Biology, Evolution, and Human Nature

Biology, Evolution, and Human Nature

Timothy H. Goldsmith

and

William F. Zimmerman

Department of Molecular, Cellular and Developmental Biology
Yale University

Department of Biology
Amherst College

John Wiley & Sons, Inc.

New York • Chichester • Weinheim • Brisbane • Singapore • Toronto

Acquisitions Editor	*Keri L. Witman/Joseph Hefta*
Marketing Manager	*Clay Stone*
Production Editor	*Sandra Russell*
Senior Designer	*Kevin Murphy*
Illustration Editor	*Anna Melhorn*
Photo Editor	*Lisa Gee*
Production Management Services	*UG / GGS Information Services, Inc.*

Cover painting by Henri Rousseau, "Le Douanier"; Art Institute of Chicago, Illinois/Super Stock

This book was set in Janson Text by UG / GGS Information Services, Inc. and printed and bound by Hamilton Printing. The cover was printed by Phoenix Color Corp.

This book is printed on acid-free paper. ∞

Library of Congress Cataloging in Publication Data:
Goldsmith, Timothy H.
    Biology, evolution, and human nature / Timothy H. Goldsmith and
        William F. Zimmerman.
        p. cm.
    Includes bibliographical references (p.).
    ISBN 0-471-18219-2 (cloth : alk. paper)
        1. Biology. 2. Evolution (Biology) 3. Human behavior. I. Zimmerman,
        William F. (William Frederick), II. Title.

QH308.2.G665 2000
570—dc21   00-021456

L.C. Call no.	Dewey Classification No.	L.C. Card No.

ISBN 0-471-18219-2

Printed in the United States of America

10 9 8 7 6 5 4 3 2 1

To
     George Williams
       William Hamilton
         Robert Trivers
           Richard Alexander

whose ideas have clarified and extended evolutionary theory into the domain of social behavior and brought us to a deeper understanding of human nature.

Preface

The mystery of existence—why and how we have come to be—is a major stimulus to religious belief and provides abiding themes in literature as well as in social and historical interpretations. This search for understanding is one of the most pervasive features of the human mind, and it lies at the core of a liberal education.

The biological sciences have some fundamentally important ideas to contribute to this search because, like all other living creatures, humans are a product of evolution. The study of evolution reveals a hierarchy of increasing complexity, reaching from the molecular structure of genes to human cognition and the diversity of human cultures. Using evolution as the unifying theme, this book traces the connections between levels of complexity, showing how both the study of other organisms and a variety of perspectives from biology, psychology, and anthropology provide complementary insights.

For whom is the book intended? The subject is broad and lends itself to interdisciplinary courses in a liberal arts curriculum where an important goal is to convey the nature of science as a powerful intellectual tool for understanding the world. There is a sufficient diversity of material that instructors can select in ways that complement their own interests.

From the perspective of traditional disciplines, the book is suitable for use in a one-semester course for students who are not majoring in biology but who wish to understand some important biological concepts that will inform their understanding of human behavioral evolution. Some of the material is not found in standard curricula for majors in biology, and the fascinated and positive response of graduate teaching assistants who have run discussion sections in the Yale course leads us to believe that the later chapters could also be used in the instruction of majors.

Because we have tried to develop all ideas from the ground up, we believe the text could also find a place in the social sciences. For example, it could be useful in college courses in anthropology, psychology, and sociology, in clinical programs in psychology and psychiatry, and in the growing number of law curricula that seek to place the study of human relationships in a broader philosophical and scientific context.

Although we assume that most students using this text have taken a course in biology in high school, in our experience students vary widely in what they have been taught and in what they remember later. Some readers who have recently had an advanced placement course in high school may find parts of Chapters 2 through 4 familiar. Others, however, may find some of the background material new or challenging. The material on evolutionary social theory and its application to human cultural evolution is increasingly taught at colleges and universities, but frequently not in biology courses.

We have tried to make this book unique in its goals and content, and a few more words about themes may be useful. There is a belief—common among students—that science is the province of an intellectual elite, dealing with the stuff that makes up the physical universe but having very little to do with human affairs. This book should help to dispel that myth.

Memorizing terminology frequently detracts from learning concepts. Clearly terms must be defined in order to develop arguments and reduce ambiguity, but we have tried to keep the specialized jargon to a minimum. We have also tried to focus on broad principles that are worth understanding—and remembering—after the course is completed. The questions at the end of each chapter were written in this spirit, and most of them can be used as topics for short essays or in discussion sections. Similarly, most of the post-chapter suggestions for further reading are general sources, usually for non-specialists, although a few provide technical detail. The last chapter, which is about human culture and is the capstone of the book, is also most likely to be the starting place for readers who want to pursue the subject further. It therefore contains more suggestions for additional reading than earlier chapters. In addition, there is a more extensive bibliography at the end of the book that includes references to the work of many of the individuals mentioned in the text.

Understanding the nature of science can be improved if students recognize that scientific discoveries and interpretations are often influenced by the social milieu in which they occur. Moreover, evolutionary theory is too often believed to be static, the work of an individual who lived over a century ago, rather than the dynamic, changing enterprise that it is. Consequently we have woven in some historical threads. Early on we introduce Charles Darwin and the understanding of biology that his insight altered forever. But Darwin's ideas were just the beginning. The twentieth century has seen landmark changes in the understanding of evolution, including the structure of genes, the role of natural variation in populations, molecular evolution, and an emerging biological theory of social evolution. Equally true, at various times evolutionary biology has also been misunderstood, even perverted.

Considering the human psyche and human behavior in an evolutionary context has frequently generated controversy. The reasons include the influence of religious beliefs and the Western philosophical tradition (partly religious in origin) of seeing culture and the human mind as beyond all reach of biological influence and therefore of biological analysis. But creating culture is a hallmark of our species' behavior. Equating "biological" with "genetic" is another major source of confusion. Compounding factors include deep misunderstandings of evolutionary theory and the multiple and complementary meanings of causation, ignorance of genes and of their complex relation to behavior, and persistent confusion about nature and nurture as alternative explanations of human behavioral development.

Unlike most of the customary descriptions of "evolutionary psychology" or "sociobiology," which rely on arguments drawn from the role of natural selection in evolutionary theory, we have embedded evolutionary arguments in a broader biological matrix. We have explained what genes are and what they do. This is common enough in introductory biology texts, but it is then generally not related to concepts of social evolution. An exploration of social insects reveals the power of kin selection, as well as a system for purposeful decision making that contrasts with the human brain. The relationships between humans and their pathogens provide a rich source of important evolutionary principles, some with immediate practical applications. We have also included basic information about nerve cells and brains in order to show that understanding how nervous systems work and develop is absolutely critical to viewing behavior in an evolutionary context.

Finally, it may be useful to clarify some of the things that the book is *not* by comparing it with others in the current market. First, this is not an introductory text designed to "cover" the sweep of biological knowledge and preview every other course in a biology major. Nor have we presented a course in "human biology" in the sense of a general text that simply uses human examples to illustrate the enormous detail that seems to characterize introductory texts. This is also not a narrow book devoted to some limited, albeit important, aspect of human evolution such as physical anthropology. Finally, this is not a book about sociobiology as that subject has come to be commonly understood, or misunderstood.

ACKNOWLEDGMENTS

We are grateful to a number of colleagues who have read all or parts of the manuscript and offered trenchant suggestions for improvement. Many of these we have gladly incorporated. Our thanks to Laura Betzig, Rodger Bybee, David Buss, John Carlson, Lee Cronk, Martin Daly, Paul Ewald, Richard Goldsby, John Hartung, Marc Hauser, Patricia Hawley, Leo Hickey, Owen Jones, Donald Kennedy, Roger Masters, Alvin Novick, David Pilbeam, Joan Silk, Jerry Waldvogel, Richard Wrangham, and an anonymous reviewer for their time and effort on our behalf. In addition Paul Ewald, Ada Lampert, Lionel Tiger, Sheldon Segal, Mitchell Sogin, and Eric Trinkaus helpfully answered specific questions. Thanks also to Napoleon Chagnon, Natalie Demong, Sarah Hrdy, Irenaus Eibl-Eibesfeldt, Steve Emlin, Paul Forscher, John Gore, Erick Greene, Michael Huffman, Steven LeBlanc, Tetsuro Matsuzawa, Gulru Necipoglu, Paul Sherman, Craig Stanford, and Franz de Waal for the use of photographs.

Brief Contents

Contents

Foundations

Earth rise photographed April 12, 1972 by the astronauts of Apollo 16 as they orbited the moon prior to landing. Notice how much of the earth is blanketed by clouds. Although life as we know it requires water and a narrow range of temperatures, many small organic molecules can be formed by natural processes, and life may very well exist other places in the universe. The distances between stars are so great, however, that we have only recently found direct evidence for other solar systems. Whether life has existed elsewhere in our planetary system is still an open question.

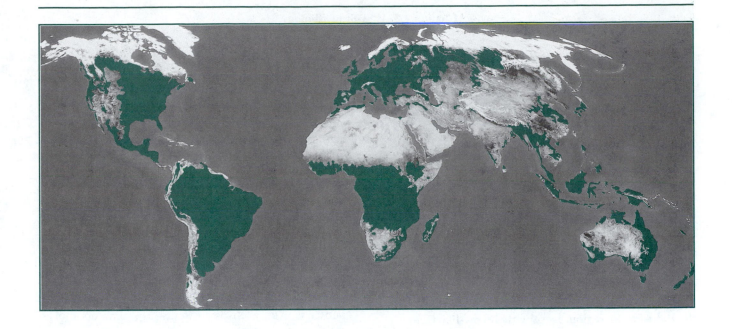

1

Why Study Biology?

It is a day in late winter on the St. Johns River in Florida, about fifty miles as the crow flies from NASA's launch site for the space shuttle at Cape Kennedy. This part of the river is the winter home to several dozen manatees, large aquatic mammals, sometimes called sea cows. Although they can weigh up to a ton and grow to twelve or fourteen feet in length, manatees are gentle vegetarians (Fig. 1.1). Florida is at the northern edge of the range of the West Indian manatee, and here they have become an endangered species. Although they have no natural enemies, their numbers have dwindled with the encroachment of human populations. Shrinking habitat is only one of their problems. About a third of manatee deaths in Florida are caused by injury from the propellers of motorboats.

But today there are no motorboats to harass these individuals, for they have entered a small, protected tributary of the river. The stream is scarcely a third of a mile in length and is fed by one of Florida's many freshwater springs. Over a million gallons of water a day well up from deep, limestone caverns. The rainwater of earlier

Photo: The atmosphere and many of the surface features of Earth have been transformed as life has evolved. This composite view of the Earth was constructed from thousands of separate images recorded by Tiros-N meteorological satellites of the National Oceanographic and Atmospheric Administration (NOAA). In a true-color image, the darker areas of the continents are green due to the presence of extensive vegetation. Note the relative lack of vegetation in a broad arid band extending east from North Africa.

FIGURE 1.1 **Mother manatee with her calf. The West Indian manatee (*Trichechus manatus*) is one of three living species of manatee, one of which is the dugong of the Indo-Pacific region. Despite their cow-like faces these aquatic mammals with broad tails are the likely source of the mermaid myth. They belong to the Order of mammals known as Sirenia. The name Sirenia refers to the sirens of Greek mythology, beasts part woman and part bird that lured unsuspecting sailors to their death by singing. Docile sea cows hardly warrant that association.**

years is now surfacing and making its way to the sea. The temperature of Blue Spring is 72° F year-round, and during the colder months of the year the manatees seek the mouth of its little stream for warmth. All species of animals and plants tend to probe the limits of their ranges and their habitats, and the manatees that reach the United States live about as far north as it is possible for this species to survive.

On this day several individuals are also probing the limits of their immediate habitat. A cow with her calf swim slowly upstream, exploring the boundaries of their world. They are not looking for food as they push outward, for relatively little grows in the first several hundred yards downstream from the site of the spring's upwelling.

Drifting downstream with the current are two Scuba divers, a man and a woman. Like the manatees, they too have been exploring the limits of their world. Equipped with air tanks, masks, and flippers for their feet they have dived a hundred feet into the depths of the spring just to see what was there. Like the astronauts who rocket into space from the launch site a few score miles away at Cape Kennedy, or the European explorers who discovered the St. Johns River several hundred years before, or the unknown adventurers who first crossed the Bering Strait from Asia and whose progeny began to populate the Americas thousands of years before them, the Scuba divers are curious about the world in which they live.

Two pairs of curious individuals move slowly toward each other, the human swimmers drifting on the surface with the current, the manatee mother and her offspring swimming languidly, easily visible in the clear water. The shrill whistle of a park ranger alerts the human swimmers to move close to the bank and continue slowly downstream. They are in no danger from the manatees; rather, it is the manatees that need to be protected, and they should not be encouraged to approach humans closely. The swimmers are clearly fascinated and try to spot the approaching manatees under the surface of the water. The manatees seem to be equally interested in these novel visitors to *their* environment. Abruptly they change course and approach the swimmers to within scarcely more than a body length. All are now motionless, the man and woman peering intently, the manatees, cow and calf looking, listening, and perhaps smelling. Then abruptly the encounter is over; each pair of curious swimmers moves on to new experiences.

A nonevent? Perhaps, to all but the participants, each of whom now has a memory, for good or ill. The calf may have learned, erroneously, that humans are harmless, an experience that at some future time could prove to be not only wrong but dangerous. For the woman and the man, the experience has also been new. Not only have they seen a rare creature living its own life in its own habitat, thereby enriching their knowledge of the world, but the encounter has also been inexplicably pleasurable. They speak to each other excitedly, comparing their sense of exhilaration. For the humans this day has brought an aesthetic as well as a physical adventure.

Why were the humans and the manatees interested in each other? What features do we humans share with these and other animals? What makes us different from them? As the encounter suggests, curiosity is not a uniquely human attribute. Whence does curiosity, or any other human feature arise? Is curiosity so widespread among humans that we can think of it as one aspect of human nature? For that matter, is it useful to think of humans as having a "nature"? If so, how might we characterize human nature? Do manatees have a manatee nature? How do we humans fit into the natural world of which manatees are so clearly a part?

These are the sorts of questions we will address in this book. These are also the kinds of questions that have interested and perplexed people throughout recorded history, and doubtless for a long time before that. Variants of these questions are central issues in philosophy, religion, and literature. They are also questions that can now be addressed by science, and that will be our approach.

That these kinds of questions should be asked at all and so frequently is a testament to human curiosity. Curiosity about other people, about the world, seems to be a universal human feature, and it takes many forms. The curiosity of the nuclear physicist about the structure of the atomic nucleus may seem at first to have little in common with the curiosity that sustains our interest in gossip about the personal problems of strangers, but perhaps these are two manifestations of a single basic human characteristic.

Curiosity about the world helps us to deal with its mysteries and uncertainties. The ephemeral nature of life is a central mystery. We are born, grow, love, nurture offspring, mature, age, and die, and while we are about it, we see the life cycles of other animals and plants and we wonder what it all means. Our ancient ancestors who hunted and gathered tens of thousands of years ago had to be concerned about why their sources of food varied in abundance, often in unpredictable ways. As later ancestors adopted a more sedentary life based on agriculture they became even more mindful of the turn of the seasons and of fluctuations in rainfall. People everywhere express these concerns in religions and seek rational explanations of natural events by postulating spirits and deities. Religions as well as secular philosophies have also tried to understand and bring order to the complex and tumultuous nature of human relationships. We suggest that these human activities and the cultural institutions to which they give rise are rooted in a curiosity that leads us to seek order in our physical and social worlds so that we can anticipate and control their effects on our lives (Chapter 14). The questions that we are posing are therefore old and enduring, but that is why they are so interesting and so important.

SCIENCE IS A MEANS OF UNDERSTANDING THE WORLD

Science is another manifestation of human curiosity, for it is a way to understand the universe. Science is also a process, a method, in which observation and controlled manipulation—experimentation—play central roles. The cultural explanations of the world that are embodied in religions (of which there are about as many as there are cultures) are passed from generation to generation, and are more symbolic or metaphorical than literal in their meaning. Science, on the other hand, has, ultimately, a disregard for traditional explanation if new observations or refined experiments reveal older understandings to be inadequate or simply wrong. We say *ultimately*, however, because scientific understanding builds cumulatively, and an understanding that stands on a firm foundation of observation and experiment is not easily shaken and not casually cast aside. But when the evidence becomes compelling, scientific understanding changes (Box 1.1).

Careful observation of nature did not start with modern science. People who depend intimately on their natural environment frequently have an extensive practical knowledge of plants and animals that is astonishing to outsiders, including scientists from western countries. What characterizes science, however, is unremitting curiosity about nature coupled to skepticism and a willingness to be guided, finally, by careful observation of nature itself and experiments that narrow the number of possible interpretations. We will give many examples of this process throughout the book.

DIFFERENT DISCIPLINES, DIFFERENT PERSPECTIVES, BUT ONE WORLD

Perhaps you wonder what human thought and behavior have to do with biology. Are not human affairs the province of psychology, sociology, and a variety of humanistic disciplines? You may also be thinking, correctly, that very practical issues involving human affairs are quite complex. For example, the ecological impact of humans on the earth, growth and distribution of food to prevent starvation, the effective delivery of health care, and the prevention of all kinds of violence are all issues that involve perspectives other than science—economics, politics, diplomacy, psychology, law, and religion to name some of the more obvious. It is also clear that the central questions about human nature and behavior have occupied the thinking of talented individuals in other disciplines well before science became a widely accepted method of studying nature. Science therefore does not exist in isolation from other human activities, either in the questions it addresses or in its practical applications. Science does, however, employ a powerful methodology, and it is playing an increasingly important role in human affairs.

HIERARCHY IN NATURE

Because of the way academic departments and curricula are organized, it is easy to gain the impression that science is the sum of separate compartments called physics, chemistry, biology, and so forth. The list of compartments multiplies as scientists find reasons to subdivide these major categories, but these fields are related to each other in important ways. One way is in sharing a common methodology, but that is not what we have in mind here.

If you take your television set or your automobile apart, in time you will have a pile of pieces—odd-shaped parts mostly made of metal, plastic, glass, and a few other materials. Even if you don't lose any parts, you nevertheless will no longer have a TV capable of creating images or a car that runs. The TV set and the automobile had properties that caused them to perform

Box 1.1
Creeping Credibility

Beginning at least with Francis Bacon in the early seventeenth century, the rough correspondence between the shapes of the shorelines of eastern South America and western Africa have suggested to some people that these continental masses may once have been joined. The hypothesis assumed a comprehensive form in 1912 in the writings of Alfred Wegener, who proposed that the present configuration of continents was the result of a slow cleaving and separation from a once contiguous chunk of land—that continents actually drift over the surface of the earth much like icebergs in the ocean, albeit much more slowly. He was led to this view not only by the distribution of similar rocks on continental faces that were presumably once in contact, but by the distribution of similar plants and animals, living and fossil, on what are now widely separated land masses.

The history of the concept of continental drift illustrates how truly revolutionary ideas in science do not instantly achieve recognition if they lack a crucial element. Until the 1960s the hypothesis of continental drift had very little support and was considered by many geologists to be fanciful. (As students, both of us remember geologists who ridiculed the idea that the continents had reached their present shapes and positions as a result of movement.) The reason for this widespread skepticism was that the hypothesis of continental drift did not include a credible mechanism that could cause the movement of great masses of land.

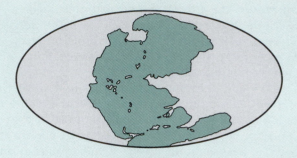

200 Million years ago (Triassic)

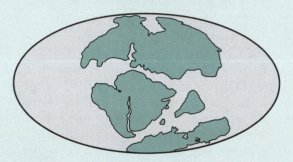

135 Million years ago (early Cretaceous)

65 Million years ago (late Cretaceous)

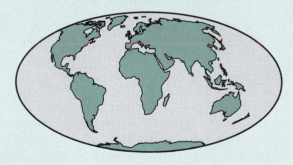

Today

FIGURE 1.2 The continents have not always occupied the same positions that they do today. This series of diagrams shows how they have moved during the past 200 million years. *Plate tectonics* is the field of geology concerned with the causes of these movements (Fig. 1–3). The present geographic distribution of many organisms has been greatly affected by such gradual continental drift (Chapter 2).

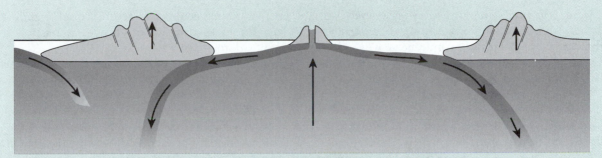

FIGURE 1.3 A simplified diagram of plate tectonics. Liquid, molten rock pushes up from deep within the earth, perforates the crust and sometimes extrudes to form volcanoes. The flowing rock also displaces old crust laterally, causing it to flow under the edge of continents and to push against their edges. The resulting compression causes horizontal shifts of the land (earthquakes) and vertical uplifting of the land (mountain building). Familiar examples are California earthquakes and the volcanoes of the Pacific Northwest such as Crater Lake, Mt. Rainier, Mt. Shasta, and Mt. St. Helen.

Today, however, the processes of *plate tectonics* (as the phenomenon is now called) are generally accepted. Wegener's hypothesis was revolutionary in its implications and required new thinking about the physics of the earth's crust. Time and the accumulation of much additional information were required in order for wide acceptance of the idea.

The evidence now includes more detailed correspondences between rocks on continental faces that are separated by wide expanses of ocean (for example, Brazil/Africa and Greenland/Norway) and the distribution of many plants and animals and their fossils. Some of this *biogeographic* evidence is described in Chapter 2, for it contributes in fascinating ways to our understanding of evolutionary history. Other, relatively recent, evidence is of an entirely new nature. The orientation of magnetic minerals locked in old, undisturbed rocks from different continents points to different positions of the earth's magnetic pole and is consistent with the continents having moved and rotated relative to each other during the last 200 million years. Other magnetic evidence (indicating reversals of the earth's magnetic field) from more recent rocks taken from the sea floor indicates that the seabed is moving several centimeters a year, scarcely faster than your fingernails grow.

Evidence of very different kinds—observations that are not readily explained by alternative concepts—has thus amassed to the point that plate tectonics is essential in understanding the history of the earth. Figure 1.2 shows how the continents were distributed at various times in the past. Furthermore, plate tectonics accounts for some important present features of the earth. Upwelling of molten rock occurs, for example, at the mid-Atlantic ridge, and new crust spreads laterally, widening (in this case) the Atlantic ocean (Fig. 1.3). Where continental masses grind against each other, the land is compressed and mountains form. For example, the Himalayas rose as the Indian subcontinent pushed into the rest of Asia. At other places, great plates are rubbing together, and one is being forced under the other. These are sites of earthquakes and volcanic activity, such as those found in Iceland and around much of the rim of the Pacific. Earthquakes are as familiar down the Andean spine of South America as they are to residents of California, and volcanoes form part of the landscape in our Pacific Northwest.

useful functions by virtue of the way their constituent parts were *organized*—the way they interacted with each other when they were properly put together. How fast the car will run and what mileage it will get are properties of the whole assembled vehicle and can only be examined when the vehicle is intact.

The same principle is at work in the way the world (or the universe) is organized. Consider an obvious example, *you*. In school you were taught that living organisms are composed of cells. The simplest organisms are single cells, but most of the organisms that are big enough to see with the unaided eye consist of many cells. Cells come in an enormous variety of shapes. Some secrete and imbed themselves in tough or hard substances, giving rise to such diverse structures as coral reefs, the trunks of trees, and the bones of your body. Other cells like the muscles of your limbs can do mechanical work, others like the cells of your liver are chemical factories, and still others like the nerve cells of your brain are in the information business. But *you*, or your cat, or the tree outside your window are more than collections of cells. Each has properties that define a person, a cat, or a tree. But if we examined many of your cells under a microscope, we would not be able to distinguish them from their counterparts in your cat. Cat blood is not very different from human blood; cat muscles work pretty much the same as our muscles; even cat neurons follow the same rules of signaling as the nerve cells in human brains. So what makes you different from your cat?

The difference arises from *organization*, the ways the basic building blocks have been put together. That comparison illustrates the enormous complexity that can arise from the different ways a large number of similar component parts can be assembled. Your humanness, like the cat's catness, *emerges* from the details of this assembly. You and the cat and the tree are each individual organisms, and as organisms you are each representatives of a level in a hierarchy of complexity of organization in which cells represent a lower (less complex) level.

At the bottom of this hierarchy (at least as we currently understand the universe) are the elementary particles that make up atomic nuclei. Atoms of the elements are assembled from these elementary particles along with surrounding arrays of electrons. Elements differ from one another in the weights of their atomic nuclei and the number of accompanying electrons. Understanding the forces that hold the nucleus together is one of the central problems of physics.

When atoms form chemical bonds with each other by sharing electrons, molecules of new substances are produced. The molecules have very different properties from the constituent atoms; they have properties that result from the way the atoms are organized. The

rules that describe chemical bonding and the properties of the molecules that result are sufficiently different that another branch of science, chemistry, exists to deal with them.

So far we have talked about elementary particles, atoms, molecules, cells, and organisms, *ordered here in a hierarchy of increasing complexity*. The complexity increases, because each higher level of entity is assembled from the stuff of lower levels. It is because of the *emergence of properties* at each level—properties that are best described in terms of that level of organization—that a branch of science (physics, chemistry, biology) exists to carry on the study at that level (Fig. 1.4). In the same way, geology and astronomy build on physics and chemistry in the study of the earth, planets, and stars. As scientific knowledge has advanced, the relationships between levels of organization have become clearer and the boundaries between the scientific disciplines have become blurred. As a consequence, disciplines with names like chemical physics, biochemistry, molecular biology, geochemistry, and astrophysics have formed at the boundaries.

Hierarchy in biological systems extends beyond organisms, for many organisms live in groups. Some species live in social groups with considerable complexity; honeybees and humans are two obvious examples. Even when social groups do not form, populations of individuals are frequently biologically important entities for reasons that we will discuss in Chapter 4. The

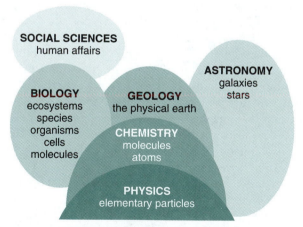

FIGURE 1.4 Nature consists of hierarchies of complexity, each level derived from those below. Each level in the hierarchy displays characteristic features that depend on the organization of its constituent parts. But each is assembled of components that are present at lower levels. The various scientific disciplines—physics, chemistry, biology, and so forth—build on each other in corresponding fashion. Science is an effort to understand both the properties at each level (as, for example, the social behavior of primate groups) as well as the connections between levels (for example, the chemistry of genes).

characteristics of social structures will occupy our attention starting in Chapter 6. Furthermore, different kinds of organisms in an ecosystem may be interdependent in complex ways, so that the introduction or removal of a single species can trigger changes in the numbers of others. We will see some examples of this principle at work in Chapter 8 in the context of human disease. Thus an ensemble of species can exhibit properties that can only be described and understood in terms of the entire ecosystem.

REDUCTIONISM

An effort to understand the properties of one level of organizational complexity in terms of the properties of lower levels is known as *reductionism*. Until about a hundred years ago, the relation of biology to chemistry was not very clear, but during the twentieth century an enormous amount has been learned about the chemistry of living cells. By breaking cells open we can study the many chemical reactions that are involved in mobilizing the energy of food, which take place in organelles called mitochondria. The process of photosynthesis, which enables green plants to harvest the energy of sunlight and which supports the life of animals, is now understood in chemical terms, and it occurs in organelles called chloroplasts. Even the transmission and expression of genetic information, which we will discuss in Chapters 3 and 4, is now known in molecular terms. At its most basic level, biology is very fancy chemistry.

Chemistry, however, does not help much in *predicting* the properties of living cells. The movement of an amoeboid cell from one place to another, the coordinated movement of chromosomes during cell division, the differentiation of a fertilized egg into many cells with different forms and functions—these and many other phenomena would be impossible to anticipate from our understanding of the fundamental rules of chemistry. They are properties of complex ensembles of molecules, properties that have appeared as a result of the organization of simpler entities.

The aim of reductionism is to analyze the complexity of a system by taking it apart, studying its components, and figuring out how they work. Reductionist explanations are extraordinarily powerful, because they generally tell us how things work in terms that are intellectually appealing. Knowing how things work satisfies curiosity, for it removes some of the mystery that is generated by complex organization. Furthermore, knowing how something works can be useful in fixing it when it is broken, an idea that is as powerful in medicine as in automobile repair. And knowing how things work often leads to new insights of understanding and novel creations. For example, knowing the detailed molecular structures of genes (Chapter 3) has greatly expanded our understanding of how genes control the growth and development of cells (Chapter 10) as well as how genes change over time (Chapters 4 and 5).

The analogy of TV sets and automobiles with nature is useful, but can be misleading. TV sets and automobiles are technological creations that are made possible by a firm understanding of electronics and heat engines. Designers of TV sets and cars are able to work upward, so-to-speak, in their design. They know what they wish to accomplish, and their task is to make parts that can serve their individual functions yet work together with the other components to form a useful product. In biology the problem is frequently the reverse. Biologists are more often confronted with a product—a cell, a brain, or an organism—and the challenge is to figure out how such complicated structures work, to strip them down to their component molecular parts and see how they function together to produce the whole.

The reductionist approach is not new, nor need it necessarily lead to an explanation in molecular terms. That outcome depends on when and where in the hierarchy the examination begins. For example, until less than 300 years ago the functioning of the human heart was not understood. In the first half of the seventeenth century, however, William Harvey, physician to the English Kings James I and Charles I, showed that the vertebrate heart is a pump with valves to keep the blood flowing in one direction through its chambers. This work is a landmark in science, not only in its reductionist approach, but because Harvey reasoned from careful observations rather than accepting the beliefs of Aristotle and the Roman physician Galen, doctrines that had been accepted for fourteen centuries. Interestingly, one important detail escaped Harvey: he could not establish the connection of blood delivered to the tissues by arteries with blood returning to the heart through veins because he had no microscope and therefore could not see the small capillaries. This gap in his understanding illustrates an additional point: advances in scientific understanding frequently must await advances in technology that allow us to observe the very small or the very distant.

The brain can also be studied by a reductionist approach, and there are now many techniques for examining the behavior of small groups of nerve cells or even single neurons (Chapter 9). A fundamental conclusion from this work is that nervous systems and the behaviors they produce obey the rules of chemistry and physics; in fact, most of the basic properties of nerve cells are shared by all other cells in the body. It is therefore possible in principle to use a reductionist approach in seeking the cause of a bird's singing or a mammal's behavior when it comes in contact with another member of its species. Which nerve cells are active? What hormones are present and influencing the behavior of nerve cells? Answers to these sorts of questions give us a picture of how at least some of the nerve cells of a brain are acting during the immediate production of a specific behavior.

MAKING CONNECTIONS BETWEEN LEVELS

Reductionism creates strands of understanding that reach downward, connecting levels in the hierarchy of organization. Reductionism therefore unites chemistry with physics, and biology with both. But reductionism does not eliminate complexity; it just helps us to see relationships more clearly. In fact, many of the properties that arise from complex organization are of such a nature that it is the *organization itself* that must remain the focus of explanation. For example, exploring how individual nerve cells signal each other can say little about how your brain is able to identify, uniquely, another person from a brief glimpse of their face. That capacity is a property of extensive networks of neurons acquiring visual information from the external world and comparing it with a stored record. Similarly, no study of an isolated chimpanzee could ever tell us about the dynamic relationships between individuals of different ages and sex that characterize a troop of these animals, for those interactions are a property of the group itself (Chapter 13). Reductionism has its role, but it cannot, alone, give us the full conceptual understanding we desire.

The study of behavior of one species, humans, is traditionally the province of the social sciences: psychology (the study of the human mind), sociology (the study of the organization and behavior of human groups, by custom generally in the dominant cultures), and anthropology (the study of human groups, by custom not the major cultures and not groups studied by sociologists). The separation of the social sciences from biology again reflects the hierarchy of complexity. Among all living creatures, humans are unrivaled in two respects: their ability to communicate with language and their intellectual capacities. These two features of humans account for an unprecedented complexity of social interactions, an ability to create cultural traditions that can be passed to others by teaching and learning, and technological achievements of ever-increasing sophistication (Chapter 14). The existence of the social sciences reflects these attributes and the additional levels of social and behavioral complexity that the human species has achieved.

But humans are biological organisms. They are similar in their biochemistry and genetics to all other organisms. Their nerve cells and their brains work on the same principles as those of other animals. These similarities reflect a common evolutionary heritage, and they explain why many human diseases can be studied successfully by using other animals as surrogates. For these reasons it is logical to expect that there are strong, downward connections between the social sciences and biology, just as understanding of much of biology is now rooted in the physical sciences. At this time, however, the connections between biology and the social sciences are weaker and more tenuous than between biology and chemistry.

Even the simplest language in everyday use complicates all discussion of the relationship between the biological and the social sciences. Consider the words so frequently employed in referring to the practice of adoption: there are "biological" parents and adoptive parents, and in cases of disputed custody, society finds itself wrestling with whether "biology" should come before whatever psychological needs the child might have in a continued relation with the adoptive parents. The word "biological" is here a surrogate for "genetic," often with the unspoken assumptions that biology is defined by genes, that genes operate independently of the environment, and that the postnatal development of the brain is not really a biological phenomenon. Such a belief in the independence of genes is naive. The nurture of offspring, physical and social, is equally a part of their biology (Chapter 11).

What is the "nonbiological" alternative to "biological"? Has it to do with socialization and learning? To alter behavior through experience is to make a change in the nervous system. Put simply, the functional properties of some nerve cells are changed in the process (Chapter 9). The attendant changes in behavior are generally considered to be "psychological" phenomena (which of course they are), but the functional properties of those changed neurons are equally biological.

Only the behavior of one species, humans, has been intellectually compartmentalized into disciplines that are often assumed to be quite distinct and separate from biology. But this tradition has the negative effect of diverting our attention from the connections between psychology and biology, or more importantly, concealing the biological roots of our nature. Throughout this book we will try to illustrate how we believe the biological and social sciences should complement each other in a coherent view of the place of *Homo sapiens* in nature.

Interestingly, there is a second, almost paradoxical feature of this example of "biological" and adoptive parents. The rights and responsibilities of individuals are not only central to our political philosophy, but early environmental influences are deemed critical in the development of human character. Thus the idea that you are what you make of yourself, and not what you inherit. Yet society invests these disputes over the custody of children with another consideration: rights of parents based on genetic continuity, sometimes, unfortunately, to the detriment of the children. We hope to provide a deeper insight as to why both the law and the emotions feel this tug of genes.

THE CENTRAL ROLE OF EVOLUTION IN BIOLOGY

Physical objects are made of parts arranged in a hierarchy of complexity, but as we move from the nonliving to the living we encounter an enormous increase in diversity and complexity. How did life become so diverse and complex? This question brings us to one of the most important findings in all of science.

At the macroscopic level—the objects that we can see and feel—there are several discontinuities between the living and the nonliving. The first is that living things perpetuate themselves, they reproduce. In due course we will modify this statement, for in sexual reproduction it is only the genes that are reproduced in the sense of making exact copies of themselves, and the organisms that appear in subsequent generations are new and unique entities. But reproduction is an important distinction between the living and nonliving.

A second distinction was mentioned before—complexity of organization. Although complexity is a matter of degree, living things are immensely more complicated in the numbers and kinds of molecular parts than anything in the inorganic world. And as we saw in the previous section, it is that very complexity of organization that leads to an obvious discontinuity that distinguishes biology from chemistry.

A third and related distinction between the living and the nonliving is that living things *appear* as if they were *designed* for something. A rock has atomic order in its structure. Indeed, some crystals of minerals exhibit a high degree of geometric order, and we may marvel at their shape or color, but there is no functional purpose to that order. (The order of course has a cause. It represents a state of low energy into which the atoms combine under particular conditions of heat and pressure.) On the other hand, living things and their constituent parts do *seem* to have a purpose. The sparrow's feathers provide a body insulation so light in weight that birds are able to fly; the sparrow's wings are *for* flying in search of food and mate; its eyes are *for* seeing the necessities of life; and its lungs, like a fish's gills, are *for* obtaining the oxygen that it needs. The sparrow builds a nest for the purpose of depositing its eggs in a safe place, and it lays eggs for the purpose of making more sparrows. In short, everything about a sparrow, from its anatomy to its behavior, *seems* to be organized for some goal or purpose. Rocks and streams, mountains and lakes, however, do not have purposes. As the Oxford biologist Richard Dawkins has put it, they just *are*.

These distinctions between the living and the nonliving have been apparent to people for millennia, and efforts to understand them in human terms—to give them sense and meaning—are found in every culture. It was not until the middle of the nineteenth century, however, that an English naturalist, Charles Darwin, brought biology into the conceptual scheme of the physical sciences by showing how the seemingly purposeful nature of living creatures reflects a natural phenomenon that is a unique property of life. Darwin called this phenomenon "descent with modification," but it soon became known as evolution.

It is impossible to overemphasize the importance of evolution as a natural process, for it provides the key to understanding the appearance of "purpose" and "design" in living organisms. We shall describe Darwin's specific contributions and the nature of evolutionary theory at greater length in later chapters, but a very brief overview of its key element is useful here.

Evolution results from a physical process of *sorting*. Whenever there are entities that are capable of replicating, and there is variation among them, some variants are more successful than others in replicating (Fig. 1.5). In the inorganic world, crystals can grow, but one crystal does not replicate itself. If it breaks, the pieces are just parts of what was there before. But the capacity of crystals to grow in a ordered way foreshadows the properties of the molecules from which genes are made. The genes of living things literally replicate—identical copies are made. Sometimes, however,

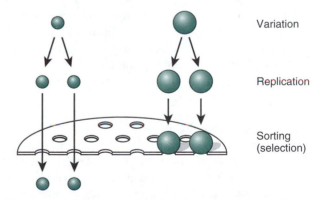

FIGURE 1.5 **A diagram showing the essential conditions that underlie organic evolutionary change: variation in a population of replicating entities, and a sorting process that removes some variants before the next round of replication. In biological evolution, the replicating entities are genes and the sorting process is *natural selection*, but the relationship between genes and selection involves the presence of organisms, as will be discussed in Chapter 5. The ultimate source of variation is mutation.**

the copies are not quite the same as the parent gene, and this leads to variation in the population—both the population of genes and the population of organisms carrying the genes. Whenever there is variation in such a population, some copies of genes are more likely to be successfully replicated in future generations than are their alternative versions. Or as seen from the vantage point of the organisms containing the genes, some individual organisms will be more successful than others in reproducing. This sorting, this *sifting of genes* in a population through successive generations is called *natural selection*, and it is the principal (but not the only) process that causes evolutionary change (Chapters 3–5).

Natural selection is a very powerful process, and it produces organisms so specialized in their structure and behavior that they appear to some people to be the result of conscious design. They seem that way because the time scale over which evolutionary change occurs is so long relative to a human lifetime that when the results of evolution are assessed by personal human experience, "design" seems to be a "commonsense" explanation. But judgments about the universe based on personal experience are not always right—in fact, they can be quite misleading—when dealing with unfamiliar dimensions of space and time. We have already given one example: the movements of the continents over the surface of the earth requiring hundreds of millions of years. Or to offer a biological example, a curtain of mystery was lifted, revolutionizing human understanding of many diseases, when microscopes revealed the presence of bacteria and their roles in infections. Similarly, the vastness of space is difficult for people to comprehend, and was not remotely realized before the invention of telescopes and modern astronomical techniques. Many people today have prescientific worldviews about biology, partly because the processes of evolution are not part of their personal experience and partly because religious belief determines their conception of existence.

We introduced the concept of natural selection by contrasting the seeming purpose we see in living organisms with its absence in rocks, lakes, and hills. Both the many kinds of organisms and the earth's physical features, however, result from natural processes that occur over long times and have contingent outcomes. In other words, both processes depend on the conditions that prevail during their operation. Nevertheless, there are important differences. Geological change is largely towards greater disorder. Streams run down hill and mountains erode. Even when a source of energy refashions the landscape, as for example during an earthquake or the eruption of a volcano, there is no particular order to the process, although the patterns

that result have natural explanations. By contrast, every time a new organism develops from an egg, energy is used to create order, and this is where the appearance of purpose arises. Organic diversity, with its attendant appearance of purpose, is the result of billions of such reproductive and developmental events winnowed by natural selection since the origin of life. The concept of purpose appears again in Chapter 11.

This evolutionary perspective points to another way in which the comparison of organisms with TV sets and automobiles is misleading and inadequate. Technological creations like TV sets are designed and assembled by humans, but the "design" and assembly of living organisms emerges from evolution, a natural process that does not require a conscious "designer" and *has no long-term goal or plan* (Chapters 4–6). Assembly is accomplished during the development of an organism from a fertilized egg, in which cells proliferate and differentiate, organs form, and the shape of the body emerges. Like the "design" that results from natural selection, this too is a natural process (Chapters 10–11).

BEHAVIOR CAN EVOLVE

Perhaps you noticed that when we described how a sparrow seems to be the result of design, we said that not only its physical appearance but also aspects of its *behavior* are the result of evolutionary change. Some people are surprised to learn that behavior is subject to evolutionary change, since they think of the physical and the behavioral as separate. Partly that is because we commonly consider our own minds to be creations of our individual experience, and we compartmentalize the study of our behavior in disciplines such as psychology and sociology. We will have much more to say about the behavior of humans and other animals throughout the book, but for the moment consider the following idea.

The behavior of an animal is caused by patterns of activity of nerve cells in its brain (Chapter 9). But brains, like other organs in the body, can be changed by evolutionary processes. Moreover, how an animal behaves can influence very directly whether it is able to survive and reproduce, so behavior contributes strongly to the sifting process we referred to as natural selection. For these fundamental reasons, we should expect that the behavior of all animals has been shaped by evolution.

It would be quite incorrect to infer, however, that there are two kinds of behaviors, those that are learned by an organism and those that result from evolution by natural selection (Chapter 11). Part of our task in this book will be to explore the relationships between changes that take place in populations of organisms in evolutionary time and changes that take place in individual organisms during their lifetimes.

THE MULTIPLE MEANINGS OF "CAUSE"

Each spring, as the weather warms, birds migrate north to breed, and the woods and fields become filled with their songs. What causes them to sing? In the autumn, as the weather starts to cool, these same birds and their young make the return journey south to warmer climates. Why do many species of birds undertake these long, seasonal migrations? What are the causes of this arduous behavior?

We could ask similar questions about you. When you eat, the food you take in is digested. You probably do not think much about that process except when you have indigestion, but that intake of food provides you with the energy you need in order to exist. Your food is fuel. How does it yield useful energy? What causes you to eat?

When you sit down to a meal, you use a knife and fork, but if you were born into another culture you might prefer other implements to manipulate your food. You also have friends. Moreover, you value their friendship, and what you believe they think of you influences how you behave toward them. There are thus things you do and don't do in your relations with others. In fact, much of your behavior is responding to something we refer to as a conscience, but how did that conscience discover what is expected of it?

You are probably thinking that last question has an obvious answer: Your parents and others taught you. But why is that conscience there in your head ready for their appeal? Or perhaps you are in love and your emotional world is suddenly topsy-turvy. But why is that other person having this effect on you? What is causing it? Why are you susceptible?

Whenever we consider what living organisms are doing, we are confronted with questions of cause, and every question has at least two different kinds of answer. Is there an orderly way to think about causation?

PROXIMATE CAUSE

What is the cause of a bird's singing or the hand's moving to the mouth with a forkful of food, or the emotional feelings of the heartsick lover? One set of answers can obviously be sought in terms of the activities of nerve cells at various places in the brain. But suppose we ask about the causes of why you use a fork at all, or why you do not steal things that you might like to have. You learned to use a fork, and you learned that stealing is not acceptable social behavior, and you remember. Actually, you probably use a fork in eating without giving the matter conscious thought. It probably does not ever occur to you to steal things, but if the suggestion is made that you might, you become keenly aware that it would not be an easy act for you to commit. In both cases the act of learning has created some alteration in your brain, some lingering change that has now become a part of you. As sparrows do not hatch from the egg in full song, perhaps they also must learn (Chapter 11). Thus a brain changes during a lifetime, and understanding the causes of behavior must therefore also include the dimension of time and the experiences of individuals.

All of these modes of explanation are instances of *proximate cause*. Each speaks to the question of *how* a process works. In each case, understanding involves taking the system apart, or at least asking questions about its individual components. In each case, explanation follows a reductionist approach. Furthermore, in the examples where active learning was part of the individual's history, explanations for behavior need to consider changes that have occurred in the brain during the lifetime of the individual.

EVOLUTIONARY CAUSE

Suppose, however, that we knew about all the individual nerve cells in the brain in enormous detail, much greater detail than now seems even conceivable. Perhaps then we could even understand how *populations* of nerve cells function to generate those aspects of mind where matters of proximate cause can only be discussed now from the vantage point of psychology, high in the hierarchy of complexity. We would have detailed explanations of how the brain works, but a curious person might nevertheless still ask Why is friendship so important to human beings? or Why do some kinds of birds migrate whereas others do not?

These kinds of questions are quite different from the How does it work? sort of question we were just considering. These *why* questions are inquiring about *purpose*, and they bring us back to the observation that organisms appear as though they are the product of design. In biology, questions about purpose and design are termed questions of *ultimate cause* or *evolutionary cause* because their answers require another mode of explanation involving evolutionary history. Explanations of evolutionary cause involve natural selection and an understanding of why some traits have been more successful than others in fostering reproductive success.

Explanations of proximate and evolutionary cause refer to different natural processes, so we should try not to confuse them with each other. However, they are frequently seen—incorrectly—as alternatives, with one precluding the other. They do not provide alternative explanations; each is capable of offering a different but valid perspective. Each refers to a different feature of nature. A common misconception is that any behavior that has been molded by natural selection—and for which we can therefore formulate a hypothesis of evolutionary cause—must be rigid and inflexible, whereas learning—an example of an explanation of proximate cause—has not been shaped or tuned by evolutionary history. In Chapter 11 we will explore these confusions in greater detail.

IS THERE A HUMAN NATURE?

Recall now the story of the human swimmers and the manatees with which we opened this chapter and the questions that were raised by watching that brief encounter in a Florida stream. When we observe manatees in their natural habitat, we can describe their behaviors and the conditions under which particular behaviors appear. If these descriptions are thorough, we can think of them as characterizing the "nature" of a manatee. They will describe the ways manatees react to each other, to their young, to possible danger, to different sources of food, in short all of the many things that manatees do. But to characterize the nature of what it is to be a manatee, or any other species of animal, is to gain insight into what has made it an evolutionary success. To describe manatee nature is thus to recount an evolutionary outcome. The reason is that behavior, no less than the shape and form of an animal, is influenced by evolutionary history. Starting in Chapter 6 we discuss in much greater detail how this idea is supported.

What about people? Is there some core of behavioral propensities, inclinations, tendencies shared by most people everywhere? Is this not what is referred to by the phrase "that's just human nature"? If you are inclined to answer "yes" (however tentatively) to these two questions, then consider where the logic of the preceding paragraph leads. Because humans, like every other living creature, are the result of evolution (Chapter 12), is it not appropriate to seek an understanding of human nature from an evolutionary perspective?

Another view of humanity, however, asserts that there is no such thing as human nature. Humans are infinitely malleable, and human cultures are virtually infinitely variable. Moreover, humans become whatever is transmitted to them by the social environment in which they develop. There is no core of humanness that remains when cultural differences are peeled away.

This, or something very much like it, is a common view in certain of the social sciences. It is a view that derives from religious and philosophical traditions in Western civilization, traditions that predate the recognition of evolution as a fundamental natural process. Consequently this view does not include evolutionary implications and it proceeds as though humankind can be adequately understood without considering this cen-

tral biological process, or indeed the rest of nature. Not only is this an overly narrow view of the world, it is quite wrong. The explanation will build throughout the book, so that by Chapter 14 we will be able to show how an evolutionary perspective actually enriches understanding of humanity.

We have now posed an issue of great philosophical moment, and this topic alone may provide ample incentive to dig into the following chapters. For some people, this subject seems to be loaded with dangerous political or ethical significance. We can assure you, however, that understanding our evolutionary heritage does not provide a rationale for throwing aside the moral teaching and the legal experience of centuries of human civilization. In fact, quite the contrary is true. Seen in an evolutionary perspective, the reasons why certain teachings and traditions are common to many cultures becomes understandable.

THE WORLD IS CHANGING RAPIDLY, AND HUMANS ARE RESPONSIBLE

Although the kinds of questions we address in this book are not new, there are some compelling reasons why they have taken on new urgency and why we should try to understand ourselves and our place in nature with greater sophistication. Here is a familiar example. One of the most dramatic events to have occurred during the 4.5 billion year history of the earth has happened during the last 200 years. Until the nineteenth century, over the hundreds of thousands of years of human existence (a mere blink of the eye when compared to the age of the earth), human population increased very slowly. As Figure 1.6 shows, by 1800 there were about a billion people on earth, but then the number began to increase rapidly.

A population will increase in size when the number of births exceeds the number of deaths, (Box 1.2) and during the present century the increase has accelerated enormously. The principal cause for this increase has been a decrease in the death rate resulting from clean water supplies, better nutrition and sanitation, immunization against infectious diseases, and the discovery and widespread use of antibiotics. (Note that all of these improvements in public health are the direct result of increased knowledge of biology.) A decreased death rate means fewer children die in infancy and more adults live to old age.

We mentioned earlier that the manatees in south Florida are threatened by the impact of human populations, but manatees are not alone. The human population of Florida increased fourfold (from 3 to 12 million) between 1950 and 1990. This increase was due more to

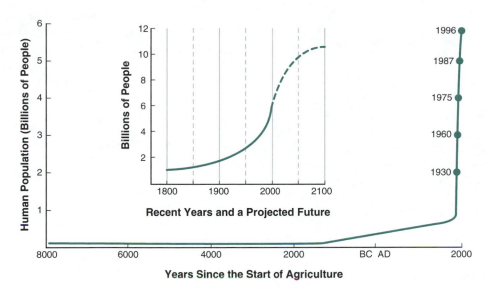

FIGURE 1.6 The increase in human population in the world since the inception of agriculture. Inset: recent growth (solid line) and the 1998 United Nations mean estimate of future growth (dashed line). It took hundreds of thousands of years for the population to reach a billion people at about 1800, but growth has been explosive during the twentieth century and the population is now almost 6 billion. Even with increasing concern about the impact of human populations on the environment and political and social efforts to control the number of births, the population may double again before the year 2100.

migration of people into the region than to a shift in the balance of births and deaths. Although this movement of people can be seen as another example of a species expanding its range, and we can point to a variety of demographic, economic, and political causes, the consequences of the increase in population have been greatly amplified by technology. As land has been drained for habitation and agriculture and the natural flow of water into the Everglades interrupted for flood control, the ecological integrity of the entire southern third of the Florida peninsula has been severely threatened (Fig. 1.8). An estimated 97% of the pine forests have been lost, and the Everglades themselves are now considered to be dying. Birds provide a sensitive indicator of ecological change. The number of wading birds like herons that nest in the Everglades has declined from about 300,000 sixty years ago to an estimated 10,000–15,000 today. Altered patterns of water flow, coupled with agricultural runoff and other contamination, have led to elevated levels of mercury as well as declines of native species of both plants and animals from mangroves to panthers. The salinity and turbidity of Florida Bay, the vast expanse of shallow water at the tip of the state, have increased, and this important nursery for a myriad of aquatic creatures, including commercially useful fish and shrimp, has been badly compromised.

Florida presents an example that is close to home, but this very recent growth of human populations worldwide and the attendant pressure on arable land threatens the continued existence of tropical rain forests. By one estimate, the tropical rain forests of the world are disappearing at a rate of nearly 1% per year. The total, worldwide area occupied by this habitat is roughly the size of the continental United States, and the annual loss is an area about the size of Ohio, or twice that of the entire Everglades. Why is this important? Tropical rain forests are the places on earth richest in the diversity of different kinds of plants and animals, and their destruction leads to an irreversible loss of living forms on a scale comparable to the most massive extinctions that have occurred in geological time. The difference is that this time the pace is faster, and the cause is the impact on the world of one species, *ourselves.*

Why should these matters concern you? Our description of the impact of human population growth and human technology sounds like a biological catastrophe well into the making. But politically and economically not everyone agrees that the results are so bad. Many economists argue that technological invention will be our salvation; that human inventiveness will continue to elevate living standards. Those who adopt this position call attention to past successes and to

Box 1.2
Populations and Bank Accounts

As long as the environment can provide the necessary support, a population of organisms can grow by the same rules that describe the accumulation of money in an interest-bearing bank account. Suppose you make an initial deposit of P_0 dollars and the account earns 4% a year. If you let the interest accumulate and do not make any withdrawals, at the end of the first year the value of your account will have increased by a factor of 1.04. In other words, your account will be worth P_1 dollars, where

$$P_1 = 1.04\, P_0$$

At the end of the second year, however, you will have P_2 dollars, where

$$P_2 = 1.04\, P_1 = (1.04)^2\, P_0$$

Look carefully at how we got from P_1 to P_2. If we extend the same reasoning we can generalize the expression and see what has happened to the value of the investment after a time, t. Because in this example the interest is recomputed annually, time is measured in years. At the end of t years (assuming you have left everything in the account), your account will be worth

$$P_t = (1.04)^t\, P_0$$

This is an example of exponential growth, and we have used only very elementary algebra to show you its logic. There is an alternative expression for exponential growth that is encountered more frequently but whose derivation is less intuitive:

$$P_t = P_0 \times e^{r \times t}$$

where r is the fractional growth in time t. (You can find the origins of this form of the equation in Box 2.3.)

How many years does it take for your money to double in value, i.e., for growth to have occurred so that $P_t/P_0 = 2$? By taking the natural logarithm of both sides of the second form of the equation and rearranging terms,

$$t = \frac{\ln(2)}{r} = \frac{0.693}{r} = \frac{69}{\rho}$$

where ρ is simply the growth rate r expressed as percentage (e.g., 4% rather than 0.04). This expression provides a ready rule-of-thumb for estimating the doubling time (in years) of a population: simply divide 69 by the percentage increase per year.

A graph of the doubling time (in years) as a function of percentage growth provides a quicker insight into the relationship between t and r than looking at the equations (Fig. 1.7). Clearly in considering finances, one would like to make investments that yield 8% or 10% interest, for otherwise there will not be a dramatic increase in principal in a few years. But if the doubling time of populations is the issue, quite another perspective is appropriate. During the 1960s the rate of world population growth reached 2.5% and was greater than 3% in some developing countries. As you can see from Figure 1.7, with an annual growth rate of 2.5%, it takes only twenty-eight years for a population to double in size. Twenty-eight years is a very short time in the span of human history.

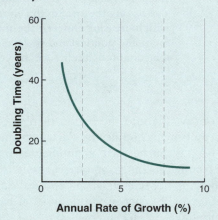

FIGURE 1.7 The time for a population to double its size (or a bank account to double in value) decreases dramatically as the annual rate of growth (or interest paid on a bank account) increases from less than 1 to several percent. A population grows when the number of births exceeds the number of deaths. This calculation assumes there is no limit on growth of the population. In nature, however, populations cannot grow indefinitely because they deplete some critical resource such as food. The calculation also assumes that the size of the population is not changing due to the arrival of new individuals by immigration or loss of members by emigration.

the generally high standard of living in developed countries.

Others are less persuaded. They question the assumptions on which these optimistic projections are based; some challenge whether economic measures of human living standards are the only values that ought to be considered. They point to the accelerating degradation of the environment, to falling water tables in many newly populated areas, to the depletion of natural resources, to lack of new land for colonization and the development of productive agriculture, to the increasing numbers of people living in poverty in the undeveloped world, to the disproportionate share of economic consumption by developed nations, to unanticipated phenomena such as damage to stratospheric ozone and the possibility of global warming, and to the relatively short

(A)

(B)

FIGURE 1.8 **(A) Saw grass prairie (*Cladium jamaicense*), Everglades National Park.**

(B) Mangrove islands in the southern end of Everglades National Park, with Florida Bay in the background.

FIGURE 1.8 (continued)
(C) Most of what remains of the everglades is now confined to Everglades National Park.

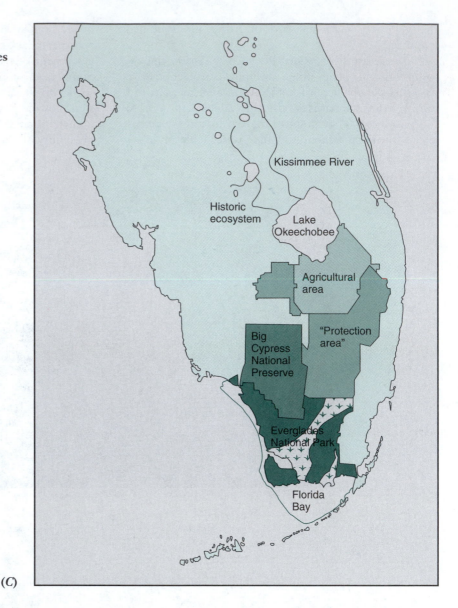

(C)

span of time over which these events have developed. In the coming years these issues will intensify and be joined by sharper debate over means to control further growth in population and the impact of development on the environment. Considerations of long- versus short-term benefits, the interplay between economics and aesthetics, and the need to resolve conflicts and competing interests between groups will further complicate the dialogue. And in addition, an impending revolution in molecular genetics will confront us with opportunities and decisions likely to shake some of our deepest beliefs. We think that a knowledge of biology is essential in these debates, and that citizens in a democratic society should be informed about these issues.

This is not, however, a book about ecology. We are raising ecological issues because they illustrate, simply and directly, that we are all part of nature. There is much more to that idea, however, than recog-

nizing our human capacity for creating ecological havoc. As with every other living organism, evolution has given shape to our biological and social worlds in ways that perhaps you have never considered. So although a rudimentary knowledge of basic biological principles will assist you in making personal decisions throughout your life, we have a deeper purpose in writing this book. We hope to enrich your comprehension of history and of our relationships with other living creatures as well as with the origins, functions, and diversity of behavior and of social structures, because this broader understanding provides insight into the complex motivations that people bring to social, economic, and political issues.

A final example may help to convey the scope of this exploration. The twentieth century has seen humans slaughtering other humans on an unprecedented scale, from the Holocaust of World War II to the

Khmer Rouge regime in Cambodia, to pick two particularly massive and brutal examples. But humans killing humans is not new. In fact much of recorded history is a saga of conflicts between groups, and even the most casual reading of today's newspaper shows that it has not abated with time. An observer from another world might well be perplexed by the paradox of how the same species (us!) can be so brutal and destructive with its own kind while simultaneously showing concern for others and displaying enormous artistic and intellectual creativity. What makes us such a welter of contradictions? Let's see.

SYNOPSIS

Humans are a part of nature in ways that are not fully explored in traditional curricula. This concept provides the rationale for what is an unconventional combination of topics in a biology text.

Human curiosity takes many forms, and one of the most important is science. Science is a way of understanding in which the human intellect is focused on timeless questions of how the world works and came to be the way it is. Science is also a method for acquiring knowledge based on careful observations and experimental testing of alternative explanations. Scientific knowledge is cumulative because each generation builds upon the discoveries and the insights of previous ones. New discoveries sometimes unsettle or overthrow previous understanding; more often they simply embellish it in new ways. Science accelerates its own progress: many advances during the last two centuries have been made possible by the invention of devices for observing and measuring very small and very distant objects, from atoms and molecules to faraway galaxies. Most importantly, science is a mode of understanding that is different from belief systems that remain unchanged and unchallenged from generation to generation.

The fields of science—physics, chemistry, biology, geology, astronomy, and the social sciences—are categories of convenience, as they reflect different levels in a continuum of increasing complexity in the ways matter is organized. Each field has its own unanswered questions and its own methods of investigation because complexity creates *emergent properties* that are not readily predictable from knowledge of lower levels. Because each level is built of components from lower levels, however, science tries to understand the downward connections—how the properties of cells are based on their constituent molecules, how the structure and behavior of organisms derive from the properties of cells, how social systems and ecosystems reflect the interactions of individual organisms, and how entities at each level interact with their surroundings. *Reductionist* ex-

planations of this sort enlarge our understanding of emergent properties.

Natural selection is an extremely important natural process that creates the diversity and complexity of life. The process is itself an example of an emergent property, which results from the ability of large organic molecules, nucleic acids, to replicate with high, but not perfect, precision. To understand life one must understand and appreciate how evolution takes place.

In biology the word "cause" has two distinct meanings. Knowing *how* cells are able to mobilize chemical energy from food or how learning changes nerve cells in a brain are examples of *proximate cause* that require reductionist analyses. Explanations of proximate cause, however, provide no insight as to *why* there are so many different kinds of beetles or why we love our children. Answers to such *why* questions must be sought in terms of evolutionary change. *Evolutionary cause* is sometimes referred to as *ultimate cause*, because it provides insights about the apparent "purposes" that biological systems display. Explanations of proximate cause and evolutionary cause are not alternatives; as we will illustrate throughout the book, they deal with different aspects of nature.

The study of human behavior and human societies has built on Western philosophical and religious traditions that distance humanity from the rest of nature. Consequently, some people have argued that human behavior is entirely shaped by the social environments into which we are born and mature, or that knowledge of evolution does not help us to understand human cultures. As we will show in detail, these views are inconsistent with present knowledge of genetics, the structure and development of the brain, and the evolutionary origins of "human nature."

There are compelling reasons why deeper understanding of these issues is an important educational goal. During the past 200 years humans have had an unprecedented impact on the earth. Although we are the most intelligent species that evolution has produced, we remain unsophisticated in our understanding of ourselves, our origins, and our capacity to plan ahead.

QUESTIONS FOR THOUGHT AND DISCUSSION

1. How would you characterize science as a way of knowing? What makes it particularly effective? Is it the exclusive province of individuals who call themselves scientists? Why?

2. What is meant by the statement that explanations of proximate and evolutionary cause are complementary and not alternatives? Illustrate with an example that is not in the text.

In what social sciences or natural sciences besides biology are historical explanations of cause useful and appropriate? What processes are involved?

3. The idea that new properties emerge as a result of complexity of organization has wide application. What examples can you draw from social, economic, or political organizations of humans? What new phenomena are thereby created?

4. The ecological changes that have taken place in Florida during the last several generations are the result of human behavior. What are some of the choices people have made, and what are possible proximate and evolutionary explanations of cause of those decisions? How might understanding evolutionary causes of human behavior be useful in thinking about public policy? Is this idea likely to generate controversy? Why?

SUGGESTIONS FOR FURTHER READING

Dawkins, R. (1986). *The Blind Watchmaker*. Essex: Longman. An evolutionary biologist skillfully addresses the issue of apparent design of living organisms, starting with the arguments of the Reverend William Paley in late eighteenth century England.

Mayr, E. (1997). *This Is Biology: The Science of the Living World*. Cambridge, MA: Belknap Press of Harvard University Press. The structure of biological science in a historical and philosophical framework, as seen by one of the foremost evolutionary biologists of the twentieth century.

Wilson, E. O. (1998). *Consilience: The Unity of Knowledge*. New York, NY: Alfred A. Knopf. A distinguished biologist provides a sweeping argument for the unity of human knowledge.

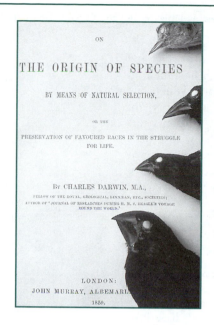

2

Charles Darwin and the Origins of Evolutionary Theory

CHAPTER OUTLINE

Human curiosity tries to make sense of the world and to account for the "whys" of events. Every culture has a traditional account of the origins of the world, of life, or of humankind. These legends reflect curiosity and wonder about the meaning of existence, and they frequently contain wisdom about human nature and the ways people behave. Each contains symbolism that is important to the members of its culture. And many of these stories are genuinely enchanting.

Science began to change our ideas about the history of the earth during the nineteenth century, and the process has accelerated with the application of new techniques from physics, chemistry, geology, and biology. As a result, the recognition of evolution and the basic understanding of how it works comprise one of a few truly revolutionary changes in scientific thought—the sort that philosophers of science refer to as a "paradigm shift."

We begin this chapter with a brief discussion of Charles Darwin, whose role in biology is comparable to that of Isaac Newton in physics. This historical ap-

Photo: In 1859 Charles Darwin suggested a natural process by which life has changed and diversified through time. Although his ideas have been refined and extended, natural selection remains one of the most important concepts of science. The finches and tortoises of the Galapagos Islands off the coast of Ecuador were among the many organisms that influenced him, but only recently have studies of "Darwin's finches" provided direct evidence of natural selection in a natural population.

proach provides a glimpse of how Darwin worked and thought, some of the evidence on which he relied, as well as the magnitude of his contributions. When one sees what critically important information was not available to him (or his contemporaries), his insights seem even more impressive. We will end with a brief discussion of some common misconceptions about evolution and how they are being exploited in contemporary American politics to the detriment of science classrooms.

Compared to ancient creation stories, scientific understanding does not diminish wonder, for in the words of Charles Darwin:

> There is grandeur in this view of life with its several powers having originally been breathed into a few forms or into one, and . . .from so simple a beginning endless forms, most beautiful and most wonderful, have been and are being evolved. [*On the Origin of Species*, p. 434]

CHARLES DARWIN

Charles Darwin (1809–1881) was the son and grandson of English country physicians. Nothing in his early life seems to have foreshadowed his later contributions. He started medical training at Edinburgh at age 16, but quickly found it not to his liking. Somewhat at a loss as to what to do with his life, he took up studies at Cambridge, where, while casually thinking about the ministry, he pursued avocational interests in geology and natural history and was befriended by John Henslow, the Professor of Botany, and Adam Sedgewick, the Professor of Geology.

THE VOYAGE OF THE BEAGLE

When he was only 23 years old, Darwin was offered the position as naturalist on *H.M.S. Beagle*, a small naval vessel engaged in charting coastal waters in distant parts of the world. Because neither his mentor Henslow nor another man felt able to accept the appointment, the offer came to Darwin quite by chance, but the impact of the journey set the course of Darwin's life. Figure 2.1 is a portrait of Darwin as a young man.

Darwin's name is linked with the concept of evolution, but to understand his contributions we should first consider how the society into which he was born saw nature. At the start of the nineteenth century, understanding of geology was very different from what it is today. Reckoning from Old Testament genealogies, the earth was commonly taken to be only several thousand years old. There was neither an appreciation of the true age of the earth (about 4.5 billion years) nor of the forces that have been and are now at work in shaping its surface. Knowledge of radioactive elements and

FIGURE 2.1 This watercolor of Charles Darwin was painted by the artist George Richmond in 1840, four years after the *Beagle* returned to England, when Darwin was 31 years old.

their use in dating rocks were yet to be discovered. Similarly, the significance of volcanic activity and earthquakes, the forces of glaciers, and the existence of plate tectonics were just dimly perceived by a few naturalists or were still completely unknown.

In Europe (as elsewhere), understanding biology was largely the domain of philosophers and theologians, professional and amateur. The Judeo-Christian tradition saw the world as the handiwork of an anthropomorphic God, a view with two important implications. First, final causes—the Aristotelian concept of purposes for which organisms exist and events occur (Chapter 1)—were ascribed to the will of God. Second (and also traceable to Greek roots), was the idea that organisms, as God's creations, all have an essential quality, defining their nature, unchanging in time, and reflecting the way God would have things be. Clearly individual organisms of the same kind frequently differ from one another, but these differences were taken to be minor perturbations without great significance.

At the time the *Beagle* sailed from England in 1831 change was in the air. Science, by appeal to observation and experiment, was assuming an increasing importance in understanding the world. In the budding science of geology, there was growing conviction that the earth was much older than Biblical accounts implied, but the church continued to influence interpretation. For example, the presence of fossils of clearly different ages led to the suggestion that there had been multiple creations. In France, Jean-Baptiste de Lamarck (1744–1829) had proposed that new forms of life were emerging all the time, either from other living organisms or from inanimate matter. This was one of several pre-Darwinian postulates of evolution, but it did not meet wide acceptance. Lamarck's countryman Cuvier countered with two arguments that continued to echo through the nineteenth century and that can still be heard from creationists to this day: current forms of life are so close to perfection that they would be seriously compromised by any change, and the fossil record fails to reveal all intermediate life forms. Biology had yet to shake free of traditional interpretations that were not drawn directly from nature.

Darwin was away from England with the *Beagle* for nearly five years. He had with him Charles Lyell's newly published *Principles of Geology*, an insatiable curiosity, a keen eye for observation, charm that enabled him to enlist assistance from both friends and casual acquaintances, and at that time in his life, considerable physical stamina and skill at riding and hunting. The *Beagle* spent over three years along the coast of South America, from Brazil south to Argentina, around Cape Horn, and north along the shores of Chile and Peru before striking west across the Pacific and circumnavigating the globe (Fig. 2.2). Its mission of charting coastal waters was prompted by the growing role of Britain as a maritime and naval power, and it left Darwin much time to go ashore for extended periods.

Three sorts of observations influenced Darwin's subsequent thinking. First, he was prepared by his earlier study of geology and by Lyell's book to see evidence that the earth was old and that it had been shaped over time by the same agents of change that can be seen at work today, including weathering, erosion, glaciers, and volcanoes. While ashore on the west coast of South America, Darwin experienced a substantial earthquake, which led him to realize how fossil beds of marine organisms he had observed earlier could have been lifted so far above the present sea level (Fig. 2.3).

Second, he also perceived that not only were the South American plants and animals different from those in Europe (Fig. 2.4, page 25), the Argentine soil contained the bones of still other creatures that were no longer living. He shipped back to England

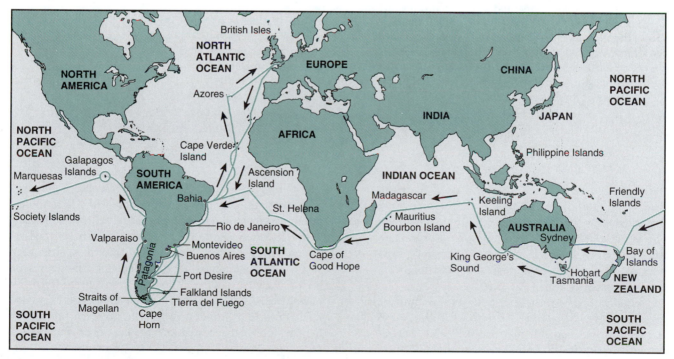

FIGURE 2.2 The *Beagle* left England on December 27, 1831 and did not return for nearly five years. During the first four years the ship sailed near the coast of South America, which gave the young Charles Darwin many opportunities to go ashore.

FIGURE 2.3 While ashore in Chile, Darwin experienced an earthquake. Although he was far from the center of destruction when the quake occurred, the experience was unsettling, and the imagery in his later description seems to foreshadow the concept of plate tectonics. "A bad earthquake at once destroys the oldest associations; the world, the very emblem of all that is solid, had moved beneath our feet *like a crust over a fluid*; one second of time has created in the mind a strange idea of insecurity, which hours of reflection would not have produced." (*Italics added.*)

This figure is from a drawing by the lieutenant of the *Beagle*, J.C. Wickham. It shows the ruins of the cathedral in Concepcion, Chile, which was close to the epicenter and was reduced to rubble in about six seconds. This experience of the earthquake, reinforced by his visit to Concepcion shortly afterward, suggested to Darwin how the great forces of geological activity could have created mountains and shifted the locations of fossil-bearing rocks.

the fossilized remains of a species of a giant ground sloth the size of an elephant that lived in South America about half a million years ago. Off the coast of Ecuador he encountered an isolated island group, the Galapagos Islands, where tortoises and birds on one island sometimes appeared to be distinct from those on other islands (Fig. 2.5).

Finally, Darwin encountered several human cultures that differed from those with which he was familiar in Europe and in the Spanish and Portuguese settlements in South America. Most notable were the indigenous people at the extreme tip of the South American continent, Tierra del Fuego, the land of fire, so named by earlier Spanish explorers who had been impressed by the many fires that these people built

every night for warmth. These observations showed him alternative ways in which people can adjust to different environments, and the observations remained with him when years later he wrote *The Descent of Man*.

When Darwin returned to England, he spent a year at Cambridge organizing the specimens he had sent back, resided briefly in London, then settled with his new wife in Down, Kent, a small town about twenty miles outside of London, where he lived for the rest of his life. He had independent means and never held a university or other position. A few years after his return to England Darwin began to suffer from a recurrent gastrointestinal problem that made him physically miserable for much of the rest of his life. He is buried in Westminster Abbey along with many of England's great.

FIGURE 2.4 Among the unfamiliar animals that Darwin encountered in South America were llamas (*right*) and marsupials like the opossum (*left*). That similar habitats in different parts of the world are frequently occupied by very different kinds of plants and animals seemed to Darwin inconsistent with the then-current concept of creation. His many observations of the diversity and geographical distribution of organisms formed part of the foundation of his concept of evolution.

Opossums belong to a primitive group of mammals called marsupials, which characteristically carry their newborn young in a pouch. Although most marsupials are found in Australia, because of the movements of continents in geological history there are species that occur today in South America and one (*shown here*) that has made its way across the Isthmus of Panama into North America. Llamas, along with guanacos, vicuñas, and alpacas are South American relatives of camels. North America was the center of evolution of this group of mammals.

FIGURE 2.5 On the Galapagos Islands about 500 miles off the coast of Ecuador, Darwin found a group of closely related birds. These birds are known from nowhere else in the world, but they are most similar to species in South America. "Why," he wrote, "should the species which are supposed to have been created in the Galapagos Archipelago, and nowhere else, bear so plain a stamp of affinity to those created in America?" His answer was that these islands had originally been populated by a few individuals (perhaps blown there by storms), and evolutionary divergence then ensued. The birds are now known as Darwin's finches. In recent years they have been studied intensively by a group of biologists at Princeton University who have documented measurable evolutionary changes occurring in the course of a few generations (Chapter 4). These three are, from left to right: *Geospiza magnirostris, G. parvula,* and *Certhidea olivacea.*

Change is an obvious feature of nature. Plants and animals grow and develop from fertilized eggs. Landscapes change as trees succeed grasses and other smaller plants after a fire, or as weathering, erosion, or cataclysmic events like volcanoes alter the terrain. Lineages of organisms evolve through many generations, and astronomers speak of the evolution of stars.

Many different processes bring about change, and because these processes operate on vastly different time scales, involve quite different physical entities, and work by different natural processes we can be easily misled by similar terminology. On the one hand, biological and stellar evolution are so different in all three respects that they can share the word evolution with little risk of confusion. In the eighteenth and nineteenth centuries, however, prior to the recognition of natural selection as a key to understanding biological diversity, the words development and evolution were frequently used interchangeably. The words now apply to two distinct biological processes. Evolution occurs on a time scale of many generations and involves a sorting of alternative genes (Chapters 3–5). Development, in contrast, involves the differential expression of genes in the life of an individual organism, directing the changes that take place in getting from a fertilized egg to an adult plant or animal (Chapter 10). A pre-Darwinian use of the word development nevertheless still occasionally enters our language. For example, the editors of biology textbooks written for the high school market have been known to employ the word development for descent with modification in a deliberate effort to avoid using the word evolution.

DARWIN'S CONTRIBUTIONS

Darwin made two major contributions to scientific knowledge. The first was to document in meticulous detail the evidence that during the history of the earth organisms have changed and that this is an important natural process. He called it "descent with modification", but we know it now as evolution (Box 2.1).

His second and more original contribution was to propose a mechanism by which evolution has occurred, *natural selection*. He was led to the concept of natural selection by reading in 1838 (two years after the *Beagle's* return) an essay that had originally been written in 1798 by an English economist, Robert Malthus. Malthus's thesis was that famine was the inevitable result of growth of human populations, but Darwin saw in that notion the germ of another and more sweeping idea. It occurred to him that variations among individuals of the same species might not be imperfections of little consequence, but could represent differences that could lead to increased or decreased reproductive success. Out of this realization came the concept of natural selection: the process through which some individuals, because of small differences, are better able to leave offspring than others.

Cognizant of the implications, Darwin worked and thought but delayed publishing his ideas about natural selection for twenty years. In June 1858, however, his hand was forced when he received a manuscript from a younger naturalist, Alfred Russel Wallace (1823–1913), who was then in the Moluccas (now a part of the Indonesian Archipelago). Wallace, also influenced by Robert Malthus, had independently arrived at the concept of natural selection. Darwin was distraught, but several of his eminent scientific friends who were aware of Darwin's years of work, urged him to present a summary of his own ideas along with Wallace's paper at the Linnean Society of London thirteen days later. Darwin then set about writing *On the Origin of Species*, which was published the following year. History has given Darwin the bulk of the credit, principally because he documented the case for evolution and natural selection far more extensively than Wallace.

In 1859 the concept of natural selection was a bold and imaginative accomplishment. The reason is that neither Darwin nor any of his contemporaries had a clear understanding about the source of the variation that makes one individual different from another. In his day, nothing was known about genetics, but that is anticipating another part of the story. In subsequent chapters we will describe how understanding of evolution has matured since Darwin's time (although without diminishing at all the importance of his insights), and we will present a more detailed account of natural selection.

A final and important point about Darwin. He was not only a keen observer and a careful and critical thinker, he also tried to test a number of his ideas. While still on the *Beagle* he conceived of atolls—those Pacific islands consisting of little more than a narrow ring of sand surrounding a shallow lagoon and flanked by coral reefs on the seaward side—as having originated by the slow submergence of volcanic islands. He realized that if the growth of coral could keep up with the changing sea level, the original volcanic center of the island could slip below the surface of the sea, leaving only the barrier reef exposed. He knew that corals are shallow-water creatures, and that if his hypothesis was correct, the outer slopes of the reefs of atolls should descend well below the deepest living polyps. He tested this deduction by lowering wax-covered weights over the side of a boat, and from impressions left in the wax he showed that the living reefs were built on a deep wall of long-dead coral (Box 2.2).

Years later at his home, Down House, and concerned with natural variation, Darwin recognized that artificial selection of domesticated plants and animals was basically similar to natural selection (although accelerated by human direction), and he immersed himself in the world of pigeon fanciers in order to measure the extent of variation displayed by pigeons, to compare juveniles and adults, and to try some breeding experiments of his own. When he turned his attention to the geographic distribution of plants and animals, he was faced with problems such as how remote oceanic islands could have been colonized or how the eggs or larvae of freshwater invertebrates could get from one lake or stream system to another. This led him to measure the number of viable seeds contained in a small quantity of pond mud, to test the viability of seeds after extensive immersion in salt water or the guts of birds, and to investigate the capacity of mollusk larvae to attach to the feet of ducks and remain alive when exposed to air. In other words, if an idea seemed plausible and testable, Darwin frequently conducted simple experiments.

In addition to *On the Origin of Species*, he wrote over a dozen books describing his observations and experiments: *The Voyage of the Beagle*, treatises on the geology of South America, *The Variation of Animals and Plants Under Domestication*, *The Descent of Man and Selection in Relation to Sex*, *The Expression of the Emotions in Man and Animals*, as well as a diversity of other subjects including barnacles; the ways in which orchids are fertilized by insects; male and female flowers on plants of the same species; cross- and self-fertilization in plants; the effect of light on the growth and movement of plants; and the role of earthworms in conditioning soil.

FIVE PILLARS OF EVIDENCE

There are five, largely independent lines of evidence that life on this planet has evolved over a span of several billion years. As we pointed about above, one of Darwin's contributions was to assemble the available evidence to form a coherent theory of change. In the intervening years, however, substantial new lines of evidence have emerged, and the *reality* of evolution as a profoundly important natural process responsible for shaping life on earth *is not a matter of scientific controversy*.

There is nevertheless a curious paradox in our culture, where the products of scientific discovery have had such profound and unprecedented technological impact on every facet of our lives from communication to transportation to medical treatment. Despite the obvious power and impact of scientific knowledge, there are still many people who not only deny the existence of evolution in their personal understanding of the world, but they object to its presence in school curricula. Because of the impact of this movement, we have chosen not only to review the scope and diversity of the physical evidence for biological evolution, we also consider some of the objections of its critics.

(1) PALEONTOLOGY AND PHYSICAL STUDIES OF THE EARTH'S AGE

Knowing that the earth is approximately 4.5 billion years old is an important piece of information. It provides the time frame over which life arose (Chapter 3) and subsequently evolved, and it demonstrates that the earth is far older than 6000–10,000 years. The universe is estimated to be 10–20 billion years old, and in that framework our solar system is a relative newcomer that formed from a condensation of dust and gas of the sort that can be observed elsewhere in the universe. The ages of rocks can be determined by measuring the ratios of isotopes (Box 2.3, page 30), and 4.54 billion years for the age of the earth is a consensus figure based on a variety of different isotopes with different decay rates as well as a study of meteorites and rocks brought back from the moon.

A variety of geological processes work to create the rocks under our feet. Material made fluid by the interior heat of the earth and solidified near the surface is *igneous* rock. Igneous rocks are used for radiometric dating. Sand and mud weathered from the highlands by wind and rain are washed to swamps, lakes, and oceans and become compacted as *sedimentary* rock. Both igneous and sedimentary rocks can subsequently be distorted by the heat and great forces associated with plate tectonics to form *metamorphic* rocks.

Box 2.2
Coral and the Formation of Atolls

Corals are marine animals in the Phylum Cnidaria (sometimes called Coelenterates) that live in warm, shallow water. Each individual or *polyp* (Fig. 2.6A) is little more than an open cup of tissue with tentacles around the rim to help capture particles of food, like a tiny, upside-down jelly fish. Corals, however, grow in massive colonies and secrete calcium carbonate. Some species grow as delicate fans, others as branched "stag horns" (Fig. 2.6B), and others as massive walls, collectively forming coral reefs along tropical shores (Fig. 2.7).

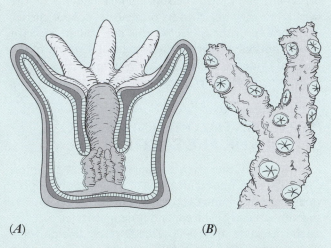

(A)   *(B)*

FIGURE 2.6 (A) A coral polyp is structurally simple, with only two layers of cells, an oral opening surrounded by a ring of tentacles, and a digestive cavity that ends blindly. (B) Twelve polyps embedded in a calcareous matrix that they have secreted.

FIGURE 2.7 Coral reefs provide a stable habitat for numerous species of coral, fish and invertebrates. In much of the world they are in severe danger from silting and pollution caused by human activity.

Sedimentary rocks—of which sandstone, limestone, and shale are examples—are deposited in layers (Fig. 2.10, page 31), and unless the terrain is later disrupted, the oldest layers are on the bottom. Moreover, sedimentary rocks frequently have preserved within them the hard remains of plants and animals that were living at the time the deposits formed. Bones of vertebrates, shells of mollusks and other invertebrates, trunks of trees, and impressions of leaves or animal feet are among the objects most likely to form such *fossils*.

Fossils therefore provide the first and most obvious evidence about the history of life on earth. Moreover, they provide the only evidence for how extinct organisms actually looked.

Among the practical problems in reading the fossil record are missing pages in the book. Sedimentary deposits usually do not accumulate on top of each other in uninterrupted sequence. Locally, layers of sediment often erode away before new deposits form, and beds of rock can be lifted up or tilted and greatly altered by

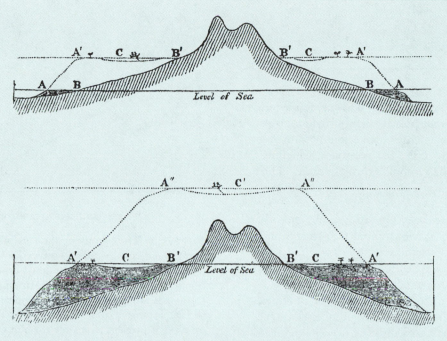

FIGURE 2.8 Darwin's diagrams showing how atolls form.

Figure 2.8 shows Darwin's drawings illustrating the formation of atolls. The upper panel is the profile of a volcanic ocean island. When the island is geologically young, the coral starts to grow between *A* and *B*, forming a fringing reef just offshore. As time passes, the land subsides, represented in these drawings by a rise in sea level. Because this geological process is so slow, the growth of the coral is able to keep up. By the time the sea has risen to the level shown by the upper horizontal line, the coral has formed a massive reef. Little islets have formed at *A'*. The main island, now much smaller in size, is surrounded by a lagoon (*C*) and a barrier reef (*A'*). Darwin was able to show that in deep water on the outer face of the barrier reef the coral had died, indicating that it had at one time been much closer to the surface.

The lower panel shows the profile when the original volcanic island has sunk below the surface of the sea. The coral has continued to grow, and a flat sandy atoll (*A''*) around the edge of a lagoon (*C'*) is all that remains. The reef is frequently broken on one side, so the atoll does not form a complete ring.

heat and pressure. In 1799 an English engineer named William Smith realized that where stretches of the record were undisturbed, fossils varied in a sequence from the oldest to the youngest rocks, and he suggested that similarity in fossils from different locations could therefore be used in interpreting the *relative* ages of rocks at sites where part of the sequence was missing. Geologists were thus beginning to use fossils in sedimentary rocks as measures of time several decades before there was a widely shared concept of evolution and more than a century before it was possible to estimate the *absolute* ages of associated igneous rocks by using radiometric methods.

But Darwin saw something new in the temporal sequences of fossils: evidence for descent with modification from earlier common ancestors. Supporting evidence for this interpretation has now accumulated for nearly a century and a half, but what features of the fossil record lead so compellingly to this conclusion? In order to answer this question we can examine a sum-

Box 2.3
Dating the Ages of Rocks

Consider a *radioactive element*, N, that becomes incorporated into a rock when the rock forms. Radioactive decay is a spontaneous process that occurs at a constant rate in which one element is converted into another element or an isotope of the same element, which we will represent by the symbol D (for daughter). Thus

$$N \rightarrow D + \text{energy}$$

Among the processes used for dating rocks and other substances are the decay of rubidium to strontium, potassium to argon, uranium to lead, and ^{14}C to ^{12}C. These elements all have different rates of decay and can be used to measure rocks of different ages.

The number of atoms of N that decay in any period of time is proportional to the number that are present. This is not a difficult idea; it is analogous to pointing out that the rate of purchase of hot dogs during a baseball game is proportional to the number of fans in the stadium. Change that follows this kind of rule is *exponential*, and you met exponential growth in Chapter 1. N decays exponentially and D increases exponentially, and Figure 2.9 shows plots of N and D as a function of time.

Those interested in more quantitative reasoning are invited to read to the end of the box. In mathematical terms $\frac{dN}{dt} = -kN$, where the proportionality constant k is different for every radioactive element. Solving for N as a function of time, measured from the formation of the rock:

$$\frac{N}{N_0} = e^{-kt}$$

where N_0 is the amount of the decaying isotope initially present. As N disappears it is replaced by D, so $D = N_0 - N_0 e^{-kt}$

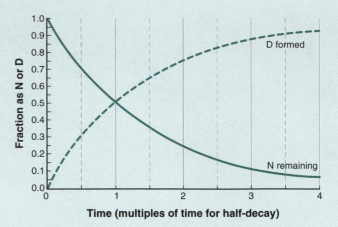

FIGURE 2.9 The exponential decay of a radioactive element (solid curve) and its conversion into a daughter isotope (dashed curve). The units on the *y*-axis are the fraction of N remaining or D formed and the units on the *x*-axis are multiples of the time required for half of the remaining N to decay. Note that regardless of how much N is present, the time for half of it to decay is a constant. Of course the time for half-decay is different for each radioactive element.

The value of N_0 is not known; one can measure only the amounts of N and D that are present now. But as $D = N_0 - N = Ne^{kt} - N = N(e^{kt} - 1)$,

$$t = \frac{\ln\left(\frac{D}{N} + 1\right)}{k}.$$

It is thus possible to calculate t from measurements of D and N and knowledge of k. The validity of this expression obviously depends on all the D that is formed remaining trapped within the rock, and elements are used in which this condition will be met. The expression also depends on there not being any D present when the rock was formed. Sometimes, however, some D may be present at the outset, but techniques are available to estimate this initial amount independently and subtract it from the total to get the amount that has arisen from the decay of N.

mary of the history of the earth as it is revealed by the presence of fossils. Table 2.1 is such a summary, and it is a more accurate record than was known in Darwin's time. Not only is it now possible to determine the ages of fossil deposits, but hundreds of thousands of additional fossils have been discovered since the middle of the nineteenth century.

Geologists divide this scale of time into four great eras, each subdivided into several periods. The names

of these geologic periods are not inviting. Some refer to places where deposits were first described—Devonian for the English county Devonshire; Jurassic for the Jura Mountains near Geneva. Others refer to characteristics of deposits—Carboniferous for the presence of coal; Cretaceous for chalk.

The principal criteria that first distinguished these geologic periods are the characteristic plants and animals found in their rocks as fossils. This is because sev-

FIGURE 2.10 Layers of sediments are deposited in lagoons, lakes, seas, and river deltas. If this continues long enough, the sediments are compressed, heated, and transformed into sedimentary rocks. If these rocks are later uplifted and exposed to erosion by wind and water, the kind of landscape shown here emerges.

TABLE 2.1 Geological Time

Era	Period	Million Years from Start to Present	What Happened
Cenozoic	Quaternary	2	Many glaciers; many mammals become extinct; humans evolve
	Tertiary	65	Adaptive radiation of mammals, birds, and flowering plants
Mesozoic	Cretaceous	144	Mammals and flowering plants starting to diversify; continents starting to separate; mass extinctions and end of dinosaurs close the period
	Jurassic	213	Many dinosaurs; first birds; gymnosperms are dominant plants; North America and Eurasia joined; South America and Africa close
	Triassic	248	First dinosaurs and mammals; rise of gymnosperms; marine invertebrates diversify; continents begin to drift apart; period ends with mass extinctions
Paleozoic	Permian	286	Diversification of reptiles; decline of amphibians; many kinds of insects; single major continental mass; glaciations; mass extinctions at end of period
	Carboniferous	360	Early vascular plants abound (including ferns and many forms extinct today); many amphibians; first reptiles; radiation of insects
	Devonian	408	Bony and cartilaginous fish originate and diversify; early arthropods (trilobites) diverse; first amphibians and insects; mass extinction near end of period
	Silurian	438	Many agnathans (jawless fishes); placoderms (early jawed fish with finlike structures) appear; vascular plants and arthropods appear on land
	Ordovician	505	Agnathans (jawless fishes) arise; diversification of invertebrates; mass extinction at end of period
	Cambrian	570	Appearance of most animal phyla; many algae
Precambrian	Vendian	670	Origin of nucleated cells and appearance of first multicellular animals
	Sturtian	800	

eral of these periods are marked by mass extinctions near their close (the Ordovician, Devonian, Permian, Triassic, and Cretaceous), in which numerous fossils simply disappear, never to be found in later strata. The following period then reveals an extensive diversification of the remaining organisms in which many new forms appear. The history of life has thus not been tranquil and has been punctuated by climate changes and asteroid impacts as well as influenced by the movement of continental masses. Many times more plants and animals have become extinct than now exist.

The age of the earth and the time scale of evolution are so vast that they challenge comprehension. Darwin, who did not have the benefit of the numbers in Table 2.1, nevertheless had a keen insight about geologic time based on measurements of the thickness of sedimentary rock strata, rates of sedimentation, knowledge that vast parts of the sedimentary record had been lost through erosion, and his own observations of the slow rates of erosion of sea cliffs. "The consideration of these facts", he wrote, "impresses my mind almost in the same manner as does the vain endeavour to grapple with the idea of eternity."

In order to recast the immensity of geologic time onto a more familiar scale, we can adapt a familiar story and collapse the 4.54 billion years of the earth's history into the six working days of the week (Fig. 2.11). On this scale, the first fossil evidence of life appears during the morning of the second day (about 3.5 billion years ago). These organisms are similar to bacteria living today, and their features will be described in the next chapter. The first cells with nuclei—cells of the kind that are present today in all organisms other than bacteria—made their appearance on Thursday (about 2 billion years ago). None of this fossil evidence was known to Darwin. It is important to realize that there are few fossils of any kind known from the Precambrian era; there were few hard structures to fossilize, and the time has been so long that erosion and metamorphosis have largely destroyed or altered the sedimentary deposits of that era. Nevertheless, from what fossils are available it appears that the evolution of nucleated cells may have been a major step that took considerable time to happen.

Shortly before dawn on Saturday, the sixth and last day, multicellular animals appear, followed by representatives of most of the present-day phyla. All of the evolutionary history of multicellular life unfolds during Saturday, and around 10 PM (end of the Cretaceous period) dinosaurs pass from the earth. The diversity of mammals evolves after 10 PM, and around 15 minutes before midnight protohumans become distinct from apes. The first signs of agriculture appear with only about 1.5 seconds remaining in the week, and the Jewish and Christian calendars represent but fractions of a single second on our 6-day scale.

FIGURE 2.11 The vastness of time over which evolution has occurred is hard to grasp. By scaling 4.6 billion years into six days, however, the tiny fraction of time that has been occupied by the human species becomes more readily apparent.

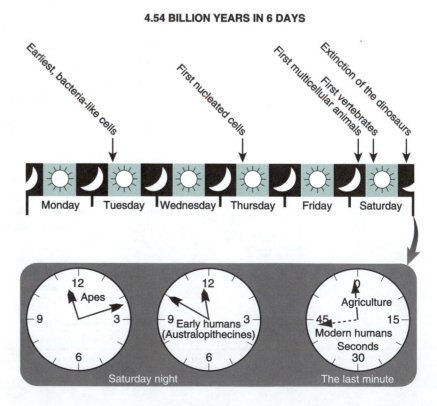

4.54 BILLION YEARS IN 6 DAYS

Box 2.4
Descent with Modification

The first Chordates (animals that share with embryos of more familiar vertebrates a supporting structure along the back called the *notochord*) appeared in the upper Cambrian period. Jawless, scaleless fish without paired fins called Ostracoderms are found in the Ordovician period (505 m.y.a.—million years ago) (Fig. 2.12A), and an extinct group of jawed fish with teeth and open gills called Placoderms arose in the Silurian period (438 m.y.a.), to be followed by bony and cartilaginous forms in the Devonian period (408 m.y.a.). The group of bony fish with fine bones in their fins that exist today—*teleost* fishes—did not arise until the Triassic period (248 m.y.a.). Familiar cartilaginous fish are sharks and rays, but these modern forms did not appear until still later, in the Cretaceous period (144 m.y.a.).

The Permian period (286 m.y.a.) saw the emergence of fish with lobed fins and internal nostrils (Fig. 2.12B–D). One group called crossopterygians was thought to be extinct until a representative called the coelacanth was fished from the Indian Ocean about 60 years ago. Amphibians, derived from crossopterygians, appeared in the Devonian period (408 m.y.a.) (Fig. 2.12E,F) and became abundant in the Carboniferous period(360 m.y.a.), then declined in numbers during the Permian period (286 m.y.a.). Some of the amphibians of the Carboniferous period were huge by present standards, reaching 15 feet in length. Amphibians are represented today by frogs and salamanders; the first fossil frogs and salamanders date from the Jurassic period (213 m.y.a.).

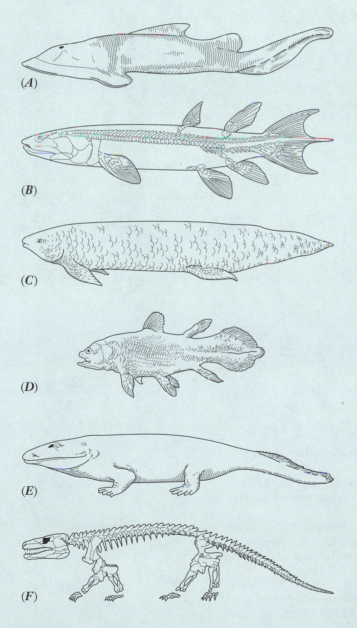

(A)

(B)

(C)

(D)

(E)

(F)

FIGURE 2.12 (A) *Acerapsis robustus*, an ostracoderm from the Devonian period. Along with hagfish, jawless, parasitic lampreys are the closest living relatives of ostracoderms.

(B) A Devonian lobefin fish, *Eusthenopteron*. The lobefins (crossopterygian fishes) gave rise to the amphibians.

(C) The Australian lungfish *Neoceratodus forsteri*, a modern descendant of crossopterygian fishes, still has a primitive lung, which was characteristic of many Paleozoic fishes. In modern (teleost) fishes oxygen is obtained through the gills, and the "lung" is present as the air-filled swim bladder, which counters the weight of the skeleton and enables fish to maintain their position in the water column without having to expend energy by swimming. In some species the swim bladder has also assumed a secondary function in producing or receiving sound signals.

(D) The coelacanth *Latimeria*, the only lobefin living today. It is found in the sea east of Africa.

(E) *Ichthyostega*, one of the earliest known amphibians, lived in the late Devonian period in what is now Greenland. It probably did most of its walking on the bottom of the bodies of fresh water where it lived.

(F) *Seymouria* was an amphibian of the Permian period of the sort that gave rise to reptiles.

(continued)

The end of the Permian period was marked by the largest number of extinctions the world has seen. The impact appears to have been greatest on marine invertebrates; by some estimates over 90% of species were extinguished. During the Mesozoic era that followed, the breakup of earth's single continent occurred, and the plants and animals on the separate land masses, be-coming isolated by ocean barriers, began to evolve in distinct ways. The presence of major barriers to migration captured Darwin's attention, although he knew nothing of the history of continental drift.

The first reptiles appeared in the Carboniferous period (360 m.y.a.), becoming more numerous as the Paleozoic era ended and the Mesozoic periods pro-

FIGURE 2.13 The Mesozoic era saw a diversity of reptiles, some quite large, with various means of locomotion and exploiting a variety of foods.

(A) A mososaur, *Platecarpus*, of the late Cretaceous period, found in what is now Kansas. Much as seals and whales of today are adapted to live in the sea, this large lizard was also a marine animal.

(B) The dinosaurs are a group of reptiles that became extinct at the end of the Cretaceous period. *Brachiosaurus* was a giant herbivore of the late Jurassic period.

(C) *Ceratosaurus* was a late Jurassic carnivorous dinosaur. It was a bipedal, with a heavy tail for balance, and was likely very agile for its size.

(D) *Monoclonis*, a dinosaur of the late Cretaceous period suggests a rhinoceros.

(E) In addition to birds, some reptiles (pterosaurs) and mammals (bats) have evolved the capability of flying. This large pterosaur from the late Cretaceous period is *Pteranodon*.

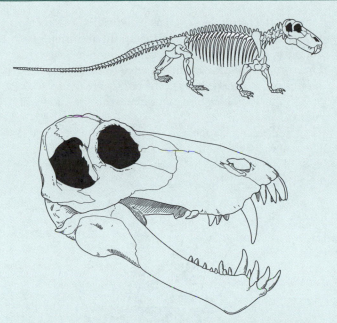

gressed. This is why the Mesozoic is sometimes called the Age of Reptiles. Reptiles are the first group of vertebrates to evolve the capacity to live on land throughout their lives. An impervious skin protected them from desiccation, and eggs with shells freed them from the need to return to water to reproduce. Dinosaurs appeared in the late Triassic period and flourished until the second largest mass extinction at the end of the Cretaceous period (Fig. 2.13A–D). The Mesozoic reptiles were a diverse lot and included separate lines that gave rise to crocodilians, turtles, lizards, birds, and mammals (Fig. 2.14).

The first evidence for mammals is found in the Triassic period (248 m.y.a.). Mammals arose from a group of reptiles called *therapsids*, but they did not start

FIGURE 2.14 *Titanophoneus* was a late Permian reptile with some features that relate it to mammals. It was over 8 feet in length.

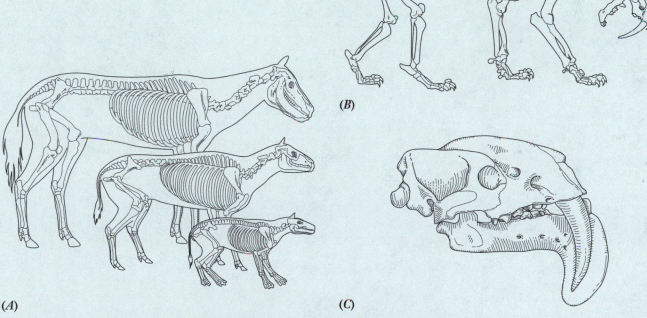

(A)   (B)   (C)

FIGURE 2.15 (A) The evolution of horses began when there was a land connection between North America and Asia. Representatives in the Old World became extinct several times but were reintroduced over the land bridge. They became extinct in North America during the Pleistocene epoch and were reintroduced by Europeans in the sixteenth century. Many fossil horses are known from several different lineages. These are three examples dating (from small to large) from roughly 50, 35, and 20 million years ago.

(B) Skeleton of the saber-toothed cat *Smilodon californicus* from the Pleistocene tar pits at Rancho La Brea near Los Angeles, California.

(C) Skull of a South American saber-toothed marsupial mammal *Thylacosmilus* that lived several million years ago. Note the similarity with the saber-toothed cat, a placental mammal. This similarity in appearance evolved independently in two different lineages and is an example of *convergent evolution*.

(continued)

to become abundant until after the close of the Cretaceous period (65 m.y.a.). The Cretaceous extinctions
appear to have eliminated virtually all terrestrial animals larger than about 50 pounds. Until this time, representatives of the mammalian lineage had been mostly
small and probably nocturnal. Since dinosaurs did not
survive to the Cenozoic era, new ecological opportunities were opened for mammals, and a profusion of new
species has evolved during the last 65 million years: rodents with teeth for gnawing, the many hoofed mammals with teeth and digestive systems for eating plant

material, the carnivores such as wolves and lions that
prey on the herbivores, bats that can fly, and whales
and porpoises that returned secondarily to the water.
This sort of evolutionary divergence to take advantage
of unfilled *ecological niches* is referred to as *adaptive radiation*. Some examples of extinct mammals are shown in
Figure 2.15.

The first birds appeared in the Jurassic period, and
like mammals, they later increased in numbers of
kinds. They appear to have arisen from a group of
dinosaurs (Figs. 2.16 and 2.17).

(*A*)   (*B*)

FIGURE 2.16 (A) Fossil of *Archaeopteryx* in Bavarian limestone about 150 million
years old (Jurassic period). It was found in 1877; other specimens also exist. The
light bones, wings, and impressions of feathers suggest flight, yet the skeleton has
features that are present in carnivorous dinosaurs. As in birds, however, the clavicles
have fused to form the "wishbone." *Archaeopteryx* is therefore believed by many paleontologists to be intermediate between one group of dinosaurs and birds. Recently, remarkably preserved fossil dinosaurs with feather impressions have been
found in China. They further support the dinosaur-bird connection and suggest that
feathers (an elaboration of scales) may have functioned first as insulation.

(B) Rendition of the same fossil

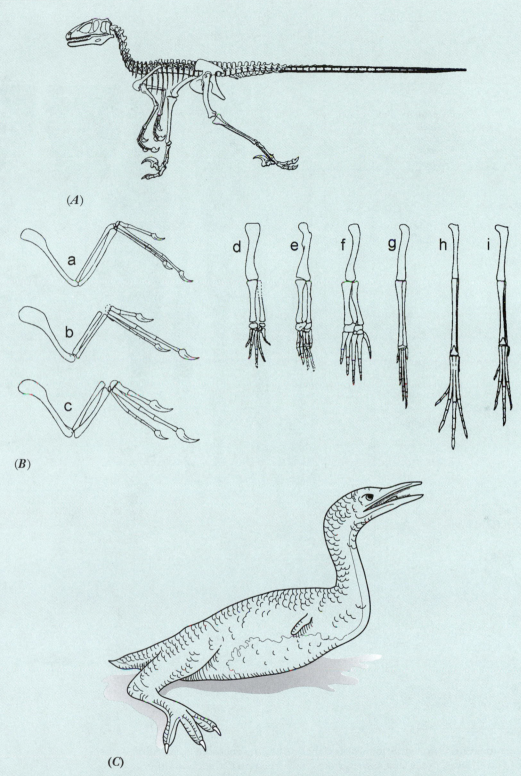

FIGURE 2.17 **(A)** A bipedal, carnivorous dinosaur (*Deinonychus*) of the early Cretaceous period related to the theropod lineage from which birds likely evolved.

    (B) The forelimb of *Archaeopteryx* (*a*) compared with the limbs of two theropod dinosaurs (*b* and *c*); *h* is the hind limb of the *Archaeopteryx* compared with a theropod (*i*) and several more-distantly related dinosaurs (*d–g*).

    (C) *Hesperornis* was a big, loonlike diving bird of the late Cretaceous period. Unlike modern birds, it still had socketed teeth.

(continued)

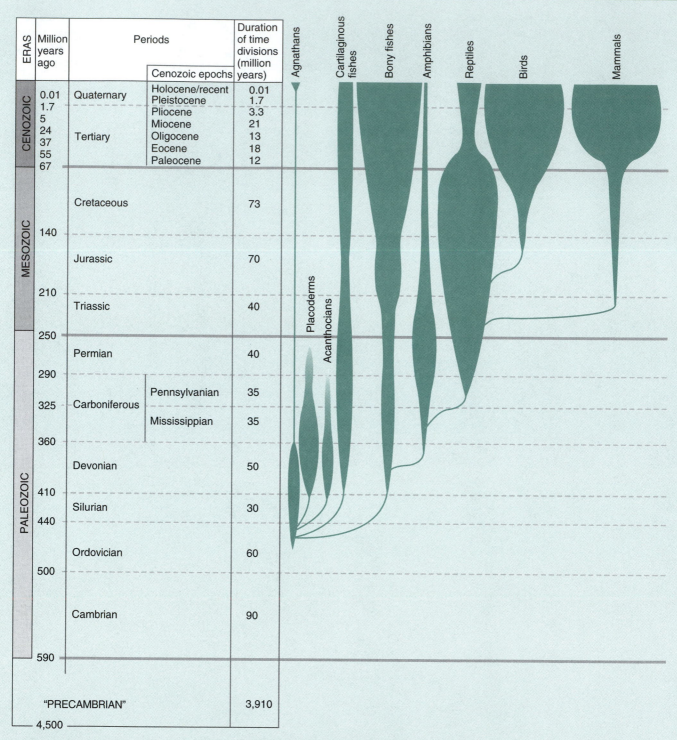

ERAS	Million years ago	Periods		Duration of time divisions (million years)
			Cenozoic epochs	
CENOZOIC	0.01	Quaternary	Holocene/recent	0.01
	1.7		Pleistocene	1.7
	5	Tertiary	Pliocene	3.3
	24		Miocene	21
	37		Oligocene	13
	55		Eocene	18
	67		Paleocene	12
MESOZOIC		Cretaceous		73
	140	Jurassic		70
	210	Triassic		40
	250			
PALEOZOIC		Permian		40
	290	Carboniferous	Pennsylvanian	35
	325		Mississippian	35
	360	Devonian		50
	410	Silurian		30
	440	Ordovician		60
	500			
		Cambrian		90
	590			
		"PRECAMBRIAN"		3,910
	4,500			

FIGURE 2.18 This diagram is an evolutionary tree that shows the relationships of the major groups of vertebrates. Note that some groups have become extinct and others are represented today by only a few species.

The important message in this brief synopsis is that the major groups of vertebrates have arisen sequentially over a broad expanse of time, and in each case the fossil record shows anatomical affinities of the earliest representatives of each group to the groups of vertebrates then existing: birds and mammals to (different) Mesozoic reptiles; the first reptiles to Carboniferous amphibians; and the earliest amphibians to fish of the Devonian period (Fig. 2.18).

Similar histories are known for invertebrates and plants (Fig. 2.19). Trilobites are marine arthropods that appeared in the Cambrian period and persisted until the end of the Permian period. Orders of insects that include present-day cockroaches, grasshoppers, cicadas, and mayflies are found in the Carboniferous period, a time of extensive swampy forests of ferns and other groups of vascular plants that are now extinct. The insect orders that include bees and butterflies emerged in the Triassic period, as did *gymnosperms*, vascular plants with seeds but not flowers and represented today by conifers and the Ginkgo tree.

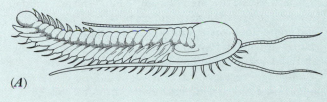

(A)

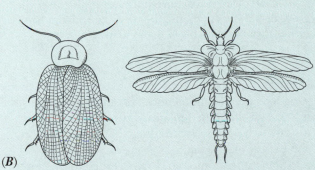

(B)

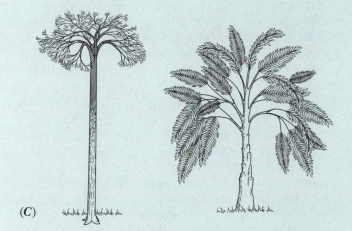

(C)

FIGURE 2.19 (A) An early arthropod. This large (18 inch) trilobite lived in the Cambrian period.

(B) Insects from the Carboniferous period. The specimen on the left is an early relative of the cockroach; the specimen on the right differs from all modern insects in having a third, small pair of wings in front of the other two pairs.

(C) Representative plants of the swamp forests of the Carboniferous period. These are both extinct.

The fossil record thus shows that the origin of life occurred relatively early in the history of the planet (i.e., during the first *billion* years!), but subsequent changes seem to have involved bottlenecks. The formation of nucleated cells may have been one, and the formation of multicellular organisms may have been another. But once those hurdles were passed, the way seems to have opened for accelerated evolutionary change, beginning about 600 million years ago. Most of our knowledge of evolutionary events is about these later stages; here is where the fossil record is richest, and here is where we find the multitude of multicellular organisms that, through time, have "descended with modification".

In summary, throughout the fossil record, groups of organisms come into existence, expand, and branch, sometimes giving rise to new groups, more frequently withering to extinction. Box 2.4 provides a more extensive view of the evidence for descent with modification with emphasis on the vertebrates. The fossil record thus provides ample evidence for evolutionary change, but it does not address the *process* of how that change occurred. That we will discuss in the following chapters.

(2) COMPARATIVE ANATOMY AND EMBRYOLOGY

We have seen that the members of the same class, independently of their habits of life, resemble each other in the general plan of their organization. This resemblance is often expressed by the term "unity of type;" or by saying that the several parts and organs in the different species of the class are **homologous** [emphasis added]. . . . What can be more curious than that the hand of a man, formed for grasping, that of a mole for digging, the leg of the horse, the paddle of the porpoise, and the wing of the bat, should all be constructed on the same pattern, and should include the same bones, in the same relative positions? . . . We see the same great law in the construction of the mouths of insects: what can be more different than the immensely long spiral proboscis of a sphinx-moth, the curious folded one of a bee or bug, and the great jaws of a beetle?—yet all these organs, serving for such different purposes, are formed by infinitely numerous modifications of an upper lip, mandibles, and two pairs of maxillae. [*On the Origin of Species*, p. 434]

This pattern of structure, to Darwin the "very soul" of natural history, begged for a causal explanation. In the living world, similarity suggests genealogical relationship, so for Darwin these similarities indicated descent with modification. The word *homology* refers to structures derived from a common ancestor and modified in different lineages of organisms by the process of natural selection. Figure 2.20 illustrates

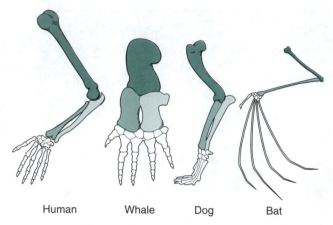

Human   Whale   Dog   Bat

FIGURE 2.20 Homology is illustrated by the front limbs of these four vertebrates. All are constructed of the same basic parts: one bone in the upper arm, two in the lower arm, and a basic five digits in the "hand." In each animal, however, the relative sizes and shapes of the parts assume different proportions during development, enabling each appendage to serve a different function. The short, massive forelimb of the whale is a flipper for swimming, whereas the delicate bones of the bat form a light, skin-covered wing for flying. The dog's padded foot enables it to use its forelimbs in running over the ground, and the opposable thumb or the human hand makes the arm effective in manipulating objects.

how the bones of the forelimbs of several mammals are variations on a common ancestral structure while serving several different functions: the human's for grasping, the whale's for swimming, the dog's for running, and the bat's for flying.

Sometimes homologies are revealed during embryonic development. Barnacles, those small, hard marine organisms that encrust pilings and rocks in the intertidal zone, were for a time hard to classify with other animals. But it was Darwin who solved that riddle by showing that because barnacle larvae resemble the larvae of lobsters and shrimp, barnacles are much-modified crustaceans (Fig. 2.21).

In 1828, three decades before Darwin published *On the Origin of Species*, a German zoologist Karl Ernst von Baer pointed out that the early embryos of dissimilar species nevertheless look very much alike (Fig. 2.22). Evolutionary theory provides an explanation. For example, that vertebrates share similarities in their early patterns of development is evidence of their descent from a common ancestor, and the characteristic differences that distinguish one group from another appear as development progresses. (We will consider some of the underlying mechanisms in Chapter 10). Thus, mammals pass through a developmental stage in which there are grooves that in fish embryos develop into gill slits. The related idea that early embryological

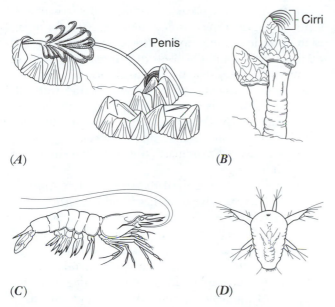

(A)   (B)

(C)   (D)

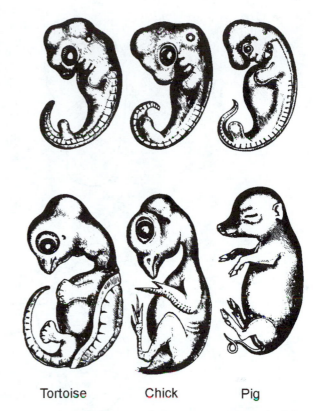

Tortoise   Chick   Pig

FIGURE 2.21 Barnacles are familiar to anyone who has visited a rocky seashore or who has looked closely at wharf pilings at low tide. There are about a thousand different species of barnacles. Some species are encased in hard, squat, conical cases and stick out their feathery *cirri* to sweep in particles of food when they are covered by the tide (A). Others are mounted on a stalk (B). Some are parasites and inhabit other animals. As the long penis in (A) illustrates, when an animal grows attached to a rock, its sex life is restricted.

Like lobsters and crabs, the shrimp in (C), with its array of paired appendages, is a more familiar crustacean. Barnacles are crustaceans too, and the *cirri* are modified appendages of the mid-part of the body, the thorax. How do we know? Barnacles, like other crustaceans, have a characteristic larva called the *nauplius* (D). The nauplius larva of a crustacean has a single median eye and three pairs of appendages. Similarities in early development frequently offer clues about evolutionary relationships.

FIGURE 2.22 During early development, several of the classes of vertebrates go through similar morphological stages. Due to different amounts of yolk and different patterns of cleavage and gastrulation (Chapter 10), however, they are less similar in the stages immediately following fertilization. The similarities suggested to Darwin descent with modification. Despite the morphological complexities revealed by more extensive study, similarities in the molecular control of embryonic differentiation (Chapter 10) point to an evolutionary history of developmental processes.

stages represent the *adult* forms of evolutionary ancestors—"ontogeny recapitulates phylogeny"—is not, however, correct. Figure 2.23 illustrates how the feet of humans and monkeys are similar early in development but diverge later, *more* so in the case of the *monkey*.

(3) BIOGEOGRAPHY

These cases of relationship, without identity, of the inhabitants of seas now disjoined, and likewise of the past and present inhabitants of the temperate lands of North America and Europe, are inexplicable on the theory of creation. We cannot say that they have been created alike, in correspondence with the nearly similar physical conditions of the areas; for if we compare, for instance, certain parts of South America with the southern continents of the Old World, we see countries closely corresponding in all their physical conditions, but with their inhabitants utterly dissimilar. [*On the Origin of Species*, p. 372]

The distribution of plants and animals over the surface of the earth is far from uniform. A major cause of this geographic variation is obviously ecological: plants that conserve water efficiently inhabit desert areas, not tropical rain forests; white bears that eat seals live in the far north, not on the shores of tropical or temperate seas. The enormous diversity of life reflects the power of natural selection, for lineages of organisms exploit virtually every ecological opportunity that becomes available. This feature of Darwin's "descent with modification" is frequently referred to as adaptation, a concept to which we will return in Chapter 5.

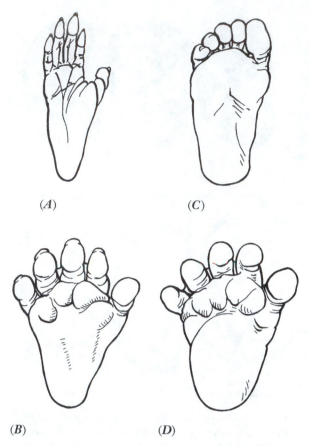

(A)   (C)

(B)   (D)

FIGURE 2.23 Adult feet of macaque monkey (A) and human (C). Early fetal feet of monkey (B) and human (D). The foot of an adult monkey is more greatly modified during development than is the human foot. The drawings are not to the same scale.

The passage that began this section points to still another feature of organic diversity. Darwin is noting that although natural selection has modified organisms to live in various environments throughout the world, similar habitats on different continents are not populated by the same organisms. For example, the deserts of the Americas have different plants and animals from the deserts of Africa. Continents usually have organisms whose shared similarities argue for descent with modification, but the great oceans of the world impose barriers to migration and mixing between continents. Similarly, the marine organisms of an ocean are more similar to each other than the marine organisms on the two sides of a continental landmass. To Darwin, these general observations suggested that the geographic distributions of organisms reflect migrations accompanied by change through natural selection, but that the migration of terrestrial organisms has been restricted by oceans and that of marine organisms, by continents.

Darwin also saw that the plants and animals indigenous to remote oceanic islands were similar to species on the nearest continents. But not all kinds of

creatures occur on oceanic islands, and Darwin presented a cogent explanation: a broad expanse of ocean is an obstacle to migration for some organisms but not others. Reptiles and the seeds of plants can float great distances on trees that wash into the ocean, but amphibians (e.g., frogs) cannot tolerate wetting with salt water, and mammals cannot survive long without food and fresh water. Birds and bats can be carried hundreds of miles to oceanic islands on the winds of storms, whereas terrestrial mammals are tied to the surface of the earth.

Darwin also recognized that historical contingencies were likely to have influenced both migrations and natural selection. As we will discuss below, he was particularly impressed by the likely effect of the southward distribution of glaciers with the accompanying change in climate. But to Darwin, the complex relationship between the geographic distribution and degree of relatedness of the major groups of plants and animals—what we call *biogeography*—was inconsistent with creationism and required a coherent explanation in terms of natural phenomena. "Why should the species which are supposed to have been created in the Galapagos Archipelago, and nowhere else, bear so plain a stamp of affinity to those created in America?" he wrote. And again: "Why, it may be asked, has the supposed creative force produced bats and no other mammals on remote islands?" He saw that migrations of organisms (albeit with some restrictions) and evolution by natural selection provided a logical and uniquely consistent explanatory framework.

Since Darwin wrote his account in 1859, many more fossils have been discovered and much has been learned about the movement of continents. Consequently, our understanding of biogeography, like much else in evolution, has been greatly enriched in the ensuing years. Following are three examples from different periods in geological time.

Africa, South America, and Australia are home to large, flightless birds: the ostrich, the rhea, and the emu, respectively (Fig. 2.24). Collectively these are known as ratite birds, and it has long been suspected that they are more closely related to each other than to other birds—in other words, that they have descended from a common ancestor. The alternative explanation is that their similarities are an example of *convergent evolution*, in which distantly related organisms acquire similar features through independent adaptations to similar ecological conditions. Some recent evidence based on similarities in the genes of these birds (the technique is described later in Box 3.2) confirms that they are in fact related, but what has enabled these lineages of birds to find homes on such distant continents? The answer to that puzzle is to be found in plate tectonics, the slow movement of great slabs of the earth's surface that was described in Chapter 1. About

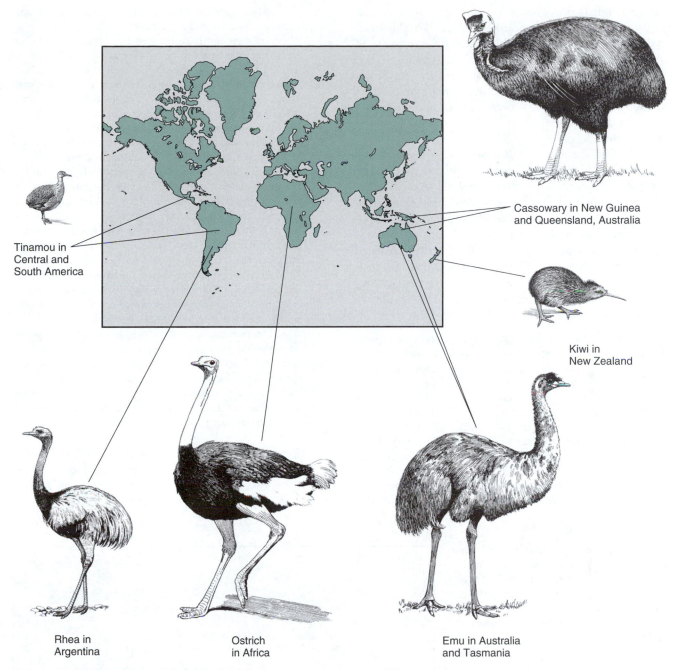

Tinamou in Central and South America

Cassowary in New Guinea and Queensland, Australia

Kiwi in New Zealand

Rhea in Argentina

Ostrich in Africa

Emu in Australia and Tasmania

FIGURE 2.24 These six birds have evolved from a common ancestor. Their similar appearance is therefore likely not an example of convergent evolution. All but the tinamou are flightless. Their occurrence on three continents is a result of continental movements during the past several hundred million years.

285 million years ago these continental masses along with what is now Antarctica were all connected, and the common ancestor of these ratite birds must have been widely distributed over this land.

Africa separated first; South America and Australia parted somewhat later. Previously you may have thought of Australia as the unique home of marsupials (mammals that carry their young in pouches), but recall (Fig. 2.4) that South America also has an extensive marsupial fauna. The reason is the geological history of the continents. One of these marsupials, the opossum, has managed to migrate north through the Isthmus of Panama (a relatively recent land connection) as far as the United States, but like all other marsupials, its presence in the Americas reflects an ancient connection between Australia and South America.

The second example is more recent and involves the evolution of mammals in the last 60 million years. Camels evoke thoughts of Saharan Africa and the Arabian Peninsula. The fossil record, however, reveals that camels originated in North America and migrated east to Eurasia and south to South America. They became extinct in North America, but their lineage exists to this day in South America as llamas and guanacos (Fig. 2.25). Elephants, by contrast, dispersed from a center of origin in Africa. They are still found in parts of southern Asia, but their representatives in Europe and in the Americas (mastodons and mammoths) have become extinct.

The last example is still more recent and involves events of the last 100,000 years. The plants that live on the tops of mountains in New England share features with the flora of Labrador and also show similarities with plants growing on the summits of European mountains. How did these pockets of plant life become isolated on mountaintops, and why are these isolated populations from distant locations related? The explanation involves extensive glaciation that took place several tens of thousands of years ago. Once again we quote Darwin, who as usual had it quite right.

> . . . we shall follow the changes more readily, by supposing a new glacial period to come slowly on, and then pass away, as formerly occurred. As the cold came on, and as each more southern zone became fitted for arctic beings and ill-fitted for their former more temperate inhabitants, the latter would be supplanted and arctic productions would take their places. The inhabitants of the more temperate regions would at the same time travel southward, unless they were stopped by barriers, in which case they would perish. The mountains would become covered with snow and ice, and their former Alpine inhabitants would descend to the plains. By the time that the cold had reached its maximum, we should have a uniform arctic fauna and flora, covering the central parts of Europe, as far south as the Alps and Pyrenees, and even stretching into Spain. The now temperate regions of the United States would likewise be covered by arctic plants and animals, and these would be nearly the same with those of Europe; for the present circumpolar inhabitants, which we suppose to have everywhere traveled southward, are remarkably uniform round the world. . . .

> As the warmth returned, the arctic forms would retreat northward, closely followed up in their retreat by the productions of the more temperate regions. And as the snow melted from the bases of the mountains, the arctic forms would seize on the cleared and thawed ground, always ascending higher and higher, as the warmth increased, whilst their brethren were pursuing their northern journey. Hence, when the warmth had fully returned, the same arctic species, which had lately lived in a body together on the lowlands of the Old and New Worlds would be left isolated on distant mountain summits (having been exterminated on all lesser heights) and in the arctic regions of both hemispheres.

> Thus we can understand the identity of many plants at points so immensely remote as on the mountains of the United States and of Europe. We can thus also understand the fact that the Alpine plants of each mountain-range are more especially related to the arctic forms living due north or nearly due north of them . . . [*On the Origin of Species*, pp. 366–367]

The distributions of animals also reflect the southward movement and retreat of glaciers. For example, there are two populations of scrub jays, one in Florida and the other in California and the Southwest. Simi-

FIGURE 2.25 The group of animals that includes camels and dromedaries of North Africa and the Middle East actually evolved in North America and spread to Eurasia, North Africa, and South America. This family of mammals subsequently became extinct throughout much of its range (hatched area), but is represented in South America today by guanacos, alpacas, and llamas.

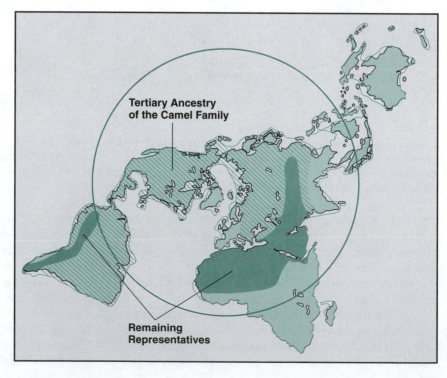

larly, the distribution of diamondback rattlesnakes in the southern United States is discontinuous, with morphologically and behaviorally somewhat different populations in Florida and Texas (Fig. 2.26).

(4) POPULATION GENETICS

The last two of our five pillars of evidence for evolution were unknown to Darwin and his contemporaries. Because both require additional background information, we will defer their detailed discussion until the next several chapters. But we can provide a sense of their importance in few words here. Darwin did not under-

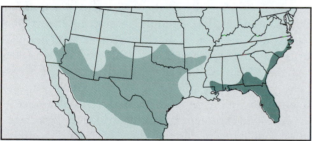

FIGURE 2.26 **The eastern and western diamondback rattlesnakes (*Crotalus adamanteus*, below, and *Crotalus atrox*, above) are two species with distinct geographical ranges. They are likely descendants of isolated populations that survived the last glacial period.**

stand the basis for the heritable variation that is required for natural selection to operate, a gap that remained a serious deficiency in evolutionary theory until the physical basis for inheritance was discovered, the rules by which genes are passed from generation to generation were understood, and the presence of genetic variation in natural populations was recognized. Only then was the process of natural selection put on a firm foundation, and it took nearly eighty years.

(5) MOLECULAR BIOLOGY

Finally, during the second half of the twentieth century the molecular structure of the gene was discovered, along with the nature of the genetic code and the means by which the code is read by living cells. With this new information and the continuously expanding technology for studying the molecular properties of genes, it has been possible to see evidence for evolution in the basic patterns of biochemistry employed by all organisms and to trace evolutionary history in the structures of genes and proteins as well as in lineages of organisms. We will return to both of these final pillars in the following chapters.

SOME COMMON MISCONCEPTIONS

EVOLUTION AND CONTEMPORARY POLITICS

Why a section on misconceptions? There are many ways in which a concept can be misunderstood, so how can we possibly anticipate the difficulties this material will present to any reader? We can't, but we have another problem in mind.

Biological evolution offers a special set of issues for many religious fundamentalists. One reason is the mistaken belief that a moral code must necessarily derive from the supernatural—that the materialism of science not only offers no insights into our ethical convictions, but may even be inconsistent with them. To quote a group objecting as recently as 1996 to the teaching of evolution in the public schools of Alabama, "If human beings are solely the product of chance meaningless forces . . . , then one can rightly question the value of man, the significance of life, and whether there is any basis for morality. If children are taught to see themselves as animals, they may well act like animals."

Understanding the history of life in no way challenges the existence or importance of codes of ethics and morals. In fact, in Chapter 14 we will see how understanding social systems in an evolutionary context provides insight as to why certain themes are repeated in legal and religious institutions in diverse cultures from around the world.

The quotation also expresses a second concern. If evolution has no goal or predetermined outcome, and if the human species is the product of chance and accident, does this not deprive us of significance and purpose? As we shall see in Chapter 4, evolution is not simply "chance and accident", but that is not the crux of the matter here. The deeper issue has to do with the difference between scientific and religious understanding. Science, by definition, can only be concerned with nature. The natural process by which evolution occurs does not have a goal, and in that sense it does not have a purpose. But as we saw in the previous chapter, organisms appear to have purposes, to which the study of evolution brings understanding.

The human mind, however, is capable of many kinds of thoughts, and when it supposes the supernatural, science has little more to say. A resolution by the Council of the National Academy of Sciences puts it in these words: "Religion and science are separate and mutually exclusive realms of human thought whose presentation in the same context leads to misunderstanding of both scientific theory and religious belief."

For most religious denominations, the reality of evolution poses no problem. Some, however, find it difficult to accommodate views that are inconsistent with a literal reading of the Bible and have tried to subvert the teaching of evolution in the public schools. Much of this goes on at the local level, but—to cite some recent examples—in 1996 the state board of education in Alabama and the state legislatures of Tennessee, Georgia, and Ohio considered changes that would compromise the teaching of science. In 1997 an anti-evolution bill passed the North Carolina house but failed in the senate. In 1999 the deputy attorney general of Nebraska, in a bizarre interpretation of constitutional law, informed the state school board that having students recognize "that the present arises from materials and forms of the past" appears to interfere with students' constitutional rights because the proposition "is phrased in terms of objective fact rather than theory." And as the end of the millennium approached, the Kansas state school board voted 6 to 4 to remove evolution from the state science standards.

In these controversies, science is misrepresented and distorted. In the early 1980s the tactic of religious fundamentalists was to demand "equal time" for "creation science." "Equal time" has an appeal to people who want to be fair, but "creation science" is a contradiction in terms. In 1987 in *Edwards v. Aguillard* the Supreme Court confirmed a lower court ruling that "creation science" is religion, not science, and consequently has no place in a *science* class. "Creation science" now travels under aliases like "abrupt appearance theory" or "intelligent design theory",

and efforts to attack the teaching of evolution have taken new forms.

One approach is to assert that evolution should be taught as theory, not as fact. This idea turns science on its head. A scientific theory is not a casual surmise, a tenuous guess, or an alternative to something that is known to be true. It is most certainly not an alternative to a fact. A scientific theory is a comprehensive explanation that is based on an enormous amount of *evidence*. Moreover, science is not the accumulation of facts; science is about assembling evidence and constructing theories. Good theories are culminations, not tentative steps along the way.

Scientific disagreements are resolved when there is consensus about what constitutes adequate evidence and how that evidence should be interpreted. Consequently, what appears to be a "fact" at one time may change in the light of new evidence. Similarly, a theory that is adequate at one time may require modification as new evidence is found. Casting the political debate over evolution as fact versus theory grossly misrepresents the nature of the scientific process as well as scientific understanding.

Misunderstanding is easy because in everyday speech the word "theory" is usually used synonymously with surmise, conjecture, or hypothesis. Even scientists frequently speak casually this way, although in writing they are usually attentive to the formal meaning. Furthermore, in other fields "theory" can have other particular meanings. For example, in law, the theory of a case is an attorney's idea of what happened and her plan for trying it. This multiplicity of meaning is the reason why the argument for "fact instead of theory" is frequently heard sympathetically by individuals who are unfamiliar with the methods and processes of science.

A second argument is that the evidence *against* evolution should be presented in the classroom. It is always appropriate to consider the evidence both for and against any scientific interpretation, so like the pitch for equal time, this tactic has an attractive sound and appears to be consistent with good educational practice. The evidence that evolution is an important natural process is so overwhelming, however, that the counter evidence offered by creationists is invariably an inaccurate observation, an interpretation that is inconsistent with other observations, a conclusion that is at odds with fundamental physical principles, or yet another appeal to the supernatural decked out in scientific-sounding jargon like "intelligent design theory." But as we pointed out earlier, a scientific theory cannot involve the supernatural. Appeal to the supernatural, by definition, lies outside of science because what cannot be observed, measured, or tested is not relevant to scientific discourse and naturalistic explanation. An alternative interpretation of scientific evidence must *itself* be based on evidence.

HOW WELL FOUNDED ARE THE CRITICISMS?

What are some of the challenges to the evidence for evolution? One is that the fossil record fails to reveal the presence of the necessary intermediate forms that should have occurred. This is an old argument, and it has its roots in Darwin's own writings. Darwin saw that it would trouble skeptics, and he was acutely aware of the fragmentary nature of the fossil record as it then existed: "Now turn to our richest geological museums, and what a paltry display we behold!" He was also aware of the reasons. Because most of the sedimentary rocks that were formed have been either metamorphosed or lifted up and eroded away before later deposits were added, the geological record is intrinsically fragmentary. Second, only part of the world had been explored by geologists in even the most rudimentary fashion. Furthermore, only a fraction of the organisms that have ever existed have parts that are likely to leave fossils, and of these only a fraction are ever likely to have existed in places where fossils could form. *On the Origin of Species* has an entire chapter devoted to what Darwin characterized as the "extreme imperfection of the geological record", and he concluded that "He who rejects these views on the nature of the geological record, will rightly reject my whole theory. *For he may ask in vain* [italics added] where are the numberless transitional links. . . ."

The argument that the fossil record does not support evolution continues to be repeated, despite the hundreds of thousands of fossils that have been discovered since Darwin addressed the problem. The evidence for the earliest cells well over 3 billion years ago, the richness of the Cambrian diversity, the presence of countless intermediates in the evolution of reptiles, birds (Fig. 2.27), and mammals, and virtually the entire fossil record of the origins of humans (Chapter 12) are all later additions to scientific knowledge. This accumulation of new data has enriched immeasurably our understanding of evolutionary history.

There is one form of criticism of the fossil record that is particularly woolly: If mammals arose from reptiles, why do we not find any intermediates between reptiles living today and mammals living today? The reason, as we suspect you are able to deduce for yourself, is that reptiles that are living today have as long an evolutionary history as do mammals that are now living. Both are descended from a common ancestor, but there is no reason *that* species looked anything like any animal now living. As is so frequently the case, Darwin understood this better than his latter-day critics:

> . . . the fantail and pouter pigeons [breeds produced by artificial selection] have both descended from the rock-pigeon; if we possessed all the intermediate varieties which have ever existed, we should have an extremely

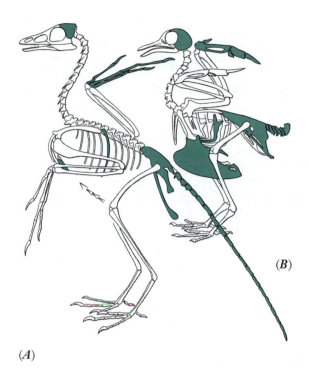

(A)

FIGURE 2.27 Comparison of the skeletons of *Archaeopteryx* (A) and the modern pigeon (B). The corresponding parts of the skeletons are readily apparent, but those structures that have been most modified in contemporary birds, and are therefore represented in a more reptilian form in *Archaeopteryx*, are shaded green. They include the enormous enlargement of the keel-shaped breastbone (sternum) to provide a surface for the attachment of powerful flight muscles, the fusion of bones, and shortening of what was originally a long tail.

> close series between both and the rock-pigeon; but we should have no varieties directly intermediate between the fantail and pouter . . .
>
> So it is with natural species, if we look to forms very distinct, for instance to the horse and tapir, we have no reason to suppose that links ever existed directly intermediate between them, but between each and an unknown common parent [parent-species]. The common parent will have had in its whole organization much general resemblance to the tapir and to the horse; but in some points of structure may have differed considerably from both, even perhaps more than they differ from each other. [*On the Origin of Species*, pp. 280–281]

A second criticism is that the formation of new species or other, more major groups has never been observed. That selection can alter characters within a species is acknowledged—for how could the results of plant breeding and animal husbandry be ignored?—but it is asserted that selection has never produced a new species. This assertion is akin to saying that growing roots are incapable of splitting rocks or glaciers are not

Box 2.5
New Species of Plants Sometimes Form Very Rapidly

In addition to forming new species through the slow accumulation of genetic differences in geographically separated populations, plants display another process that can operate very rapidly. *Spartina* is a genus of common salt-marsh grasses. About 1800, *S. alterniflora*, which is native along the eastern coast of North America, was introduced in the marshes of Great Britain, perhaps inadvertently carried by ship. It quickly spread, existing with the much shorter local species *S. maritima* from which it is easily distinguished. In 1870 a hybrid was discovered, and although the hybrid was sterile, like many other plants it was able to propagate by making rhizomes, underground horizontal stems that send down additional roots and send up green shoots. About twenty years later, it generated a variant that was able to reproduce sexually, i.e., make seeds. This new form spread rapidly, remained distinct from the earlier forms, and consequently was recognized as a new species.

What had happened? *S. maritima* has 30 pairs of chromosomes, *S. alterniflora* has 31, and the sterile hybrid, 31. The new species, however, has 61 pairs. It is thus *polyploid*, with multiple copies of chromosomes. The formation of polyploids and the hybridization of related species with the same or different numbers of chromosomes is a frequent occurrence in plants. Wheat, which is one of the most abundant food crops in the world, and the wild grass from which existing strains of wheat arose by hybridization, exist as several species with 7, 14, and 21 pairs of chromosomes. When fertile hybrids or mutational variants appear with different numbers of chromosomes from their parents, new species can result. The assertion that no one has observed the formation of a new species is therefore not correct.

---

responsible for carving U-shaped valleys because a human observer cannot stand still and watch it happen. Natural events occur on characteristic time scales, and the accumulation of heritable differences sufficient to separate two populations into distinct species ordinarily requires time well in excess of a human life. We say ordinarily, because in plants separate species can arise on the basis of small numbers of heritable differences, and the formation of new species has been observed (Box 2.5). Furthermore, examples are known of populations of animals that look different and have the properties of distinct species because they do not interbreed, yet each is connected to the other through a series of populations of intermediate and interbreeding variants (Fig. 2.28). Examples like these show that new species do represent the accumulation of heritable differences of the same kind that are observed when a population is subjected to either natural or artificial selection. We will return to the question of how major anatomical differences might have evolved when we discuss development in Chapter 10.

It has been said, incorrectly, that evolution violates an important physical principle known as the Second Law of Thermodynamics. This rule states that physical systems tend to a disordered condition: the milk mixes with your coffee (albeit slowly) even if you do not stir; mountains wear away to sand and gravel; a tree that dies rots and returns to the soil. But living organisms are the antithesis of disorder. Are they violating the Second Law of Thermodynamics?

They are not. Order can be created if energy is supplied. Boiling seawater separates the water from the dissolved salts; mountains can be created with energy from deep within the earth; and an oak grows from an acorn, utilizing the energy of sunlight, mobilizing it through the process of photosynthesis. The assertion that life violates the Second Law of Thermodynamics is a misunderstanding of physics.

Another objection takes the following form: "I don't see how blind chance could produce something as complicated as a tree or a human being." This statement contains a deep misunderstanding of evolution. As we will explain at length in Chapter 4, chance plays a number of roles in evolution, but natural selection is not based on chance. This objection also invokes what the Oxford biologist Richard Dawkins has called the "The Argument from Personal Incredulity" as a criterion for whether a proposition could be correct. (Dawkins coined the phrase in reference to the Bishop of Birmingham's inability to imagine why natural selection made polar bears white, considering that they have no natural enemies.) Personal incredulity may have a lot to do with whether a particular individual understands or believes a scientific concept, but it has very little bearing on whether the concept is valid. Evolution takes place on a time scale that is so much longer

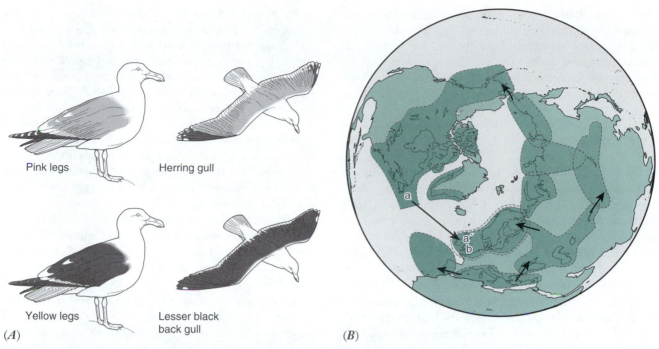

Pink legs

Herring gull

Yellow legs

Lesser black back gull

(A)

(B)

FIGURE 2.28 **(A) The herring gull (*Larus argentatus*, above) is common throughout North America and has spread to Europe where it coexists with another species, the lesser black-back gull (*Larus fuscus*). The herring gull has a gray back and pink legs and feet, whereas the lesser black-back gull has a black back and yellow legs and feet.**

(B) The two species are connected across Europe and Asia through a series of recognizably distinct populations of gulls that interbreed where they come in contact. Such populations are sometimes referred to as subspecies. These subspecies evolved from smaller populations that became isolated from one another during the last Pleistocene glaciation. When the glaciers melted, the ranges of these smaller populations expanded. As the time of separation had not been long enough for new species to have formed, interbreeding started to occur when adjacent populations came into contact. Individuals from widely separated populations still rarely encountered each other, so nonadjacent populations slowly diverged. When in recent times herring gulls at one end of the chain (a) spread eastward across the Atlantic to Europe (a'), they were sufficiently different from the lesser black-back gulls at the other end of the chain (b) that the two now coexist where they meet and are recognized as separate species. The existence of this ring of populations demonstrates directly how geographical separation can lead to the formation of new species.

than our personal experience prepares us to contemplate that "commonsense" notions of what is likely or possible largely fail us. Evolutionary biology is not unique in this regard; astronomy offers comparable challenges.

Sometimes one encounters the criticism that if evolution is just "survival of the fittest", the entire concept reduces to a tautology. Those that survive are deemed the fit; to say that the fittest survive is therefore to say nothing. What is the problem with this reasoning?

Bankruptcies are common events in the world of commerce; those businesses that are more fit continue while those that are less fit fail. But what makes one en-

terprise succeed while its competitors are forced out of business? There are a number of possible differences: better labor relations, lower costs of production, better access to raw materials, closer markets, better marketing, or better management, to name some of the more obvious. To be useful, any economic analysis of failures and successes must consider such factors.

Similarly, to understand how natural selection leads to reproductive success biologists must study organisms with the same attention to detail. Is one animal better able to utilize a wide variety of foods during a drought, or better able to survive harsh winters, or is it more effective in courtship, or more resistant to disease, or have any of a number of other possible advantages? Put another way, understanding evolutionary processes is in principle subject to experimental analysis. The phrase "survival of the fittest" is thus a cliché that conceals the deeper truth that every claim of evolutionary advantage is a hypothesis that either has been or needs to be tested by observation and experiment.

SYNOPSIS

In this chapter you met the nineteenth century naturalist Charles Darwin and saw how his travels as a young man not only shaped the rest of his life but also led to a profound change in the way scientists understand the biological world. Darwin did not set out to accomplish such a change. He was a keen and very thoughtful observer, and with time and experience he became broadly knowledgeable of both geology and biology. His conclusion that evolutionary change was an important historical process came to him slowly and was the result of years of both observation and experimentation. Like all good scientists, he was always aware of the need to consider alternative interpretations of what the evidence seemed to say. Not only did Darwin make a convincing case for the reality of evolution, he proposed that natural selection was the likely mechanism responsible.

The evidence that living organisms are the product of a very long evolutionary process driven by natural selection rests on five different lines of evidence. Darwin knew about three of them, and all have been greatly strengthened since *On the Origin of Species* was published in 1859.

The *fossil record* provides the only direct evidence about how earlier plants and animals looked. The record is inherently incomplete, partly because uplift and subsidence of the earth's crust ensures that sedimentary rocks are frequently weathered and eroded before new sediments are deposited on top of what remains. Nevertheless, the hundreds of thousands of fossils that have been discovered since Darwin's time strengthen his interpretation that all organisms are modified descendants of previously existing forms. Thus mammals and birds arose from different lineages of reptiles, reptiles from amphibians, amphibians from a group of ancient crossopterygian fishes, and equivalent findings for invertebrates and plants.

The fossil record also documents major extinctions in which entire groups of organisms were greatly reduced in number or totally annihilated. Each of these cataclysmic events provided ecological opportunities for the evolution of new forms from surviving lineages. For example, the extinction of the dinosaurs about 65 million years ago opened the way for the diversification of mammals.

A second line of evidence for evolution is the *fundamental similarity of body structures* that have entirely different functions in related groups of organisms. For example, the forelimbs of mammals that fly, swim, run, and grasp are built from comparable sets of bones by modifying their shapes and proportions. To Darwin this fundamental similarity of structure suggested genealogical relationship, and structures that are similar by reason of descent from a common ancestor are referred to as *homologous*. Embryological development provides supporting evidence. For example, the early development of vertebrates passes through some similar stages, and the changes in later development that lead to the characteristic adult forms represent evolutionary specializations.

The third line of evidence comes from *biogeography*, the distribution of plants and animals over the earth. Different groups of organisms arose at different times in different places and dispersed from their centers of origin, encountering barriers to movement. Land animals are stopped by oceans, and marine organisms are stopped by continents. The movements of large land masses (plate tectonics), the formation or elimination of land bridges caused by changes in sea level, and alterations in climate associated with periods of glaciation have all influenced the present distribution of organisms.

The fourth and fifth lines of evidence were unknown to Darwin. Because they require a basic knowledge of the molecular structure of genes and of population genetics to understand, we have deferred them to the next two chapters. Collectively, however, these five lines of evidence document a rich and detailed understanding of the history of life.

Despite this extensive evidence, some people do not believe in the reality of biological evolution because they feel it conflicts with their religious beliefs. But "science and religion are separate and mutually exclusive realms of human thought." Failure to understand this distinction leads to confusion about the roles of each. Science can only address questions about nature, so it has nothing to say about spirits and deities that are not part of the observable, measurable universe. Conversely, scientific understanding cannot be refuted by assertions that are based on personal faith, particularly when belief is regularly contradicted by observation and experiment.

QUESTIONS FOR THOUGHT AND DISCUSSION

1. In November 1995, the Board of Education of the state of Alabama adopted the following disclaimer on evolution to be affixed to every biology textbook used in the state schools.

A Message from the Alabama State Board of Education

This textbook discusses evolution, a controversial theory some scientists present as a scientific explanation for the origin of living things such as plants, animals and humans. No one was present when life first appeared on earth. Therefore, any statement about life's origins should be considered as theory, not fact. The word "evolution" may refer to many types of change. Evolution describes changes that occur within a species. (White moths, for example, may "evolve" into gray moths.) This process is microevolution, which can be observed and described as fact. Evolution may also refer to the change of one living thing to another, such as reptiles into birds. This process, called macroevolution, has never been observed and should be considered a theory. Evolution also refers to the unproven belief that random, undirected forces produced a world of living things.

There are many unanswered questions about the origin of life which are not mentioned in your textbook, including:

—Why did the major groups of animals suddenly appear in the fossil record (known as the "Cambrian explosion")?

—Why have no new major groups of living things appeared in the fossil record for a long time?

—Why do major groups of plants and animals have no transitional forms in the fossil record?

—How did you and all living things come to possess such a complete and complex set of "instructions" for building a living body?

Study hard and keep an open mind. Someday you may contribute to the theories of how living things appeared on earth.

Analyze this disclaimer. What errors and misconceptions are you able to identify? What can you say about the authors' understanding of the scientific process? If there are issues in this disclaimer that you feel were not discussed in Chapter 2, identify them and consider what you think might be an appropriate response. You may find it interesting to look at the disclaimer again after reading Chapters 3–5.

SUGGESTIONS FOR FURTHER READING

Browne, J. (1996). *Charles Darwin Voyaging: A Biography*. Princeton, NJ : Princeton University Press. A recent and very readable account of Darwin's life up to 1859.

Carroll, R. L. (1977) *Patterns and Processes of Vertebrate Evolution*. New York, NY: Cambridge University Press. Rates, directions, and patterns of vertebrate evolution and the relationship of microevolution to macroevolution are discussed in the light of modern knowledge of population and developmental genetics.

Darwin, C. (1859). *On the Origin of Species by Means of Natural Selection or the Preservation of Favored Races in The Struggle for Life. A Facsimile of the First Edition. 1964*. Cambridge, MA: Harvard University Press. Well worth examining to appreciate the detail with which Darwin assembled evidence for "descent with modification." The page numbers cited in Chapter 2 refer to this facsimile edition.

Darwin, C. (J. Browne, ed.) (1989). *The Voyage of the Beagle: Charles Darwin's Journal of Researches*. New York, NY: Penguin. A reprinting of Darwin's own description of his observations and experiences while he was traveling with *H. M. S. Beagle*, compiled after his return to England but well before his ideas about natural selection had crystallized.

Steering Committee on Science and Creationism. (1999). *Science and Creationism: A View from the National Academy of Sciences, 2nd ed*. Washington, DC: National Academy Press. This pamphlet explains the distinction between scientific understanding and belief in the supernatural. The National Academy of Sciences is the most distinguished honorific society in the profession, and this document was written in response to efforts to introduce creationism into science classrooms in public schools.

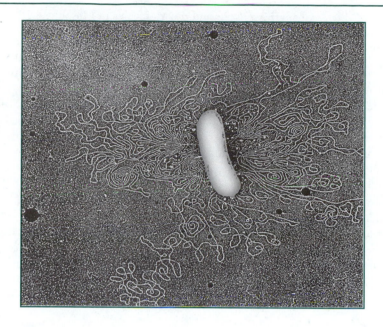

3

Cells and the Molecules of Life

Much of this book is about things you can see and with which you have at least a passing familiarity—animals and people, how they appear, the things they do, how they relate to each other, and how they reproduce. But in order to understand what is readily observable, we have to expand our sensory capacities with instruments such as telescopes and microscopes and enlarge our concepts of space and time beyond our everyday experience. In this chapter we consider some things that are too small to see with the unaided eye: *cells*, of which all

Photo: This bacterial cell (*Escherichia coli*) was treated with an enzyme to weaken its cell wall, then placed in water causing its DNA to be ejected. The double helix of the DNA molecule appears in this scanning electron micrograph as thread-like loops and tangles. The length of DNA is about 1.5 mm, more than 1000-times longer than the cell from which it came. The length of DNA in the nucleus of a mammalian cell is about a meter, approximately 700-times longer.

organisms are composed, and smaller still, some of the important *molecules* from which cells are constructed. Our emphasis will be on *structure* and how the shapes and arrangements of molecules and parts of cells tell much about their *functions*. This is essential background for what follows later in the book, because to understand complex entities it is necessary to consider the building blocks from which their complexity arises. Exploring how structure and function are related at the level of molecules and cells gives a greater appreciation of how natural selection operates, directly and indirectly, at every level of biological organization.

It is unnecessary to memorize innumerable chemical formulas in order to develop a generalized picture of a protein that will help in understanding the multiple roles that they play in living systems. And understanding the geometrical structure of genes is particularly rewarding, because the organization of this molecule holds the key to how it encodes information and how it replicates between generations.

LIFE IS ORGANIZED IN COMPARTMENTS CALLED CELLS

Next to evolution, the most important unifying principle in biology is that all organisms consist of *cells*, whether single-celled organisms like bacteria or multi-

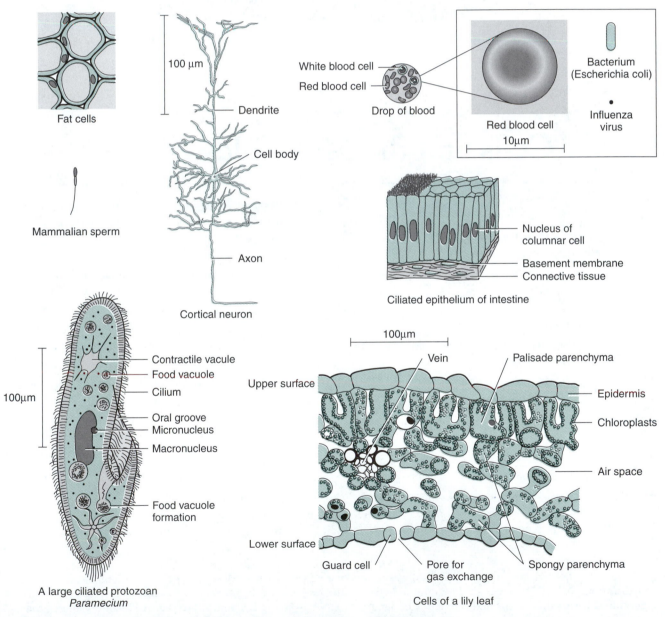

FIGURE 3.1 Some representative cells illustrating a diversity of sizes and shapes. The box in the upper right corner is at a 10-fold larger scale and shows the relative sizes of a bacterium and a virus. The protozoan in the lower left corner is just visible to the naked eye, but the largest cells are the eggs of big birds such as the ostrich.

cellular organisms like ourselves. In multicellular organisms, groups of cells specialize to perform distinct functions and are organized into tissues and organs. Not only are all organisms fashioned from cells, all cells arise by division of preexisting cells. By extension, all living organisms are descended from single-celled organisms that existed very early in the history of life, approximately 3.5 billion years ago.

The finding that all life consists of structural units called cells and that all cells arise from other cells is referred to as the *cell theory*. Like evolution, this is a scientific generalization that is based on a vast amount of observation and for which there are no known exceptions that would require the concept to be modified. Although the first evidence for the existence of cells dates from the creation of the first microscopes in the seventeenth century, the evidence for the cell theory was persuasively summarized by the German scientist Rudolf Virchow in 1855, four years before Charles Darwin published *On the Origin of Species*.

Some cells are shown in Figure 3.1. The formation of the first cells very early in the evolution of life represents a major evolutionary step, for a cell creates a tiny environment, separated from the rest of the world, in which local chemical reactions can occur and the participating molecules cannot drift away. But the interior of the cell is not sealed behind an impenetrable boundary through which nothing can pass. Instead it is contained within a thin membrane composed principally of lipids only two molecules thick (Fig 3.2). The cell membrane has several functions. It regulates the passage of molecules in and out

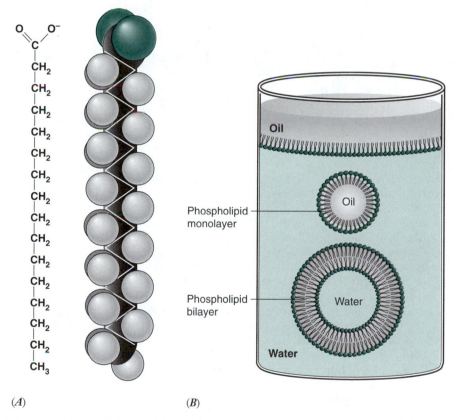

(A)   *(B)*

FIGURE 3.2. **(A) A fatty acid molecule. The chemical formula is on the left and a model showing how the molecule fills space is on the right. (Refer to Box 3.1 for more background on chemical bonding.) The *carboxyl group* at the upper end of the molecule is *acidic*: it loses a hydrogen ion (that's what makes it an acid) and takes on a negative charge, becoming an anion. The carboxyl end of the molecule is therefore *polar*, and it is soluble in water. The remainder of the molecule consists of carbon atoms with hydrogen atoms attached. This *hydrocarbon* tail is *nonpolar* and has very different solubility properties. It associates readily with solvents such as gasoline or oils, which are themselves composed largely of other hydrocarbons.**

(B) If fatty acids or phospholipids are placed on the surface of water, they orient with their polar heads down and their hydrocarbon tails up. In the presence of a pool of oil floating on water, phospholipids will form a layer at the interface between the polar and nonpolar solvents. If the oil is broken into droplets by vigorous shaking, the phospholipids will surround the droplets with their hydrocarbon tails sticking inward and their polar heads in the surrounding water.

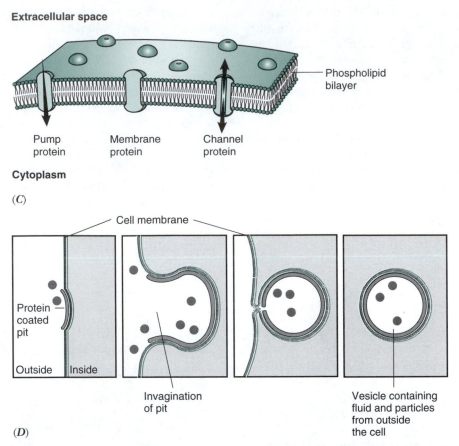

(D)

FIGURE 3.2. (continued) **(C) The interior of cells is aqueous, and the surrounding layer of phospholipids forms a bilayer two molecules thick. The hydrocarbon tails of each layer face each other, and the polar head groups face the water on the inside and the outside of the cell, as in the lower diagram in (B).**

The lipid bilayers of cell membranes have associated proteins that perform specialized functions. Some are channels allowing ions to pass through the membrane, others are pumps that use metabolic energy to transport sugars and other substances across the membrane, and some are receptor molecules that detect the presence of signal molecules on the outside of the cell and cause the cell to respond by altering its growth, changing its shape, or responding in a great variety of other ways.

(D) Cells can engulf foreign objects or small volumes of fluid by *endocytosis*, which occurs at pits in the cell membrane that are coated with a protein. Similarly, vesicles within the cell can fuse with the membrane and secrete proteins or small quantities of other molecules by the process of *exocytosis*.

of the cell, acquiring and retaining those that are needed within, disposing of unwanted by-products, and secreting molecules that the cell has manufactured for use elsewhere in the body. Transport through the membrane involves a number of processes, most of which are mediated by large molecules (*proteins*—whose structure is described below) embedded in the lipid matrix of the membrane. Some proteins form pores that allow water and small ions such as Na^+ and K^+ to pass through the membrane, and in nerve and muscle cells these channels can be rapidly opened and closed so that the fluxes of ions generate electrical signals. Using other membrane proteins, sugars, which are useful sources of energy

for the cell, can be transported into the cell, even against a concentration gradient. And larger objects can be engulfed within a pinched-off pocket of membrane for either uptake or secretion (Fig. 3.2D).

Cells also respond to specific molecular signals such as hormones. The cell detects these signal molecules when they react with other membrane proteins that serve as receptors for the signals. Cells can respond by moving, changing shape, altering their rate of growth, and changing the chemical reactions that they are performing. Signal molecules can also affect a cell's genes, as we will describe below.

The brief review of chemical bonding in Box 3.1 provides background for the rest of this chapter.

Box 3.1
Making and Breaking Chemical Bonds

When atoms share electrons they are held together by strong, *covalent bonds*. Some examples are pictured in Figure 3.3A, where a single line means that a pair of electrons are shared and a double line indicates two pairs. Bonds are strong and stable when their formation fills the outer shell of *valence* electrons. This number is 2 for H and 8 for C, N, and O. When an atom combines with several others, the bonds form at characteristic angles to each other (Fig. 3.3B).

Adjacent atoms influence the kinds of bonds that are made. Although C-H and O-H bonds are usually strong and stable, when in the presence of water the H atom in a *carboxyl group* readily dissociates as a hydrogen ion, leaving the $-COO^-$ anion (Fig. 3.3C). Carboxyl groups are therefore *acids*. Or consider the electrons around the nitrogen in an $-NH_2$ group. Two of the electrons are not paired with another atom, but in the presence of water they can be shared with a hydrogen ion to make the $-(NH_3)^+$ group a *cation* (Fig. 3.3C). Cations (with a positive charge) and anions (with a negative charge) are readily attracted to each other, particularly if each is not surrounded by water molecules.

Covalent bonds are strong in the sense that energy is required to make and break them. Furthermore, compounds with many $-CH_2-$ and $-CHOH-$ groups are stores of energy that can be released when the bonds are broken. That is what happens when gasoline burns in an automobile engine; heat is generated and the explosion drives the pistons, thus converting *chemical energy* into *mechanical energy*. This is also what happens when you metabolize the food that you eat, thereby obtaining energy to go about your daily activities.

For covalent bonds to break and reform, energy must be put into the system. This is called the *activation energy*, and in an automobile engine it is supplied by a spark. A *catalyst* is a substance that associates temporarily with the reacting molecule and decreases the activation energy that is required in order for the reaction to occur. At a given temperature, more of the participating molecules therefore have the requisite activation energy, and the reaction goes faster. *Enzymes* are biological molecules, almost always proteins, that function as catalysts in cells (Fig. 3.4), enabling reactions to occur rapidly at normal body temperature. In other words, they permit us to burn the food we eat and obtain energy from the $-CHOH-$ of sugars and

(A)

Single and double bonds have different spatial arrangements.

(B)

$$RCOOH \longleftrightarrow RCOO^- + H^+$$
$$RNH_2 + H^+ \longleftrightarrow RNH_3^+$$

(C)

FIGURE 3.3 (A) Carbon has four valence electrons and forms single covalent bonds with four other atoms. Hydrogen has one electron to share, so four hydrogens can bond with one carbon. In the second row, a double bond has formed between carbon and oxygen. Ordinarily it is too much trouble to show the electrons, and bonds are represented by short straight lines. Sometimes compounds are represented in an even more condensed notation, as in the examples on the right.

(B) When an atom bonds with several other atoms, the bonds form at different angles. Molecules therefore have characteristic three-dimensional shapes.

(C) An *acid* is a substance that gives up a hydrogen *ion* in aqueous solution. An ion with a positive charge is called a *cation*. The carboxyl group $(-COOH)$ is an acidic group commonly found in carbon compounds. (The "R" in the figure represents the remainder of the molecule). The positively charged hydrogen ion differs from a hydrogen atom because it has left its valence electron behind. This leaves a negative charge on the COO^-, which is therefore an *anion*.

Amino groups with the structure $-NH_2$ are common constituents of important biological molecules. They are *bases* because they can attract a hydrogen ion to an otherwise unshared pair of electrons. The resulting $-NH_3$ is a cation because it has a net positive charge associated with the nitrogen atom. (continued) the $-CH_2-$ of fats without the application of high heat.

Another kind of bond is much weaker than covalent bonds but is enormously important in the structure of genes and proteins. It is called the *hydrogen bond*. The easiest place to visualize hydrogen bonds is in water molecules (Fig. 3.5). Although water molecules do not have a net charge, the oxygen atom has a greater attraction than does hydrogen for the pair of electrons that they share. The result is that water molecules are *polar*, slightly more negative on the oxygen side than on the hydrogen side. Because of this polar structure, water molecules are attracted to and surround ions.

Hydrogen bonds form when a hydrogen on one water molecule is attracted to the oxygen on a different water molecule. This occurs in liquid water, but because the attractions are weak, they are constantly breaking and reforming at room temperature. They are important, however, in determining the physical properties of water. For example, water has a much higher boiling point than many small molecules like alcohol because to bring molecules of water from liquid to vapor phase it is necessary to break many hydrogen bonds.

In proteins and nucleic acids, hydrogen bonds form between the oxygen of a C=O group and a hydrogen atom in an $-NH_2$ or $-NH-$ group as well as other places where a hydrogen atom is placed near an

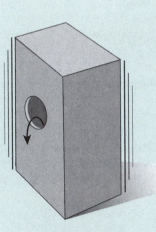

Uncatalyzed

FIGURE 3.4 A simple model of the role of enzymes in chemical reactions. Imagine a ball in a box like that shown in the figure. The floor of the box is higher than the surface of the surrounding table. The ball therefore has potential energy by virtue of its more elevated position, and kinetic energy because the box is being jiggled. If there were no barriers to the ball's movement, the jiggling of the box would cause the ball to fall down onto the table. If the kinetic energy imposed by shaking the box is great enough, there is a chance the ball will get over the barrier at a low point and drop onto the table. If the barrier lowers, the shaking will not have to be as vigorous.

A sugar molecule has greater potential energy stored in its bonds than does an equivalent amount of carbon and hydrogen in CO_2 and H_2O, but at room temperature the kinetic energy of molecular motion is insufficient to break and rearrange covalent bonds. Bonds can be broken, however, if the temperature is raised high enough. In more familiar terms, if you heat sugar on the stove you can burn it. The role of enzymes is equivalent to lowering the barriers on the box with the ball. By decreasing the activation energy of the reaction, many more molecules will have sufficient kinetic energy at body temperature to "escape from the box." In other words, an enzyme causes the number of molecules reacting in a given interval of time—the rate of the reaction—to increase.

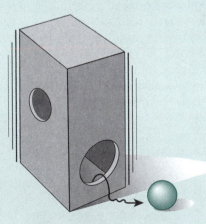

Enzyme catalysis

atom that is more negative than its immediate molecular environment (Fig. 3.5). As in water, these hydrogen bonds are about 5% as strong as covalent bonds, so they are easily broken by molecular motions at room temperature. But where they are numerous and the donor groups cannot move relative to each other because they occupy fixed positions as part of a large molecule, hydrogen bonds can have an enormous cumulative influence in determining the final shape of the molecule. Specific examples are present in the discussion of proteins and genes.

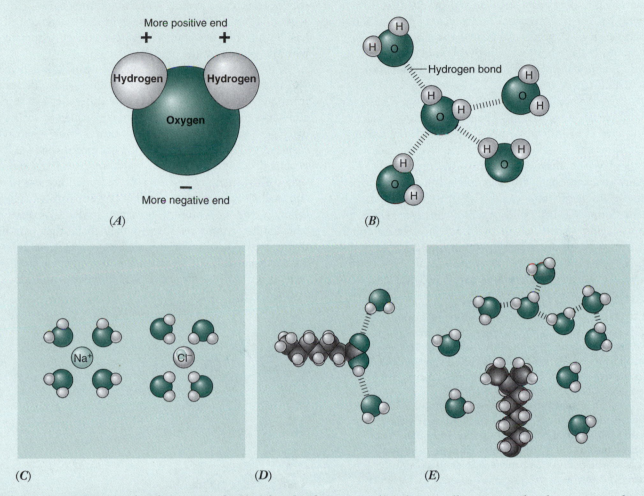

FIGURE 3.5 (A) Because the electrons in the H-O bonds of water are shared unequally, water molecules are somewhat more negative on the oxygen side of the molecule. This *polar* nature of water has a number of important consequences.

(B) Water molecules are attracted to each other through the weak forces of *hydrogen bonds*, as shown here. Liquid water therefore has an internal structure that is constantly changing as H-bonds break and reform.

(C) When salts (consisting of ion pairs of cations and anions) come into contact with water they dissolve because the individual ions are surrounded by clouds of polar water molecules.

(D) Water molecules are also attracted to polar groups on organic molecules, with which they form hydrogen bonds, thus increasing their solubility in an aqueous medium.

(E) A nonpolar region of a molecule, like the hydrocarbon tail on the fatty acid in Figure 3.2, interrupts the structure of the surrounding water without attracting any of the polar molecules of water. Nonpolar molecules are therefore not soluble in water.

PROTEINS AS MOLECULAR MACHINES

As you can see, we have been unable to get beyond the surface of the cell without encountering proteins. In nature, the only source of proteins is cells, and proteins are responsible for the shapes and so many of the activities of cells that in an important sense life is the expression of protein diversity. Moreover, there is a direct relationship between genes and proteins that is not found with other important molecules such as carbohydrates and fats. For these reasons we need to spend some time describing what proteins look like and how their functional properties are related to their molecular structure.

Proteins are examples of *polymers* (from Greek for many parts), which simply means that they consist of long chains of individual units, like beads on a string. The individual units in proteins are amino acids (Fig. 3.6), of which there are twenty different kinds. The chains of amino acids do not branch, but proteins differ greatly in length and can be several hundred amino acids long. Clearly with twenty different kinds of amino acids with which to build and no fixed number that must be used for every protein, the possible number of different sequences is astronomical. This is why there are so many different kinds of proteins and why

proteins are able to play so many different roles in the life of the cell.

All amino acids share the following feature: on the first (or α) carbon atom they have an *amino* ($-NH_2$) group and an acidic (i.e., *carboxyl*, $-COOH$) group. The carboxyl group is acidic because it readily gives up a hydrogen ion and takes on a negative ionic charge (Fig. 3.2). Conversely the amino group can take up a hydrogen ion and assume a positive ionic charge. Consequently, an amino acid in solution characteristically has both a positive and a negative charge associated with the groups on the α-carbon.

The remainder of the molecule, which is labeled "R" in Figure 3.6, is what makes amino acids different from one another. There are twenty different R-groups, and some of these are simply chains or rings of carbon, hydrogen, and nitrogen whereas others bear a second carboxyl group or amino group. This means that the R-groups differ greatly in their solubility in water. Those that consist solely of carbon and hydrogen are *nonpolar* and will associate with a nonpolar environment like lipids if the opportunity presents itself. Conversely, where ionic groups are present they make that part of the molecule more polar and readily soluble in an aqueous environment. These features of amino acids are important for both the structures and functions of proteins.

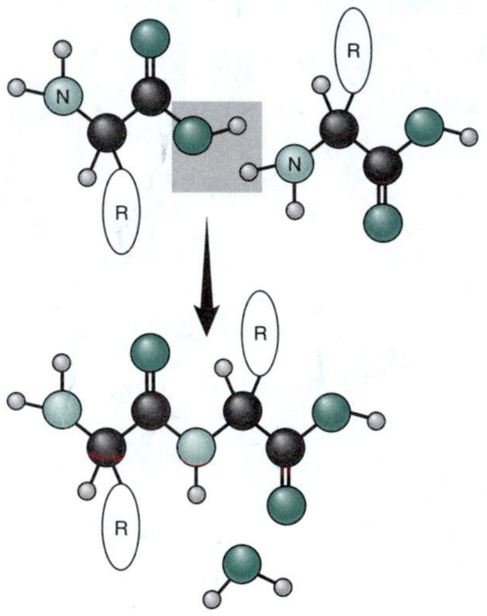

FIGURE 3.6 (A) A protein is a polymer of amino acids strung together by strong bonds between the carboxyl group of one α carbon and the amino group of the next. This bond is called a *peptide bond*. Notice that when it forms a molecule of water is also generated.

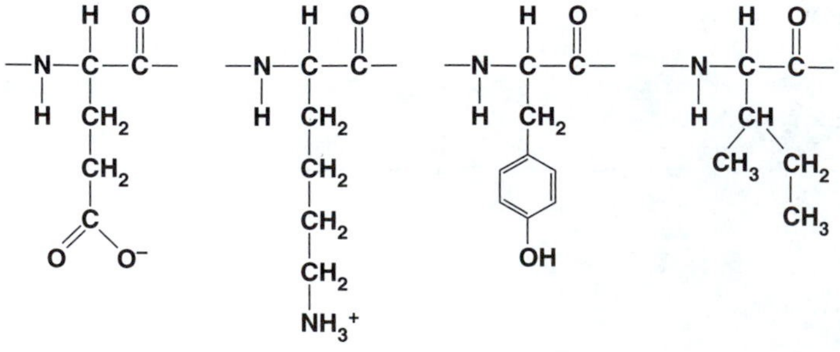

(B) These are four of the twenty amino acids, illustrating that the R-groups can be either polar (with carboxyl and amino groups) or nonpolar. The shapes and functions of proteins depend in critical ways on the presence of these side chains. (The structural formulas are drawn as though each is part of a protein.)

(A)

(B)

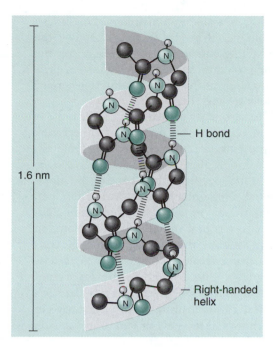

1.6 nm

H bond

Right-handed helix

(A)

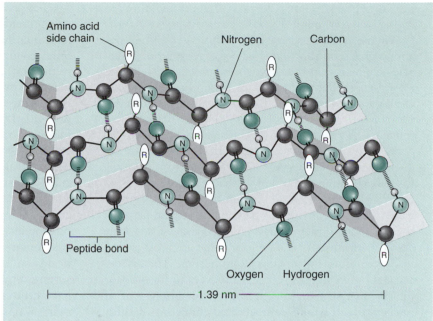

Amino acid side chain

Nitrogen

Carbon

R

Peptide bond

Oxygen   Hydrogen

1.39 nm

(B)

Polar side chains   Nonpolar side chains

Hydrogen bonds can form to polar side chains on the outside of the molecule

Hydrophobic core region contains nonpolar side chains

Unfolded polypeptide

Folded conformation in aqueous environment

(C)

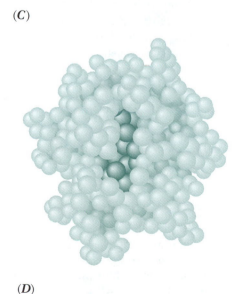

(D)

FIGURE 3.7 **Some principles that determine the three-dimensional structures of proteins.** (A) Domains can coil into an *α-helix* that is stabilized by the presence of hydrogen bonds between α-carbon-NH-groups of one amino acid and a −C=O several amino acids further along the backbone of the molecule. (B) Or sections of the chain of amino acids can curve back and hydrogen bond with other regions, forming a *β-pleated sheet*. Both the α-helix and the β-pleated sheet are examples of *secondary structures* involving domains of the protein. A protein molecule can have domains of this sort separated by portions of the amino acid chain that fold in less regular ways.

(C) The *tertiary structure* of a protein refers to the way in which the various domains interact with each other to form the final, three-dimensional, folded configuration. This panel illustrates another principle. If the protein is soluble in water, it folds so that polar side chains are exposed on the external surface of the molecule. On the other hand, if the protein is to exist in a nonpolar environment like the phospholipid bilayer of the cell membrane, it folds around its polar side chains and presents nonpolar *R*-groups on its external surface.

(D) This is a space-filling model of a protein, illustrating how a small substrate molecule or coenzyme that participates in an enzymatic reaction binds at a specific site on the protein. Specificity is determined by the particular R-groups in the binding region and the three-dimensional shape of the protein.

In proteins, amino acids are connected through strong (*covalent*) bonds (Box 3.1) called *peptide bonds* between the α-carboxyl on one amino acid and the α-amino group on the next (Fig. 3.6). (A string of amino acids connected in this way is sometimes referred to as a *polypeptide*.) Because these are strong bonds, enzymes are required to form them, and we shall see later in this chapter how that process occurs. The particular linear sequence of amino acids in a protein is known as the *primary structure* of the protein. If you knew that sequence, you could write down the structural formula for the protein in the same way that we wrote the structures for the amino acids in Figure 3.6. That exercise would be about as exciting as copying several pages of the telephone directory, and it would convey very little about the features of the molecule that are most critical in determining its functional properties. That is because proteins fold in very specific ways to form three-dimensional structures with pockets for binding other molecules and whole domains that differ in solubility and other physical properties. So knowing the primary structure of a protein is only the first step in understanding its folded structure.

The folded shapes of proteins are stabilized by a variety of forces that (with a few important exceptions) do not involve additional covalent bonds. For example, a positive charge on one R-group may be attracted to the negative charge on another, even though the two amino acids are widely separated in the primary structure of the protein. A second form of weak attraction is the *hydrogen bond*, whose importance is so great for understanding the structure of both proteins and genes that you should refer again to Box 3.1 if you passed over that material on first reading.

Regions of a protein can coil or fold in recognizable patterns stabilized by hydrogen bonds. Examples of such *secondary structures* are shown in Figure 3.7. The entire molecule then folds further into its *tertiary structure*, stabilized by ionic attractions and numerous weak forces acting at very specific sites to produce, for example, a molecule capable of functioning as an enzyme, with a crevice into which a specific substrate molecule will fit to have its bonds broken and reformed. Regions of a protein molecule with compact sub-structures form *domains*. Some proteins require the interaction of two or more different polypeptide chains (*subunits*) to achieve their final functional form. The three-dimensional arrangement of such subunits is referred to as *quaternary* structure.

Many proteins function as receptors for small molecular signals. Sometimes these receptors are embedded in the cell membrane and sometimes they are located inside the cell, but like enzymes they must have *binding sites* that can recognize the appropriate signal molecules and then respond with changes in shape that can be recognized by still other proteins (Fig. 3.7).

If a protein is to reside in a membrane, its surface will present nonpolar R-groups in that region of the molecule that will be surrounded by lipids. If, on the other hand, the protein is soluble in the aqueous environment of the cell or in the blood, its external surface must expose polar groups (Fig. 3.7).

Because the secondary and tertiary structures of proteins depend on hydrogen bonds and other weak forces, most proteins cannot tolerate high temperatures. When a solution of proteins is heated, hydrogen bonds break and the specific folded shape of the molecules is disrupted. New interactions form at random, and the proteins are said to be *denatured*. A familiar example occurs when you boil an egg. The thick solution of proteins that compose the "white" of the egg congeal into a solid mass. Because proteins are characteristically so sensitive to heat, most organisms cannot survive high temperatures. However, some bacteria have evolved heat-resistant proteins and can thrive in thermal hot springs and in vents of very hot water on the ocean floor.

THE INTERNAL STRUCTURE OF CELLS REFLECTS DIFFERENT FUNCTIONS

Life requires a source of energy. Ultimately the energy is supplied by the sun and is captured in chemical bonds by the photosynthesis performed by green plants. This series of reactions can be summarized by the following shorthand:

$$CO_2 + H_2O \rightarrow -CHOH- + O_2$$

In other words, carbon (from carbon dioxide in the air) is *reduced* (has hydrogen atoms added) to form sugars (which you can think of as molecules containing repeats of $-CHOH-$), and oxygen is released as a by-product. Energy of sunlight is thus captured and transformed into potential energy in sugars and other carbohydrate molecules.

Animals, however, are totally dependent on plants for energy, which they consume either directly by eating plants or indirectly by eating other animals. When animals (and plants too) need energy, they tap into the supply that was generated by photosynthesis. They oxidize sugars and other molecules that have been made from sugars.

$$CHOH + O_2 \rightarrow CO_2 + H_2O$$

Each cell is responsible for mobilizing the energy it needs for growing, moving about, doing the mechanical work of muscle, and synthesizing proteins and other molecules that it uses itself or secretes. This process of burning sugars is equivalent to the combustion of a log in your fireplace or gasoline in your automobile in that oxygen is required and carbon dioxide is produced. There are two important differences, however. First, the process occurs at the normal temperature of the cell because it proceeds through a large number of chemical steps, each mediated by an *enzyme*. [These enzymes are proteins that function as catalysts, making chemical reactions go faster than they otherwise would without raising the temperature (Box 3.1).]

Second, some of the energy locked in the chemical bonds of the molecules that are being utilized for food is recaptured in the bonds of another molecule known as ATP (Fig. 3.8). In order to minimize the fraction of the energy lost as heat and to maximize the fraction that is recaptured as ATP, the metabolic burning of sugars and other foodstuffs requires many chemical reactions occurring in sequence. The details of the chemistry are therefore numerous, and they not going to concern us here. The important point is that ATP serves as a sort of currency for energy exchange within the cell. For example, if the cell needs energy to make more membrane so that it can grow larger, it makes ATP by burning sugars or other foodstuffs, then uses the ATP as a source of energy to synthesize the lipids and proteins that it needs for growth. When you take a walk, lift a box, or otherwise put your muscles to work, you are paying with ATP (Fig. 3.9). In this case, ATP reacts with the contractile proteins of your muscle cells. The ATP loses its terminal phosphate group (thus some potential energy), and the proteins in their turn shorten (thus the muscle does mechanical work).

Except in bacteria and other primitive cells, many of the chemical reactions associated with the burning of sugars (collectively known as *oxidative metabolism*) occur

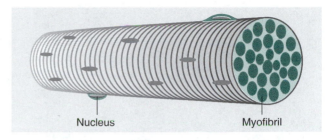

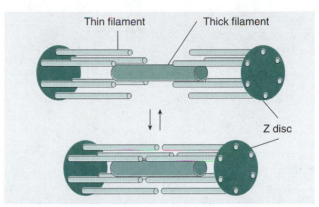

FIGURE 3.9 Muscles illustrate how cells use proteins to perform mechanical work.

Above A muscle cell is a long tube containing many nuclei. In a human arm or leg it can be 50 cm long and only 50 μm in diameter—10,000 times longer than its width. It contains bundles of thick and thin filaments of protein.

Below When a muscle is stimulated to contract, the sets of thick and thin filaments slide together and the muscle shortens. The sliding is caused by breaking and reforming bonds between tiny arms on the thick filaments (not shown in this version) and sites on the thin filaments. Changing these sites of attachment requires ATP, which loses its terminal phosphate in the process.

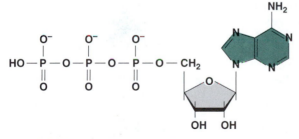

FIGURE 3.8 This is ATP, the molecule that serves as the principal mode of energy exchange in cells. ATP is made by metabolizing sugars and other foods. It can donate some of its internal potential energy when the loss of its terminal phosphate group is appropriately coupled to other energy-requiring reactions.

ATP consists of a nitrogen-containing base, a 5-carbon sugar, and three phosphate groups. ATP is similar to the molecules from which genes are constructed, to substances that are involved in molecular signaling, and to some of the coenzymes that participate in enzymatic reactions.

in the cell within structures called *mitochondria* (Fig. 3.10). Note that the internal structure of mitochondria consists largely of membranes. These provide surfaces in which enzymes reside in set relationship to each other, enabling the chemical reactions in which ATP is synthesized to proceed in appropriate sequence.

Much of the interior of such cells (the *cytoplasm*) is also occupied by an array of membranes known as the *endoplasmic reticulum*, and in places these membranes are studded with dense clusters of small bodies called *ribosomes*, together forming rough endoplasmic reticulum. Ribosomes are the sites in the cell where proteins are synthesized. We will consider their function in a few pages when we discuss the relationship between genes and protein synthesis. In other places the endoplasmic reticulum is smooth; these are sites where lipids are manufactured. The *Golgi apparatus* is a series of adjacent, flattened membrane sacs where molecules destined for export from the cell are packaged.

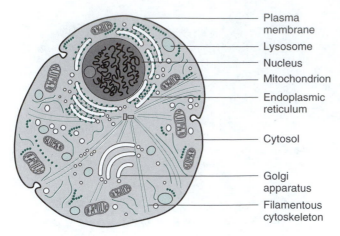

FIGURE 3.10 The parts of a typical animal cell.

Within multicellular organisms, cells assume many shapes, and there is an internal architecture of filaments and tubules called the *cytoskeleton* that is responsible for both shape and movement. These filaments and tubules consist of specific proteins, and like the membranes and other constituents of the cell they are synthesized as the cell grows.

Lysosomes are organelles that dispose of large molecules that are not needed by the cell. Like a mitochondrion, a lysosome is surrounded by a lipid bilayer membrane that keeps its contents separated from the rest of the cell. This is particularly important for lysosomes because they contain a variety of soluble enzymes that break down proteins and other macromolecules. Once the membrane of a lysosome is either ruptured or fused with the membrane of another cellular structure, the degradative enzymes are released to attack the macromolecules.

The nucleus of the cell is bounded by a membrane that is continuous with the endoplasmic reticulum. During cell division, material in the nucleus condenses into discrete bodies called *chromosomes*, which contain the genes in association with protein. Genes bring us to a second class of large molecules, the nucleic acids, of which there are two kinds, DNA and RNA.

THE STRUCTURE OF THE GENES REVEALS THE BASIS FOR THEIR REPLICATION

The discovery that DNA (*deoxyribonucleic acid*) is the substance of which genes are made is one of the most important findings in all of biological science. Shortly before 1900 it was known that hereditary information was associated with the chromosomes that appear and replicate during cell division, but the molecule that carries the genetic information remained a mystery for an-

other half century. Of all the molecules in the cell, only proteins seemed to have enough variety and complexity to account for the myriad of features that genes must control. Proteins are assembled from twenty different amino acids, but DNA, although largely localized in the chromosomes, consists of long chains of only four different subunits.

Serious attention became focused on DNA in 1944 when Oswald Avery, who was interested in the microorganism responsible for pneumonia, showed that DNA extracted from one genetic strain of the bacterium could be taken up by another strain and cause a genetic change that was then inherited by daughter cells. That experiment showed that DNA contained genetic information. Several years later, James Watson and Francis Crick asked whether the secret of DNA might be found in the structure of the molecule. Their answer, which came in 1953, had revolutionary consequences for the science of biology.

The four building blocks of DNA are called *nucleotides* (Fig. 3.11). Each consists of three simpler molecules linked together by covalent bonds: a nitrogen-containing, ring-shaped *base*; a phosphate group (PO_4); and a five-carbon sugar, *deoxyribose*. We shall refer to the four nucleotides as A, T, C, and G, which stand for the chemical names of the four bases, *adenine*, *thymine*, *cytosine*, and *guanine*. However, it is possible to understand everything you need to know about the structure and function of DNA without memorizing these four names.

Notice in Figure 3.11 how the carbon atoms in the sugar are (arbitrarily) numbered $1'–5'$ and how the phosphate group is attached to the $5'$ carbon. In

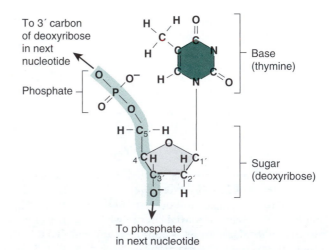

FIGURE 3.11 Nucleotides are the building blocks of the nucleic acids DNA and RNA. A nucleotide consists of a nitrogen-containing base, a 5-carbon sugar, and a phosphate group. (If you compare this with Figure 3.8, you will see that ATP is one of these nucleotides to which two additional phosphate groups are attached.)

DNA the nucleotides are in turn linked together by strong covalent bonds between the phosphate group of one nucleotide and the 3' carbon on the next nucleotide.

The critical discovery of Watson and Crick was that two long strands of DNA are wound together in a *double helix* (Fig. 3.12). The two backbones of the helix consist of alternating sugar and phosphate residues on the two strands of nucleotides, and the bases on each strand project inward toward the axis of the helix. The key to this structure of DNA—and the ability of DNA to replicate —lies in the precise molecular shapes of the bases (Fig. 3.13). A and T can simultaneously form two hydrogen bonds with each other, whereas G and C can simultaneously form three hydrogen bonds with each other. Moreover, because of their molecular shapes, no other pairwise alignments of these bases allow the formation of more than one hydrogen bond at a time and fit along the axis of the helix.

The consequence of these relationships is that once the sequence of nucleotides on one strand is de-

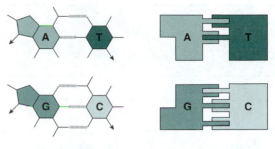

Structural formulas   Prongs and sockets

FIGURE 3.13 The shapes of the nucleotide bases, with the positions of their polar groups, make complementary pairs in which T hydrogen-bonds with A, and C pairs with G. A familiar analogy is the many different shapes of electrical plugs and sockets that exist around the world. The distribution of prongs and sockets in the model corresponds to the array of H atoms and the negative sites to which the hydrogens are attracted.

termined, the sequence of the *complementary* strand is also specified. An A on one strand will pair with a T on the other; a C on one strand will pair with a G on the other (Box 3.2). Although the two strands are not identical, they contain equivalent information. What are the implications of this relationship?

Every time a cell divides to form two daughter cells, its DNA is replicated. The double helix starts to uncoil, exposing the sequence of bases on each strand. An enzyme, *DNA polymerase*, then goes to work. Starting at the 3' end on one strand, free nucleotides begin to pair with their complementary bases on the intact strand, and as they line up in sequence, DNA polymerase connects them together by catalyzing the formation of covalent bonds between the phosphate group on the newest nucleotide and the sugar of the last nucleotide in the growing chain (Fig. 3.15). The enzyme then moves along the growing strand and incorporates the next base into its length.

If you examine Figures 3.11, 3.12 and 3.15 closely, you will see that the two strands of the double helix run in opposite directions with respect to the linkages in their backbones. DNA polymerase must start at the 3' end of the template strand, and it therefore constructs a complementary strand from the 5' end. As DNA replicates prior to cell division, the enzyme must assemble a new strand from both templates; consequently, as you can see in Figure 3.15 one of the new strands must be assembled in short segments, which are then connected by another enzyme, *DNA ligase*. This detail takes on additional interest, because DNA ligase is one of several enzymes that have become important research tools in manipulating DNA in genetic engineering.

In summary, the structure of DNA and its special ability among biological molecules to serve as the

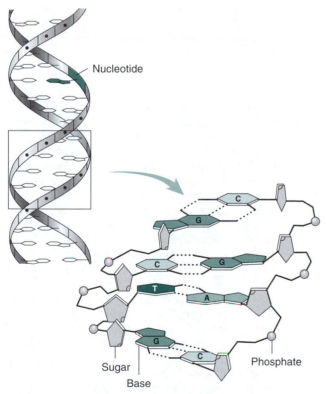

Nucleotide

Sugar | Phosphate
Base

FIGURE 3.12 The structure of DNA is key to understanding how this molecule is replicated. DNA is a double helix. Each strand of the helix is a string of nucleotides connected to each other through strong (covalent) bonds between the phosphate group of one and a hydroxyl group (OH) on the sugar residue of the next nucleotide. The bases project inward, toward the axis of the helix, where they form hydrogen bonds in complementary pairs, A with T and G with C.

BOX 3.2
Melting the Double Helix

DNA can be extracted from broken cells as long threads of double helix. Although hydrogen bonds are very weak compared with the covalent bonds that form the backbone of each strand, the helix remains intact partly because there are so many bases and so many hydrogen bonds. If the temperature is raised close to the boiling point of water, however, the hydrogen bonds break, and the two strands separate. The DNA is said to have "melted" because the solution becomes less viscous.

For a rough analogy, think of a "Velcro" fastener on your clothes. Each half of the fastener is sewn to the clothing, making strong connections. When the two halves engage, many tiny hooks on one side entangle with an irregular mat of fibers on the other. Individually, the connections are weak, but the fastener holds because there are so many hooks. In this analogy, pulling the two halves apart is equivalent to heating a solution of DNA; all of the weak connections (hydrogen bonds) break well before the sewn seams (the covalent bonds in each strand) begin to part.

The melting of a solution of DNA is like the denaturation of protein molecules in that a large number of weak forces between very specific places in the molecule are broken, and profound changes in molecular shape then occur. There is a very important difference, however, if the solution is then cooled. Protein molecules remain denatured; weak interactions form, but all specificity in their placement is lost. Parts on one molecule become attracted to places on other molecules, and the result is a massive tangle that does not remain in solution. With DNA, something remarkably different happens: regions of complementary bases find each other, and double helices reform (Fig. 3.14).

This feature of DNA has a number of practical uses, one of which we will mention here. If single strands of DNA from different species are mixed together in solution, they will base-pair and form double helices whenever and wherever there are extensive regions of complementarity in base sequence. The more closely related two species are, the more base-pairing there will be, and the higher the temperature must be raised in order to melt the hybrid DNA. This is the technique used to establish the evolutionary relationships of the ostrich, rhea, and emu discussed in Chapter 2. It was also used to create an evolutionary tree of relationships of the major groups of primates, as described in Chapter 12.

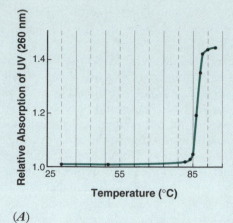

(A)

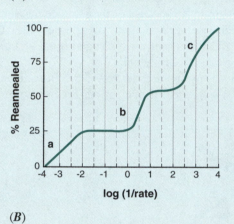

(B)

FIGURE 3.14 (A) When a solution of DNA is heated, it "melts" as the hydrogen bonds break and the two strands of the helix come apart. The change can be monitored by measuring the absorption of ultraviolet light by the bases, which increases as the strands separate. In this example the DNA was from a bacterium. All the base sequences were the same, so the melting curve was very sharp.

(B) When a solution of DNA is slowly cooled, the helices reassemble. In DNA from organisms other than bacteria, some DNA helices are present in many copies, some in a few copies, and some in only a single copy. If such a solution is melted and then allowed to reanneal, the DNA that is present in multiple copies finds complementary strands faster than the DNA that exists as only one or a few copies. (If you go to a football game alone and look for a friend with whom to sit, you can locate a companion much faster if you have many friends in the crowd.)

In this figure the fraction of DNA that has reassociated as double helices is plotted as a function of the logarithm of 1/*rate* of annealing. (Faster rates are on the left, and because the axis is logarithmic, it spans a range of 10^8 i.e., 100 million-fold.) The curve has three distinct regions: *a*, the DNA that reassociates fastest is the DNA for which there are many copies; *b*, the middle section of the curve corresponds to DNA present in a few copies, and *c*, the last and slowest DNA to anneal, is the DNA present as a single copy.

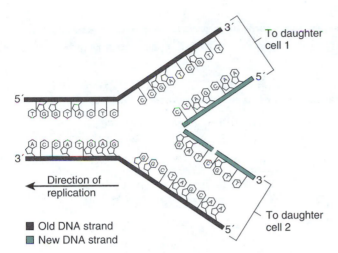

FIGURE 3.15 Prior to cell division, DNA replicates. The double helix opens, and each strand serves as a template for the synthesis of a complementary strand. The enzyme that performs this task works from the 3′ end of the template strand, but because the two strands of a double helix are oriented in opposite directions, one of the new strands has to be assembled in short pieces. The result, however, is two double helices, each identical to the parent helix, each destined for one of the two daughter cells that results from cell division, and each containing one of the strands of DNA originally present in the parent helix.

	Second Base			
	U	**C**	**A**	**G**
U	UUU ⎤ Phe UUC ⎦ UUA ⎤ Leu UUG ⎦	UCU ⎤ UCC ⎥ Ser UCA ⎥ UCG ⎦	UAU ⎤ Tyr UAC ⎦ UAA ⎤ STOP UAG ⎦	UGU ⎤ Cys UGC ⎦ UGA STOP UGG Trp
C	CUU ⎤ CUC ⎥ Leu CUA ⎥ CUG ⎦	CCU ⎤ CCC ⎥ Pro CCA ⎥ CCG ⎦	CAU ⎤ His CAC ⎦ CAA ⎤ Gln CAG ⎦	CGU ⎤ CGC ⎥ Arg CGA ⎥ CGG ⎦
A	AUU ⎤ AUC ⎥ Ile AUA ⎦ AUG Met	ACU ⎤ ACC ⎥ Thr ACA ⎥ ACG ⎦	AAU ⎤ Asn AAC ⎦ AAA ⎤ Lys AAG ⎦	AGU ⎤ Ser AGC ⎦ AGA ⎤ Arg AGG ⎦
G	GUU ⎤ GUC ⎥ Val GUA ⎥ GUG ⎦	GCU ⎤ GCC ⎥ Ala GCA ⎥ GCG ⎦	GAU ⎤ Asp GAC ⎦ GAA ⎤ Glu GAG ⎦	GGU ⎤ GGC ⎥ Gly GGA ⎥ GGG ⎦

(First Base shown along left side.)

FIGURE 3.16 The genetic code found in all organisms. There are sixteen ways in which one of the bases (the first) can be combined with another (the second). But there are sixty-four possible combinations if a third base is added. There are thus sixty-four three base *codons*. As there are twenty amino acids (here represented by their abbreviations), some amino acids are represented by more than one codon. Three of the sixty-four codons are termination (STOP) signals that indicate the end of a coding region.

template for its own replication are due to the complementary hydrogen bonding between specific nucleotides.

DNA AS CODED INFORMATION

The hereditary information contained in DNA is the information necessary to specify the primary structures of all the proteins that an organism will need and make in its life. In other words, by employing just four nucleotide bases, DNA codes for the sequence of amino acids that are the building blocks of each and every protein in the body. There are twenty amino acids, so the alphabet of four nucleotide letters must code for twenty different words. There are $4^2 = 16$ ways in which four nucleotides can be combined in pairs, but sixteen is four short of what is required. The evolutionary solution has been to employ *triplets* of nucleotides, for which there are $4^3 = 64$ different possible combinations. The nucleotide triplets and the amino acids they designate are shown in Figure 3.16.

Each triplet of nucleotides is called a *codon*, and sixty-four codons are many more than required to designate the twenty amino acids. Some amino acids are therefore represented by more than one codon. (A code with this kind of redundancy is said to be *degenerate*.) You can think of a gene as a stretch of DNA that con-

tains a *sequence* of codons corresponding to the sequence of amino acids in the primary structure of a protein. Note in Figure 3.16 that three of the codons have a different meaning; they are used as "stop" signals to indicate the end of a coding region. A coding region within a gene does not have any other punctuation; it does not need any, because all of the codons are the same length, and so all that is necessary is for the cell to "read" them sequentially.

The formal relationship between a piece of a gene and a short section of the corresponding protein is shown in Figure 3.17. We will shortly consider how a gene directs the synthesis of a protein, but it is first important to understand the consequences of a mistake in the replication of DNA. If one nucleotide is substituted for another during the replication of DNA, this mistake constitutes a *copy error* and results in a *point mutation* (Fig. 3.18). A point mutation is obviously the simplest possible alteration in a gene; we will see examples of more extensive mutations later in this chapter.

What are the consequences of a point mutation? The chart of the genetic code (Fig. 3.16) shows that some mutations, most of which are in the third position of the codon, will not result in a codon for a different amino acid. This is because although the codon has been altered physically, the amino acid that it designates has not been changed. Such copy errors are *silent mutations*;

FIGURE 3.17 This diagram shows the relationship between genes, codons, amino acids, and proteins. A gene is a stretch of DNA that codes for a particular protein. It consists of a linear sequence of codons, each composed of three nucleotides. Each codon represents an amino acid, and the linear sequence of codons corresponds to the sequence of amino acids in the primary structure of the protein.

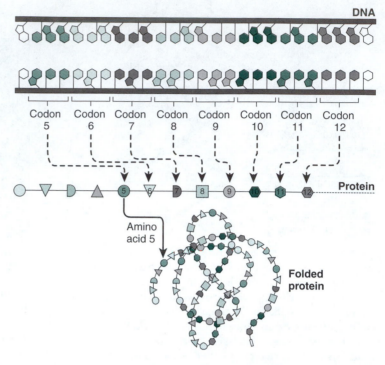

they are evident in the gene, but they do not change the gene product, the protein for which the gene codes.

Other point mutations do lead to changes in the protein, with a wide range of consequences. At one end of the spectrum, if the new amino acid is chemically similar to the original and if the substitution occurs at a site that is not critical for the function of the protein, the change may be inconsequential. In other words, the new protein may be able to do its job as well as the original version, and in such a case the mutation is said to be *neutral*—neutral, that is, to natural selection. Of course a mutational change that is neutral in one environment might prove to be either advantageous or disadvantageous under other circumstances. Consequently, a mu-

tation that is neutral when it occurs may subsequently become an object of selection in a later generation.

Other amino acid substitutions can have more severe consequences (Box 3.3). The protein may be altered functionally in modest ways, or the substitution may be catastrophic, preventing the protein from folding properly, altering an essential binding site, or otherwise destroying its ability to function. Which of these outcomes happens depends on the structure of the protein and whether it can accommodate the particular amino acid being inserted.

How often do mistakes occur? The copying machinery can be astonishingly accurate, with only one error every 10^9 bases. To do as well, a typist would have

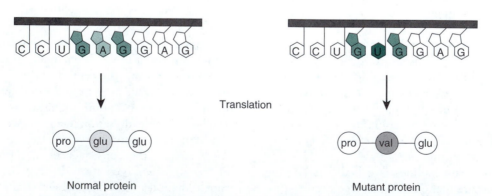

FIGURE 3.18 A point mutation is a single base substitution in the code. If it changes the codon so as to specify a different amino acid, the result is a change in the primary structure of the protein. Whether or not an amino acid substitution alters the function of the protein, however, depends on differences in the side chains of the original and the replacement amino acids.

Box 3.3
Sickle-Cell Hemoglobin

Sickle-cell anemia is an inherited condition in which the red blood cells tend to lose their round form and collapse into sickle shapes that do not pass freely through capillaries (Fig. 3.19A). The immediate trigger is a shortage of oxygen, brought about, for example, by exercise or by flying to high elevations in an unpressurized cabin.

Hemoglobin is the protein in red blood cells that carries oxygen. When the blood passes through capillaries in the lungs, the hemoglobin binds oxygen. When the blood arrives in the tissues where there is less oxygen, the bound oxygen is released. Sickle-cell anemia is caused by a point mutation in the gene for hemoglobin that substitutes an amino acid with a non-polar side chain for one with a polar group. This change alters the physical properties of the hemoglobin molecule in such a way that when it is not binding oxygen it comes out of solution (precipitates) more readily than normal hemoglobin. As the mass of precipitated hemoglobin grows, the red blood cell assumes a sickle shape, which blocks narrow capillaries and starves the surrounding tissue of oxygen. Consequently, the symptoms become acute when an individual with sickle-cell anemia is having trouble getting enough oxygen.

The complexity of organisms can conceal the interdependence of parts. Figure 3.19B lists some of the manifestations of sickle-cell anemia. As cells sickle, they frequently clump, leading to a local blockage of blood supply, which can damage organs like the kidneys, the lungs, or the brain. Depending on where the damage is greatest, the patient may suffer from kidney failure, a lung infection like pneumonia, or paralysis. Or the loss of functional red blood cells can lead to anemia, which in turn leads to weakness, or in young children, poor development. In the absence of information about an individual's hemoglobin or red blood cells, all of these symptoms have alternative explanations. These many possible outcomes, however, are all due to a point mutation in a single gene.

In the United States, sickle-cell anemia is found among African-Americans. The reason is interesting from an evolutionary perspective. The gene is present in indigenous populations of tropical West Africa where the incidence of malaria is high. Malaria is caused by a single-cell parasite that infects red blood cells. Individuals whose red blood cells have a tendency to sickle have some resistance to malaria, possibly because sickled cells tend to be preferentially removed from the circulation by the spleen. A shorter lifetime for red blood cells means a shorter stay in the body for the *Plasmodium* that

(A)

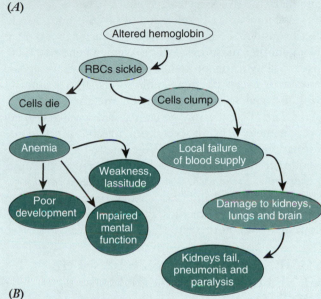

(B)

FIGURE 3.19 **(A) A normal red blood cell (left) is round, biconcave, and flexible. Hemoglobin in the red blood cells of people with the sickle cell mutation is more likely to crystallize when oxygen is scarce, and the cell becomes sickle-shaped and less flexible (right).**

(B) A point mutation that changes a single amino acid in a protein can have many consequences for the organism. This is illustrated here by the example of sickle-cell hemoglobin.

causes malaria and a decreased opportunity for it to multiply and cause bouts of high fever.

The distribution of the gene for sickle-cell hemoglobin illustrates two related principles. First, whether a gene confers a reproductive advantage—and therefore whether it is maintained in a population—can depend upon the organism's environment. Where there is no malaria, the sickle-cell gene offers no advantages whatsoever. Second, in the tropics of Africa and Asia, there are advantages and disadvantages to this gene. Individuals with one normal and one sickle-cell hemoglobin gene are more resistant to malaria than individuals with two normal copies. About 9% of African-Americans are carriers of the mutant allele; in the absence of medical care individuals with two mutant copies, however, usually do not survive to reproduce. Natural selection often leads to genetic compromises of this sort.

to transcribe about 800,000 double-spaced pages without making a mistake! The error rate in replicating DNA would be about a thousand times higher, however, if there were not proofreading. Proofreading is possible because there is a correct copy of the information available in the template. Consequently, the DNA-copying enzymes are able to recognize and replace any base that distorts the helix as it forms. Other proteins are able to examine a newly synthesized double helix, recognize distortions due to improper base-pairing, and make the necessary correction.

Accurate as it is, replication of DNA is not perfect. Let's consider the implications of one error in a billion from another perspective. If an average gene has 1000 bases coding for its protein product, there will be about one error every 10^6 gene replications. In a mammal, with 10^5 genes, one sperm in ten will therefore have some sort of copy error. Actually it is more likely to be every gamete, because several cell divisions precede the formation of a sperm cell. When viewed as the likelihood that any single gene will be different in the next generation, a mutation is a relatively rare event. Considered as the probability of change somewhere in the entire genome of a mammal's eggs or sperm, however, copy errors are not so rare.

SOME ANALOGIES BETWEEN THE GENETIC CODE AND HUMAN LANGUAGE

A simple comparison between words and their meaning illustrates by analogy the different effects of point mutations on the content (the meaning) of the genetic code. Consider the following sentence, composed of three-letter words ("codons"),

... THE PUB GOT ITS NEW RUG WET ...

which suggests that a lot of beer was spilled in the tavern last night. (The sentence is written with punctuation (spaces) after each word to help you read it, but with knowledge that all the words have only three letters you could read the sentence without having the spaces present.)

After a "copy error" in the third position of the second "codon" the message retains meaning (it keeps a "function")—

... THE PUP GOT ITS NEW RUG WET ...

—but the meaning seems quite different (the "function" has changed). This "mutation" is thus analogous to an amino acid substitution that leaves the gene product (the protein) functional but with altered properties.

The analogy with an amino acid substitution that changes the functional properties of the protein is apt unless the name of the only pub in town happens to be *The Pup*, in which case the "mutation" was neutral.

As with DNA, most of the possible "point mutations" in the sentence will leave it meaningless, without function. For example,

... THE PUB GOT ITS NEW RZG WET ....

Codons are embedded in very long sequences of nucleotides, and in transcribing genes there have to be signals to initiate the beginning and the end of the coding region. You met the three codons that signal "stop," but the signals to begin reading consist of *start sequences* of bases. With such a code, if reading does not begin at precisely the right point, the *reading frames* are shifted, with the likely result that the gene product will be totally wrong. In the analogy of our simple sentence,

... THE PUB GOT ITS NEW RUG WET ...,

a shift of one "nucleotide base" in the sequence of reading produces

... T HEP UBG OTI TSN EWR UGW ET ...,

which contains sense but conveys nonsense.

HOW THE GENETIC CODE IS TRANSCRIBED

How does the information in the genetic code, present in the nucleus, become expressed in the synthesis of proteins, a process that takes place in association with *ribosomes* in the cytoplasm? This occurs in two steps. First, the information in a gene is *transcribed* by copying it into a molecule of another nucleic acid called *messenger RNA* (mRNA). The mRNA then makes its way through pores in the nuclear envelope to the site of *translation* on the ribosomes. We will consider each of these steps in turn.

First a word about RNA (*ribonucleic acid*). The building blocks of RNA differ from those of DNA in two ways. First, the sugar in RNA, *ribose*, differs from the sugar in DNA, *deoxyribose*, in having an additional hydroxyl group on the 2′ carbon. The second difference is that the base *uracil*, designated U, substitutes for the base designated T in DNA (Fig. 3.11).

There are four major kinds of RNA performing different functions. Unlike DNA, they are all single-stranded. We shall introduce them as needed.

The first is *messenger RNA* (mRNA). When a gene becomes active, the portion of the DNA corresponding to that gene uncoils and exposes the bases on the individual strands. A strand of RNA is then made by a process analogous to the replication of DNA. This region of DNA serves as a template for the synthesis of complementary mRNA: free ribose-containing nucleotides pair with the complementary bases on the DNA and are then attached to each other, phosphate-to-sugar, by the enzyme *RNA polymerase* (Fig. 3.20).

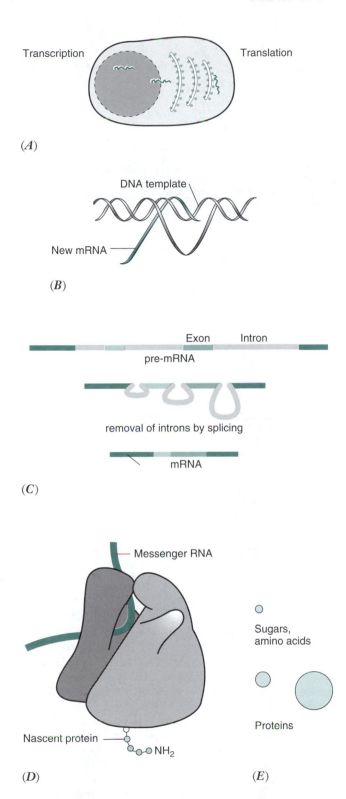

Transcription   Translation

(A)

DNA template

New mRNA

(B)

Exon   Intron

pre-mRNA

removal of introns by splicing

mRNA

(C)

Messenger RNA

Sugars,
amino acids

Proteins

Nascent protein

NH₂

(D)   *(E)*

FIGURE 3.20 (A) Genes are copied into mRNA in the nucleus (*transcription*); the mRNA then moves through nuclear pores to ribosomes in the cytoplasm where it provides the information for the synthesis of proteins (*translation*).

(B) Transcription is similar to replication of DNA in that a portion of one strand of the double helix serves as a template for the synthesis of mRNA. Different enzymes are involved, and mRNA does not form a double helix like DNA.

(C) Because DNA contains coding regions (*exons* interrupted by stretches of non-coding nucleotides (*introns*), the pre-mRNA must be "edited" before it is ready to leave the nucleus as mRNA. The introns are excised and the ends of the exons spliced together.

(D) Ribosomes are large molecular complexes of ribosomal RNA and protein. During translation they read the mRNA from one end to the other, and the polypeptide chain grows by the addition of amino acids.

(E) Sizes of small molecules and proteins relative to ribosomes.

In the genes of organisms other than bacteria there is a curious but very important complication to this otherwise simple account of *transcription*. The genes do not consist of uninterrupted sequences of codons corresponding to the amino acids in a protein. Within single genes, sequences of codons which contain the information specifying the primary structure of a protein (called

exons) are interrupted by sequences of nucleotides (*introns*) that do not appear to be coding for anything. Introns are sometimes referred to as "junk DNA," but they can contain sequences that bind proteins that are important in activating specific genes, as described in a later section. Although there is much variation, there are about eight introns in a typical gene, accounting for up to 90% of the nucleotides (Fig. 3.20),

There are several reasons for characterizing introns as "junk." Proteins that perform the same function in different species have similar structures. They have similar sequences of amino acids because their genes have descended from a common ancestral gene, and they are thus *homologous*. If we compare the sequences of nucleotides in the genes we find that whereas the exons have been *conserved* (i.e., show relatively little variation between closely related species), the introns vary much more. The great evolutionary divergence of homologous introns suggests that they cannot be coding for anything. Second, introns can vary in length and sometimes may even be lost, with no apparent consequences for the host organism. In Chapter 5 we will discuss a reason why cells carry all this "excess baggage."

The single strand of RNA transcribed from a gene is *pre-mRNA*. To become mRNA, the portions of pre-mRNA that are introns have to be recognized and removed, a process known as *RNA splicing*. But how are introns recognized? There are a few nucleotides at the ends of introns that do not vary and serve as recognition sites. (Such short stretches of nucleotides that are *conserved* in evolutionary time and can be recognized by other molecules are called *consensus sequences* and can be found in both DNA and RNA.) The ends of introns are recognized by *spliceosomes*, catalytic bodies consisting of a number of proteins and small nuclear RNAs (*snRNAs*). The snRNAs of the spliceosome are critical for recognizing the junctions between introns and exons, and together with their associated proteins they cut out the introns and rejoin the exons, sugar-to-phosphate. The result is a single strand of mRNA, now consisting of contiguous exons, and a number of excised loops of RNA corresponding to the former individual introns. The mRNA then moves through the pores of the nuclear membrane to the sites of translation on the ribosomes, and the RNA of the introns is broken down so that the nucleotides can be reused.

TRANSLATION: THE BUILDING OF PROTEINS

The mRNA brings to the ribosomes the information necessary to assemble amino acids in proper sequence, but how is that information utilized? There are two aspects to the problem: recognizing the information, and forming covalent (peptide) bonds to connect the amino acids together. As with gene splicing, the chemistry is catalyzed by complex structures consisting of proteins and RNA. Here, however, the structures are ribosomes containing *ribosomal RNA (rRNA)*. As previously, we will pass over the details of the chemistry and focus attention on how molecules recognize information. In this case the problem is particularly interesting because it involves the transfer of sequence information between molecules of two different kinds.

The solution is provided by a fourth kind of RNA, *transfer RNA (tRNA)*, shown diagrammatically in Figure 3.21. Notice how tRNA folds into characteristic cloverleaf structures by complementary base-pairing between different parts of the molecule. There is at least one tRNA for each of the twenty amino acids, each differing in the details of its three-dimensional shape. On one end of its folded structure, a molecule of tRNA has a sequence of three nucleotides complementary to a codon in mRNA. This *anticodon* enables the tRNA to recognize and base-pair with the complementary codons on mRNA. At the other end of the tRNA molecule there is a site to which the amino acid corresponding to the codon and anticodon can be attached covalently. Each kind of tRNA has a shape that is recognized by a specific enzyme that attaches the appropriate amino acid. Note that the specificity of this reaction depends on the ability of the enzyme to recognize *both* the tRNA and the amino acid and then catalyze the formation of a covalent bond between them. These enzymatic reactions are thus "interpreters" that relate the nucleotide sequences of codons to specific amino acids.

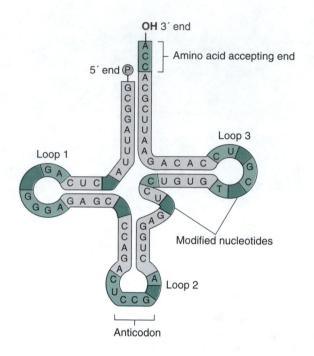

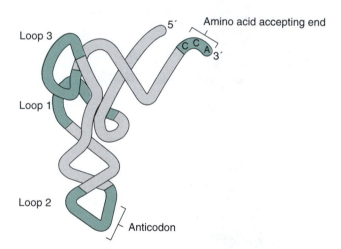

Figure 3.21 Transfer RNA (tRNA) is responsible for relating the genetic information contained in individual codons to specific amino acids. Molecules of tRNA have a basic cloverleaf pattern (*above*) that is stabilized by regions of complementary base pairing. They then fold into specific three-dimensional shapes stabilized by additional hydrogen bonds. There is at least one tRNA for every amino acid, and each is recognized by a particular enzyme that attaches the appropriate amino acid covalently to a nucleotide on one end of the molecule. On the other end of the folded molecule are three nucleotides (the anticodon) that are complementary to the mRNA codon for that amino acid.

A growing chain of amino acids is shown in Figure 3.22. The mRNA lines up on a ribosome, and the anticodon of a tRNA that has been primed with its amino acid finds its place at the head of the first codon. A second tRNA then occupies the adjacent codon, and the

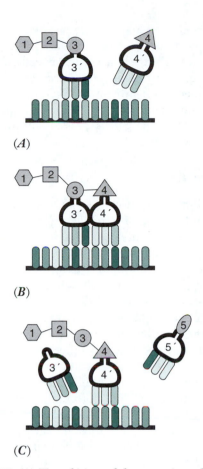

FIGURE 3.22 (A) Translation of the genetic code on ribosomes. A section of mRNA is shown at the bottom. The first three amino acids have been linked, and the third is still attached to its tRNA (3′), which is in turn H-bonded through its anticodon to a codon on mRNA. The next amino acid (4) is finding its place as the appropriate tRNA brings its anticodon into alignment with the next codon on the mRNA.

(B) Enzymes now break the bond between tRNA and the carboxyl group on amino acid 3. The carboxyl group on amino acid 3 then forms a peptide bond with the amino group on the α-carbon of amino acid 4.

(C) tRNA 3′ has finished its job, and it leaves the scene. The polypeptide chain has grown by one amino acid residue, the mRNA has slipped along by one codon, and another tRNA (5′) is delivering the next amino acid. This cycle will be repeated until a "stop" codon is encountered. The polypeptide chain may then be several hundred amino acids long. It then comes off the ribosome and folds into the tertiary structure characteristic for that protein.

catalytic machinery of the ribosome transfers the carboxyl group of the first amino acid from its bond with tRNA, forming a peptide bond with the second amino acid. The first tRNA, having now completed its job, leaves its place on the mRNA to be recycled. The mRNA slips along the ribosome like videotape through a player, and new tRNAs deliver the appropriate amino

acids in sequence. When a stop codon is encountered, the polypeptide chain is complete and is severed from the final tRNA.

Thus the genes, through mRNA, specify the primary structures of proteins—the sequence of amino acids in the backbone. How does a newly synthesized protein find the final tertiary structure that is essential for an enzyme or receptor to function? Interestingly, in most cases this folding just happens. The secondary and tertiary structures represent favored, low-energy configurations into which proteins fold spontaneously. In some cases involving large proteins, folding may be assisted by other proteins called *chaperones*.

NOT ALL GENES ARE ACTIVE AT ONCE: CONTROL OF TRANSCRIPTION

One of the most fascinating features of life is the capacity of a fertilized egg—a single cell—to give rise to an entire functioning organism like yourself. In an adult there are many different kinds of cells performing different biochemical tasks, but all cells contain copies of the same genes that were present in the fertilized egg from which it developed. If every cell in the body is genetically identical to every other, what makes liver cells biochemically distinct from muscle or nerve cells, each with different but overlapping arrays of enzymes?

The answer is that not all the genes in a cell are active simultaneously. A gene can be turned on—that is, it can become active in transcribing mRNA—in some tissues but not in others. Or a gene in one type of cell may be active at some times but not at others. The activation of particular genes is one of the factors responsible for the selective formation of specialized tissues such as nerve or muscle during development.

Selective activation of genes is actually a very basic phenomenon even displayed by single-celled organisms like bacteria. For example, many bacteria obtain their energy from sugar molecules that occur in their environment. There are many different sugars, however, and each requires unique enzymes to transport it into the cell and to metabolize it. Bacteria are able to detect the presence of specific sugars in their environment and activate those genes that code for the specific proteins that are needed. Synthesizing proteins costs energy, so with the ability to activate genes selectively, bacteria do not waste energy making enzymes until they need them.

Abnormalities in the control of gene activity can have serious consequences for human health. For example, when specific genes that control the mitotic cycle of the cell escape normal regulation, the cell divides too often and a cancer results. The isolation and characterization of such genes thus has considerable medical significance, and their investigation is underway in many laboratories.

How are genes regulated? Regulation can occur during either transcription or translation, but we shall illustrate with the former. Let us then revisit the process of transcription, providing some detail that we omitted on first telling. In order for mRNA to form, the enzyme RNA polymerase and several associated proteins called *basal factors* must bind to the DNA at a *promoter region* ahead of the coding region in the first exon (Fig. 3.23). Together these proteins comprise the general-purpose molecular machinery for transcription, but by themselves they do not control which genes will be read. Other proteins called *activators* must also be present. Activators bind to the DNA at sites called *enhancers*. The enhancer sites may lie many basepairs distant from the promoter region, and the DNA then loops so that the activators can also come into physical contact with the transcription machinery through the participation of still other proteins. Only when this complex of molecular interactions has formed, is the RNA polymerase able to start making mRNA. The specificity of gene activation lies in the great variety of enhancer sites, activator molecules, and other participating proteins that can be combined to control an individual gene.

Several other features extend the complexity of this process. First, in addition to activator proteins binding to enhancer sites on the DNA, there are *repressor* proteins that bind to the DNA and prevent the action of RNA polymerase. Second, the action of activators or repressors can be under control of small molecules like steroid hormones. Thus the presence of a molecule like the hormone testosterone can have a direct effect on the expression of genes in particular cells. And third, activators, being proteins, are coded for by genes. A gene for an activator protein can therefore control the activity of other genes. These mechanisms play a critical role in differential gene expression during development (Chapter 10).

SPLIT GENES

Although introns do not code for proteins, they can provide binding sites for regulatory proteins. Moreover, the presence of noncoding regions between exons may have been exploited during evolution in another way. Many protein molecules have different *domains*, differing in the nature of their secondary and tertiary structures and associated physical properties such as solubility in an aqueous environment (Figure 3.7). Where proteins have this complexity, the different domains are frequently coded by different exons. Because in the gene, exons are separated by noncoding introns, there is the possibility of combining exons in different ways during RNA splicing. For example, the same nuclear gene can give rise to more than one protein depending on which of its exons are incorporated into mRNA. One gene is known that contains the information to make twenty different proteins, all controlled during the final splicing of mRNA. There are even examples of mRNA that are made from exons from the genes for different proteins. This finding requires a refinement to how one might define a gene. More importantly, it suggests that some exons originated early in the evolution of life and were then used to compose a variety of new proteins.

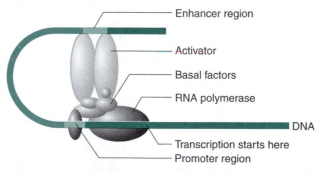

FIGURE 3.23 The control of transcription involves several proteins. RNA polymerase is the enzyme that catalyzes the addition of nucleotides to form mRNA. It does not act, however, without the presence of a squad of other proteins called basal factors, one of which binds directly to the promoter region of the DNA upstream from the first exon. Control is further enhanced by the need for other transcription factors—activator proteins—binding to an enhancer region of the DNA at one end and the complex of basal factors at the other. Genes can be kept from transcribing by other transcription factors, repressors.

Labels in figure:
Enhancer region
Activator
Basal factors
RNA polymerase
DNA
Transcription starts here
Promoter region

THE BEHAVIOR OF GENES DURING CELL DIVISION

GENES AND CHROMOSOMES

We return to DNA replication, just before cell division, now viewing the process at the level of the whole cell. View is the right word, because as cells divide, the DNA in the nucleus—the equivalent of about a three-foot length of double helix in every mammalian cell—condenses into structures called *chromosomes*, which are readily visible in optical microscopes. (Chromosome simply means "colored body," and the name refers to the fact that these structures are easily stained with dyes available to nineteenth century microscopists.) Each cell in an organism that reproduces sexually has a double set of genes, one from each parent, and therefore two sets of chromosomes (and is thus *diploid*). Different species of organisms have different numbers of chromosomes; humans have twenty-three pairs, the fruit fly *Drosophila* has four pairs.

The first step in preparation for cell division is for each strand of the double helix of DNA to serve as a template for the synthesis of a new strand. As the DNA condenses into visible chromosomes, each chromosome appears as a double structure consisting of two *sister chromatids*, and each has an identical copy of the other's DNA. The nuclear envelope dissolves and the chromosomes line up in a central plane. The sister chromatids of each chromosome are then drawn apart by threadlike strands of a special protein, forming two diploid sets of chromosomes at opposite poles of the cell. A nuclear envelope forms around each complete set of chromosomes, the material of the chromosomes (DNA and protein) decondenses in each new nucleus, and cell membranes form across the cytoplasm to complete the process of cell division (Fig. 3.24).

The process of nuclear division is called *mitosis*. It has traditionally been divided into several stages with Greek names, but it is more useful to think of mitosis as a smooth, continuous process that has the effect of

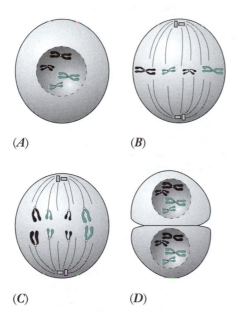

(A)   (B)

(C)   (D)

FIGURE 3.24 **When cells divide, the genetic material replicates and identical copies are distributed to each of the two daughter cells in a process called mitosis.**

(A) Prior to cell division the DNA replicates and condenses with protein as chromosomes visible in the light microscope. The nuclear envelope then dissolves.

(B) The chromosomes, each now consisting of a pair of sister chromatids, line up in the middle of the cell.

(C) The sister chromatids are drawn to opposite poles of the cell by fibers of protein.

(D) A plasma membrane forms across the original midplane of the cell, nuclear membranes appear around each of the two sets of chromosomes, and the chromosomes decondense to form an irregular mass of DNA and protein called chromatin.

producing two nuclei where there was one before, each daughter cell possessing an exact copy of the DNA that was present in the parent cell.

GENES AND THE FUNDAMENTAL MEANING OF SEX

Sex is a reproductive process in which completely new combinations of genes are assembled in every generation. That is the basic significance of sex, and all of the social and psychological ramifications of sexual reproduction (Chapters 6, 13, and 14) are informed by an evolutionary perspective that starts with the behavior of DNA.

THE NUMBER OF COPIES OF GENES IS REDUCED BY HALF IN THE FORMATION OF EGGS AND SPERM

A fertilized egg (a *zygote*) is *diploid* because it contains a set of maternal chromosomes (from the female's egg) and a paternal set (from the sperm or pollen). Egg and sperm are therefore each *haploid* (containing a single set of genes). But how does the diploid number of genes in a parent become halved in the formation of *gametes* (eggs and sperm)? This is accomplished by a special sequence of two divisions known as meiosis (Fig. 3.25). Meiosis is a process that goes on only in the germ line, the cells that are destined to become eggs or sperm.

Meiosis begins like mitosis with all the DNA replicating, but the next event is what distinguishes the two processes from each other. As the chromosomes (now double because of the synthesis of new DNA) line up on a plane in the middle of the cell, *homologous chromosomes pair with each other*. In other words, each chromosome pairs with its equivalent partner that originally came from the other parent. Now as the protein fibers of the meiotic spindle draw the chromosomes apart, *homologous pairs are separated*, and each daughter cell from this first meiotic division receives a single chromosome of each kind.

Each daughter cell of this first meiotic division is thus haploid with respect to the number of chromosomes it possesses, but it is diploid with respect to its content of genes. This is because the DNA replicated prior to the first meiotic division, and the chromosomes in the daughter cells of the first meiotic division are double, consisting of two sister chromatids.

The process of meiosis is then completed with a second division in which the sister chromatids separate from each other and the cytoplasm divides to form two daughter cells. Each of the resulting gametes has a single set of chromosomes and the haploid complement of genes.

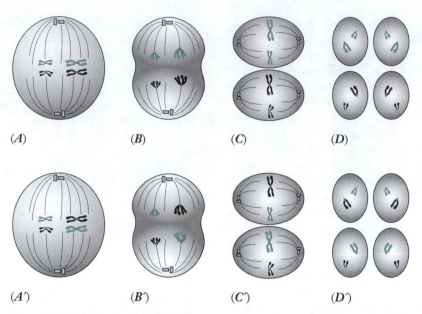

(A)   (B)   (C)   (D)

(A′)   (B′)   (C′)   (D′)

FIGURE 3.25 Cells that are to become eggs or sperm undergo a pair of divisions called meiosis in which the chromosome number is halved.

(A) DNA replicates and the chromosomes line up in the center of the cell, but unlike in mitosis, homologous chromosomes pair with each other.

(B) Instead of sister chromatids separating, the chromosomes of each homologous pair are then drawn to opposite poles of the cell.

(C) When the first division is complete, each cell has a haploid number of chromosomes (one from each pair). But because the DNA replicated prior to the first division, each chromosome consists of a pair of sister chromatids.

(D) The second meiotic division separates the sister chromatids. Each of the four cells contains a single set of chromosomes.

The paired chromosomes line up independently of each other at the start of the first division (Compare A with A′). Consequently, in this example in which the diploid cell has two pairs of chromosomes, four different mixtures of parental chromosomes are possible in the haploid gametes (two in D and two in D′).

As with mitosis, the continuous process of meiosis can be described with a platoon of Greek names applied to stages along the way. None of this terminology is important in grasping the consequences of these two meiotic divisions. The first consequence is obviously that eggs and sperm have a single copy of each gene, so that following fertilization the zygote will be diploid. But there is another consequence of equal importance, one that is responsible for the whole point of sexual reproduction.

NONHOMOLOGOUS CHROMOSOMES ASSORT INDEPENDENTLY

When we described the array of paired, homologous chromosomes during the first meiotic division, we left an important point to your imagination. For example, consider an organism with only two pairs of chromosomes. At the first meiotic division do the chromo-

somes from the father always go to one daughter cell and those from the mother to the other daughter cell? The answer is no: the chromosomes have an equal chance of going to either daughter cell (Fig. 3.25). In other words, *the alignment of any pair of homologous chromosomes at the first meiotic division is independent of the alignment of all the other pairs*. The result is that nonhomologous chromosomes sort themselves into the gametes in a totally random manner.

The consequence of this *independent assortment* of chromosomes is that sexual reproduction creates new ensembles of genes with every fertilization. Remember that mutation can produce different variants of the same gene; these different copies are called *alleles* of the gene. For many genes an offspring will have inherited a different allele from each parent. When the offspring in turn makes eggs or sperm, the independent assortment of chromosomes scrambles the genes from the two parents so that the genes (alleles) on one chromosome become associated with different alleles of other genes on the other chromosome.

We can readily calculate the extent of *recombination* that results from the independent assortment of chromosomes. In our hypothetical organism with two pairs of chromosomes, independent assortment led to four different kinds of gametes: one with only paternal chromosomes, one with only maternal chromosomes, and two with the two possible mixtures of maternal and paternal chromosomes (Figure 3.25). Now consider an organism with three pairs of chromosomes, with one member of each pair from each of two parents. We leave it to you to convince yourself that eight different combinations of three haploid chromosomes can form as a result of independent assortment. In other words, during meiosis this organism will generate eight genetically distinct classes of gametes.

On the basis of these two examples we can formulate a simple expression for the extent of this form of recombination. If n is the number of pairs of chromosomes, recombination due to independent assortment will generate 2^n genetically distinct haploid gametes. But you have twenty-three pairs of chromosomes, so by this process you can produce 2^{23} (8,388,608) different kinds of eggs or sperm. Your future mate will be generating an equivalent number of different combinations of parental chromosomes, all of which, like yours, have a very small probability of ever having existed before. No wonder we all are different!

CROSSING OVER FURTHER EXTENDS THE AMOUNT OF RECOMBINATION

A calculation of recombination based on the independent assortment of chromosomes during meiosis still seriously underestimates the possible number of new combinations of alleles that are formed during sexual reproduction. That is because DNA in one chromosome can exchange with DNA in its homologous partner that came from the other parent. During the first meiotic division, when the homologous chromosomes have paired, a length of chromatid from one parental source may interchange with the corresponding piece of a chromatid from the homologous chromosome. This process, known as *crossing over*, is diagrammed in Figure 3.26. The number of potential crossover points is in principle as large as the number of base pairs in the DNA of the chromosome, so crossing over provides an astronomical number of additional possible recombinants.

Sometimes the crossover points on the two chromatids are not quite in register, and one of the resulting chromosomes has a *deletion* of DNA whereas the other has an *insertion* of some extra DNA. Deletions are likely to have adverse consequences for the organism that in-

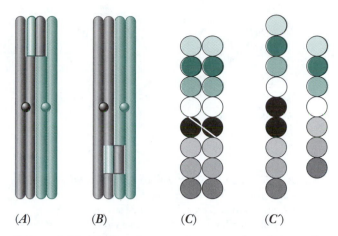

(A)   (B)   (C)   (C′)

FIGURE 3.26 (A) Crossing over is a mutual exchange of portions of sister chromatids from homologous chromosomes early in the first meiotic division. Crossing over greatly extends the amount of genetic recombination that occurs during meiosis.

(B) Multiple crossovers are frequent. This is the result of a pair of crossovers during the same meiotic division.

(C) An enlarged diagram of a pair of sister chromatids. The circles represent sections of DNA corresponding to individual genes. As shown by the diagonal line, sometimes crossovers do not take place at precisely the same point on the two sister chromatids.

(C′) The result of such an *unequal crossover* is an *insertion* of extra DNA in one chromatid and a *deletion* from the other. As drawn in this example, the insertion has resulted in a *gene duplication* in the chromatid on the left. As described in the text, gene duplications open evolutionary opportunities. Deletions, on the other hand, are most likely bad news for the organism that inherits that chromosome. Deletions and insertions can involve longer stretches of DNA than single genes.

Think of deletions and insertions as mutations. They are random events, but different from the point mutations that occur during replication of DNA. They are examples of chromosomal mutations.

Because crossovers can occur virtually anywhere in the DNA, they need not take place between genes, as in these diagrams. They can happen between exons or within exons, with a variety of consequences for the protein for which that gene codes. For example, the insertion or deletion of a single nucleotide can generate a shift in the reading frame.

herits such a chromosome; one or more genes may be missing, and unless their functions can be covered by the genes provided to the zygote by the germ cell of the other parent, the mutation may even be lethal. Insertions, on the other hand, provide an evolutionary opportunity. A particularly interesting case occurs when there has been a *gene duplication*. Having an extra copy of a gene may be initially redundant, but mutations that subsequently occur in the extra copy may be subject to their own selection. The result is that in time organisms can have two closely related genes, both of

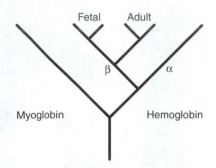

FIGURE 3.27 Gene duplications can lead to the evolution of proteins for related functions, leading in turn to organisms with families of related genes. For example, many vertebrates have two distinct oxygen-binding proteins: hemoglobin and myoglobin. Hemoglobin is found in red blood cells, whereas myoglobin in located in muscles. (The difference between dark meat and white meat in a chicken or turkey is the presence of myoglobin in the dark meat.) Hemoglobin and myoglobin have evolved from a common genetic ancestor, and their seperate evolution likely started with a gene duplication. Hemoglobin consists of four polypeptide chains, two α chains and two β chains. Furthermore, the β chains are different in fetal and adult hemoglobin. Different chains are coded by different genes. The evolutionary relationships are shown in this diagram.

whose products are used in similar but importantly different ways (Fig. 3.27).

THE ORIGIN OF LIFE

Where did the first genes and cells come from? The earth is about 4.54 billion (4.54×10^9) years old, and virtually all of the kinds of organisms more complex than a single cell have appeared during the last 14% of that time (650 million years). But there are possible traces of life that date perhaps as much as 3.8 billion years, to a time when the earth was less 1 billion years old. All organisms, living and fossil, consist of cells, so the formation of these membrane-bound, reproducing entities was a landmark event in the history of life. The oldest fossil cells (Fig. 3.28B) are about 3.5 billion years old.

All life shares a common chemistry: it is built around the activities of proteins, it is based on a genetic code that uses DNA and the same codons to represent the twenty amino acids, it utilizes RNA in intermediary steps in protein synthesis, and organisms everywhere employ similar chemical reactions for mobilizing energy from food. As this chemistry characterizes all life, its basic features must have come into existence before the first cells. Tracing its origins, however, is not easy, partly because we do not understand well the physical conditions that existed on the earth at that time.

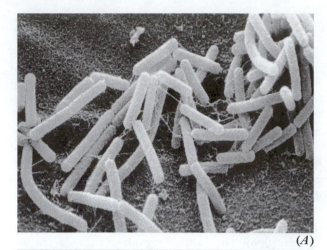

(A)

(B)

FIGURE 3.28 (A) These rod-shaped objects, viewed here with an electron microscope, are an example of a prokaryote. This is the hay bacillus *Bacillus subtilis*. It is sometimes pathogenic in humans and causes severe eye infections.

(B) Stromatolites are formed of calcium carbonate secreted by cyanobacteria. These present day stromatolites are in Shark Bay, Australia. Similar masses have been found as fossils 2–3 billion years old.

But there is a second reason why this problem is perplexing. In life today, DNA and RNA are required in order to make proteins, but proteins are required to synthesize the nucleotides that make up DNA and RNA as well as to catalyze the formation and replication of nucleic acids. Here in the realm of molecules is the classical dilemma of which came first, the chicken or the egg? There is as yet no complete answer to this riddle, but there is fascinating evidence from which the broad outlines can be discerned.

The first replicators may have been molecules of RNA, not DNA, and early in the evolution of life, molecules of RNA may have functioned as catalysts and were responsible for making the first proteins. In this view, the coding function was subsequently handed to a more stable molecule, DNA, and most of the catalytic functions were assumed by proteins. Although at present, RNA molecules can neither replicate themselves nor direct the synthesis of proteins without the aid of preexisting proteins, some of them have properties that suggest that both of these abilities were present in the earliest RNAs. First, RNA is the only biological molecule other than proteins known to have enzymatic activity. Until about fifteen years ago it was generally thought that all enzymes were proteins, but in 1983 it was shown that some RNA can cut and splice together other RNA as well as itself. Second, very recent experiments have shown that rRNA is the catalyst for peptide bond formation, and the ribosomal proteins perform supporting roles. Additionally, a number of enzymatic reactions in the cell require the catalytic participation of small molecules called *coenzymes*, some of which are nucleotides or closely related molecules. All of this suggests that early in the evolution of life the first self-replicating molecules may have been part of an early "RNA world" in which RNA or some similar molecule could act both as templates and catalysts for its own replication as well as messengers and catalysts for the synthesis of proteins. In this view those nucleotides that today function as coenzymes represent molecular fossils, remnants of catalytic processes that now also engage proteins.

But where did the first RNA come from? During the first billion years of its existence the earth was very different from how it is today, with more volcanic activity, numerous impacts of asteroids, and possibly many electrical discharges of lightning. Furthermore, there was little if any free oxygen in the atmosphere, which probably contained much carbon dioxide and nitrogen (from volcanoes), water vapor, and possibly hydrogen cyanide (HCN). With a scarcity of oxygen there would have been no shield of ozone in the upper atmosphere to filter out the far ultraviolet rays of the sun, light that is so energetic that it is capable of breaking and forming covalent bonds. Furthermore, with a relative paucity of oxygen, compounds of carbon, hydrogen, and nitrogen are more stable because they are less likely to be broken down (oxidized) to CO_2 and water.

Under these conditions, many of the small organic molecules like sugars, nucleotides, amino acids, and fatty acids can be formed by natural processes such as electric discharges and short wavelength radiation and without the aid of catalysts. They are present in carbonaceous meteorites and comets, their formation can be replicated in the laboratory, and they very likely formed on the surface of an early earth. Furthermore, nucleotides of RNA can polymerize. From here, however, the specific details are more speculative, but they hinge on the known properties of RNA.

In principle, RNA can serve as the template for the synthesis of a complementary strand, which in turn could provide the mold for a copy of the original strand. This does not happen under present conditions without the help of proteins, but with the right catalyst it might well occur. Some have speculated that mineral surfaces could have catalyzed this reaction, but there is another possibility in which another molecule similar to RNA (but no longer in existence) catalyzed its own formation. Central to this idea is the realization that an RNA capable of replication has the properties necessary to launch evolution. Any variant that was more successful in replicating would quickly appear in greater quantity and would enjoy a considerable selective advantage in a primordial soup of molecules. Two features necessary for this to have occurred are present in RNA—a template for replication, and the capacity for catalysis.

Several additional changes are necessary to bring forth life in a form that we recognize today. Compared to RNA, proteins can assume a vastly greater number of three-dimensional shapes and engage in a correspondingly larger number of enzymatic reactions. In this view of events, RNA "discovered" how to make polypeptides and proteins, and natural selection further amplified the possibilities. Proteins thus became the agents of RNA, better able than RNA to manipulate other molecules. Second, RNA that could make proteins that could in turn catalyze the formation of lipids had a potential advantage, for as we saw in Figure 3.2, lipids tend to form bilayers. An RNA-protein complex that surrounded itself with a lipid membrane would hold its diffusible components together and become a more efficient replicator. Third, at some stage in the process, RNA began to use DNA as a storage site for what was by then the universal code for amino acids. Having one less hydroxyl group on its sugar residues than RNA (Fig. 3.11), DNA is more stable. Furthermore, DNA forms spontaneously into a double helix, a closed book that can only be read or copied through the intervention of RNA and proteins.

What remains of this original role of RNA is an array of intermediary supporting activities in which genes are replicated in each cell cycle and the information in genes is transcribed and translated into a vast variety of proteins. Chemists have not yet discovered how to fill in all the details. Perhaps there were even earlier polymers than RNA that have left no trace and from which RNA subsequently evolved. Our present ignorance of these early events poses a number of interesting challenges, but because these processes occurred so long ago in a world vastly different from conditions today, a detailed understanding of the origin of life will be difficult to achieve in the near future. The conceptual model that we have just outlined, however, has an appealing elegance that will drive scientific inquiry for years to come.

We need to make one final point about the origin of life. The question of how life arose is different from understanding how it has changed during the last billion years. Because evolution is going on around us, we know a great deal more about how lineages of organisms change over time than we know about the origins of the first cells. We have drawn on the current understanding of evolution to place the speculations about life's origins in a common conceptual framework, but uncertainties about events that occurred on the early earth do not diminish the strength of evolutionary theory as an explanatory concept for how the world has been shaped during the last billion years.

TWO KINDS OF CELLS: PROKARYOTES AND EUKARYOTES

With the origin of cellular structure, evolution created entities capable of mobilizing energy and capturing materials in the service of replication. Here in a single complex package was the information necessary not only to specify more of itself, but also to engineer its successful reproduction into the future. Present were the code for all the enzymes necessary to copy DNA, to make the molecular machinery for harvesting energy, synthesizing a vast array of molecules needed for growth, organizing the division of one cell into two daughter cells, and enabling the cell to move about and respond to signals from its environment.

Individual cells have organizational complexity that far exceeds anything encountered in the nonliving world. This intricacy of function is reflected in the structure of cells, with their numerous internal compartments (organelles) and specialized membrane surfaces for coordinating enzymatic reactions. Nature, however, reveals two stages in the evolution of this complexity. *Prokaryotic* (*pro* before, *karyon* nucleus) cells are simpler and more ancient. They are found today in bacteria (Fig. 3.28A) and blue-green algae.

They have a relatively small amount of DNA, and as the name suggests, they lack an organized nucleus surrounded by a nuclear membrane. *Eukaryotic* cells (*eu* true) represent a substantial increase in structural complexity. Not only do they have a nucleus surrounded by a membrane, much more DNA, and protein in their chromosomes, their cytoplasm contains the array of specialized compartments—organelles—described earlier (Fig 3.10).

Prokaryotes probably existed for a billion years before the first appearance of eukaryotes. In fact, mitochondria and chloroplasts are believed to have arisen as prokaryotic invaders living originally as intracellular symbionts and now completely integrated into the life of eukaryotic cells (Box 3.4). The major reason for this interpretation is that both mitochondria and chloroplasts contain some DNA that codes for several of their proteins.

The relative structural simplicity of prokaryotes is in one sense misleading. As a group, bacteria have metabolically diverse means for obtaining energy and tolerating extremes of temperature. Their evolutionary success has been undiminished by the presence of eukaryotes; in fact, many bacteria such as the common intestinal bacterium *E. coli* exploit larger organisms for a habitat.

SYNOPSIS

Cells are the basic structural unit of all organisms, and the appearance of the first cells about 3.5 billion years ago represents a watershed event in the evolution of life. Cells are small compartments enclosed within a thin membrane of lipid molecules that not only prevents the contents of the cell from diffusing away but actively regulates the passage of molecules between the cytoplasm and the external environment. Each cell is an autonomous unit that contains the molecular equipment for obtaining energy, moving, growing, and reproducing. Each cell also contains the genetic information required for synthesizing the molecules it needs for these activities.

The simplest (and oldest) kind of cells, *prokaryotes*, are represented today by bacteria and blue-green algae. They lack an organized nucleus and have little structure to their cytoplasm. All other organisms, *eukaryotes*, have cells with *nuclei*. Their *ribosomes* (sites of protein synthesis) and *chromosomes*, are more complex than in prokaryotes, and they also contain specialized organelles such as *mitochondria*. Both *chloroplasts* (the light-harvesting organelles of green plants) and mitochondria have some genetic material of their own and are believed to have originated as symbiotic invaders early in the evolution of eukaryotic cells.

Box 3.4
The Early Evolution of Cells

The prokaryotes are actually a very diverse array of organisms. Comparisons of their ribosomal RNA suggest that they comprise two groups (the Archae and Bacteria), as distinct from each other as each is from eukaryotes. Figure 3.29 is an evolutionary tree of relationships of contemporary organisms as it is believed to have unfolded more than 600 million years ago, before the Cambrian period.

The prokaryotes have diverse ways of obtaining energy. Some are photosynthetic but instead of using water as a source of electrons for reducing carbon dioxide as green plants do, they employ other sources. For example, one kind uses hydrogen sulfide and generates elemental sulfur. Other bacteria perform still other reactions that are unknown in eukaryotes. For example, different species of bacteria get energy by oxidizing sulfur (S) to sulfate (SO_4^{2-}), ferrous iron to ferric iron, or ammonia (NH_3) to nitrite ions. This evolutionary diversity has no counterpart in eukaryotic organisms. The "discovery" of how to use water as the electron donor in photosynthesis (as occurs in higher plants) was a later innovation that not only opened new evolutionary opportunities, but was instrumental in creating the present atmosphere with about 20% oxygen. This in turn enlarged the evolutionary opportunities for organisms to obtain their energy from the metabolic labors of others, by exploiting this oxygen to oxidize carbon compounds that had been made by photosynthetic plants.

Figure 3.29 also indicates the likely interactions of pro- and eukaryotes that led to the presence of mitochondria and chloroplasts as organelles of eukaryotic cells. Current evidence suggests that the progenitors of both mitochondria and chloroplasts were prokaryotes that entered the cells of eukaryotes, existed in symbiotic relationship, and eventually lost their capacity for an independent existence. Chloroplasts have independently entered more than one lineage of eukaryotes, leading to brown, red, and green algae plus "higher" plants.

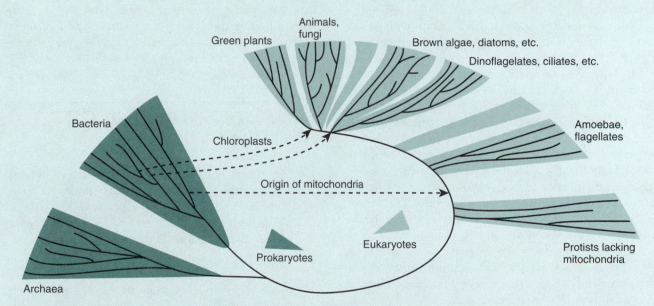

FIGURE 3.29 An evolutionary tree of relationships based on differences in ribosomal RNA and indicating separations that took place very early in the history of life. The origin of mitochondria and two of the origins of chloroplasts are shown by the broken arrows.

Both the structures and the numerous activities of cells depend on *proteins* working with a supporting cast of nucleic acids, lipids, carbohydrates, and a variety of other smaller molecules. Proteins play many roles: for example, enzymes catalyze chemical reactions, tubules and filaments determine the shapes and movements of cells, receptor molecules recognize molecular signals from other cells, channels and pumps regulate the passage of substances through the cell membrane, and transcription factors control the expression of genes.

Proteins are linear chains of twenty different kinds of *amino acids*. Each kind of protein has a unique sequence of amino acids, which defines its *primary structure*. Protein molecules fold into more compact, three-dimensional shapes (*secondary* and *tertiary structures*), stabilized by several kinds of forces including ionic attractions and *hydrogen bonds*. Some proteins require the combination of two or more polypeptide chains (*quaternary structure*) usually coded by different genes.

An average protein is about 300 amino acids long, and there is an astronomical number of ways that twenty different kinds of amino acids can be strung together in long chains. As each chain can fold into a unique three-dimensional shape, there is a vast number of shapes, and therefore functions, that proteins can perform.

Proteins have a special relationship with *genes*: genes code for proteins. The genetic information resides in the structure of a long, ladderlike polymer, DNA (deoxyribose nucleic acid) twisted into the shape of a helix. The two sides of the ladder consist of linear chains of four different nucleotide bases (symbolized by A, T, G, and C) linked together in specific sequences. Each rung of the ladder consists of a pair of *complementary bases* forming hydrogen bonds with each other, A with T and G with C. Because of complementary base-pairing, the sequence of bases on one side of the ladder determines the sequence on the other side. This structural feature of DNA is the basis of its ability to replicate prior to every cell division. The helix opens, and each exposed strand acts as a template for the synthesis of a new complementary strand. A point mutation is a copy error and occurs when a wrong nucleotide base is inserted during DNA synthesis.

The sequence of bases in DNA specifies the primary structure of proteins. Three nucleotide bases in a row constitute a *codon*, and each of the twenty amino acids is represented by one or more codons. In eukaryotic cells the DNA that codes for a protein is first *transcribed* into a single strand of messenger RNA (*mRNA*) within the nucleus, and the mRNA then moves to *ribosomes* in the cytoplasm where the protein is assembled (*translated*) with the help of transfer RNA (*tRNA*). A point mutation can lead to the substitution of one amino acid for another during translation.

In eukaryotes, regions of DNA that code for protein (*exons*) are separated by intervening base sequences with no coding function (*introns*). To make mRNA, the base sequences corresponding to the introns are removed, and the regions corresponding to the exons are spliced together.

Which genes are being transcribed at any time is under the control of specific proteins (*transcription factors*) that bind to the DNA upstream from the first exon and activate or inhibit the enzymes that make mRNA. These binding proteins are also coded for by genes, and this is the mechanism by which regulatory genes control the expression of other genes.

Some of the RNA in present-day cells has enzymatic activity, and RNA works together with protein enzymes in transcription and translation. These roles suggest that RNA (or some precursor to RNA) may have had more numerous catalytic functions very early in a pre-cellular stage in the evolution of life.

In *sexual reproduction*, two *haploid* gametes (each with a single set of genes) combine at fertilization to form a *diploid* offspring (with two sets of genes). In this process, an offspring frequently receives different copies (*alleles*) of many of its genes from each parent. The function of sexual reproduction is to recombine the assemblages of parental genes when the offspring makes its own eggs or sperm. This *recombination* is accomplished by the *independent assortment* of *homologous chromosomes* during *meiosis* and by *crossing over*, the exchange of pieces of *homologous chromatids*.

QUESTIONS FOR THOUGHT AND DISCUSSION

1. What does it mean to say that molecules (nucleic acids) contain information in the form of a code? How is that information expressed? Do protein molecules contain information? Does information require the presence of human minds?

2. Three-base codons drawn from a code with four bases allow for sixty-four different codons, several times more than enough to specify the twenty amino acids. How long would the codons have to be if there were only two bases in the code? What is wrong with the idea of a total of three-bases in the code?

3. There are currently plans to look for evidence that life has evolved on Mars. What sorts of evidence do you think would be useful?

4. What are some emergent properties of cells that are not readily predictable from the properties of genes and proteins? In answering, consider what events in the evolution of cells occurred before the

appearance of prokaryotes and between the appearance of prokaryotes and eukaryotes (Figure 2–11).

Do cells contain information that is not present in genes? If your answer to this question is "yes," from where did that information come?

5. The Alabama textbook disclaimer (see the question at the end of Chapter 2) asks "How did you and all living things come to possess such a complete and complex set of 'instructions' for building a living body?" What insights does science offer in answer to that question?

SUGGESTIONS FOR FURTHER READING

Alberts, B.; Bray, D.; Lewis, J.; Raff, M.; Roberts, K.; and Watson, J. D. (1994). *Molecular Biology of the Cell, 3rd ed.* New York: Garland Publishing, Inc. An excellent book for details on the topics in Chapter 3.

Dricla, K. (1992). *Understanding DNA and Gene Cloning: A Guide for the Curious, 2nd ed.* New York, NY: John Wiley & Sons, Inc. A basic and easy-to-understand introduction to procedures for identifying and manipulating genes.

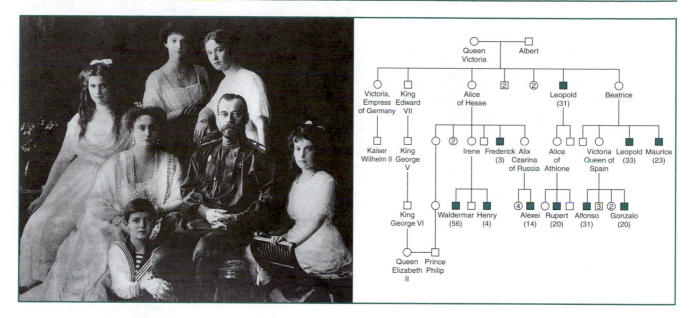

4

Genetic Continuity and Change: Organisms, Environments, and Microevolution

AN OLD MYSTERY

The discovery of the structure of DNA and its role in coding hereditary information, all accomplished in the last fifty years, is one of the intellectual triumphs of human history. But humans have known for thousands of years that the traits of plants and animals can vary and that these variations can be passed to their offspring. The archaeological record shows that many of the domestic breeds of plants and animals that we currently use for

Photo: *Left.* The last of the Russian Czars, surrounded by his wife and children, photographed about five years before their murder. The young prince Alexi suffered from hemophilia, a genetic disorder in which the blood fails to clot. The gene for hemophilia is recessive but is located on the X chromosome. The condition is therefore much more common in sons than in daughters.

Right. Alexi inherited his gene from his grandmother, Queen Victoria, along with nine other males in the royal families of Europe. The ages at which these individuals died are shown in parenthesis. Their mothers were all carriers, but as the gene is recessive the mothers all had normal blood clotting.

food, sport, protection, and companionship are the products of a long history of selective breeding by humans.

Prior to 1900, the most common view of inheritance was that parental traits were blended in the offspring. Even after the invention of optical microscopes in the seventeenth century and the realization in the mid-nineteenth century that living organisms are not only constructed of cells but that all cells come from other cells, there was no accurate concept of how physical traits of adults could be represented in eggs and sperm. The science of chemistry was still rudimentary, there was no idea of proteins or nucleic acids, and most of the techniques for studying living cells had not yet been invented. Consequently the only means for understanding how traits come to be inherited was simple observation of their expression across generations. The concept of "blending" inheritance was a logical inference from such observations, because most observable traits of organisms—height, weight, strength, color of hair or skin—do vary continuously (from tall to short, strong to weak, dark to light), and offspring are usually intermediate between their parents in most such respects.

But there were some troubling features of this view of inheritance that were not easy to understand. Sometimes brown-eyed parents would have blue-eyed children, sometimes a new variant of a dog or a pigeon or a rose would arise and "breed true"—that is, not be diluted out by breeding with other individuals that did not possess the trait. So besides continuous or blending inheritance there were examples of inheritance that appeared to be discontinuous.

Between 1856 and 1866 an Austrian monk named Gregor Mendel performed some simple experimental crosses of pea plants in the garden of his Augustinian monastery, work that launched the science of genetics. He discovered the basic rules of heredity that resolved the confusion posed by blending and discontinuous examples of inheritance. His findings were published in 1866, but in a curious turn of fate their significance remained unappreciated for thirty-four years. Then in 1900 equivalent experiments were repeated independently by three different botanists. When his work was finally recognized, it soon became apparent how Mendel's rules of inheritance reflect the behavior of chromosomes in meiosis, and still later his discovery led to an understanding of how natural selection causes the changes in gene frequencies that are the basis of biological evolution. But we are getting ahead of the story.

WHAT MENDEL FOUND: TWO SIMPLE EXAMPLES, SOME IMPORTANT RULES, AND A FEW TERMS

Mendel was able to discover the basic laws of heredity because he chose to experiment with the inheritance of traits that are easy to recognize and trace, namely discontinuous traits. He obtained inbred (self fertilized) strains or varieties of garden peas that consistently showed either red or white flowers, long or short stems, either round or wrinkled seeds, and yellow or green seeds. He then crossed these strains by fertilizing the flowers of one with pollen from another.

The results of one of Mendel's crosses—between a lineage of peas with red flowers and one with white flowers—are shown in Figure 4.1. The parental plants, termed the P (parental) generation, were cross-pollinated, and the flowers of all of their offspring, termed the F_1 (first filial) generation, were red. These F_1 plants were then allowed to pollinate each other, and the resulting offspring, termed the F_2 generation, included red and white flowered plants in the approximate ratio 3 to 1.

The most remarkable result of this cross was that although the F_1 plants did not show white flowers, white flowers reappeared among the F_2 plants. Mendel realized that the F_1 plants, even though they have red flowers, must be carrying some hereditary factor for white flowers. His explanation of these results, which is diagrammed to the right of his results in Figure 4.1, contained the following elements:

- In each plant a *pair of hereditary factors* controls flower color.

- During the formation of gametes (eggs and pollen), the paired factors separate or *segregate* from each other so that each gamete receives only one factor.

- Consequently, in the resulting cross of plants with red and white flowers, each parent contributes one of each of the factors to the offspring.

The factors for red flowers and white flowers are different. They are actually *alternative* forms of the same factor, and the red form is *dominant* over the white one. That means that in the F_1 plants with both forms of the hereditary factor for flower color, the red color is expressed, and the presence of the alternative factor is not apparent from the color of the flowers.

A contemporary understanding of this experiment follows the same diagram; we need only change a few of the words. The hereditary factors of Mendel are of course *genes*, regions of nuclear DNA that code for a specific protein. The gene for flower color resides at a specific *locus* on one of the chromosomes, but the gene can exist in more than one form, coding for slightly different versions of a protein. The alternative forms of a gene are referred to as different *alleles*, and the two alleles in this example are represented by the letters "*C*" (for colored, i.e., red) and "*c*" (colorless, i.e., white). Although we see only two alleles in the example for this trait, in nature and for many traits there may be *multiple alleles* available in a population of organisms. Of

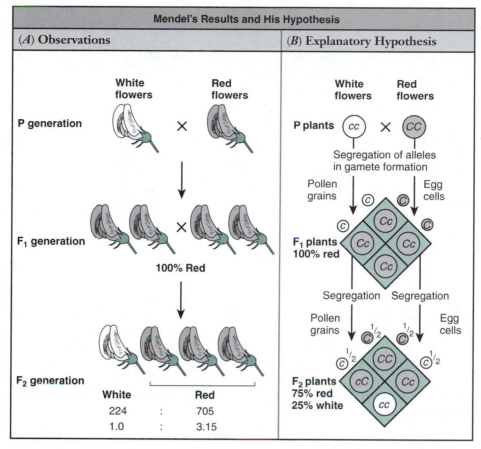

FIGURE 4.1 *Left*: The results that Mendel observed on crossing inbred strains of red and white peas. In the F_1 generation, all the flowers were red, but when the F_1 plants were crossed with each other, the next generation (F_2) had both red and white flowers in a ratio of approximately 3:1.

Right: The basis for this result in terms of a pair of alleles at a single locus. In the first cross the parental strains were homozygous for *CC* (colored) and *cc* (white) flowers. (The terminology is explained in the text.) *CC* parents give rise to haploid gametes of genotype *C*, and *cc* parents to *c* gametes. When the gametes from the two parents fuse, all of the F_1 offspring are heterozygous and have genotype *Cc*. Because *C* is dominant, the F_1 phenotype is colored (red).

The lower table shows an analysis of the second cross where all the parents (drawn from the F_1 plants) are of genotype *Cc*. Each parent makes haploid gametes *C* and *c* in equal numbers, and these combine randomly at fertilization to produce *CC*, *Cc*, and *cc* offspring in a ratio of 1:2:1, respectively. Because of dominance, however, the ratio of phenotypes in the F_2 generation is 3 colored to 1 white.

course any individual organism will have no more than two alleles at a given locus, because it gets one on the chromosome that it receives from each parent.

Note that sometimes "gene" refers to the DNA at a particular locus, as in "the gene for flower color" or "the gene for hemoglobin." When "gene" is used this way it is inclusive and does not seem to recognize the possible existence of different alleles. But sometimes the word gene can refer specifically to different alleles, as when we speak of "natural selection as a process of sorting genes." Don't be confused by these slightly dif-

ferent uses; the precise meaning is usually quite clear from the context.

If both alleles at a locus are identical, the organism is referred to as *homozygous* ("like paired") at this genetic locus. If the alleles are different, the condition is *heterozygous* ("other paired"). Thus individuals that are *CC* or *cc* are homozygous, and those that are *Cc* are heterozygous.

This example of discontinuous inheritance also illustrates the distinction between how an organism appears, its *phenotype*, and its hereditary constitution, its *genotype*. The F_1 hybrids in Mendel's cross were all red, thus they were identical in phenotype. But when they were inbred, some white flowers appeared in the F_2 generation. This means that the *c* allele must have been lurking in the F_1 plants. Obviously phenotype does not necessarily reveal genotype. In this example the gene for red flowers (the allele for red!) was *dominant*, and the red phenotype can have either one of two genotypes—homozygous for the dominant allele or heterozygous. The *c* allele was *recessive*; the appearance of white phenotype required the *cc* genotype.

Alleles at the same locus (such as *C* and *c*) differ from each other in base sequence. In the heterozygous state, the gene product of one allele may be solely responsible for determining that aspect of the phenotype that is under the control of genes at this locus. That is what makes one allele dominant over another. We will shortly show why the gene for red flowers is likely to be dominant over the gene for white.

Mendel tried his method of analysis on the more complex case of the inheritance of two distinct traits, seed texture and seed color. A strain of peas with smooth, yellow seeds was crossed with a strain with wrinkled, green seeds. All of the F_1 hybrids had smooth, yellow seeds, but if these F_1 hybrids were allowed to cross-pollinate, their F_2 offspring were found to include all combinations of the parental traits in the approximate ratio: 9 smooth, yellow; 3 smooth, green; 3 wrinkled, yellow; and 1 wrinkled, green. Mendel realized that the results of this cross could be explained by assuming that the hereditary factors for seed texture and color are separate and that they *segregate from one another independently* during the formation of gametes (Fig. 4.2).

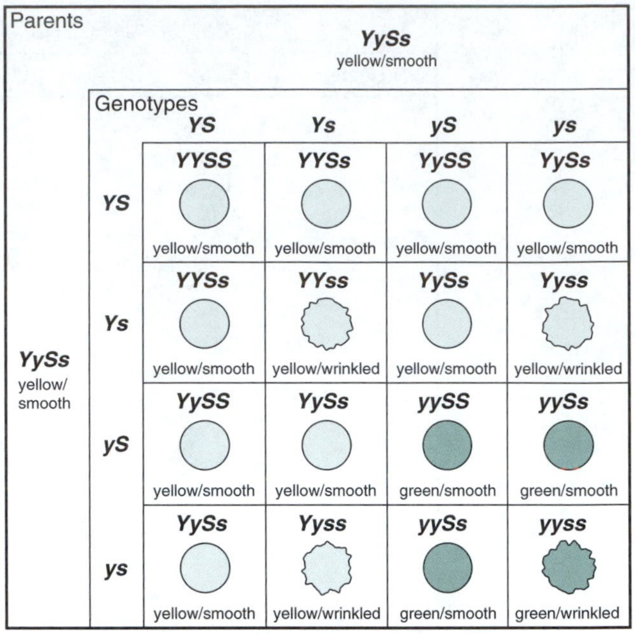

It was a large step in the understanding of inheritance when Mendel's observations on the independent segregation of traits could be put together with knowledge of the segregation of chromosomes during meiosis that was obtained by looking at cells under the microscope (Chapter 3), for this correlation made it clear that genetic information must reside in the nucleus of the cell and travel with the chromosomes during cell division. We described that latter process in Chapter 3 with no reference to the history of discovery, but reflect for a moment on the significance of this correlation in the context of the early years of the twentieth century. At that time there was no knowledge of DNA nor of how genes function in the cell. But to be able to focus attention on the cell nucleus as the place where genes must reside was a very substantial discovery.

It may have occurred to you that had Mendel chosen traits whose genes were located on the same chromosome, the rules of inheritance would not have been so obvious or appeared so simple. Genes that are *linked* together on the same chromosome will not segregate with the same degree of independence as the chromosomes themselves, but as we also described in Chapter

FIGURE 4.2 Independent assortment of genes for the texture and color of pea seeds, in a classical experiment of Gregor Mendel. Because of dominance, the ratios of phenotypes are 9:3:3:1, but the ratios of genotypes are different. Examine this diagram to be sure that you understand the origins of the different phenotypes and genotypes. How does the construction of this diagram reflect the independent assortment of alleles and the random nature of fertilization?

(For the curious: When there is a single gene with two alleles, the number of genotypes is given by the coefficients of the binomial expansion $(x + y)^2 = x^2 + 2xy + y^2$, i.e., 1*CC*:2*Cc*:1*cc*. Those who like to think in numbers may want to explore how the various genotypes in this dihybrid cross relate to the coefficients in the binomial expansion of $(x + y)^4$.)

3, the process of recombination of genetic material is not restricted to segregation of entire chromosomes. The process of *crossing over* greatly enhances the shuffling of genes, but it also complicates somewhat the rules of genetics. If two genes do not segregate independently, it is because they are located on the same chromosome. Moreover, the closer together two loci are on the same chromosome, the smaller will be the probability of crossing over between them. Conversely, the further apart they are, the more likely there will be recombination due to crossing over.

BLENDING INHERITANCE REVISITED

It did not take long after the rediscovery of Mendel's work around the turn of the century before geneticists realized that the concept of genes as discrete entities was completely compatible with the more common continuous variation found in nature. To see how this is possible, we will illustrate with two very different examples.

INCOMPLETE DOMINANCE

First, consider a cross between a parent that is homozygous for a dominant allele and another parent homozygous for a recessive allele at the same locus, just as in Figure 4.1. In some cases the phenotype of the F_1 individuals is truly intermediate in character between the phenotypes of the two parents. In the example of Figure 4.3 we are supposing that red flowers crossed with white flowers yield individuals with pink flowers. (These could be geraniums rather than the peas that Mendel used.) The genotype of the F_1 plants is heterozygous, but in this example the gene for flower color exhibits *incomplete dominance*. To extend the model we proposed to account for the dominance that Mendel observed in his observations on flower color, incomplete dominance will be found in cases where full phenotypic expression requires the presence of two copies of the dominant allele (Fig. 4.4). For example, if the dominant allele codes for a protein (enzyme) involved in the synthesis of red pigment, there are two possibilities relating the *amount* of this enzyme to the *amount* of pigment. If one copy of the gene is sufficient to make enough enzyme for full pigmentation to develop, the gene exhibits dominance, as in the case of Mendel's peas. On the other hand, if in the presence of only one copy of the gene a reduced amount of enzyme leads to a reduced level of pigmentation, the F_1 flowers have a lower concentration of pigment in their petals. They appear pink instead of red, and the gene is said to exhibit incomplete dominance.

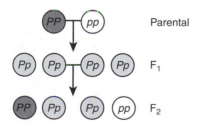

FIGURE 4.3 The F_1 generation is produced by crossing pigmented and white flowers, each homozygous for an allele of the gene controlling the synthesis of pigment. Members of the F_1 generation are all heterozygous, and the pale color is intermediate between the pigmented and white parents. When the F_1 generation is inbred to produce the F_2 generation, pigmented, intermediate and white flowers appear in an approximate ratio of 1:2:1. Do you understand how this ratio arises? And why we say it is *approximate*?

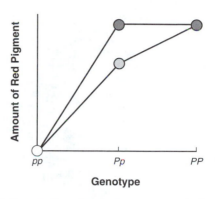

FIGURE 4.4 Incomplete dominance can result from differences in the amounts of gene product. *Upper curve*: Complete dominance, in which enough gene product (enzyme) is made by one copy of the *P* allele that a normal amount of pigment is synthesized in the flower. *Lower curve*: Incomplete dominance, in which not enough gene product is made by a single copy of the *P* allele to make up for the deficit in enzyme resulting from a dysfunctional *p* allele.

As an exercise, convince yourself that if the F_1 plants in Figure 4.3 are crossed with each other, the F_2 plants will be red, pink, and white in a ratio of 1:2:1.

POLYGENES

There is another explanation for why many traits of offspring appear intermediate between the traits of the parents. Suppose that height of a plant is controlled by a single genetic locus, *A*, with two alleles, *A* and *a*. Rather than thinking of these alleles as having dominance-recessive relationships to each other, suppose that each makes an additive but slightly different contribution to height: *A* contributes 6 inches to height and *a* contributes 4 inches to height, so that *AA* individuals are 12 inches tall and *aa* individuals are 8 inches in height. Clearly, the F_1 hybrids between these individuals will be 10 inches in height, intermediate between the parents. Furthermore, if we carry out an F_1 cross, we expect the F_2 offspring to be in the ratio: 1 8-inches:2 10-inches:1 12-inches. (If you do not understand these ratios, go back to the preceding paragraph and work through the example of the F_2 generation of red, pink, and white flowers.)

Now suppose there are *two* gene loci, *X* and *Y*, that control height. (In order to maximize their independent assortment for the purposes of this example, we will assume that they are on separate chromosomes, but of course in nature they need not be.) *X* and *Y* each contribute 3 inches to height and *x* and *y* each contribute 2 inches. *XXYY* individuals will be 12 inches in height, *xxyy* individuals will be 8 inches in height, their F_1 offspring will be 10 inches in height and the F_2 offspring, in turn, will include individuals of 8, 9, 10, 11, and 12 inches in the ratio 1:4:6:4:1 (Fig. 4.5). The dif-

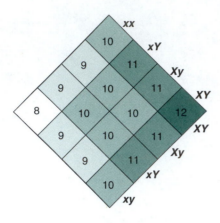

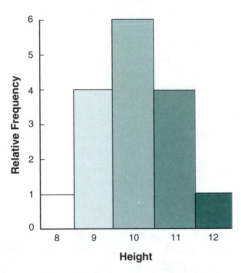

FIGURE 4.5 *Above:* Matrix of phenotypes (heights) resulting from a dihybrid cross in which height is controlled by alleles at two gene loci, as described in the text. Genotypes of parental gametes are shown outside the box. The heights of the resulting offspring are shown inside the box.

Below: This graph (a frequency histogram) shows the relative number of individuals of each height. If more than two loci contributed additive effects, the distribution of phenotypes would become broader and smoother.

ference between this hypothetical example and the previous one is, of course, that we have postulated more genes contributing to the trait, and as a result there are more intermediate genotypes and phenotypes.

These examples illustrate that the more genes there are contributing to a trait, the more genotypic and phenotypic categories there are, and the smoother the distribution of phenotypes becomes. In fact, most phenotypic traits are under the control of several genes and are thus *polygenic.* The examples that we have used to illustrate this point, however, are very simple in that

the effects of genes at different loci are independent and additive. In nature such simplicity is infrequent. For example, the effect of genes at one locus may depend on the presence of particular alleles at another locus.

GENES AND ENVIRONMENT

One of the central points that we wish to convey in this book—one that will arise repeatedly in many different contexts—is that genes are helpless without an environment in which to act. This dependence starts in the *internal* environment of the *cell.* In order for a gene to direct the synthesis of the protein for which it codes, that gene must be part of a living cell with all of the associated metabolic machinery for protein synthesis (Chapter 3).

Phenotypic expression of complex traits is also influenced by the *external* environment of the *organism.* For example, plant height is affected by several features of the environment including light, temperature, water, and mineral content of the soil. (For simplicity, we deliberately ignored this role of the environment in introducing polygenes in the previous section.) Moreover, it is a matter of chance where a plant grows, for its seeds can fall into environments that range from very favorable to very unfavorable. Environmental interactions with different arrays of polygenes may result in many more different phenotypes than there are genotypes—even as many phenotypes as there are individuals in the population!

This principle is hardly limited to plants. The physical growth and appearance of a person is no less dependent on environmental factors such as optimal nutrition, and as we discuss later, the sensory and social environment are critical players in the development of the brain and the generation of behavior.

The flowers of tobacco plants provide a good example of complex gene-environment interactions in development. In one species of tobacco these flowers vary more than twofold in length, and this phenotypic character, which is easy to measure, is under polygenic control. The upper panel of Figure 4.6 shows the distributions of flower length for two highly inbred strains of this plant. In other words, plants were previously selected for either long or short flowers, and plants in each population were self-fertilized. From the progeny of the short-flowered plants, individuals with the shortest flowers were inbred, and this procedure was repeated for several generations. Similarly, plants with longer flowers were subjected to the same sort of selection. As a result of several generations of selection, two strains of plants were produced with distinctly different lengths of flower.

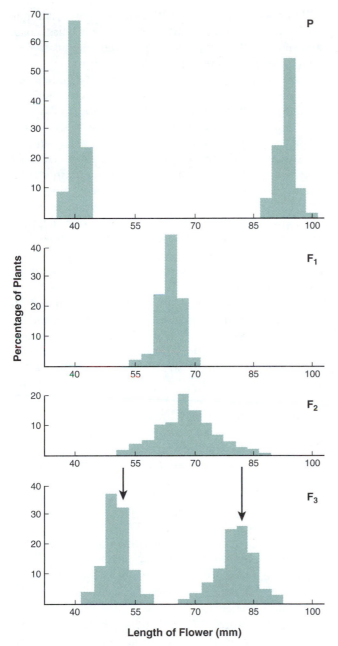

Length of Flower (mm)

FIGURE 4.6 Polygenic control of floral length. The upper panel shows the distribution of length of the flowers in two inbred strains of tobacco plants. When these homozygous strains were crossed, the F_1 plants were all heterozygous at the several loci controlling floral length, and their flowers were intermediate in length between those of the parental strains. The phenotypic variation in these two generations was due to the environment. When the F_1 plants were intercrossed, the F_2 flowers showed greater variation because recombination had increased the genetic variation. Finally, selection for length among the F_2 plants led to two distinct populations of floral length among the F_3 generation. Refer to the text for a further description of this experiment.

This procedure is *artificial selection*, meaning that the reproductive advantage of particular phenotypes is imposed by human intervention. The process is in principle exactly the same as natural selection, and the realization that plant and animal breeders can generate new phenotypes in a few generations was one of the observations that led Darwin to propose *natural selection* as a basic evolutionary process.

Because the two strains of tobacco flower were so highly inbred for this characteristic, each of the several genetic loci that contribute to length of flower had approached homozygosity. Consequently, after inbreeding the amount of variation in the length of flower within each strain was largely due to different environmental influences on the growing plants.

These two strains were then crossed with each other, producing an F_1 generation with flowers intermediate in length between those of the parents. Because these plants were identically heterozygous at every genetic locus contributing to length of flower, this F_1 population was also genetically homogeneous. If you examine the distribution of floral length in the second panel, you will see that it is approximately the same width as the broader of the two parental distributions. Most of the phenotypic variation in the F_1 generation was therefore also due to different environmental influences on the plants.

The F_1 generation was then self-pollinated to produce an F_2 generation. At this point you can anticipate the result. Because of the segregation of alleles during meiosis and the resulting recombination of genes in the F_2 plants, the distribution of genotypes, and thus phenotypes, in the third panel is substantially broader than in either the F_1 or parental generations. Clearly environmental differences contributed to this distribution just as in the two previous generations, but some of the phenotypic variation in the F_2 was now due to genetic differences among the plants.

Note in the distribution of floral length in the F_2 plants that there are no plants with flowers as long or as short as those in the two parental strains. This might be because there are so many contributing genes that the probability is very small of recovering individuals homozygous at all loci in the sample size under study. But it might also be caused by a more complex relation between the participating genes than a simple additive contribution to floral length. For example, alleles present in the very shortest plants might actually inhibit growth, and vice versa. There is also a third potential source of variation in phenotypes. Different genotypes might vary in the way they interact with their environments; for example, two plants might achieve the same floral length in a favorable environment but differ in length in unfavorable growing conditions. For an analogous example in humans, consider that some people get fat from eating potato chips and drinking beer whereas others do not. If we were going to study the control of floral length at great length, we would want to know much more about how these polygenes are interacting with each other and with the environment.

Nevertheless, when artificial selection of the F_2 plants was imposed by inbreeding individuals with the phenotypic characters above the two arrows, two different F_3 strains were produced, each with an average floral length close to the extremes of the distribution of the F_2 plants used as parents. These selective crosses are the first steps toward recovering the genetic strains that appear in the top panel.

HERITABILITY

This example of tobacco flowers illustrates an important general principle about populations of organisms. There is always phenotypic variation in a population; no two individuals are exactly alike. Variation means how much members of the population differ from the average. Variation could be measured as the average difference between individuals and the mean, but statisticians calculate the average of the *square* of the difference, called the *variance*.

As we just saw, organisms can differ from each other because they have different genes and because they have experienced different environments. If we look at any *individual organism*, however, we cannot say how much of the phenotype is caused by genes and how much by environment; the two influences are usually coupled (Chapters 10 and 11). But if we consider the expression of a phenotypic character in a *population*, we can ask how much of the variation *among organisms* reflects differences in genotypes and how much is caused by different environmental effects.

If all the individuals in the population have the same genotype, all of the phenotypic variation is caused by environmental differences. On the other hand, if the environment is completely uniform and the organisms differ in the alleles that are responsible for the phenotypic character, all of the variation is genetic. *Heritability* (designated h^2) is simply the fraction of the variation that can be attributed to genetic differences among individuals. In symbolic notation, $h^2 = V_G/(V_G + V_E)$, where V_G and V_E are measures (technically the *variances*) of genetic and environmentally-caused variation, respectively. Heritability is a fraction that varies from 0 to 1.

Heritability is thus a *population statistic*. It expresses the extent to which phenotypic differences in a trait are influenced by genetic differences among individuals. The importance of heritability is that it measures whether natural selection is able to act on the trait. For example, in the limiting case where heritability is 0, differential reproductive success of some of the variants will not change gene frequencies, and natural selection will not occur because all of the phenotypic variation is due to environmental differences. In other words, natural selection for a trait can only act on a population in which there is genetic variation for that trait.

Heritability is an important concept to understand because it has appeared in science writing with political implications (for example, a recent book on alleged racial differences in IQ), and it can be easily misunderstood. First, any measure of heritability applies only to a particular environment and the particular ensemble of genes that were present when the measurement was made. Estimates of heritability are therefore not transferable from one set of conditions to another, and they must always be interpreted in the context in which they were made. Second, although the techniques for measuring h^2 are beyond the scope of this account, realize that estimates for complex phenotypic traits generally have a large uncertainty.

Finally, heritability is not the same as *heritable*. To say a trait is heritable means that an aspect of its expression in under genetic control. But in a genetically homogeneous population, a heritable trait will have a heritability of 0. Confusing? You're right! But all that statement means is that in a genetically homogeneous population there can be no variation among individuals caused by the presence of different alleles.

Consider Figure 4.6 again. The heritability of flower length is small in each of the parental populations as well as in the F_1. In the F_2, however, it is significantly larger. If we assume that the heritability of flower length is 0 in the F_1 (due to extensive inbreeding of the parental populations), all of the variation is due to environmental differences. V_E can be calculated from the frequency distribution of phenotypes and is about 11. In the F_2, however, the variance is $V_E + V_G$ and is about 47. If we assume that V_E is the same in the F_2 as in the F_1, the heritability of flower length in the F_2 is approximately $(47 - 11)/47 = 0.77$.

The relevance of heritability to the way in which organisms respond to natural selection is illustrated in Figure 4.7. The two upper panels show the frequency distributions of egg weight and egg number for chickens. Each distribution is approximately bell-shaped, with the most common values around the mean. The heritability of egg weight is fairly high (somewhere between 0.4 and 0.8 depending on the strain of chickens and the conditions under which they are reared). In other words, a substantial fraction of the variation in egg weight is due to the presence of different alleles. Consequently when the flock is subjected to artificial selection by breeding hens that lay eggs at the extremes of the distribution (arrows), the frequency distributions of the eggs laid by their progeny are quite different from that of the parental flock (lower panel).

Egg number has a much lower heritability, and efforts to increase or decrease egg number by selective

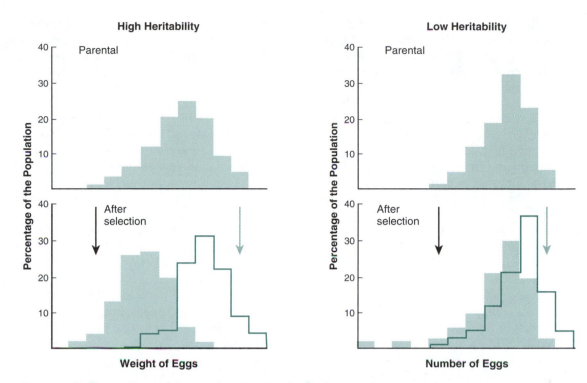

FIGURE 4.7 Comparison of the results of artificial selection on phenotypic characters exhibiting relatively high heritability (*left*) and low heritability (*right*). Response to selection for a phenotypic trait is larger when there is substantial genetic variation for the trait present in the population—that is, when heritability is high.

breeding are much less successful (lower panel, right). This is because most of the variation in egg number laid by the parental population was due to environmental factors, and there was little genetic variation on which selection could operate. In qualitative terms these two experiments show how the speed with which selection can operate on a trait depends on the heritability of the trait.

THE HARDY-WEINBERG LAW

Biological evolution consists of changes in the genetic makeup of populations over time: mutations create new genes or modify old ones, and with time these genetic novelties either increase, decrease, or remain stable in frequency. In order to understand how and why biological evolution occurs, we need to consider changes in the frequencies of alleles at the level of populations of organisms.

We introduced the Mendelian rules of inheritance with examples involving genes at one or two loci and random mating between parents. Let us now build on those concepts by exploring what we might expect to happen to the frequencies of alleles and genotypes if we mix two genetically different populations of *different*

sizes. Imagine two populations of sea urchins that differ in both number and in color. One population consists of 80 *RR* individuals (40 males, 40 females) that are red, and the other consists of 20 *rr* individuals (10 males, 10 females) that are white. We now place the two populations together in the same large seawater tank. At the time of mating in their natural habitats, sea urchins shed their gametes (eggs, sperm) into the water synchronously, and we shall assume that under the conditions of our thought experiment, fertilizations are completely random.

Male and female sea urchins shed millions of sperm and hundreds of thousands of eggs, respectively. What is the fraction of sperm with the allele *R* and what fraction with *r*? As all of the *R* sperm come from the *RR* animals and all of the *r* sperm come from the *rr* animals, the relative proportions of the two classes of sperm are the same as the relative proportions of the two parental genotypes, 0.8 for *R* and 0.2 for *r*. Exactly the same argument holds for eggs, even though there are fewer eggs produced than sperm: 0.8 *R* and 0.2 *r*.

We can now consider the gametes in this aquarium as a "gene pool" in which the frequency of the *R* allele is $p = 0.8$ and the frequency of the *r* allele is $q = 0.2$. What will the frequencies of the genotypes be when gametes fuse together to form zygotes?

The expected frequencies of the genotypes are found by calculating the probabilities that the two kinds of gametes, R and r, will randomly combine to form the zygotes, RR, Rr, and rr. This is easy if you remember that the probability of observing two independent events is the product of their individual probabilities. For example, when flipping a coin the probability of tossing two heads in a row is $0.5 \times 0.5 = 0.25$. By the same reasoning, the chance of observing a fertilization of an R egg by an R sperm is the product of their individual probabilities of occurrence among similar gametes in the tank, which is 0.8 for each. Consequently, the probability in any fertilization of forming an RR zygote is $p \times p = p^2 = (0.8)^2 = 0.64$. Similarly, the probability of forming an rr zygote is $q^2 = (0.2)^2 = 0.04$. Whereas these homozygous genotypes, RR and rr, can result from only one type of "gamete collision"—that between gametes bearing the same allele—the heterozygous genotype, Rr, can result from two kinds of encounters. An R sperm may combine with an r egg with a probability of $p \times q = 0.8 \times 0.2 = 0.16$, or an r sperm may combine with an R egg with the same probability, $q \times p = 0.16$. The sum of these probabilities is then $2pq = 0.32$. Every fertilization must be one of the types for which we have just calculated probabilities. Consequently, the sum of the three frequencies of expected genotypes is $0.64 + 0.32 + 0.04 = 1.0$.

You will likely recognize that we can calculate the expected frequencies of genotypes from the frequencies of the alleles by using the binomial expansion $(p + q)^2 = p^2 + 2pq + q^2 = 1.0$. (Remember, $p + q = 1$, so $(p + q)^2 = 1$.)

We now carry this quantitative reasoning one step further and ask what the frequencies of the alleles and genotypes will be in the sea urchin population after another generation of random mating. If each parental genotype is equally successful in producing offspring (as we have just assumed), 0.64 of our F_1 generation are RR, 0.32 are Rr, and 0.04 are rr. Using the same reasoning that we employed when we considered the first set of parents, we can calculate the proportion of gametes of each haploid genotype that will be produced by this F_1 population. In a random sample of 100 sperm (or eggs), there will be 64 R gametes from the RR urchins and 4 r gametes from the rr urchins. But what about the gametes from the Rr animals? Each gamete has a 50% chance of being either R or r. Consequently in our random sample of 100 sperm produced by the whole population there will be 16 R and 16 r sperm produced by the heterozygous urchins. The frequency of R gametes is therefore $0.64 + 0.16 = 0.8$, and the frequency of r gametes is $0.16 + 0.04 = 0.2$.

Astonishing? Even though the original population of red sea urchins outnumbered the white ones by 4 to 1, after two generations of random matings the proportions of the two alleles in the population has not changed. These calculations demonstrate the most basic tenet of population genetics, the Hardy-Weinberg Law, discovered independently in 1908 by the English mathematician Hardy and the German physician Weinberg. *Under certain conditions (to be described presently), in populations of sexually reproducing organisms allele frequencies, and therefore genotype ratios, remain constant from one generation to the next.* Such a population is in Hardy-Weinberg equilibrium. In these conditions it doesn't matter what the relative sizes of the initial populations of red and white animals are, no loss of genetic variation and no change in allele frequencies will take place after they have been mixed. A population at Hardy-Weinberg equilibrium is therefore not evolving. Conversely, if a population is not at Hardy-Weinberg equilibrium, evolutionary changes *are* taking place.

Shortly, we will consider what might be driving such evolutionary changes, but let us first see how the Hardy-Weinberg Law can be used to determine whether a population is actually at equilibrium with respect to the alleles at a particular genetic locus. Consider a population with the following numbers of individuals of each of the three genotypes:

	AA	Aa	aa	sum
observed	65	10	25	100

Is this population at Hardy-Weinberg equilibrium? From the numbers of individuals of each genotype, we can calculate the frequencies of the two alleles in the population, and from this information we can calculate the *frequencies at which we should expect each genotype to occur at equilibrium.*

For A the frequency $p = 0.65 + (0.1 \times 0.5) = 0.7$

For a the frequency $q = 0.25 + (0.1 \times 0.5) = 0.3$

As a check, note that $p + q = 1$, as it should. Now from these allele frequencies we can calculate the frequencies of the diploid genotypes that should exist in the population at equilibrium: $p^2 + 2pq + q^2 = 0.49 + 0.42 + 0.09 = 1.0$. (Note again, as a check on the accuracy of your arithmetic the three frequencies of genotype must sum to 1.) In a population of 100 organisms, the expected numbers of each genotype are therefore:

	AA	Aa	aa	sum
expected	49	42	9	100

which differ substantially from the numbers that were observed. This population therefore does not appear to be in Hardy-Weinberg equilibrium with respect to the alleles at this genetic locus. We'll soon see the possible reasons why.

GENETICS AND EVOLUTIONARY CHANGE

MICROEVOLUTION AND MACROEVOLUTION

Evolutionary events taking place within populations, phenomena that can be characterized in terms of specific genes, are referred to as *microevolution. Macroevolution*, by contrast, refers to the emergence of categories of organisms more major than species. Macroevolution builds on microevolutionary processes, but it is enriched by historical contingencies such as physical events that cause major extinctions. Recently, study of the roles of genes in development has begun to contribute importantly to a richer understanding of macroevolutionary changes (Chapter 10). There is nevertheless a persistent misconception (nurtured by creationists) that macroevolution is not a natural phenomenon that can be understood by the scientific process.

We can now place the discussion of the Hardy-Weinberg Law in a larger context of microevolutionary processes.

DEPARTURE FROM HARDY-WEINBERG CONDITIONS

We suggested above that when a population is not in Hardy-Weinberg equilibrium, evolutionary changes must be occurring to shift the distribution of genotypes from the expected frequencies. What are the "certain conditions" to which we alluded at the outset that must be met to maintain Hardy-Weinberg equilibrium? Or put the other way around, what factors generate microevolutionary changes in populations of organisms?

We have already provided a clue, for more than once in discussing the Hardy-Weinberg Law we emphasized the assumption that the sea urchins were mating randomly. Clearly, if they were not mating randomly, our procedure of multiplying the frequencies of the genotypes of the gametes would not have been valid. But what causes nonrandom mating? And what other factors could cause departures from Hardy-Weinberg equilibrium?

There are four kinds of phenomena that lie behind microevolutionary change.

RANDOM GENETIC DRIFT

If a population is very small, the frequency of an allele can increase or decrease over time entirely by chance. If the probabilities of various matings are chance events, when there are only a few occurrences there will likely be departures from the calculated probabili-

ties. If you flip a coin four times, you expect two heads and two tails. Try a series of four-flip tosses and see how many yield two heads and two tails, or try the more elaborate genetic model in Box 4.1. The point is that with small numbers of organisms, chance events may either "fix" an allele in the population or move it to extinction.

The importance of this *genetic drift* in evolution will vary with circumstances. It may play a significant role in changing the frequency of alleles that are little affected by other factors, especially in small populations. For example, a random sample of a few dozen or a few hundred humans is likely to have, by chance, a higher or lower frequency of individuals with some genetic mutation than the larger population from which they came. If this small sample of individuals then founds a much larger population, the larger population can have the high incidence of the allele too. Such *founder effects* are thought to account, for example, for the relatively high incidence of Tay-Sachs disease, a hereditary degeneration of nerve cells, that is found among the Ashkenazi Jews. On the other hand, it is unlikely that genetic drift affects the evolution of most adaptations. This is because adaptations are usually controlled by many genes, and the more genes that influence a trait the less likely it is that all of the alleles will randomly "drift" in the same direction at the same time, thus causing the phenotype to change directionally.

GENE FLOW

The example of our sea urchins was convenient and somewhat artificial because the possible matings were constrained by the walls of the aquarium. In other words, the population was discrete; there was no possibility of urchins wandering out of the tank or other urchins with a different frequency of alleles wandering in. Of course in nature populations of organisms differ greatly in their degree of genetic isolation. Some populations may be truly isolated for extended periods of time; more frequently, however, organisms come and go, bringing or taking their genes with them. This movement of alleles between populations is what is meant by *gene flow*. The term may evoke an image of liquid being poured, but the process of gene flow is the movement of genes in organisms, or in their gametes, or in seeds. Gene flow can be an important player in microevolution, tending to neutralize changes in a local population that are driven by natural selection.

MUTATION

As discussed in Chapter 3, mutations are introduced into genetic information at several levels, from the substitution of one nucleotide for another when DNA

Box 4.1

A Simple Experiment Illustrating Genetic Drift

You can demonstrate genetic drift for yourself with the following experiment. In nine cups place a hundred or more small tokens (beans, peas, marbles, "M&Ms") of two colors. Make the frequency, p, of one color equal to 0.1 in the first pot, 0.2 in the second, and so on to $p = 0.9$ in the last pot. Now you can simulate fluctuations in gene frequency due to genetic drift by drawing different numbers of peas (gametes) of different colors (alleles) from the pots over time. For example, start with a population size of 5 individuals with p and $q = 0.5$ by drawing 10 peas (gametes) without looking from the pot in which $p = q = 0.5$. If you drew, say, 3 peas of one color and 7 of the other ($p = 0.3$, $q = 0.7$), then you have already demonstrated genetic drift: p changed from 0.5 to 0.3 in one generation due to sampling error. You can go on to the next generation by returning the 10 peas to the pot in which $p = q = 0.5$ and drawing the next 10 peas from the pot in which $p = 0.3$; you will find that eventually one color (allele) will go to extinction (p or $q = 0$). The effect of population size on the magnitude of random fluctuation (due to sampling error) in gene frequency can be investigated by increasing the numbers of peas you draw each generation.

Readers who like to play with computers might find it more interesting to forget the teacups and the beans and program this exercise as a game. That will make it very easy to explore the effect of population size on genetic drift.

---

replicates (copy error), to duplications, inversions, and translocations of genes or of entire pieces of chromosomes. As we also discussed earlier, many nucleotide substitutions in DNA have little effect on the phenotype because they either do not result in the replacement of one amino acid for another, or if they do, the new amino acid does not have a consequential effect on the function of the protein.

It is not surprising that the majority of mutations that do have phenotypic consequences are detrimental. This is because random changes in the components of a complex structure are much more likely to harm it than improve it. In fact organisms have mechanisms to insure that such errors are seldom made during the replication of genetic information (Chapter 3); in most cases the rates of mutation are around one in a million gene replication events.

If mutations with phenotypic effects are rare and usually harmful, how do they affect evolution? Mutation is the ultimate source of new variation, and the random background rate of mutation provides the raw material on which natural selection feeds. Out of the many mutations that occur, a few will lead to some change that enhances the reproductive success of the organism in which the allele resides. Natural selection can rapidly increase the frequency of the allele.

NATURAL SELECTION

Natural selection is the most important factor in causing short-term microevolutionary change, but of course natural selection is ultimately dependent on mutation to supply the grist for its mill. We discuss natural selection at considerable length throughout this book, and in the next chapter we will raise the question of what actually gets selected. We emphasize again, however, that natural selection leads to changes in frequencies of alleles due to the effects of those genes on the abilities of their host organisms to reproduce.

Recall from an earlier page the example of a population of organisms that we showed was not in Hardy-Weinberg equilibrium because the frequencies of the several genotypes were not what was expected from the frequencies of alleles:

	AA	Aa	aa
expected	0.49	0.42	0.09
observed	0.65	0.10	0.25

It is readily apparent that there are fewer heterozygous individuals in the population than expected, somewhat more individuals homozygous for the A allele, and many more homozygous for the a allele. In the absence of genetic drift and gene flow, the likely explanation is that Aa individuals are at a selective disadvantage and the reproduction of aa individuals is being favored. Further observations and experiments would be necessary to test this hypothesis. For example, one might look for differential mortality of the heterozygous individuals to discover the cause of this observed distribution of genotypes.

Whether an allele enhances the reproductive success of the organism in which it occurs can depend on a variety of other conditions. It may be influenced by whether the allele is in the homozygous or heterozygous state, by the activities of genes at other loci, and by the environment. None of these outcomes is ordinarily predictable in advance, and all are matters for investigation. What we are describing in this elementary account of microevolution is a reformulation of Darwin's theory of natural selection in terms of the distinction we now understand between an organism's phenotype and its genes.

What sorts of evolutionary change do we expect at the level of phenotypes if phenotypic differences are both heritable and influence reproductive success? Natural selection will have three broad classes of effect (Fig. 4.8).

1. The most common selective outcome is *stabilizing*. Stabilizing selection tends to keep the population close to the average because individuals at the extremes of the distribution are at a reproductive disadvantage. Such selection therefore tends to reduce the amount of genetic variation for the trait in question. Stabilizing selection is so common because natural selection is rapid compared with the rate of mutation. This is another way of saying that organisms are usually well adapted to their environments, that most mutations will be detrimental, and that lineages of organisms frequently remain unchanged for long periods of time.

   Figure 4.9 illustrates stabilizing selection for birth weight in human infants. Note that the average birth weight (histogram) is very close to the

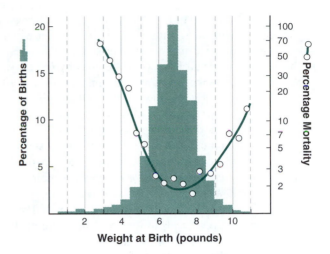

FIGURE 4.9 Birth weights and early mortality for 13,730 children. The optimum birth weight corresponds to the the weight at which mortality is lowest. Note that the vertical axis for mortality is logarithmic.

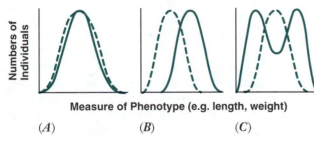

Measure of Phenotype (e.g. length, weight)

(A)   *(B)*   *(C)*

FIGURE 4.8 Natural selection can be *stabilizing* (A), *directional* (B), or *disruptive* (C). Each plot shows the numbers of individuals as a function of some phenotypic parameter such as height. In each panel the initial distribution is shown by the dashed line, and the distribution after a period of selection by the solid line. (A) Stabilizing selection holds the population around some optimal mean, perhaps narrowing the distribution of phenotyes slightly. (B) Under directional selection the distribution shifts, as individuals with phenotypic characters at some distance from the original mean have greater reproductive success. (C) In disruptive selection more than one phenotype is favored, and the distribution of observed phenotypes can be increased. Remember that in order for selection to work as described in these examples, the various phenotypes must be based on heritable differences, i.e., genes.

minimum of the curve for early infant mortality. Large infants have difficulty traversing the mother's birth canal, and smaller infants may have insufficient fat reserves or be too underdeveloped to sustain themselves.

This figure warrants additional thought, however, because not every factor that contributes to birth weight is necessarily heritable and thus subject to natural selection. For example, during a famine birth weight might be low because of poor maternal nutrition. On the other hand, can you think of reasons why poor maternal nutrition might have a heritable basis?

2. Obviously not all selection can be stabilizing, or evolutionary change would not be observed. If circumstances change—either because a mutational event has provided a new opportunity or an environmental change has presented a new challenge—some individuals not previously favored may become the more successful reproducers. The responsible genes and the more effective phenotypes with which they are associated will start to increase in frequency. This is *directional* selection. Directional selection is assumed to be the cause of most morphological evolution—for example, increases or decreases in the size of the organisms and their parts, increase in complexity of sense organs, and the evolutionary transformation of fins into feet or arms into wings.

   Because the volcanic Galapagos Islands, 500 miles off the coast of Ecuador, are isolated, subject to dramatic changes in rainfall, and contain endemic species, they provide a natural laboratory for documenting microevolutionary changes. Among the animals unique to the islands are several species of finches, (Fig. 2.5) named after Charles Darwin, who was the first to collect them. These finches have been observed for several decades by the evolutionary ecologists Peter and Rosemary Grant and their students. They showed that bill size in one species, *Geospiza fortis*, is closely adjusted to the size of the seeds that are available for food. As a result of a draught in 1976–1977, there was a shortage of small seeds, and selection rapidly favored larger birds with larger bills that could exploit the bigger seeds with harder coats that were still available (Fig. 4.10). When an El Niño event in 1983 caused exceptionally prolonged and heavy rainfall, smaller seeds were superabundant for eight months, and between 1984 and 1985 there was selection for birds with smaller bills (Fig. 4.10).

These observations illustrate the rapidity with which evolution can occur in response to changed environments, and they provide insight as to how morphological changes can arise by microevolution. In the Galapagos Islands, however, this directional selection is being caused by irregular weather patterns and is not sustained for more than a year or two. Consequently, over a decade or so, the selection on these remote islands is effectively stabilizing.

Because microorganisms reproduce so quickly, directional selection can be particularly rapid in increasing their resistance to antibiotics. Rapid evolution of the pathogen is part of the reason why it has been so difficult to stop the ravages of the virus responsible for AIDS (Chapter 8). The quantitative interplay of selection and mutation is illustrated further in Box 4.2.

3. Diversifying or *disruptive* selection occurs when individuals with two or more different phenotypes have a reproductive advantage over individuals with intermediate phenotypes. Figure 4.8 shows that such selection can increase the phenotypic variation in a population and ultimately lead to separation into two different forms. Figure 4.7 also illustrates disruptive selection.

CHANCE AND CONTINGENCY IN EVOLUTION

Consider the role of chance events in the four processes of microevolution. Mutations are chance occurrences, whether copy errors, recombination events, or induced environmental damage (for example, UV solar radiation). Genetic drift by definition is due to chance, and there is an element of chance associated with gene flow.

Natural selection, however, is the antithesis of chance. Differential reproductive success due to heritable differences that endow some individuals with relative advantage in a given environment is a process of great power. It is the major factor at work in producing biological "design." But natural selection is a statistical process. The pair of song sparrows whose nest is parasitized by a cowbird and who spend the season raising the young of another species may have simply had bad luck compared with the song sparrows in the next yard that produced several young of their own. The rabbit that falls prey to the fox may be no more "fit" than its sibling in a neighboring burrow; it may simply have been in the wrong place at the wrong time.

It is sometimes argued that natural selection cannot lead to complex structures because the product of the probabilities of the necessary mutations is infinitesimally small. The process is likened to the probability of a tornado sweeping through a junkyard and assembling a Boeing 747. This is faulty reasoning on two counts. First, life uses energy to create order; meteorological and geological processes don't. Second, the selective advantage of any mutation depends on the genome in which it occurs and the environmental conditions that exist at the time. Consequently, evolutionary change is cumulative, and every step is dependent on what had occurred previously. By analogy, based on your behavior last year, the probability of your eating salmon on Friday night may be 0.01

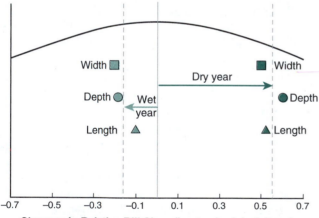

FIGURE 4.10 Selection for bill dimensions in the Galapagos finch *Geospiza fortis* resulting from an especially wet year and a drought. Many small seeds were available during the wet year, and birds with smaller bills were favored by natural selection. Only larger seeds with tougher coats were available as a result of the drought, and selection favored birds with stronger bills. The actual changes in bill size were only a few percent. They are shown here in terms of changes in the average values before and after a season of selection, plotted in units of a statistical parameter called standard deviation. The curve at the top of the figure is the crest of a normal distribution with a standard deviation of 1.

Box 4.2
Natural Selection, Mutation, and Uncommon Alleles

I—WHY RARE, LETHAL RECESSIVES AREN'T ELIMINATED

The genes of all species include a number of relatively rare recessive alleles that have adverse phenotypic consequences when homozygous. The hemoglobin allele responsible for sickle-cell anemia is an example in the human genome. Even when the homozygotes are lethal, however, such alleles persist in the population, present in individuals who have only a single copy (i.e., are heterozygous at the locus). Such people are "carriers" of the gene, but they may suffer no ill effects. Why does natural selection not eliminate such troublesome alleles entirely?

There are several reasons. Suppose that one person in a thousand is heterozygous at such a locus, and that everyone homozygous for the recessive allele fails to reproduce. Applying the Hardy-Weinberg Law, where p and q represent the frequencies of the dominant and recessive alleles, we can estimate q from the frequency of heterozygous individuals. In our hypothetical example the frequency of heterozygotes is 0.001, so

$$2pq = 2(1 - q)q = 0.001,$$

from which we can calculate that $q \approx 5 \times 10^{-4}$. The homozygous recessives in the next generation will be much rarer than the carriers and can be estimated to be $q^2 \approx 2.5 \times 10^{-7}$, or about 1 person in 4 million. These are the individuals who do not reproduce and are thus the focus of natural selection against the recessive allele.

From the Hardy-Weinberg Law we can also calculate p and q over successive generations and thus the rate at which the frequency of heterozygotes declines (Fig. 4.11). Even after 250 generations, the frequency of heterozygous individuals has fallen from 1 in 1000 people to about 0.89 per 1000, a decrease of only 11%. In other words, natural selection acts slowly against recessive alleles if they are rare and deleterious only when homozygous.

Furthermore, this rate of selection is sufficiently slow that it can be offset by the rate of mutation, μ, at the same locus. How mutation and selection against recessives interact can also be derived from the Hardy-Weinberg Law, and the result is disarmingly simple. When the increase in q by mutation is equal to the decrease in q through selection, the rate of mutation $\mu = sq^2$, where s, a coefficient of selection, is the frac-

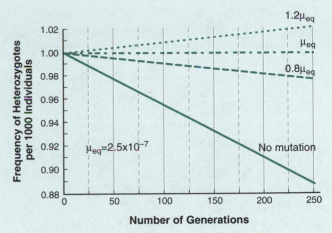

FIGURE 4.11 Selection works slowly against a deleterious recessive allele that is present at low frequency. Elimination by selection can be offset by mutation.

tion of homozygous recessives that do not reproduce. In this example, none reproduce, so $s = 1$ and $\mu = 2.5 \times 10^7$. If a rare mutation is not expressed until after its bearer has reproduced, the allele will experience less adverse selection (s will be smaller than 1), and the necessary mutation rate for replacement will be even lower.

The horizontal dashed line in Figure 4.11 shows that when selection is just balanced by the rate of mutation (μ_{eq}), the frequency of carriers remains constant through 250 generations. If the mutation rate is 20% less or 20% greater, the frequency of heterozgotes falls or rises slowly.

Is this value for mutation rate reasonable? A gene is likely to mutate about once in every million cell generations (Chapter 3), and the maturation of sperm involves several cell generations (successive cell divisions). We can therefore expect that as a combined result of mutation and the ineffectiveness of natural selection in reducing rare recessive alleles, heterozygous carriers can in fact be present at rates approaching 1 in 1000 individuals.

II—HETEROZYGOTE ADVANTAGE

A second reason why detrimental recessive mutations may be maintained in the population was illustrated by sickle-cell hemoglobin (Box 3.3). Historically, few individuals with two copies of the sickle-cell allele lived to reproduce, so selection against these homozygous individuals was strong. About 9% of African-Americans are carriers, reflecting what were likely even higher frequencies in West African populations several hundred years ago. There is positive selection for the allele when it is heterozygous and strong negative selection when homozygous. However, the *heterozygote advantage* of this allele clearly occurs only in environments where malaria is prevalent.

Consider a rare allele (A', not necessarily recessive) that is present at the same frequency as the lethal allele in the first example, $q = 5 \times 10^4$. We will assume that selection against the homozygotes, A'A', is strong: $s = 1$. But suppose that there is now modest selection that favors the heterozygotes, AA', over the homozygotes, AA. As we follow the population for many generations, the frequency of individuals with the AA genotype decreases and the frequency of the AA' genotype increases until the population comes into Hardy-Weinberg equilibrium.

Figure 4.12 shows two examples. When the selection coefficient (s) against AA' is 0, that against AA is 0.05, and that against A'A' is 1 (i.e., all die), change in the frequencies of AA and AA' genotypes is slow and modest (dashed curves). If the selection coefficient against AA is increased to 0.25, however, selection is much more rapid (equilibrium is reached in fewer than 40 generations), and there is a larger change in the frequencies of AA and AA' in reaching Hardy-Weinberg equilibrium.

There are other recessive alleles that cause serious disease when homozygous but whose frequencies in the population suggest that they confer some advan-

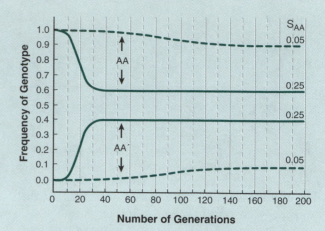

FIGURE 4.12 Heterozygote advantage of AA'. Approach to Hardy-Weinberg equilibrium from an initial very low frequency of the deleterious allele A'. Results are shown for two strengths of selection against the homozygotes AA.

tage when heterozygous. The proximate causes of heterozygote advantage are frequently not well understood. Fragile X (a gene that affects the X chromosome) causes mental retardation in males but seems to bestow greater reproductive success on women who are carriers. The gene for Tay-Sachs disease (fatal at an early age) is hypothesized to make its carriers less susceptible to tuberculosis. The gene for phenylke-

and the probability of your going to the fish market this week may be 0.3. But there is no point in multiplying the two numbers together. Eating salmon on Friday is dependent on going to the fish market first.

Chance, contingency, and circumstance have played important roles in macroevolution. Whole populations may be compromised by a serious drought or several severe winters. The flourishing or extinction of major groups has frequently been caused by climatic changes, and asteroid impacts have dramatically altered environments. Such an event is now believed to have been important in the extinction of the dinosaurs about 65 million years ago, opening possibilities for the adaptive radiation of mammals.

WHY THE THEORY OF INHERITANCE OF ACQUIRED CHARACTERISTICS FAILED

We will close this chapter by returning to the core feature of microevolution that evaded Darwin's understanding. In his time it was commonly thought that

modifications acquired by an organism during its lifetime could be passed on to its descendants. This belief—referred to today as the inheritance of acquired characteristics—implied that the strong muscles of a blacksmith's arms, the thick fur of a mammal exposed to cold, and behaviors and skills acquired by learning could appear in offspring without the experiences or conditions that caused them to appear in their parents. Two observations made this a reasonable belief. First, the blacksmith's arms do strengthen by virtue of his work at the anvil, and the environmental conditions do alter the appearance of plants and animals. Second, even the most casual observation reveals that offspring tend to resemble their parents. It seemed reasonable, then, to conclude that acquired characters were inherited.

The basis for the variation among organisms that is a condition for natural selection puzzled Darwin, and he never solved the problem. Nevertheless, he was quite aware of some of the problems posed by an inheritance of acquired characteristics. For example, the behavioral and morphological adaptations of worker bees and soldier ants could not have evolved by the inheritance of acquired characteristics because these individuals do not reproduce (Chapter 7).

tonuria (described in Chapter 11) and the gene for childhood diabetes may give heterozygote advantage to the fetus by decreasing the chance of a miscarriage. When heterozygous, the allele that causes cystic fibrosis may confer some protection against water loss caused by diarrhea.

III—WHEN RARE ALLELES OFFER AN ADVANTAGE, SELECTION INCREASES THEIR FREQUENCY DRAMATICALLY

How do rare alleles become established and spread through the population? Suppose a rare allele (A′), perhaps a new mutation, actually confers an advantage. For comparison with previous calculations, we assume that initially $q = 10^{-4}$. For purposes of calculation we have to postulate how selection is acting on each of the three genotypes. Let's suppose first that selection is modest: for example, $s = 0.1$ against AA, 0.05 against $AA′$, and 0 against $A′A′$. Selection then increases the frequency of individuals with the $A′A′$ genotype until it approaches 1 (Fig. 4.13, green curves on the right). At the same time, the AA genotype, originally accounting for 99.9% of the population, is eventually driven close to extinction in 250 generations. Notice how the frequency of heterozygotes rises and then falls. It reaches a maximum when $p = q$ (when the frequencies of the A and $A′$ alleles are equal).

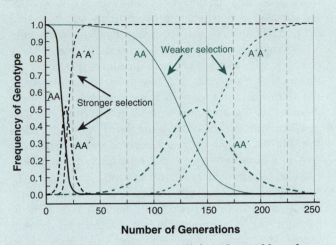

FIGURE 4.13 When a recessive allele is favored by selection, it increases in frequency, even if initially present at very low frequency. The approach to Hardy-Weinberg equilibrium is faster the stronger the selection.

If selection is stronger (s against AA is 0.5 and s against $AA′$ is 0.25), the outcome is the same, but the changes occur in fewer than 40 generations (Fig. 4.13, heavier black curves on the left).

This last pair of calculations contrasts markedly with the very small change in allele frequencies caused by selection against homozygous recessives (Fig. 4.11), and they illustrate the power of natural selection to shift gene frequencies many times faster than mutation.

Although the rules by which genes are expressed from one generation to the next only began to be widely understood in the first decades of the twentieth century, in 1883 August Weismann made a discovery that seemed to preclude the inheritance of acquired characteristics. In early stages of embryonic development of mammals and other vertebrates, those cells that are destined to give rise to eggs and sperm are set aside. They form the *germ line* (or *germ plasm*), a lineage of cells that does not participate in the making of bone or muscle or any of the other *somatic* (body) structures of the animal. They simply wait (in testes or ovary) until the animal reaches sexual maturity, and then they form sperm in males and eggs in females. Weismann reasoned that any changes that took place in somatic cells during the life of the organism were affecting cells on a dead-end road of development and evolution, as only the undifferentiated cells that formed gametes would make a contribution to the next generation.

Weismann's discovery and his reasoning were seen by most biologists as quite persuasive against the inheritance of acquired characteristics in mammals and other vertebrates, but some organisms do not sequester the germ plasm in this manner. For example, many plants form their reproductive organs (flowers) on the ends of growing shoots and refashion them from undifferentiated cells every growing season. Although the idea of an inheritance of acquired characteristics quickly receded in importance in the community of scientists, it lingered into the 1930s and 1940s in the Soviet Union as political doctrine because it suggested the possibility of permanently modifying human behavior in directions dictated by communist ideology. Its effect in that nation was to impede the science of genetics.

If characteristics acquired during the lifetime of an organism could be inherited by its offspring, there would have to be mechanisms for producing environmentally directed changes in the genome. The failure of the misguided theorizing of the Soviet geneticist Lysenko to boost agricultural production provides evidence that such a mechanism does not exist. Much stronger evidence, however, comes from an understanding of the relationship between genes and proteins and from the nature of mutations.

First, although mechanisms exist for reading the nucleotide sequence of RNA back into DNA (this is the trick employed by retroviruses discussed in Chapter 8), there is no known way for a protein to translate its primary structure back into RNA. Thus there is no way for the environment to change the primary structure of a protein in a heritable manner *except* by changing a gene.

Mutations can occur in somatic cells, and these changes will affect the gene products produced by those cells or any cells that arise from them by mitosis, but they will not appear in the next generation of organisms. For a mutation to find its way into the next generation, it must occur in the germ line of a parent. But can mutations of genes be directed to particular environmental ends, as any mechanism of the inheritance of acquired characteristics requires?

Genes would have to change in ways that improve the organism's ability to utilize (if favorable) or resist (if harmful) the environmental agent that caused the mutation. Mutations, however, are random with respect to what is good or bad for organisms, and we have seen why most mutations are harmful. There is no known mechanism by which the effects of changed environments, including agents that increase the rates of mutation—high temperature, ultraviolet radiation, certain chemicals—can cause genes to change in an adaptively advantageous way. Mutations are mistakes of the moment, and whether they are ultimately favorable depends on the subsequent working of natural selection.

SYNOPSIS

The resemblance between parents and offspring began to interest humans long before recorded history: the first archaeological evidence for domesticated animals and plants shows that they had already undergone many generations of selective breeding.

Prior to the work of Mendel in the middle of the nineteenth century, inheritance in sexually reproducing organisms was commonly viewed as a blending of parental traits. Mendel experimented with traits that are inherited as distinct alternatives, which enabled him to discover that heritable traits are determined by stable paired factors, one of which comes from each parent. We now understand these factors to be *alleles* (slightly different copies) of genes that travel with the chromosomes during cell division. They are paired because the cells of each individual carry two complete sets of chromosomes and therefore two copies of each gene.

When one allele is *dominant* over the other, *heterozygous* individuals (carrying two different copies) have the same *phenotype* (appearance) as individuals *homozygous* (carrying identical copies) for the dominant allele. When dominance is incomplete, the phenotypes of heterozygous individuals are intermediate between those of homozygous individuals. Another genetic mechanism for such intermediacy is control of the trait by many genes (*polygenes*) whose individual effects are small and additive.

Variation in phenotypes is found in virtually all natural populations of sexually reproducing organisms. Some of the variation is genetic, caused by the presence of two or more different alleles of the genes that affect a given trait, and individual organisms differ from each other in which of the alleles they have. Some variation, however, is caused by the environment, because different external conditions can affect growth and development differently, even of organisms with the same genotype.

Heritability is a measure of the fraction of phenotypic variation *in a population* caused by genetic differences among individuals. It is an important concept because natural selection can only change a trait if there is genetic variation for the trait in the population. A measure of heritability, however, applies only to the particular array of genotypes and environmental conditions present when the measurement was made.

The *Hardy-Weinberg Law* is at the foundation of population genetics. It shows why the frequencies of two alternative alleles in a population will remain constant in successive generations unless evolution is occurring.

Phenotypic evolution is caused by changes in allele frequencies that can be induced by four different *microevolutionary processes*.

1. When a population of interbreeding organisms is small, random fluctuations in allele frequencies may take place from one generation to the next by *genetic drift*.

2. When organisms or their gametes move between populations, frequencies of alleles can change by *gene flow*.

3. When DNA changes due to copy errors, or when genes are lost or duplicated during the formation of gametes, genes change by *mutation*. Most mutations are detrimental, but depending on the environment, some may enhance an organism's reproductive success. Mutations are the ultimate source of variation on which evolution depends. They are random events. Useful phenotypic features that develop during an organism's lifetime, however, cannot be inherited because they do not alter the organism's genotype.

4. *Natural selection*—the changes in allele frequencies caused by heritable differences in the abilities of organisms to reproduce—is the most powerful process at work in evolution. *Stabilizing selection* keeps phenotypes the same, reflecting the fact that organisms are usually well adapted to their environments. But if conditions change or new genotypes appear, *directional selection* can shift the distribution of phenotypes toward a new average. *Disruptive selection* occurs when individuals with divergent traits leave more descendents than those with intermediate phenotypes.

In contrast to microevolution, which occurs over time spans of a few generations and can be studied by geneticists, *macroevolution* refers to larger phenotypic changes that occur over longer time periods (thousands to millions of generations). Macroevolution reflects the cumulative effects of microevolutionary changes.

Mutation, genetic drift, and gene flow all introduce *chance* into evolution. These random events can influence the life of any individual. The fates of populations or even major groups of organisms, such as dinosaurs and mammals, can be forever altered by random environmental changes, from a particularly severe winter, to continental drift, to an asteroid impact. Natural selection, however, is not a chance process. On average, those individuals that have greater reproductive success are evolutionary winners *because* of the alleles they possess. Natural selection is the physical process that produces the elaborate and wondrous evolutionary outcomes that we know as plants and animals.

QUESTIONS FOR THOUGHT AND DISCUSSION

1. Comment on the following assertion. If 50 blue-eyed people immigrate into a population of 950 brown-eyed people and intermarry with them, blue-eyed people will eventually disappear because there are so many more people with brown eyes. For the purposes of this question, make the simplifying assumption that eye color is controlled by a single genetic locus with two alleles and brown is dominant to blue. In your answer, what additional assumptions are you making about selection?

2. Why is the distinction between genotype and phenotype important?

3. The legal scholar Philip Johnson, author of *Darwin on Trial* and skeptic of evolutionary theory, echoes a nineteenth century criticism when he suggests that the breeding of dogs produces enormous phenotypic variation, but the results are always dogs. He concludes that selection will inevitably exhaust genetic variation before a new species can be formed.

   Analyze the validity of this conclusion, drawing on knowledge of the rates of mutation, selection, and speciation.

4. The heritability of various human traits has been estimated by studying monozygotic twins. (Sometimes called identical twins, these pairs of individuals result from a single fertilization. Early in development the embryonic cells separate into two groups, each of which forms a complete fetus.)

   If you were planning a large study of some specific phenotypic trait, how might you use twins in order to estimate what fraction of the variation in the general population is due to genetic differences among individuals? In other words, in your experimental design, what different pair-wise comparisons among people would you want to make? How would each add validity to your conclusions?

5. Of the several processes that contribute to microevolutionary change, why is natural selection deemed most important? Explain whether each of the other processes is essential to microevolution? Is each inevitable? Why?

SUGGESTIONS FOR FURTHER READING

Hartl, D. and Clark, A. G. (1997). *Principles of Population Genetics*. Sunderland, MA: Sinauer Associates. A comprehensive introduction to modern population genetics.

Mange, E. J. and Mange, A. (1999). *Basic Human Genetics*. Sunderland, MA: Sinaur Associates. An excellent and extensive introduction to basic genetics and human genetics.

5

After Darwin: Molecular Evolution, Selection, and Adaptation

CHAPTER OUTLINE

- *The "modern synthesis:" Bringing natural selection and genetics together*
    - *Genetic variation in natural populations*
    - *The concept of species and the process of speciation*
- *Evolution as seen through the structure of molecules*
- *Levels of selection*
    - *Replicators and vehicles*
    - *Genes or organisms?*
    - *The concept of "selfish DNA"*
    - *Cells in an organism must cooperate*
    - *Group selection*
    - *Asexual reproduction*
    - *Two perspectives on evolutionary change*
    - *Species selection*
- *Adaptation*
    - *Criteria for recognizing an adaptation*
    - *Does adaptation optimize?*
    - *Adaptation and behavior*

- *Adaptation is only one component of evolutionary change*
    - *Adaptation and group selection revisited*
- *Synopsis*
- *Questions for thought and discussion*
- *Suggestions for further reading*

THE "MODERN SYNTHESIS:" BRINGING NATURAL SELECTION AND GENETICS TOGETHER

The rediscovery of Mendel's experiments and the recognition of abrupt and discontinuous heritable changes—mutations—did not immediately clarify how evolution works. In fact, quite the opposite happened. Around 1900, species were still considered by many to be distinct categories within which individual variation was deemed unimportant, so many biologists who were

Photos: Adaptation by natural selection produces (in Darwin's words) "endless forms, most beautiful and most wonderful." *Left*: This walking stick insect (family Pasmatidae) nearly disappears as a twig. The capacity of animals to blend with their surroundings or to mimic other animals as either threat or lure presents a virtually endless array of adaptations. *Right*: The tubular shape of the flowers of *Penstemon barbatus* and the long bill of the black-chinned hummingbird *Archilochus alexandri* are coevolutionary adaptations of plant and pollinator.

convinced of evolutionary change invoked the newly discovered phenomenon of mutation as a mechanism for the origin of new species. They suggested that only one or a few mutational changes could produce individuals of a new species. If new species were originating by such single mutational leaps, natural selection would be unnecessary.

This idea proved to be wrong: it seriously misconstrued both the role of mutation in evolutionary change and the processes by which new species are formed. It is nevertheless ironic that at a time in history when the genetic basis and importance of natural selection were still not fully clear to biologists, Herbert Spencer and his followers invoked natural selection to account for cultural changes in human societies. This serious misstep, known as Social Darwinism, is discussed at greater length in Chapter 14.

It was not until the 1930s and 1940s that the science of genetics—the relation of genes to discrete phenotypic features, the role of mutations, and the rules by which genes are passed to the next generation—was merged with Darwin's idea of natural selection. This conceptual step, referred to as the "modern synthesis," provides the foundation of current evolutionary theory. It is based on a fundamental change in the concept of species from a fixed "type" with unimportant variation to a *population* of potentially interbreeding organisms *characterized* by heritable variation. In other words, heritable differences among individuals came to be viewed not as an inconsequential characteristic of groups of related organisms but as fundamentally important in providing the basis for evolutionary change. In the last chapter we led you from basic Mendelian genetics, through the Hardy-Weinberg equilibrium, to the four processes that underlie microevolutionary change. This was an efficient way to present these ideas, but historically their relationships emerged from the "modern synthesis."

GENETIC VARIATION IN NATURAL POPULATIONS

The basis for heritable variation in natural populations was unknown to Darwin and his contemporaries. Direct evidence for genetic variation began to emerge in the 1930s when geneticists showed that wild populations of the fruit fly *Drosophila* contain many mutations that are lethal when homozygous. The presence in natural populations of genetic variation for a variety of other phenotypic traits therefore seemed likely, but evidence was needed. Confirmation came in a flood starting in the mid-1960s, first by studying specific proteins (Box 5.1), and more recently by the direct study of the genes themselves. These sorts of results have established at the molecular level an essential precondition for natural selection to operate: the existence in natural populations of extensive amounts of heritable variation.

THE CONCEPT OF SPECIES AND THE PROCESS OF SPECIATION

In Darwin's words, the great diversity of plants and animals is the result of "descent with modification." But natural relationships of organisms were apparent even before there was a concept of evolution. The Swedish physician and botanist Carolus Linnaeus (1707–1778) introduced a system of classifying plants and animals into nested categories based on degrees of similarity, and, with modification, his system is used to this day. The major categories are:

> *Phylum*
>> *Class*
>>> *Order*
>>>> *Family*
>>>>> *Genus*
>>>>>> *Species*

For example, all the vertebrates from fish to people are grouped together in the same phylum, mammals comprise a class, the primates an order, the great apes a family, and the two kinds of chimpanzee a genus. Linnaeus's concept of species draws on the commonsense perception that the world is populated with different "kinds" of plants and animals: horses are one thing, zebras are another. But how are species defined, and how do they arise?

The modern synthesis clarified how species of animals can be formed by the operation of microevolutionary processes described in Chapter 4. A lineage of organisms can change with time so that a paleontologist viewing early and later fossils would conclude that the two samples are so different that they should be recognized as different species. No branching of the evolutionary tree is involved, only slow change over time. Alternatively, one species can split into two by the gradual accumulation of many small genetic changes in an isolated subpopulation. In time this population may become so different from the parent population that if the two groups come into contact they no longer interbreed. Any of a number of *isolating mechanisms* can have arisen that then keep the populations genetically separated. For example, eggs and sperm may be genetically incompatible, the animals may breed at different times of the year, or the songs or coloration that serve to attract mates to each other may have become different. Two species then exist whereas there was only one before. Understanding of this evolutionary process has led to the concept of a species as a population of potentially interbreeding organisms.

Box 5.1
Looking for Heritable Variation in Proteins

One approach to assessing heritable variation in natural populations has been to look at the immediate gene products, proteins. A variety of soluble proteins with enzymatic activity can be readily identified. Recall from Chapter 3 that soluble proteins have ionic charges on the surfaces of the molecule that they present to the surrounding water. Moreover, some of the many possible point mutations that can occur lead to amino acid substitutions that change the charge on the molecule. Many amino acid substitutions do not affect the overall ionic charge of the protein, but those that do can be detected by a simple and efficient assay. If a few drops of a solution of a soluble protein are placed on a mat of moist, porous material such as wet filter paper or a gel of starch and a voltage applied from one end of the gel to the other, the protein molecules will begin to move. The speed of movement depends on the net charge on the molecule. Figure 5.1 shows an example of such an experiment.

Remember that this technique can detect genetic variation of a restricted kind (a subset of possible amino acid substitutions) in a limited class of genes (those coding for identifiable, soluble proteins). Consequently it provides only a sample of possible genetic variation. In separate studies of humans and the fruit fly *Drosophila*, 30% of the loci were found to be polymorphic, i.e., with more than one allele present in the population. In a population of the fruit fly *Drosophila*, there can be 2–6 different alleles at a locus. Similar studies have now been done on many different organisms, and equivalent polymorphism is typically observed. In consequence, 3–15% of an organism's loci are heterozygous, and for any two organisms in the same population, the percentage of their loci that have different alleles is even greater.

The recent development of rapid techniques for isolating individual genes and for sequencing the DNA bases has made it potentially possible to measure all ge-

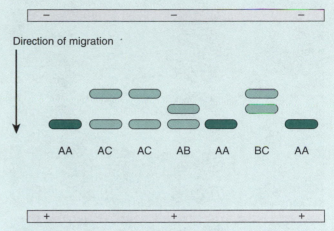

Direction of migration

AA   AC   AC   AB   AA   BC   AA

FIGURE 5.1 The migration of soluble proteins in an electric field depends on the numbers of ionic charges on exposed side groups. In this experiment, samples of a specific enzyme taken from seven individual seaside mussels (*Mytilus edulis*) were placed on a starch gel substrate and caused to migrate in an electric field. After a time the voltage was removed and the proteins were stained with a dye so that their locations on the gel were visible. This protein exhibits *polymorphism* (*poly*, many; *morph*, forms). That is, different individual animals from the same population have different variants of the molecule. The presence of the several variants is revealed by small differences in ionic charge, but whether or not the several forms of the molecule have functional differences is not addressed by this kind of measurement.

The interpretation of this experiment in terms of genetic differences is quite straightforward. There are three alleles in the population, referred to as *A*, *B*, and *C*. Three of the animals were homozygous for *A*, two were heterozygous for *A* and *C*, one was heterozygous for *A* and *B*, and one for *B* and *C*. Notice that the single spot of protein from the homozygous individuals stains darker than the other spots. That is because it contains twice as much protein as the other spots.

netic variation. Studies of variation in base sequences of the same genes from many individuals leads to the same conclusion as analysis of the amino acid sequence of their protein products: *natural populations of plants and animals have ample variation for natural selection to work.*

Bullocks oriole (*Icterus bullockii*, found in the western U.S.) and the Baltimore oriole (*I. galbula*, in the east) differ in a number of features, most conspicuously in plumage and song (Fig. 5.2). About thirty years ago hybrids were found in the Great Plains in a region where human modification of habitat had allowed the two species to come into contact. For a few years it therefore appeared that their separation had not proceeded far enough that they should be considered distinct species, and taxonomists lumped them together with the common name of Northern oriole. Further study, however, showed that the occasional hybrids

FIGURE 5.2 Baltimore (*below*) and Bullock's orioles. Hybrids occur, but they are confined to a narrow geographic zone in the Midwest where Bullock's and Baltimore orioles have recently come into contact. Studies show that there is no significant "gene flow" between the two species.

were confined to a narrow geographic region, and consequently there was negligible mixing of genes between the parent populations. The two populations are again considered to be distinct species.

This account of the relationship between two newly formed species illustrates the dynamic nature of speciation. Geographic separation (a result of glaciation and the formation of the Great Planes) had allowed the two populations to evolve into distinct forms that did not interbreed because they did not come in contact. When habitat change subsequently brought them into contact, biologists had to study the fate of hybrids in order to establish that there was little if any gene flow. Had the Bullocks and Baltimore forms been separated for a longer time, it is likely that they would now not hybridize at all. Conversely, if the populations had come back into contact sooner, the hybrids might be more numerous, gene flow greater, and the two populations would today be classified as geographical variants of a single species (i.e., subspecies).

EVOLUTION AS SEEN THROUGH THE STRUCTURE OF MOLECULES

In 1953 James Watson and Francis Crick discovered the structure of DNA, a finding that revolutionized the science of biology, opening new vistas, not only on how genes function, but on the history of evolutionary change and the evolutionary process itself. All of the knowledge of DNA, its relation to protein synthesis, and the structure of the gene that we summarized in Chapter 3 has been achieved in the last fifty years, and it is impossible to overstate the importance of these findings for the advance of biology and medicine. The information and the techniques of "molecular biology" (as this field is now called) are growing rapidly and are contributing daily to an enhanced understanding of evolution. We will refer later to some of these methods in specific contexts, but at this point we want to draw attention to the new line of evidence for evolution that molecular biology has provided: evolutionary tracks left in the structures of proteins and their genes. These discoveries could not have been anticipated in the nineteenth century, before there was knowledge of either genes or protein molecules.

We saw in Chapter 3 that some mutations can have inconsequential effects on the function of the protein for which the gene codes. Such *neutral mutations* will accumulate over long periods of time (measured in millions of years). Now consider a protein such as hemoglobin that occurs in many kinds of animals from fish to people. The evolutionary separations of fish, amphibians, reptiles, birds, and mammals, as well as the subgroups of each of these classes of vertebrates, took place at various times in evolutionary history, so neutral mutations have been accumulating in the hemoglobin gene independently in each of these lineages of vertebrates for various periods of time. It stands to reason that the more ancient the separation of two lineages of organism, the larger the number of neutral mutations that will have accumulated in their hemoglobin genes, and the larger the number of inconsequential differences there will be in the amino acid sequences (i.e., the *primary structures*) of their hemoglobins. If we know from the fossil record that two lineages separated about 100 million years ago, and we count n amino acid differences in their hemoglobins, we can conclude that the average rate at which neutral mutations are accumulating in the hemoglobin gene is $n/200$ million years. We double the time because those differences have been accumulating independently in the two lineages for 100 million years. And of course if we had looked at differences in the nucleotide bases in the hemoglobin gene, we would also have seen some *silent mutations* that do not lead to amino acid substitutions in the protein (Chapter 3).

Figure 5.3 shows a plot of the changes in the primary structure of three proteins—hemoglobin, fibrinogen, and cytochrome *c*—as a function of time, based on the kind of measurement that we described in the preceding paragraph. The most noteworthy feature of these plots is that they are straight lines. This means that the *rate* of evolutionary change in each of these proteins has been constant. One might expect that if mutations are random events, their average rate of occurrence would be constant for all proteins, but the slopes of the three lines in Figure 5.3 are very different; the three proteins have been changing at very different rates. Each protein is a molecular clock, but the clocks are ticking at three different rates.

Why do these rates differ? The explanation may have already occurred to you. A mutation that is neutral in one protein may have severe consequences in another. The rate at which amino acid changes accumulate thus depends not only on the rate at which mutations occur but also on whether natural selection then retains or eliminates the new variants. Since the rate of mutation is roughly the same in genes for different proteins, the three slopes in Figure 5.3 must reflect different effects of selection. But why is natural selection a more severe censor of amino acid changes in some proteins than in others? The answer to that question will become clear if you think about the different functions that proteins must perform.

Consider the three examples in Figure 5.3. Cytochrome *c* is a team player. It must fit into the mito-

chondrial membrane and interact in very specific ways with other membrane proteins in electron transport and in the synthesis of energy-rich molecules (Chapter 3). The structure and function of cytochrome *c* are so finely adjusted that most of the amino acid changes resulting from mutations will disrupt the function of the protein and consequently impair the organism's chances of survival and reproduction. In other words, natural selection eliminates most of the mutations that occur.

Fibrinogen represents the other extreme of constraints on protein structure and function. This polypeptide participates in blood clotting, and all it has to do is precipitate from solution at an open wound. This relatively simple function is compatible with many amino acid changes, and consequently during its evolution natural selection has been less of a censor.

Hemoglobin is intermediate in its functional and structural constraints. This oxygen-transporting protein in red blood cells consists of four subunits—two sets of two peptide chains that must bind to each other in specific ways in order to capture oxygen in the lungs and release it in the tissues. Unlike cytochrome *c*, hemoglobin works alone and does not have to interact with membranes and other proteins. Nevertheless, there are more requirements on its shape and chemical properties than there are for fibrinogen. Hemoglobin is therefore intermediate between cytochrome *c* and fibrinogen in the rate at which amino acid changes have accumulated.

We first introduced protein evolution in Chapter 3 (Fig. 3.26) when we described the evolutionary opportunities that are presented when a gene is duplicated by an unequal crossover during meiosis. We saw how evolutionary divergence of different copies of the same gene gave rise to myoglobin and the several polypeptide subunits of hemoglobin. Every vertebrate animal has thereby come to possess a family of related globin genes coding for different polypeptides with somewhat different oxygen-binding functions. These evolutionary relationships are therefore seen in proteins from individuals belonging to the same species.

Homologous proteins from individuals belonging to *different species* also reveal evolutionary history. If we examine the amino acid sequence in cytochrome *c* from a number of different species, we can draw a phylogenetic tree for this molecule based on the number of amino acid differences that are found in organisms living today (Fig. 5.4). The rationale for drawing such a tree is that the more similar in amino acid sequence two variants of the protein are, the more closely they are related to some common ancestral form. And the more differences there are, the longer it has been since they had a common ancestral gene. Note, however, that the tree is drawn without reference to an absolute measure of time; the lengths of the branches between

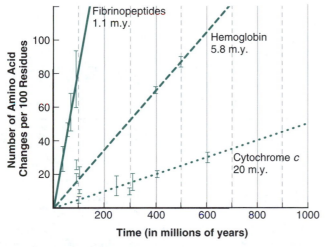

FIGURE 5.3 Evolutionary change—i.e., the accumulation of amino acid differences—occurs at different rates in different proteins. Mutational rates are probably similar in the genes for these three proteins, but as explained in the text, the mutational changes are subject to different stringencies of selection in each of these molecules. The numbers by each name are the times in millions of years (m.y.) required for a 1 percent change in amino acid composition.

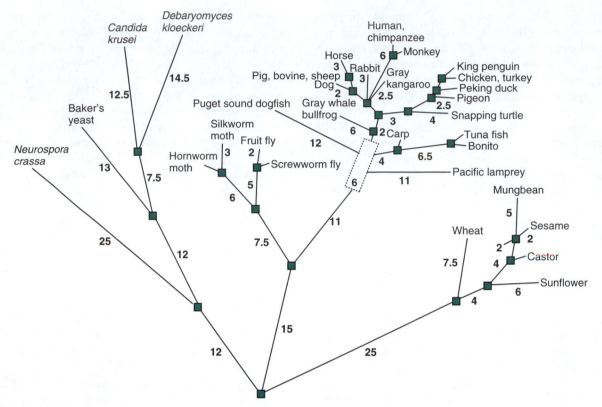

FIGURE 5.4 A phylogenetic tree of one protein, cytochrome *c*, based on amino acid differences. Protein was sampled from different organisms, and the structure of the tree is based on the reasonable assumption that the most likely evolutionary changes have involved the minimum number of amino acid substitutions. Nodes in the tree represent likely ancestral forms of cytochrome *c*. There is a close correspondence between the shape of this tree and the phylogenetic tree for these living organisms which has been inferred from the fossil record and comparative anatomy.

branch points are based only on the number of amino acid differences between the cytochrome *c* from organisms living now. But if you look at the labeling of the ends of the branches you will see that the tree comes very close to describing the evolutionary relationships—known from anatomical and paleontological evidence—of the *species* from which the cytochrome *c* came. More careful examination of the tree will reveal a few misplaced limbs, but this is not surprising: no *single* protein should be expected to track precisely the evolutionary history of the parent organisms with their thousands of proteins.

What makes this kind of comparison so interesting is that it provides a reasonably accurate record of the evolutionary history of life preserved in the structures of individual proteins and the genes that code for them. This is a convincing confirmation of evolution that is completely independent of all of the evidence that Darwin and many others have accumulated by studying the phenotypes of whole organisms, living and fossil. Both the existence of Darwinian "descent with modification"

and understanding of the processes of evolutionary change have been greatly augmented by the introduction of molecular techniques.

LEVELS OF SELECTION

What does natural selection select? This may seem like a trivial question, but it is not. At various times people have proposed that selection takes place at different levels of organization, such as genes, cells, individual organisms, groups of organisms, and species. Not all of these ideas are equally useful, and there are some semantic issues involved, so the question has both subtlety and substance.

Recall that evolution by natural selection is a consequence of three properties: (*i*) individual entities are copied, using energy and materials from their environment. Genes fit this description, but perhaps other things do too. Let's reserve judgment for the moment. (*ii*) Variants of these heritable entities occasionally occur. And *(iii)* some variants are consistently more successful than others in achieving reproductive success. To complete the Darwinian argument, the entities that reproduce more than others will increase in frequency over time. Stated in this general way, evolution automatically occurs among entities that have these three properties. Wherever entities with such properties exist, evolution by natural selection is therefore inevitable. Given the immensity of the universe and the now-known existence of other solar systems (based on the detection of stars that have orbiting planets), it seems likely that evolution has occurred elsewhere. But direct knowledge of evolution is of course currently limited to what we can observe on earth.

Life evolves, but what features of life exhibit these three requisite properties? In other words, which entities in the hierarchy of complexity—from molecules to systems of many organisms—replicate (and thus persist through time), exhibit variation, and are selected? Are they genes, genotypes, cells, organisms, groups of organisms, species, or some combination? Do all of these categories display all three properties? And which of these entities can be said to evolve? To explore answers to these questions, we will start with what is now familiar—genes and sexually reproducing organisms like ourselves—and build on the ideas discussed in the previous chapters.

REPLICATORS AND VEHICLES

The Oxford biologist Richard Dawkins has introduced the words *replicator* and *vehicle* to capture an important distinction between genes and the individual, sexually reproducing organisms that contain unique combinations of genes. *Genes are replicators.* They make (usually) exact copies of themselves by the process of base-pairing and DNA synthesis described in Chapter 3. To be more precise, they are replicated by the cells in which they occur. As replicators they propagate through successive generations. Any gene of yours was present in either your mother or your father and has a 50% chance of appearing in any one of your children. The process of replication is very accurate, so genes usually retain their identity across generations. There is a very small probability, however, that a gene may change (mutate) to a new allele before it is transmitted to one of your children. The slow and steady introduction of new alleles through the process of mutation and the winnowing of alternative alleles by natural selection cause populations of genes to change their composition through spans of time measured in many generations. *Populations of genes therefore evolve.*

Now consider yourself, a person, a *vehicle* for genes. Your genotype—the particular ensemble of alleles in your genome—is unique because half of your genes came from your mother and half from your

father. Furthermore, you and your siblings are each genetically unique because each of you received a different combination of your parent's alleles. (Identical twins are an exception: they have the same genotype.) Similarly, when you and your mate reproduce, each of your children will have different combinations of alleles (recall the section on meiosis in Chapter 3). Moreover, crossing-over during meiosis changes the array of alleles on individual chromosomes, so chromosomes, like entire genotypes, are not passed with fidelity from one generation to the next. Individual, sexually reproducing organisms and their unique complements of genes therefore do not replicate. Consequently, neither your chromosomes, genotype, nor phenotype have continuity from one generation to the next as your individual genes do. Replication of genes is therefore basically different from sexual reproduction of individual organisms. Thus, *you* (neither genotype nor phenotype) cannot evolve through time, because you did not exist in a previous generation and you will not exist in the next. But *lineages of organisms clearly evolve.* Evolutionary change is therefore not confined to replicators. Let's see why.

GENES OR ORGANISMS?

The *process* by which the replicators (genes) are sifted through the generations is based upon the reproductive success of their vehicles (organisms). The genes depend totally upon the individuals whose phenotypes they help to specify during development (Chapter 10). Some individuals leave more offspring than others do, so the alleles of their genes will be more frequent in the next generation. Note, however, that the fates of all the genes traveling together in a particular organism (its genome) are bound together, and whether or not a particular allele reaches the next generation depends not only on what it does but also upon what all the alleles of its companion genes do. Of course, copies of an organism's genes are present in other individuals, so if one individual organism fails to reproduce, that failure dooms only those copies of genes that it possesses. Populations of organisms reflect their genes. Because populations of genes evolve, lineages of organisms also evolve.

Another concept that may help you to think about this issue was introduced by the evolutionary biologist George Williams. He suggested that one can think of natural selection as occurring in domains of *information* and of *matter*. This may seem unnecessarily abstract, but the distinction addresses the issue we have been exploring. Genes are information. Genes, you may suggest, are also matter, but in this context that is incidental. The physical form of information can be changed without changing its content. For example, the printed information in this book also exists as magnetized particles on computer discs, and the information in the

Box 5.2
Extra-nuclear DNA

At first encounter it may seem that there is an enormous conceptual leap from the idea of functional genes to that of "parasitic" DNA residing within genomes. The gulf, however, is not so wide as it may first appear. Mitochondria and chloroplasts contain some functional (coding) DNA, and we saw in Chapter 3 that this indicates that these cellular organelles originated early in the evolution of cells as prokaryotic invaders. How these associations started, whether through ingestion, as symbiotic associations, or through some unknown process is not clear. At present, however, these or-

ganelles are so completely integrated into the cell cycle that their evolutionary origins were well concealed until their DNA was discovered. Moreover, the small amount of DNA that is present in these organelles codes for some of their proteins and is therefore essential to the life of the cell.

The small proportion of the DNA that is present in mitochondria is inherited from the mother (in the cytoplasm of the egg), and because the mechanisms for proofreading during replication are not as stringent as in the nucleus, it evolves faster than much of the rest of the genome. For these reasons it has been used extensively for evolutionary studies, including the origin of humans (Chapter 12).

---

base sequences of genes is also present in the amino acid sequences of the proteins they encode. In the domain of *information*, alternative forms of genes (alleles) are selected, and consequently evolution is a slow change in the information residing in genes.

This is a useful concept because the mechanism by which genetic information changes over time involves the differential reproductive success of those *material entities*, the individual organisms (vehicles) in whose bodies (phenotypes) that genetic information is expressed. This is what is meant by the statement that natural selection takes place "at the level of the individual," a phrase that draws attention to the *process* that leads to differential reproductive success. To say that natural selection takes place "at the level of the gene," on the other hand, is to draw attention to the *consequences* of selection: changes in the frequencies of different alleles over time.

To summarize to this point, genes replicate. The presence in a population of different alleles provides genetic variation, and as mutations occur and frequencies of the various alleles change with time, genes evolve. Individual organisms, on the other hand, do not replicate. They do, however, display different degrees of reproductive success that depend on differences in heritable characteristics (i.e., the presence of different genes), and this is the mechanism by which the underlying genes are selected. Populations of organisms are thus molded and shaped by natural selection, and a lineage of organisms, like the genes it contains, also evolves with time.

THE CONCEPT OF "SELFISH DNA"

The chicken is the egg's way of making another egg. This old saying unwittingly anticipated that organisms are DNA's way of making more of itself, a perspective

that provides an interesting view of evolution. The propensity of DNA for replication has led to the evolutionary emergence of vehicles (organisms) with a multitude of capacities for mobilizing energy from the environment and ensuring genetic continuity through time. Through this process like begets like with such focus and capacity that DNA has been referred to metaphorically as "selfish."

But we saw in Chapter 3 that only a small fraction of the DNA in the nuclei of eukaryotic organisms codes for proteins. Vast stretches of the double helices of DNA, regions known as *introns*, seem to have no clear function in directing the development and sustaining the life of the organism. There is other seemingly nonfunctional DNA, not embedded in coding regions, much of it present in multiple copies. There are also quasi-autonomous pieces of DNA known as *transposable elements* that can move about in the genome and influence the expression of genes. One hypothesis to account for the presence of such DNA is that it is simply along for the ride, hitchhiking on the evolutionary road. DNA that does not have a gene product could nevertheless be acted on directly by natural selection.

The concept of "selfish DNA" thus not only expresses a proclivity of DNA to thrive in environments (living cells) where replication is possible, but for some DNA to propagate even though it makes no contribution to the life of the host organism (Box 5.2). In this view, the long-term persistence of both the extra DNA and the host depends on the former not preventing the survival and reproduction of the latter. In other words, the selfishness of all hitchhiking DNA must be constrained (by natural selection) to the point that it does not interfere with natural selection at the level of the host organism.

CELLS IN AN ORGANISM MUST COOPERATE

Cells are the building blocks of organisms, intermediate in complexity between genes and entire plants and animals. But are cells the objects of natural selection? In many sexually reproducing organisms the cell lines destined to become eggs and sperm—the *germ cells*—are set aside early in development and the body is constructed of other cells—*somatic cells*—with no reproductive prospects. Virtually all cells in the body of a mammal (exceptions below) have exactly the same complement of genes, as all arise by mitotic divisions of the fertilized egg. The genes in the somatic cells must achieve their long-term evolutionary success via identical copies residing in eggs and sperm. Consequently the somatic cells are constrained to cooperate in creating a functional and reproductively successful organism, and there is no differential selection of somatic cell lines that would interfere with this process.

There are two notable exceptions, one pathological and the other normal and useful. Sometimes mutations in somatic cells cause the daughter cells to divide more frequently than they should, producing cancerous tumors. If cancer cells detach from the tumor and migrate throughout the body, new tumors form (a process called *metastasis*), and unless stopped, the errant DNA of malignant tumors usually causes death. The second exception involves clones of antibody-producing cells in the immune system, but we will hold that longer story until we discuss the evolution of disease in Chapter 8.

GROUP SELECTION

Natural selection is not an autonomous force of nature that can "do" something at different levels of biological organization. It is simply a physical process of sorting over time that occurs automatically given the right conditions. We have seen that in sexually reproducing organisms, natural selection occurs among individuals and brings about changes in the frequencies of genes in populations. Is it also useful to think about natural selection working at higher levels of biological organization, between groups of organisms, or species, or even larger taxonomic groupings? This issue is beset with confusion, largely because it can be approached from several different perspectives.

In microevolution, a focus on individual organisms is pivotal to understanding the process of natural selection. Simply put, organisms behave in ways that maximize their own reproduction and do not try to increase that of others if it imposes an increased cost on their own. Organisms do not sacrifice their own prospects or otherwise behave "for the good of the species."

The idea that individuals might curtail their reproduction in the interests of the larger group is referred to as the "group selection fallacy." A common form of this idea is that when some resource limits reproduction—food, for example—individuals are presumed to forego their own reproduction in order to limit their group's size, thus avoiding overexploitation of the resource and promoting the long-term reproductive success of all.

The killing of young of the same species (infanticide) by male langur monkeys provides an example. Langur monkeys form troops of a dozen or more adult females, their young, and a single dominant adult male (Fig. 5.5). Other "bachelor" males from all-male troops will try to displace the dominant male of the troop, and when a newcomer is successful he frequently attempts to kill the young monkeys then in the troop and infants that are born during the first several months after his takeover.

One early interpretation was that the male engages in this behavior to regulate the size of the troop. This is not plausible, because the females typically resist efforts to kill their offspring, and it is not apparent why the population should need regulation only at the time of a takeover and why only the new male should have the insight to see how the group would benefit from being smaller. A second interpretation of this infanticide was that it is pathological, but that reflects how unpleasant the behavior seems to human observers rather than an analysis of its evolutionary implications. Pathological individuals who kill young indiscriminately should have lowered reproductive success, but these males do not kill indiscriminately, and the infanticide increases their reproductive success. In other words, infanticide serves the genetic interests of the new male. By killing the youngsters sired by his predecessor he shortens the time before lactating females will come into estrus and be reproductively receptive to him. In fact, this sort of male behavior is observed in a number of other species and a variant is known in humans (Chapters 13 and 14).

As Robert Trivers has pointed out, hypotheses that individuals are genetically inclined to behave for the

FIGURE 5.5 A female langur monkey with her young.

good of the group usually suffer from a logical flaw that can be exposed with an appropriate question. For example, which male langur is likely to have the greater lifetime reproductive success: a male who kills the young fathered by his predecessor, or one who waits and runs the risk of being displaced by another male or killed by a predator before he is able to inseminate females?

This reasoning may seem too facile. Although it seems to fit langurs, people and even nonhuman animals do not always behave in selfish ways. That is right, and the next chapter will explore the various circumstances in which one creature might help another. To anticipate the general point that is relevant here, natural selection will not favor behaviors that incur reproductive costs to an individual unless those costs are likely to be compensated by equal or greater return benefits to the individual's reproduction.

The term "group selection" also has a larger connotation—the differential reproductive success of one group relative to others. For this to be closely analogous to natural selection of individuals, *logically the two groups should differ in some heritable way whose phenotypic expression is the basis for differential reproductive success.* Groups can reproduce by budding or splitting, but the idea of group selection ordinarily means that the individuals that comprise one group have greater reproductive success than the members of another group do. Thus one group becomes evolutionarily more successful than the other. Greater reproductive success of one group could be based on its more effective utilization of resources, and in principle, competition between groups might or might not involve them in direct physical conflict.

If the groups are to retain their genetic identities, however, there cannot be significant gene flow (migration) from one to another. For between-group selection to be an important force in evolution, the disappearance and formation of groups must occur at a rate comparable to the rates at which individuals reproduce. Otherwise selection among individuals will change the distribution of genes within groups faster than new groups can form and old groups can be driven to extinction by a process of selection among groups.

The problem with the hypothesis that selection takes place at the level of groups is that the conditions under which it can occur are infrequently found in nature. In the realm of microevolution, entire interbreeding populations of organisms are seldom entirely eradicated or accurately reproduced as wholes (but see the following section on species selection), and emigration and immigration have the same mixing effects on groups that crossing over and independent segregation of chromosomes have on chromosomes and genotypes. Furthermore, even if groups are isolated and do retain their unique gene frequencies, they are likely to be stable much longer than the organisms' generation time. Under these conditions, selection among individuals will occur more frequently than selection among groups.

Humans are very efficient in killing other creatures, including other groups of humans (Chapter 14). Does this lead to effective group selection? First, only rarely are all the members of a group annihilated. More importantly, although human conflict can bring about local and modest changes in the human gene pool, these outcomes are not the result of heritable differences between groups. Access to resources, an advantage in technology or leadership, or different numbers of combatants can be causal agents, but none of these reflects group differences in genes. The vast majority of human genetic diversity occurs throughout our species and is not associated with particular groups (Chapter 12). That the culture of one group can thrive at the expense of another is a different issue, and it is confusing to refer to it as group selection.

ASEXUAL REPRODUCTION

There is one rather special circumstance in which selection occurs among groups because all the members of a group are genetically identical. Some plants and animals can reproduce asexually: their cells do not undergo meiosis and form gametes. Their eggs are therefore diploid, do not require fertilization, and develop into offspring that are genetically identical to the parent. A group of genetically identical progeny is called a *clone*. Clones arising from different parents, however, may be genetically different, so selection can occur among clones. The situation can be further complicated in species (for example, certain lizards) in which intervals of sexual and asexual reproduction alternate.

Clones would seem to be efficient reproducers, as it is unnecessary to put any resources into males. In the presence of sufficient resources, individuals who can rear two offspring asexually will see their clone double in size in each generation. A female who produces two offspring by sexual reproduction, however, is just replacing herself and her mate.

There are other costs of sexual reproduction besides this twofold decrement in the potential rate of reproduction: advantageous combinations of genes are broken up by recombination, and (Chapter 6) in many species a great deal of energy that could be invested in producing offspring is used to make antlers, horns, tusks, large canines, and conspicuous displays and use them in competition and courtship, some of which attract the attention of predators as well. If sexual reproduction bears such costs, why is it much more common than asexual reproduction?

A number of reasons have been proposed, but we will mention only a couple. In asexual reproduction mutations will be passed to successive generations, and harmful mutations can therefore accumulate in a clone. Recombination, however, provides a way of generating gametes that have eliminated harmful mutations. Furthermore, among a large number of genetic recombinants, a few may actually offer an advantage. Recombination is particularly important because the most unpredictable aspect of the environment is presented by parasites and prokaryotic infectious agents that have short generation times and are capable of rapid evolution (Chapter 8). Sexual reproduction, with its attendant genetic recombination, offers the possibility of much more rapid evolutionary response than is possible with asexual reproduction.

TWO PERSPECTIVES ON EVOLUTIONARY CHANGE

The accumulation of genetic differences that occurs in microevolution can lead to macroevolutionary changes over tens and hundreds of millions of years. The evolution of humans from a common ancestor shared with apes (Chapter 12), the evolution of primates from an earlier mammalian ancestor, and the evolution of mammals from reptiles all involved speciation events summed over vast periods of time. The process can be likened to the growth of a wide and bushy tree. Twigs (species) grow at the tips of branches, and as more of them sprout, the branches thicken (genera, families). But most of the twigs do not survive. Many twigs, and even entire limbs, are lopped off by extinctions.

The fossil record gives a view of what the bush looked like at various times in the past. As discussed in Chapter 2, however, it is not a complete record, partly because only some kinds of organisms are able to fossilize, and partly because many rock strata become modified or destroyed by erosion, volcanic activity, or other geological processes. Furthermore, the temporal record laid down in rocks is not a clock with hour, minute, and second hands. For a paleontologist to be able to resolve differences in time of less than hundreds of thousands of years is unusual.

Phenotypes do not evolve at constant rates. For example, the mammalian lineage has undergone a substantial evolutionary radiation since the extinction of dinosaurs about 65 million years ago. Other groups of organisms have changed but little during the same span of time. For example, dragonflies, those large, helicopter-like insects that cruise along the shores of ponds and streams during the summer, have easily recognizable forebears as fossils from a time scores of millions of years before the earliest mammals (Fig. 5.6).

(A)

(B)

FIGURE 5.6 **Some lineages of organisms have changed relatively little over long periods of time. (A) These spawning horseshoe crabs (*Limulus polyphemus*) are quite similar to fossils dating from the Ordovician period (505-440 mya). (B) Relatives of contemporary dragonflies are known as fossils dating from the Carboniferous period (320-280 mya).**

One of the common findings in the fossil record is that a lineage of organisms frequently remains unchanged for millions of years, then a new form appears abruptly. This pattern of evolution is referred to as *punctuated equilibrium*. It is important to realize, however, that what appears as an abrupt change in the fossil record can occupy a much longer time than natural selection requires in order to generate visible phenotypic change. As a dramatic example, in response to extremes in annual rainfall and thus the relative availability of different kinds of seeds, the bills of one of the species of Galapagos Island finches have been observed to change in size by about 15% in just a few generations (see also Fig. 4.10). Even with much more conservative values for the rate of phenotypic change produced by natural selection, new forms arising by gradual, microevolutionary change would appear "abruptly" in the fossil record. Most intermediate forms simply would not be resolved in the record of the rocks. The discovery of intermediate forms is even more unlikely when speciation occurs in relatively small, isolated populations, which then spread rapidly.

Thus geneticists, studying the geographic distributions and short-term changes of gene frequencies in populations of living organisms, provide one window of understanding on biological evolution. Paleontologists, studying the fossil record, peer through another window. What they see is a much coarser record of distributions of organisms and long-term changes in the major kinds of phenotypes. The first approach yields information about the microevolutionary processes that lead to organic change; the second approach tells us about the appearance of macroevolutionary changes and the sweep of the historical record. As is frequently the case in science, different perspectives summon different language to characterize the observations and to formulate hypotheses about the underlying processes. With this background we can now pose the question Are species objects of selection?

SPECIES SELECTION

The replacement or succession of one recognizable phenotype by another as seen in the fossil record suggests to some that species can be units of selection. In the informational domain of genes, populations of genes are indeed replaced by other populations of genes as entire lineages expand, ebb, and become extinct. In that sense, species are selected. Logically, however, this differential success of groups of genes is not limited to species. It involves gene pools within both species and taxonomic groups larger than species. It is another question, though, whether the mechanisms involve anything more than the cumulative effect of natural selection on the reproductive success of individual organisms, often accelerated by climatic and other environmental change.

By analogy with the idea of group selection, one might ask whether species exhibit any collective features that are more than the aggregate traits of their individual members. One species might then be more successful than another by virtue of such a feature. Genetic variability is an obvious possibility. Genetic variability is not a characteristic of an individual organism; it is a property of a group of organisms—a population, a species, or some other grouping.

Imagine two similar species, one of which (A) exhibits greater genetic variation whereas the other (B) has less genetic variation, is phenotypically more specialized, and is possibly better able to deal effectively with the current environment. Now suppose that the environment changes in some manner that puts members of species B at a disadvantage. In time species A expands its numbers and species B declines, perhaps going extinct. What has happened?

This is species selection in the sense of describing an *outcome*. In other words, the genes of species A have increased their numbers whereas those of species B have declined. Furthermore, this evolutionary change occurred because of the greater genetic variability of species A. But the *process* by which that occurred involved the greater reproductive success of a subset of individuals of species A that were able to cope more efficiently with their new environment(s). Here again it is useful to distinguish between the results of natural selection and the detailed process by which the sifting occurs. In this example the process may have involved nothing more than natural selection working at the level of individuals. What happened at the level of species—one persisted and the other became extinct—was therefore not due to competitive interactions between the species as units, but was caused by the summed effects on individuals and their descendants.

ADAPTATION

Change that occurs by natural selection and leads to improvement in reproductive success is *adaptation*. The word adaptation can refer either to the *process* of change or to an *outcome*, a feature or state in which the organisms seem to be "matched" to their environment. For example, organisms that inhabit deserts have adaptations for conserving water. Cactus plants have a compact structure lacking broad, flat leaves with much surface area through which water would be readily lost by transpiration (Fig. 5.7). And kangaroo rats have very efficient kidneys that greatly reduce their need to drink water as well as powerful hind legs that enable them to move rapidly in an open environment with little cover as they escape from predators. These adaptations permit these organisms to live in a demanding environment that would be impossible for an oak tree or a forest mouse to exploit.

(A)

(B)

FIGURE 5.7 (A) This cactus *(Melocactus bahiensis?)* has several adaptations for life in very dry habitats. The compact shape allows the plant to store water, and the sharp spines provide protection from browsing animals in search of either food or water. The absence of thin leaves prevents water loss, and the bright desert sunlight supports sufficient photosynthesis in the greatly reduced surface area of chlorophyll-containing tissue. **(B)** The rock ptarmigan *(Lagopus mutus)* is a bird of the north. During the Arctic winter it trades its mottled brown feathers for white and blends with the snow.

The *mechanism* by which evolutionary adaptation occurs is natural selection. Features of cactus plants and kangaroo rats that enable them to conserve water therefore emerged over time as ancestors of these present-day organisms came to inhabit, i.e., became adapted to, deserts. But the word adaptation has other uses as well (Box 5.3).

Plants and animals often appear to be exquisitely crafted; indeed it was this seeming perfection that fathered the pre-Darwinian explanation for biological diversity as the handiwork of an omniscient deity. This feature of adaptations—finely tuned adjustments between organisms and their environments—provides impressive testimony to the power of natural selection as a formative process. But all useful features of organisms need not be adaptations. To pick a famous example to which Darwin himself drew attention, the seams between the bones of the mammalian skull allow a flexibility that enables the large heads of newborns to pass through the pelvic region of the mother. But this is not an adaptation for this function, because birds and reptiles also have skulls that are constructed in parts in much the same fashion. As flexibility of the skull is not necessary in order to hatch from an egg, this morphological feature cannot be an adaptation of mammals for internal gestation and birth (Fig. 5.8). On the other hand, the size and flexibility of the human birth canal have likely adapted as the human brain enlarged.

CRITERIA FOR RECOGNIZING AN ADAPTATION

How do we decide whether a useful feature of an organism is in fact an adaptation? In other words, how do we determine that it arose by natural selection for the function that it seems to serve? Evidence for the involvement of natural selection is critical in this decision, but because adaptation is a historical process, the argument must often be indirect. Evolutionary biologists who have thought about the problem have suggested several useful criteria. Sometimes *homology of structure* makes the history clear. For example, that the flippers of seals are adaptations for swimming and the wings of birds and bats adaptations for flying is clear from their obvious homology with the forelimbs of other vertebrates and the uses to which they are put (Fig. 2.20). Each structure is a modification of a basic plan for a particular use, and natural selection is the only plausible mechanism for how these modifications came into existence.

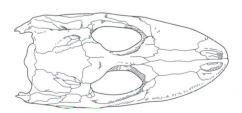

Alligator

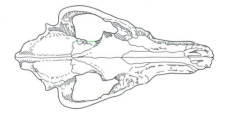

Dog

FIGURE 5.8 The skulls of both reptiles (e.g., an alligator) and mammals (a dog) consist of many bones. The sutures of the mammalian skull are thus not an adaptation to facilitate birth.

BOX 5.3
Physiological Adaptation

Adaptation by natural selection is an evolutionary process that brings about gradual change in a lineage of organisms. The word adaptation is also used in biology to describe physiological changes that are experienced by *individual* organisms over much shorter spans of time and which do not involve changes in the frequencies of genes. For example, people who go from home at sea level to visit much higher elevations in the mountains may find that moderate exertion leaves

them gasping. In a few weeks, however, their bodies adapt by producing more red blood cells so that their blood becomes more efficient in delivering the limited available oxygen from the lungs to tissues throughout the body.

When going from bright sunlight to a dimly lit room, several minutes pass before your eyes increase their sensitivity and adjust to the new level of illumination. Conversely, on stepping outdoors on a very bright day, your eyes will be dazzled for a few moments. These adjustments in the sensitivity of the eye to new ambient conditions are also examples of physiological adaptation, in this case involving dynamic properties of nerve cells.

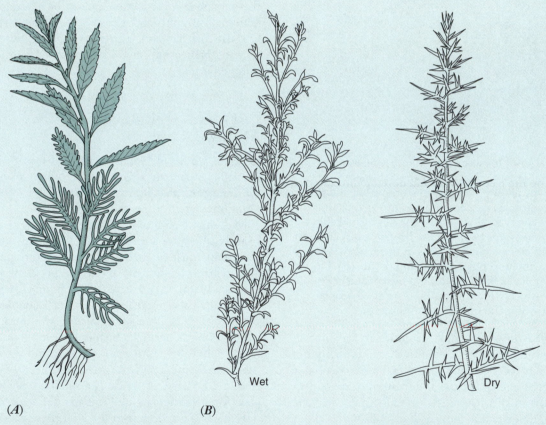

(A)   *(B)*

FIGURE 5.9 **(A) The mermaid weed (*Proserpinaca*) grows partly submerged in water. The leaves that are above the surface are not divided and present a greater surface area for gas exchange. (B) When it grows in dry places (*right*) gorse appears to present greater defenses against herbivores than when it grows in wet habitats (*left*).**

Another criterion for an adaptation is *complexity*. An image-forming eye or an ear that can analyze elaborate sounds are not only morphologically complex structures, they are obviously suited for particular functions. An eye has optics that form an image on a layer of light-sensitive receptor cells in the retina. The receptor cells,

in their turn, signal a layer of nerve cells, and the nerve cells report to the brain where additional neurons analyze the information in terms of shapes, edges, colors, and brightness as well as the changes in these properties as the eye moves or the scene changes. Ears detect complex patterns of pressure in the surrounding air or water,

 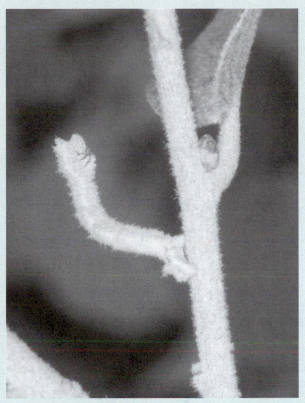

FIGURE 5.10 The physical appearance of the caterpillars of the moth *Nemoria ari-zonaria* depends on diet. If eggs hatch on oak trees in the early spring, the larvae feed on the tree's flowers and develop into caterpillars that look like the flowers. If eggs hatch later in the summer after the flowers are gone the larvae feed on leaves and develop into caterpillars that look like twigs. Each phenotype is adaptive in helping to conceal the larvae from predators.

Organisms also frequently display physiological adaptations in growth and development (Figs. 5.9 and 5.10). For example, a poison ivy plant can grow as a shrub or as a vine, depending on opportunities presented by the environment. Obviously, to be a vine is easier if there is a tree or fence on which to climb, but in the absence of such a possibility, poison ivy can assume another form and thrive. The idea that phenotypes of organisms with the same genotype can vary depending on the environment was introduced earlier (Chapter 4), but as the examples in Figures 5.9 and 5.10 illustrate, the differences can involve shape and form as well as just size and weight.

Functionally useful physiological or developmental responses to particular environmental challenges can thus be features of an organism's phenotype. The ability of the body to respond to high altitudes by increasing the manufacture of red blood cells, the capacity of the eye to adjust its sensitivity automatically to increases and decreases in ambient light, and alternative forms of growth of insects and plants are themselves the result of evolution. The *capacity* of organisms to respond with appropriate physiological changes is therefore the result of evolution by natural selection. Adaptation (by natural selection) thus begets adaptation (of the physiological sort).

and in the case of our ears, these patterns are transformed into the meaning of speech or the beauty of a Mozart concerto. Complexity of structure and suitability for a specific function are always found together in living things, and natural selection is the only known mechanism for molding this association.

A third line of evidence involves *comparisons* of closely related species that have exploited different habitats. The kidneys of kangaroo rats that we mentioned above are much more efficient in retaining water than the kidneys of closely related rodents that don't live in deserts. The correlation of this feature with habitat is strong presumptive evidence that the difference is an adaptation.

Sometimes, however, useful features are simply exploited, as in the example of the sutures in the mammalian skull. Moreover, a structure that has a specific function at one time may later be modified by natural selection for an entirely unrelated function. Such opportunistic selection has taken place many times in the evolution of flight; forelimbs were modified into wings in the dinosaurs that gave rise to birds (Fig. 2.17), in the reptiles that gave rise to pterosaurs (Fig. 2.13E) and in the mammals that gave rise to bats (Fig. 2.20).

DOES ADAPTATION OPTIMIZE?

Despite the impressive and elaborate array of adaptations displayed by organisms, adaptations are not necessarily optimal in their design. There may be *constraints* in the building materials; for example, the lenses of vertebrate eyes are not well corrected for chromatic aberration, and different wavelengths (colors) of light are therefore not brought to a focus at precisely the same distance from the front of the eye. Because of constraints, some adaptations have never arisen. For example, partly because of the difficulty of dealing with the continuity of nerves and blood vessels, the front limbs of vertebrates have never become wheels.

Furthermore, adaptations are frequently *compromises*. Bright colors may attract predators as well as mates; becoming too gaudy may be an invitation to become dinner. Large antlers may make a male deer or elk effective in confronting other males and ensuring his reproductive success with females, but as they require energy to grow and to carry about they come at a cost (Fig. 5.11).

If a lineage of organisms lacks the appropriate genetic information on which mutation and selection can act, certain adaptations may have a vanishingly small probability of occurring. For example, it is unlikely in the extreme that dogs will develop chlorophyll, chloroplasts, and the capacity to make their own food by means of photosynthesis. Put another way, major evolutionary changes are effectively irreversible, and the vertebrate lineage has long since lost the opportunity to be independent of plants for sources of energy.

The concept of adaptation is sometimes compared with an irregular landscape of peaks and valleys in which the tops of the hills and mountains represent effective adaptive outcomes (Fig. 5.12). The process of adaptation is therefore movement from a valley to a peak, but natural selection tends to find the closest peak, not necessarily the highest. Consequently, many adaptations appear to be *adequate but not optimal*, for

one can find other organisms with seemingly better adaptations. For example, for millions of years the chambered nautilus, a marine invertebrate with a chambered shell, has had an optically simple eye that is structurally like a pinhole camera. Our own eyes are much more sophisticated, with far greater resolving power, so there are clearly much higher peaks on the adaptive landscape of eyes. During its evolution, the eye of the nautilus moved to the top of a low hill and remained there, its capabilities adequate for the nautilus but otherwise quite modest (Fig. 5.12).

Incidentally, in human engineering, historical contingency can also lead to continued use of suboptimal design. Perhaps the most familiar example is the inefficient layout of keys on computer keyboards. As computers came into wide use, they copied the "QWERTY" layout of keys that people had learned in using typewriters. But that layout had been deliberately designed to be inefficient because skilled typists would otherwise type so fast they would jam the mechanical works on early typewriters. That problem does not exist on computers, but most people making the transition from typewriters were not interested in dealing with an unfamiliar keyboard. That would have been the technological equivalent of moving from a low to a higher adaptive peak. It is not often done because the intermediate states have very low fitness.

ADAPTATION AND BEHAVIOR

The concept of adaptation is particularly tricky when we turn to examples of behavior. Many of the things that animals do have direct consequences for reproductive success, and this makes behavior very sensitive to natural selection. But behaviors are not physical enti-

FIGURE 5.11 Adaptations are generally compromises, involving both benefits and costs. The size of a male elk's antlers are important in securing mates (Chapter 6). However, antlers are costly: they can be heavy and awkward to carry, and are shed and regrown each year. This is a reconstruction of an Irish elk, now extinct.

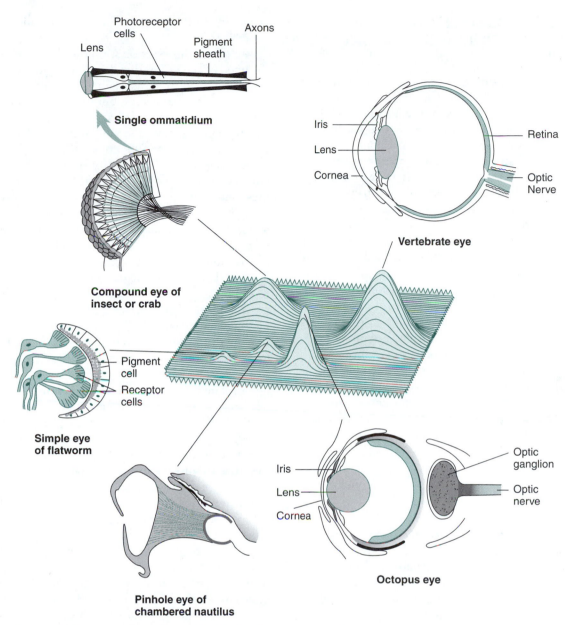

FIGURE 5.12 A much simplified adaptive landscape of eyes. Eyes have evolved independently scores of times. The large image-forming eyes of vertebrates and cephalopod mollusks (octopus, squid) are an example of convergent evolution and are represented in this diagram by two high peaks on the adaptive landscape. The simple eyecup of the flatworm is represented by a foothill.

ties like morphological structures; they are actions. Moreover, they can change quickly, and as a rule they depend on circumstances. Behaviors reflect patterns of activity of nerve cells in brains, and as we shall see in later chapters, the interplay of genes and environment in the development and function of the brain requires its own discussion. Brains are part of the package—the organism, the vehicle—that must reproduce successfully if its genes are to be present in future generations,

and brains have therefore been molded by natural selection. We should thus expect to find that *patterns* of behavior are functionally appropriate for specific developmental stages and environmental circumstances in which an organism is likely to find itself. We can also anticipate that in each species these behaviors are based on an evolved, adaptive architecture of the brain. We shall return to these matters when we discuss the concept of instinct in Chapters 9 through 11.

ADAPTATION IS ONLY ONE COMPONENT OF EVOLUTIONARY CHANGE

Evolution is a historical process, and in all historical analysis it is easier to look backward and interpret events than it is to predict the future. About 65 million years ago something happened that led to the extinction of the dinosaurs and opened evolutionary possibilities for that group of small, nocturnal creatures that gave rise to the great variety of present-day mammals. One hypothesis is that these events were triggered by the impact of an asteroid 10–15 kilometers in diameter that is known to have struck the earth near what is now the tip of the Yucatan peninsula. It threw a great deal of dust into the atmosphere and likely caused significant changes in temperature. There is in fact evidence that many such collisions have occurred during the history of the earth with consequences for living organisms. The point, however, is that major shifts in the pattern of evolution caused by climate changes and other irregular events makes the subsequent appearance of any particular form of life utterly unpredictable. Our arrival over 60 million years later, in the company of bats and squirrels, giraffes and whales, need not have happened. Evolution has no preordained goal, but that does not make adaptation any less important or the present any less wonderful.

ADAPTATION AND GROUP SELECTION REVISITED

The concept of adaptation has been entangled with the "group selection fallacy," which we discussed above. Presumed adaptations for the group have influenced many introductory accounts or popular films on animal behavior. Because at one time this confusion was widespread and has considerably influenced the thinking of individuals in other disciplines, we will examine a couple of simple examples.

The first is the familiar behavior of a small, burrowing, arctic rodent, the lemming. These little mammals undergo wide fluctuations in numbers from year to year, and when they reach a critical population density, large numbers emigrate at the same time.

Some people have suggested that this synchronous movement of lemmings and their occasional death by drowning are behavioral adaptations whose function is to relieve overpopulation. The idea is that by leaving and committing suicide the emigrants prevent their natal population from overexploiting their environment and starving. If these are the evolved functions of the behaviors they certainly would have to be called "group adaptations," because survival of the group is promoted at the expense of the genes that

programmed these behaviors to occur under crowded conditions.

In a preceding section we showed how behavior for the good of the group that comes at genetic cost to self is very unlikely to be sustained by natural selection. Let's now apply that reasoning to this example. First, consider whether these behaviors make sense as group adaptations. If emigration and suicide are for the good of the group, we would expect the process to work most efficiently if the majority of the emigrants were females in their reproductive prime, since the loss of these individuals would slow population growth more than the loss of either males or pre-reproductive females. In fact, the majority of the emigrants are juvenile males and females. So the sex and age composition of the emigrants is not what we would expect on the hypothesis that this suicidal behavior is an adaptation to foster the well-being of the group.

As George Williams suggested, the emigration can be explained by considering this behavior as an adaptation in the genetic interests of individuals. Consider first the reason why many Americans are descended from Irish people who immigrated here during the potato famines of the nineteenth century. If they had not emigrated, many of them would have starved to death. If their natal areas regularly become so overcrowded that the resident lemmings have a negligible chance of surviving and producing offspring, natural selection could reasonably be expected to have generated compensatory behavior. Thus when faced with starvation, lemmings might well be expected to look for greener pastures elsewhere, even if emigration holds only a small and uncertain chance of success and only a few of the emigrants manage to reproduce. All that is required for natural selection to have fostered this behavior is that the average reproductive success of emigrants be slightly greater than that of the lemmings who stay behind. Consistent with this explanation, the emigrants do show symptoms of poor nutrition and stress from crowding.

Why are the emigrations synchronous? Do they represent some sort of "mass suicide hysteria"? A plausible explanation consistent with adaptation is that this small, slow-moving mammal, which is easy prey for hawks, owls, and foxes when it is caught in the open, is doing what many other animals do when they must carry out activities that make them vulnerable to predators—they use the group for cover. Flocking of birds or schooling of fish reduce the individual's risk of being eaten. Not only is an individual in a crowd less likely to be singled out by a predator, but by synchronizing their activity, vulnerable animals saturate their predators with more food than can be consumed at one time. Sometimes humans also use such safety in numbers to carry out activities that they would otherwise not do alone. Looting and speeding on the highway are examples.

Why do some of the lemmings hurl themselves off cliffs or drown at sea? Isn't this suicidal? Like other animals, lemmings are adapted to the conditions they are most likely to encounter in their environment. Small ravines, streams, and ponds are the common obstacles that this small mammal can and does get across. When some emigrating lemming populations encounter cliffs and the sea at the periphery of their geographic range, their response is the one that is appropriate throughout most of their range—they simply try to get to the other side. In other words, death under these dramatic but exceptional circumstances is a local and maladaptive by-product of a behavior that has an adaptive function most of the time.

Note as an aside that humans are trapped with many such maladaptive behaviors in recently created environments that are very different from the one in which we spent 99% of our evolutionary history. Our tastes for sweets, fat, and salt evolved to guide our intake of energy-rich foods, vitamins, and minerals in environments where these items were scarce. Our preferences can be maladaptive, however, when we find ourselves in the presence of an easy abundance—kitchen, candy store, or "fast food" restaurant. The occurrence of behavior that seems to have no adaptive significance or is even maladaptive is a phenomenon we will encounter again in later chapters when we will be discussing human behavior more directly.

Another example of "group adaptation" thinking is frequently seen in nature films that show predators in action. Several lions are pulling down a zebra, and they begin to eat their catch while it is still alive. The horrified viewer is consoled with the suggestion that this is all part of the balance of nature; lion predation keeps zebra populations in check, thus preventing overgrazing of the environment and its ruin for future populations of zebras. Furthermore, we are assured, because the lions single out old, sick (and slow!) individuals, the zebra population is kept free of disease and maintained fleet of foot.

Are the lions actually doing the zebras a favor? Is the function of predation to keep animal and plant populations from overexploiting their environment and thus provide a stable and balanced ecosystem? A quick answer to this question follows from the same kind of reasoning we used in analyzing the behavior of lemmings: the function of predation is to enable predators to reproduce. If zebra populations are kept in check and healthier, this is an *effect* of lion predation, not its function. A perfectly reasonable explanation involving only natural selection for what is good for individual lions is sufficient.

SYNOPSIS

Darwin's proposal that natural selection is the primary cause of evolution remains the core insight of evolutionary theory. But like all scientific understanding, evolutionary theory has changed during the last century. The "modern synthesis" (in the 1930s) made clear that genetic variation in natural populations provides the variation on which natural selection operates. With the subsequent discovery (1953) of the structure of DNA it became possible to examine the distribution of genes in natural populations as well as study evolutionary history by determining the base sequences in genes or the primary structures of proteins for which the genes code. These molecular approaches have greatly strengthened our understanding of evolution and the mechanisms by which it operates.

What gets selected by natural selection? Alternative alleles of *genes* are selected, but in sexually reproducing species the process of selection involves the differential reproductive success of individual *organisms*. It is thus useful to think of selection as taking place in the domain of information (genes) as well as in the material domain of organisms (individuals).

Individuals generally reproduce faster than the groups to which they belong. For this and other reasons, selection does not ordinarily take place at the level of groups, and individuals characteristically do not behave "for the good of the group" if such behavior comes at a cost to their own reproductive success.

The fossil record provides a view of evolutionary history and the diversity of life that complements the understanding of evolution provided by the study of genes and natural selection in populations of living organisms. The perspectives are very different. Long periods of the fossil record show little change in fossil organisms and are interrupted by shorter periods of more rapid evolution, a phenomenon known as *punctuated equilibrium*. The periods of rapid change, however, are very long relative to the time needed for natural selection to act. Macroevolution reflects the accumulation of microevolutionary changes.

Adaptation by natural selection is a formative process of great power and is responsible for the myriad of features of organisms that seem to be the product of "design" for particular functions. There is more to evolution, however, than adaptation. Constraints on starting materials and on developmental pathways can restrict evolutionary outcomes, and environmental changes, catastrophic or minor, can set evolution on new trajectories. Evolutionary change is thus contingent on circumstances and is therefore unpredictable. In this regard, evolutionary history, like the social/cultural/political history of human societies, provides few clues as to what the future may hold.

QUESTIONS FOR THOUGHT AND DISCUSSION

1. How has evolutionary theory changed since *On the Origin of Species* was published in 1859, and what led to those changes? What unresolved issues do you think will likely lead to further change in evolutionary theory during the next several decades?

2. How would you decide whether two populations of similar, contemporary organisms should be considered the same or different species? Is it possible for a single lineage of organisms to change over time and produce a different species? In this case, how would you decide whether the ancestral and later populations were the same or different species?

SUGGESTIONS FOR FURTHER READING

Dawkins, R. (1989). *The Selfish Gene*. New York, NY: Oxford University Press. A skillful writer for a general audience focuses attention on the role of genes in evolution.

Ridley, M. (1996). *Evolution*. Cambridge, MA: Blackwell Science. A good place to turn for further understanding of evolutionary theory.

Williams, G. C. (1966). *Adaptation and Natural Selection: A Critique of Recent Thought*. Princeton, NJ: Princeton University Press. A clear and rigorous analysis of evolutionary processes and the language used to describe them. This classic book focused attention on how natural selection acts.

6

Evolutionary Social Theory

CHAPTER OUTLINE

- The rudiments of evolutionary social theory
- Altruism toward genetic relatives and the concept of kin selection
    - Shared genes
    - Some common confusions about kin selection
    - Nepotism and self-interest I: The alarm calls of ground squirrels
    - Nepotism and self-interest II: Helpers at the nest and in the den
- Reciprocal altruism: Return effects in the absence of close genetic relatedness
    - Sharing of food by vampire bats
- Parent-offspring conflict
    - The evolutionary basis of weaning conflict
    - Psychological adaptations of offspring and parents
    - Social behavior of offspring toward near relatives
    - Dependence on the mating system
    - Conflict during pregnancy
    - Reproductive future of the parent

- Sexual Selection
- The role of parental investment in sexual selection
    - Minimal parental investment by males is frequently associated with polygynous mating systems
    - Large parental investment by males is often associated with monogamous mating systems
    - Species displaying reversed sex roles have polyandrous mating systems
    - Some further behavioral consequences
- Manipulating the sex ratio
    - Opossums manipulate the numbers of sons and daughters
    - Red deer also manipulate the sex ratio
    - Female dominance enhances the reproductive success of daughters
- Synopsis
- Questions for thought and discussion
- Suggestions for further reading

Photo: Sexual selection for female choice can take different forms. The peacock (*left*) grows his gaudy and unwieldy tail, but the satin bower bird (*right*) builds his display of grass and then decorates it with brightly colored objects that he brings to the site. How such seemingly non-utilitarian features could have evolved by natural selection was a source of controversy for many years after Darwin's death.

Some animals such as leopards and orangutans live alone and come into contact with members of their own species only at the time of mating. Other animals live together as mated pairs—or even, in the case of certain birds, wild dogs, and wolves, small families—and cooperatively feed and protect their young. Still other species—ground squirrels, elephants, many kinds of primates—live in larger groups with variable ties of genetic relatedness but with considerable social interaction among individuals. And some species—schools of fish, flocks of birds, herds of wildebeest and zebra—congregate to migrate or breed in groups that range in size from a few individuals to thousands or millions. Although the interactions between individuals in large aggregates are often inconsequential, in some species cohesion and organization are important. For example, the social insects—bees, ants, termites—are obligatorily bound together by ties of genetic relationship in colonies containing tens of thousands to millions of individuals.

The social interactions of animals are among their most interesting phenotypic features. Why is there such variety in the sizes of groups as well as in the extent to which individuals recognize, communicate, and cooperate with each other? What special evolutionary issues are raised when the capacity for communication enables animals to behave in an organized, cooperative way? Clearly such social groups are more complex and more interesting than mere aggregations of individuals that have come together because of some mutually attractive stimulus such as food but display few meaningful interactions among themselves. This is the same distinction that makes a football team more interesting to watch than the crowd of spectators that have come to see the game.

Probably because we are ourselves a highly social species, we find the social interactions of other animals among the most interesting manifestations of their behaviors. However, the study of animal behavior played a much less important role in the early development of evolutionary biology than did studies of anatomy, embryology, and paleontology. There were understandable reasons for that neglect (Chapter 11), but beginning about forty years ago the social structures of animals began to receive concerted attention from a few evolutionary biologists who provided a theoretical framework for understanding social evolution. In this chapter we will examine this relatively new and important body of evolutionary social theory, which builds on and extends Darwinian thinking. We will consider some special issues raised by the social insects in the following chapter, and in later chapters we will explore how evolutionary social theories can be used to illuminate human behavior.

THE RUDIMENTS OF EVOLUTIONARY SOCIAL THEORY

We have seen how natural selection is a process of differential replication of genes caused by the effects of genes on the relative reproductive success of the individuals (vehicles) carrying them. Individual organisms are thus the elemental "atoms" of social behavior, and

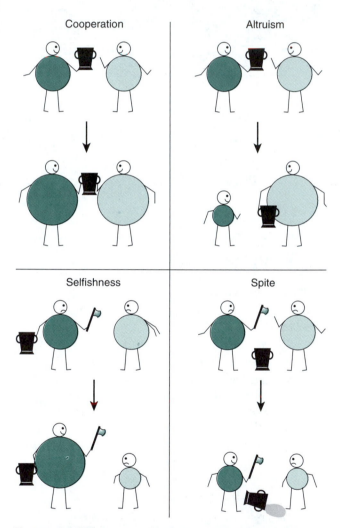

FIGURE 6.1 When actor (*darker green*) interacts with recipient (*lighter green*) there are four classes of behavior, whose consequences are depicted by the arrows. *Cooperative* acts are beneficial to both, whereas *altruistic* acts bestow a benefit on the recipient but at a cost to the actor. *Selfish* acts retain a benefit to actor but at a cost to the recipient and *spiteful* acts are costly to both. In evolution, costs and benefits are measured in terms of *reproductive success*, which is represented in these cartoons by increases and decreases in size. As explained in the text, behavior that appears to be altruistic can be evolutionarily beneficial.

reproductive success is the measure that should be used to assess the evolutionary consequences of behavior.

Let's consider the simplest possible social interaction: the behavior of one individual, the "actor," toward another individual, the "recipient." The reproductive success of the actor and recipient can either benefit (*B*) from the interaction or suffer costs (*C*), measured as increments and decrements in the numbers of surviving offspring. Since there are two possible effects on reproduction (*B* or *C*) for both actor and recipient, there are four categories of behavior, as shown in Figure 6.1.

If benefit accrues to both actor and recipient, the effect of the behavior is *cooperative*; if costs are incurred by both, actor's behavior can be characterized as *spiteful*. When actor's behavior benefits himself and comes at a cost to recipient, the behavior is *selfish*. Conversely, if costs are borne by actor while recipient benefits, the behavior is *altruistic*.

The words "cooperative," "spiteful," "selfish," and "altruistic" are used here to describe evolutionary outcomes and do not imply anything about the actor's conscious motives or the recipient's feelings. Actor and recipient might both be ants, zebras, or any other creature between A and Z. For the limited task of categorizing behavior and describing its reproductive consequences, it is not necessary to understand how their nervous systems process information and generate behavior. It is useful, though, to describe the several possible behavioral outcomes without creating specialized jargon.

These categories of behavior and their one-word definitions are, of course, simplifications. Social interactions of animals result from many behaviors—aggression, play, courtship, submission, food begging, food sharing, alarm calls, and cooperative hunting, to name just a few. Also, the roles of actor and recipient may be repeatedly reversed; the interactions may be "multiparty" (involving many actors and recipients); and the benefits and costs may be indirect or delayed in time. Nonetheless, these categories of behaviors and their effects describe recurring categories of interactions that have reproductive consequences in all social animals.

ALTRUISM TOWARD GENETIC RELATIVES AND THE CONCEPT OF KIN SELECTION

SHARED GENES

We expect reproductively selfish behavior to be favored by natural selection: if one lion repeatedly steals food (benefits) from another with little effort (cost), the replication of the genes underlying such behavior is likely to be enhanced. A similar argument can be made for cooperative behavior: if two lions collaborate in wresting food from a third at little cost and then share it, the genes underlying this behavior are favored. Spiteful behavior could be favored by natural selection if the cost to the actor is less than the cost to the recipient, but as spite incurs costs to both individuals, spiteful behavior is not common in animals.

Behavior that appears to be altruistic seems to present a paradox: if one hungry lion shares its small catch or helps fend off attacks of outsiders, it can be benefiting another's reproduction at a cost to its own. And yet such altruistic behavior is common in nature. How can this be?

The answer lies in inheritance: copies of one's genes reside in near relatives. As you received half of your genes from your mother, your degree of relatedness *r* to your mother is $\frac{1}{2}$. As the probability that any particular maternal gene was passed on to you is also 0.5, your mother's degree of genetic relatedness to you is also $\frac{1}{2}$. Similarly, your *r* with your offspring is $\frac{1}{2}$. You are related to full brothers and sisters through two parents. Examine Box 6.1 to see why your *r* to your siblings is $\frac{1}{2}$, the sum of $\frac{1}{4}$ through your father and $\frac{1}{4}$ through your mother.

This pattern of inheritance means that it is possible to further the replication of your alleles not only by having your own children but also by enhancing the reproduction of near relatives. Your degree of relatedness to your child is $\frac{1}{2}$, but to a niece or nephew it is only $\frac{1}{4}$. Consequently, in terms of the propagation of your alleles you can achieve the same reproductive success by rearing two nieces or nephews as by rearing one daughter or son. In general, by influencing the reproductive success of close relatives, an individual can influence the replication of identical copies of its genes that reside in close kin. This kind of natural selection—in which genes change in frequency because of an individual's influence on its relatives' reproduction—is called *kin selection*. This important idea was first developed in 1964 by the evolutionary geneticist William D. Hamilton, who showed how kin selection could account for the evolution of some instances of altruistic behavior.

The theory of kin selection predicts that whatever genes cause an actor to benefit the reproduction of relatives will increase in frequency if the benefits exceed the costs. Benefits and costs to whom, however, and how should they be reckoned? To illustrate with an extreme example, suppose an individual who is genetically disposed to behave altruistically toward relatives sacrifices his life in saving three siblings from certain death in a boating accident. Copies of those "altruist" alleles are present in each sibling by reason of common descent with a probability that is simply *r*, the degree of genetic relatedness of actor to each sibling. Each of the

Box 6.1
Calculating Degrees of Relatedness

The degree of relatedness r between any two individuals is the probability that they share copies of a gene that are identical by virtue of descent from a common ancestor. To calculate the degree of relatedness between one individual (darker green) and another (lighter green), draw arrows through the common ancestors as shown in Figure 6.2.

(A) For two siblings (sisters in this example), r is $\frac{1}{2}$. The probability that the first sister has received any one of her genes from her mother is $\frac{1}{2}$ (meiosis, Chapter 3), and the probability that same gene in her mother was passed to the sister is also $\frac{1}{2}$. These two probabilities are independent of one another, so r through their mother is $\frac{1}{2} \times \frac{1}{2} = \frac{1}{4}$. By the same reasoning, r through their father is also $\frac{1}{4}$. Because the genetic paths through the mother and father are separate and distinct, we add their individual probabilities to get $r = \frac{1}{4} + \frac{1}{4} = \frac{1}{2}$.

(B) The degree of relatedness of a sister with her nephew (her sister's son) is calculated in the same way. The only additional step required is to extend it to the next generation by multiplying the r for the sisters ($\frac{1}{2}$) by the probability that the nephew's mother has given him one of her genes, $\frac{1}{2}$. Therefore r for the first sister with her nephew is $\frac{1}{2} \times \frac{1}{2} = \frac{1}{4}$.

(C) Convince yourself that the degree of relatedness between first cousins is $\frac{1}{8}$.

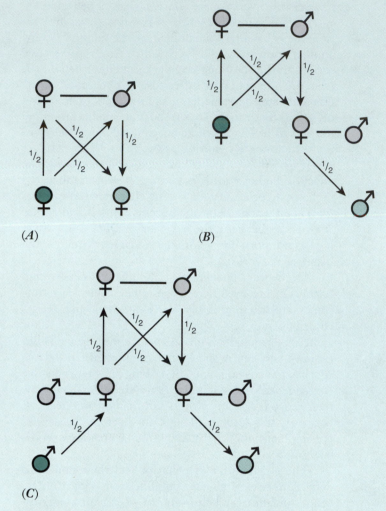

FIGURE 6.2 Tracing degrees of genetic relatedness.

actor's alleles is lost when he perishes, and so the reproductive cost, C, is 1. But as copies of those alleles are present in each sibling with probability $r = \frac{1}{2}$, the devalued benefit $B \times r = 3 \times \frac{1}{2} = 1.5$. Consequently in this example, $B \times r > C$.

Let's now state the rule more precisely. Actor's alleles that enhance the reproduction of relatives will be selected if the reproductive benefits to relatives, discounted by each relative's degree of relatedness to actor, are greater than the reproductive costs to actor. Kin selection is thus a process by which alleles change in frequency because of the behaviors they engender in an individual that in turn influence the reproduction of identical copies of those alleles in genetic relatives. The effect, however, is to promote (or

detract) from the propagation of *all* of the actor's alleles in individuals with whom he shares them by common descent.

It is important to recognize that the proximate mechanisms by which genes influence behavior is through the assembly of the brain (Chapter 10). Furthermore, how the brain filters information and weighs consequences of alternative behaviors is an emergent property of neural tissue (Chapter 9) that varies among species. For kin selection to exist, brains must have the property of directing "altruistic" behavior to genetic relatives. For this property of the phenotype to have evolved means that natural selection has favored some alleles and discarded others. That is the sense in which one can speak of "altruist" alleles.

The process of kin selection has to be considered ultimately one of differential gene replication because in our example the behavior fostered by the "altruist" alleles actually prevented their vehicle (actor) from reproducing while promoting the survival of identical copies of themselves in other vehicles. This also brings us to the reason why we have put the term "altruist" within quotation marks. At the level of the phenotype, actor was altruistic in the usual sense of the word: he died so that three siblings might live. From the viewpoint of gene action and replication, however, actor's behavior was not altruistic: as a consequence of the influence of the "altruist" alleles on his behavior, more copies of those alleles survived than if he had lived and his three siblings had drowned.

SOME COMMON CONFUSIONS ABOUT KIN SELECTION

Before considering examples of kin selection at work in nature, we will discuss some of the criticisms and confusions that surround this concept. One argument is that because most organisms in a social group have the same alleles at most of their genetic loci, beneficence of an individual toward *all* members of the social group—indiscriminate beneficence—would help the same genes and therefore be favored. To see the problem with this line of reasoning, consider how indiscriminate altruism would be vulnerable to a genetic change that directs benefits *specifically* to near relatives. Imagine a population in which *all* individuals *do* carry identical copies of a gene, X, for indiscriminate beneficence—for example, a population of zebras in which lactating mothers allow each other's young to nurse at will, irrespective of genetic relatedness. Because the genes for such behavior in mother zebras promote survival of identical copies in all baby zebras, we expect those genes to persist. And they will persist until mutations or recombinations produce a new genotype that leads to a nervous system that causes the mother to allow only its own young (with $r = \frac{1}{2}$) and those of its sisters (nieces and nephews with $r = \frac{1}{4}$) to nurse. This new allele, X_1, by causing the mother to direct benefits (milk) only to close genetic relatives, would benefit its own replication more often than it would the original, alternative allele, X. However, the original allele, X, would benefit X_1 and other copies of X equally. As a result, the baby zebras carrying X_1 would get a greater share of milk than those carrying X, most likely grow faster and stronger, be less susceptible to predation, and therefore leave more descendants. Eventually X_1 would replace X in the population. In other words, the X allele for indiscriminate altruism is unstable and open to invasion by the X_1 allele for nepotism, whereas the reverse is untrue.

The process of kin selection would not stop here. Consider what would happen if another mutation, X_2, arose, which caused the mother to allow only *its* young and *not* its nieces and nephews to nurse. Again, the X_2 allele would benefit only its own continuance in the mother's young ($r = \frac{1}{2}$), but its propagation would also continue to be enhanced by the less discriminating behavior caused by X_1. So eventually X_2 would replace X_1. These two examples show that kin selection will increasingly favor beneficent behavior directed to nearest kin, and beneficence toward more distant kin will become relatively disfavored. *All that is required is a heritable mechanism for recognizing and favoring kin when directing benefits.*

A common difficulty in understanding how kin selection works is just this matter of recognizing kin. Isn't it asking a bit much of the intelligence of a zebra, or for that matter any animal other than humans, to be aware of degrees of genetic relatedness? We certainly do not expect nonhuman animals to do cost-benefit analyses. However, an animal need not understand genetic relatedness in order to favor kin, any more than a bird need understand the aerodynamics of flight in order to fly, or astronomy in order to navigate by stars, or biochemistry in order to digest its food. It is sufficient that the animal behave *as if* it understood. Humans, too, favor kin in cultures around the world (Chapter 14), but notions of genetic relatedness could not have been discussed before there was knowledge of genes.

In the case of a mother zebra, allowing only its offspring to nurse requires two behavioral mechanisms for which there is much evidence in many species: (1) the mother be programmed to learn and remember the combination of visual and olfactory cues unique to the infant to which she gave birth, and (2) only this combination of stimuli will subsequently elicit her nursing behavior. It may be useful to describe this behavior as "recognizing her own young," but this phrase does not imply that zebras have humanlike mental representations of their young or that they have the intelligence to develop the concepts of "mine" and "others." We shall return later to this problem of animal cognition (Chapter 11).

NEPOTISM AND SELF-INTEREST I: THE ALARM CALLS OF GROUND SQUIRRELS

Our so far hypothetical examples show how natural selection might shape the behavior of animals toward others of the same species. But the principles can be made much more convincing if we are able to observe these predictions at work in field studies of real animals.

Many social animals give alarm calls when they see a predator, thus alerting other members of the group. The first impression is that this is altruistic behavior: the caller seems to have helped others but jeopardized its own safety by drawing the predator's attention to itself. In fact, this assumption is implicit in calling these vocalizations "alarm calls"! There are, however, other conceivable interpretations of these calls. (*i*) Perhaps calling diverts the predator's attention to other prey. This could be possible if the alarm call caused other members of the group to run or fly quickly for cover and the predator's attention were drawn to the movement of individuals it had not seen previously. This could decrease the chance that the caller would be attacked. (*ii*) Or maybe calling discourages the predator by informing it that it has been detected and that a successful attack is unlikely. This might be effective if the animals that call were particularly fleet or evasive and could only be caught by surprise or ambush. (*iii*) Calling could possibly discourage the predator from making future attacks. That is, the risk of giving away one's position by a call might be offset by the benefit of warning other members of the group, which then denies the predator a meal and discourages it from returning to attack the same group and the caller itself. (*iv*) Calling might warn others of the group who are likely to reciprocate. The cost to the caller (increased risk of attack) could be less than the benefit to the alerted individuals (decreased risk of attack), who later reciprocate under similar circumstances. Over time the benefits of such reciprocity to each member of the group exceed the costs. We will say more about such "reciprocal altruism" shortly. (*v*) Finally, giving alarm calls could alert relatives and be favored by kin selection. The callers jeopardize their own survival but increase the probability that alerted relatives will survive, so that $B \times r > C$.

Considering the difficulties of making the required observations and measurements, it is little wonder that only a few animals have been examined in sufficient detail to distinguish among all of these hypotheses. But Paul Sherman and his coworkers have successfully studied the function of alarm calls in a population of Belding's ground squirrels (Fig. 6.3). These squirrels live in underground burrows in alpine and subalpine meadows in the far western United States and are active during the day. Females that successfully rear young retain their burrows from one year to the next, and their daughters mature and breed in their natal area. Males, however, emigrate and breed elsewhere. During their period of sexual receptivity, females mate with an average of two males, but they subsequently give birth and rear their young alone. Some males are highly polygynous (i.e., one male characteristically mates with more than one female). As a consequence, 20% of the males can account for up to 60% of the copulations, so the females of a given area are likely to be related to each other through a common father as well as by being sisters or first cousins on their maternal side.

Most of the members of a ground squirrel population at Tioga Pass Meadow in the Sierra Nevada

FIGURE 6.3 Ground squirrel *Spermophilus beldingi* on alert (*left*) and high alert (*right*).

Mountains of California were marked over a period of seven years (almost 1900 squirrels, altogether), which enabled Sherman and his associates to know their exact ages and family relationships through common female ancestors. During 3,100 hours of observation over two summers, five different species of terrestrial predators (weasels, badgers, martens, coyotes, dogs) stalked members of the population 102 separate times and killed nine animals. Upon seeing one of these terrestrial predators, a Belding's ground squirrel sits up, remains stationary, and gives a multiple-note trill. Upon hearing an alarm call, other squirrels immediately either sit up and look for the danger or run to the nearest rock and peer about.

Observations of the squirrels and their predators make clear that calling is a nepotistic act. First, uttering an alarm call increases the risk of being killed; ground squirrels that called were chased or stalked significantly more often by all five of their predators than were squirrels that remained silent. This is because the calls are acoustically easy to locate and are uttered repeatedly—an average of twenty-seven times over six minutes. On this basis, calling therefore appears to be an altruistic act.

Second, squirrels were more likely to call when there were close relatives in the vicinity. This is why females called nearly ten times as often as males. Furthermore, females that were reproductively active or that had relatives (mother, sisters, daughters) nearby were more likely to call than other females.

This sort of discriminative nepotism requires, of course, that female ground squirrels be able to recognize their female relatives. There is other evidence that this is possible. Female ground squirrels chase out of their territories nonkin and kin of $r < \frac{1}{4}$ significantly more often than they chase away relatives of $r = \frac{1}{2}$ (mothers, daughters, sisters).

The willingness of female Belding's ground squirrels to risk their own lives by warning near relatives of terrestrial predators is just the sort of behavior that is predicted from the theory of kin selection. Humans show the same kind of behavior: across all cultures concern with the fate of others in times of danger or disaster increases directly with genetic relatedness. In our own culture, the Carnegie Hero Commission has bestowed some 8,000 awards since 1904 for acts of heroism in helping others in danger, but rarely is such heroism recognized if it involved the rescue of relatives. Helping relatives is considered normal.

Interestingly, Belding's ground squirrels give another call—a single-note whistle—when hawks fly overhead, and the response of other squirrels is also different. Sherman observed fifty-eight encounters between wild hawks and ground squirrels, and he staged six hundred more with hawks that had been trained to fly directly over the colony without attacking the animals. The first caller was usually the one in greatest danger because it was close to the predator and often far from cover. Instead of remaining where it could be seen by other squirrels, the caller scurried to the nearest burrow or bush as it sounded the alarm. The other animals responded by similarly calling and running immediately to the nearest shelter, even if it was another animal's burrow. In striking contrast to the alarm trills for terrestrial predators, the tendency to whistle-alarm was not correlated with age, sex, reproductive condition, or the presence or absence of kin. And in contrast to the callers for terrestrial predators, the callers for hawks were significantly less often killed than the noncallers.

These findings leave little doubt that whereas the function of alarm calling for terrestrial predators is to alert relatives, the alarm calling for flying hawks seems to be for self-preservation. Unlike the stealthy approach of a coyote, the swoop of a hawk gives its prey little time to escape, and when the squirrels signal the presence of a hawk they do not linger on the spot. Their call in turn triggers pandemonium in the colony, with an explosion of calling, scurrying animals.

Why should ground squirrels bother to give hawk-warnings at all if not to benefit kin? For many of the animals there are likely to be relatives around, but is there an explanation that does not require either the presence of close kin or an appeal to the implausible argument of group selection? One possibility is that in the confusion of many scampering ground squirrels triggered by the first call, the hawk's attention may be distracted and the caller's chances of escape momentarily enhanced. If this explanation is correct (and it is consistent with a reason why birds are believed to flock and fish to school), the vocalization given when a hawk appears is a "manipulative call." It is a striking example of the power of natural selection that these two kinds of predators, presenting somewhat different dangers, elicit distinctly different warning calls and adaptively appropriate behaviors on the part of both caller and listener.

NEPOTISM AND SELF-INTEREST II: HELPERS AT THE NEST AND IN THE DEN

We think of "helping" as any behavior that appears to be helpful, no matter who the actors and recipients are. Behavioral biologists have come to use the term in a more restrictive way: "helping" is parent-like behavior extended to young that are not the direct offspring of the helper, exclusive of adoption (uncommon in non-human vertebrates), parasitism (as in cowbirds and European cuckoos, who lay their eggs in the nests of other species), and the mixing of broods by communally breeding parents (e.g., certain kinds of birds). Helping behavior is now known in over one hundred species of birds as well as jackals, African hunting dogs, and a few other mammals.

Unlike the social insects (Chapter 7), helpers among birds and mammals are fully capable of breeding on their own. So why do they spend a year, a few years, or their entire lives feeding the young of others, protecting them from predators, and defending the feeding territory from intruders? How could natural selection, acting on individuals, cause the evolution of this seemingly altruistic behavior? As in other cases of behavior that appears to be altruistic, we need additional information in order to understand these breeding systems. Under what circumstances does this behavior occur? Are the breeders measurably benefited by the helpers? What benefits accrue to the helpers? And what are the relationships between the breeders and helpers?

Helping behavior has been studied extensively in birds. With colored plastic bands placed on their legs, individual birds can be recognized over a period of years and the roles and relationships of each individual understood. Helping behavior occurs in species in which a breeding pair requires territory to supply food for themselves and the growing nestlings but available territory is limited by the presence of other breeding pairs in the surrounding area. In other words, the size of the population is close to the *carrying capacity* of the environment. Under these conditions, birds one or two years of age, although physiologically capable of producing and rearing young, may not be able to do so because they cannot establish their own territories.

These same studies answer the questions of who the helpers are and whether their assistance is valuable. Helping helps: 20–250% more young are produced when helpers are present, although the average increase in the reproductive success of the breeders (across many species) is less than a factor of two. Almost invariably the helpers are earlier offspring of the breeders or, much less commonly, they are siblings of one of the breeders. In long-term studies of the Florida scrub jay (Fig. 6.4), biologist Glen Woolfenden and his associates have shown that helpers may aid parents, grandparents, uncles, or half-brothers, but it is usually one or both parents that are helped. In this study, the average degree of genetic relatedness of 180 helpers and the offspring they helped was 0.41.

The genetic relatedness between helpers and beneficiaries is strong evidence that kin selection has played a role in the evolution of helping behavior. For full siblings, $r = 0.5$ between helper and beneficiary, the same as between a helper and its own offspring. If in a given year the chances of finding a territory and successfully reproducing with a mate is very small, a young adult can enhance its reproductive success by helping its parents. Its reproductive success will be increased as long as the number of *additional* young that its parents produce (with its help) is greater than the number of young it might have produced in its own territory with its own mate.

If we refer to the increment in the number of its siblings as a benefit (B) and the number it failed to produce with a mate as a cost (C), natural selection will

FIGURE 6.4 The scrub jay (*Aphelocoma coerulescens*) is an example of a species in which young offspring frequently delay their own reproduction and help rear siblings. This behavioral option of scrub jays is taken when there is no available territory in which the helpers can seek their own mates and rear their own offspring. From Wilson, 1975, reprinted with permission of Harvard University Press.

foster helping as long as $B > C$. Now if the helper is assisting in rearing half-siblings, his r with them is only 0.25, and the cost of foregoing his own reproduction is twice as great. Under these conditions his helping behavior must therefore have twice the impact as in the previous example in order for benefit to exceed the cost. To state the relationship in symbolic terms, benefit must be devalued by the degree of relationship the helper has with those helped ($r = 0.25$), and the cost by his degree of relationship with his own offspring ($r = 0.5$). Thus for helping to increase the reproductive success of the helper, $0.25B > 0.5C$, or $B > 2C$.

Helping behavior in these birds evolves when the reproductive prospects of young adults are restricted by the near-term opportunities for obtaining territory of their own. If their chances of breeding are close to zero, virtually any increase in the number of near relatives that is due to their help means that $B > C$. Viewed this way, we can understand why helping has evolved, although, on average, it increases the reproductive success of the breeding pair by less than a factor of two.

But in the case of the Florida scrub jay, hanging around and helping benefits the young adults in another way: they stand a better chance of acquiring a territory, either by inheriting it or, with the aid of siblings, by expanding the family territory into that of neighboring birds and then claiming a piece as their own.

Helping in birds is a behavioral alternative that makes the best of a bad situation and is only done when necessary. Successful breeding is likely to lead to greater reproductive success than helping, particularly if the helper's degree of relatedness to the beneficiaries is less than $\frac{1}{2}$, and in birds, helpers take every opportu-

FIGURE 6.5 Hunting (or wild) dogs (*Lycaon picus*) are native to the savannas of Africa. They are a true, wild species and not feral animals that have escaped domestication. As the name suggests, they are proficient hunters. They are also an example of a mammal in which helpers assist in rearing younger siblings. As is also the case with jackals, the litters are large, and reproductive success of parents is significantly greater if older offspring are present to help hunt and protect the newest litter. In this picture, an adult has just returned to the den with food and is about to regurgitate to feed the eager, begging pups. From Wilson, 1975, reprinted with permission of Harvard University Press.

nity to breed on their own. Breeding status is achieved through active, aggressive defense of territory, and a breeding pair may continue to act aggressively to suppress the breeding efforts of other group members.

Among mammals helping behavior analogous to that in birds—in which individuals postpone breeding and help rear the young of their social group—is rare. Most of the known examples are found among the canids or dog family. Among the canids, wolves, jackals, and the African wild dog hunt cooperatively in packs and share what they capture with each other and the young back at the den (Fig. 6.5). The packs usually

consist of a dominant breeding pair and helpers, which are previous offspring and/or siblings. In these species the presence of one or more helpers is usually crucial to rearing offspring, and the more helpers there are, the more offspring the breeding pair are able to produce.

The fact that helpers among the canids are usually close relatives is presumptive evidence that kin selection has been important in the evolution of this behavior. Additional factors may be at work, however. Because the litters are large and hunting is cooperative, successful breeding may be more dependent on delayed reproduction and helping than in the examples of birds described above. In other words, under most circumstances helping may be virtually obligatory and an ordinary component of the life history of these mammals. One might also hypothesize that the costs to the helpers of postponed breeding are more than offset by the later enhancement of their own reproduction resulting from the experience in hunting and rearing young they gain by helping. If this were true, though, we would expect nonrelatives to be helpers more often, and we would not expect significant numbers of individuals to spend their entire lives as helpers.

RECIPROCAL ALTRUISM: RETURN EFFECTS IN THE ABSENCE OF CLOSE GENETIC RELATEDNESS

The concept of kin selection provides an explanation for how phenotypic altruism directed toward genetic relatives—nepotism—can evolve. But many societies, and particularly human societies, contain many interacting individuals who are not closely related. The evolutionary success of such societies must therefore be based on some processes in addition to kin selection. In 1971, the biologist Robert Trivers provided a key insight in solving this problem by showing how natural selection can lead to altruistic exchanges between individuals, irrespective of their degree of genetic relatedness. The process is termed *reciprocal altruism*, or sometimes simply as *return effects*. Trivers' argument is easy to understand because this sort of behavior is familiar to us in our everyday lives, and without a doubt it also played an important role in our behavioral evolution (Chapter 13).

If in a population it is common for two or more individuals to exchange altruistic acts reciprocally and repeatedly and in such a way that the summed benefits to each of them exceed the costs, in succeeding generations the genes underlying these behaviors will increase in frequency. For example, imagine that at a time of food shortage Ed gives Sam enough food to keep Sam and his family from starving, but the gift does not endanger Ed's own survival. Or Ed, upon seeing a preda-tor, gives a warning cry that greatly increases Sam's chances of escape but only slightly increases Ed's own danger of being killed. If these hypothetical situations are later reversed—Sam helps Ed and the benefit to Ed exceeds the cost to Sam—then, over time, for both individuals, the summed benefits of these exchanges exceed the costs. Selection for this kind of behavior can operate whenever and wherever two or more individuals stand to gain more than they lose by such exchanges, even though an individual might lose more than it gains during some of the transactions (Box 6.2).

Reciprocal altruism requires several conditions. First and foremost is opportunity. Consequently, reciprocal altruism will be most likely to evolve in long-lived species in which individuals live in close proximity, providing occasions for repeated contacts. Furthermore, individuals must be able to recognize one another as individuals and be able to do things that benefit each other. A principal obstacle to the evolution of reciprocal altruism, however, is its vulnerability to cheating. Natural selection should favor participants who attempt to maximize benefits to themselves by not fully reciprocating. Following this line of reasoning, Trivers further suggested that selection should engender behavioral traits that will maximize benefits to self and minimize the costs of cheating by others. For example, individuals not only need to recognize reciprocators, they require cognitive mechanisms for (*i*) keeping track of their own credits and debits and of those individuals with whom they interact, (*ii*) for concealing their intentions to cheat, and (*iii*) for detecting cheating in others. There is thus a tendency in reciprocal exchanges for the donor to proffer aid when the benefit to the recipient exceeds the cost to self and to extend aid to individuals who have recently aided them.

As we have described reciprocal altruism, with actors Ed and Sam and with familiar cognitive requirements and psychological features, it seems to be a very human practice. Natural selection has indeed honed and embellished it in our species (Box 6.3), and we examine some of the implications in later chapters. But reciprocal altruism is not a uniquely human phenomenon. For example, vampire bats are reciprocal altruists, and their behaviors show many of the features expected of such exchanges.

SHARING OF FOOD BY VAMPIRE BATS

Vampire bats engage in reciprocal altruism by sharing food at the roost (Fig. 6.6). The stable social unit of vampire bats is a group of eight to twelve adult females with their dependent offspring that roost together during the day in hollow trees. Females, but not males, remain in their natal group. As biologist Gerald Wilkinson has shown, these associations are stable; an average of one unrelated female moves into the group every two years. The result of this pattern of birth, dispersal, recruitment, and movement between groups is

Box 6.2
The Prisoner's Dilemma

	He cooperates	He defects
I cooperate	**3** (We both cooperate)	**0** (I'm a sucker)
I defect	**5** (I take him for a sucker)	**1** (We both defect)

Suppose two individuals have been arrested because they are suspected of collusion in committing a crime. If neither confesses, there will not be enough evidence to convict, and both will be released. They are kept apart, and each is told that if he implicates the other, he will not be prosecuted and will receive a modest reward. On the other hand, if he remains silent and his partner implicates him, he will receive the maximum sentence. If each implicates the other, both will be jailed, but not for the maximum time.

Here is the dilemma that each faces. If neither implicates the other, both will go free. But can either count on his partner? If my partner decides to go for the reward by implicating me and I remain silent, I will be played for a sucker! So maybe I should implicate him. But if we both do that, we will both end up in jail! What should I do? What will he do?

This problem has no solution in a single instance, but Robert Axelrod has used computer simulations to determine what strategy should be played if two individuals interact in a series of encounters where each has the possibility of either cooperating or defecting. For each encounter, there are four possible outcomes. The relative reward a player receives at each encounter is shown in the following payoff matrix (where the value for "both cooperating" is greater than the average of the two "sucker" outcomes).

Computer simulations show that in a long series of prisoner's dilemma interactions one strategy is su-

perior to any other: "tit-for-tat." In other words, if two strategies are in competition, over a couple hundred encounters tit-for-tat will displace an alternative. The tit-for-tat strategy can be stated simply: Be cooperative on the first encounter, but thereafter treat the other player just as he treated you in the previous encounter. The success of tit-for-tat is enhanced if players can recognize who is playing by the same rule and interact with them preferentially and cooperatively.

In reciprocal altruism, cooperation is mutually desirable, but there is also a temptation to cheat and a severe cost to being cheated. The games of prisoner's dilemma therefore model reciprocal altruism, and the computer simulations indicate that tit-for-tat should be favored by natural selection and thus displace other strategies. Put in terms of human cognition, the computer simulations suggest that humans have not only evolved to seek cooperation, but also to retaliate when cheated, and not to hold a grudge if the other person shows a willingness to cooperate. The validity of such a general cognitive rule is not disproved if some people fail to follow it, or if most people occasionally violate it. Cognitive processes are flexible and contingent (Chapters 9 and 11).

---

that the females in a roost group are related to each other by $r \leq 0.1$. This low degree of genetic relatedness, coupled with the possibility of recognizing individual bats by placing small, numbered bands on their legs, made it possible for Wilkinson to distinguish between nepotism, brought about by kin selection, and reciprocal altruism, brought about by selection for the ability to engage in reciprocal exchanges.

The feeding behavior and food requirements of the bats Wilkinson studied suggest that reciprocal altruism must be crucial to their survival. At night the bats fly out to find large mammals, use their sharp teeth to cut exposed skin, and then lap the blood. On a given night about one third of the bats less than two years of age and 7% of older animals fail to feed. Other than inexperience, no other variable was found to be correlated with failure to feed; success seems to be a matter of luck, so all individuals are likely to fail a number of times. Moreover, it does not take many failures to get the bats

in trouble. If they do not feed for three nights, they starve to death! The likelihood that a given bat will fail two nights in a row is $1/3 \times 1/3 = 1/9$ for the younger bats and $0.07 \times 0.07 = 0.0049$ for the older bats. In other words, on any given night about 11% of the younger bats and 0.5% of the older bats are within a day of death by starvation. Finally, starving bats lose weight more slowly as their weight decreases, so that the longer a bat is deprived of food, the more a given amount of food increases its likelihood of survival. This means that the benefit of blood-sharing to a recipient is always greater than its cost to the donor, assuming only that the recipient has been deprived longer than the donor.

Wilkinson banded bats and observed their roosting associations and food exchanges; he also removed and returned bats to their roosts in order either to feed them or deprive them of food. He found that the sharing of ingested blood was quite specific with regard to the donors, the recipients, and their associations and

Box 6.3
Social Reasoning and Detecting Cheaters

The psychologist Peter Wason designed tests to explore whether the kind of thinking we frequently use is really the same as employed in scientific reasoning—the search for evidence that falsifies hypotheses. He devised a task that required subjects to choose information that could be used to determine whether a conditional rule—"if P, then Q"—had been violated. This rule is only violated when P is true and Q is false, so the logically correct way to check for violations of the rule is to determine whether "P" and "not Q" are found together.

As an example, suppose you are a bartender who must enforce drinking-age laws. The rule is "if a person is drinking beer, then they must be at least 21 years old" (if P then Q). Now suppose further that in order to be served, patrons must present a card that says either "drinking beer" or "drinking soda" on one side and has their age on the back. A group of four patrons places their cards on the bar, and you see "drinking beer," "drinking soda," "23," and "18." Which of the cards must you turn over in order to see if that patron is violating the rule?

If you are like most people who take this test, you chose "drinking beer" and "18." The logic is clear: one has checked for the association of "P" (drinking beer) and "not Q" (less than 21), and it doesn't matter what is on the reverse side of the other two cards.

The most interesting feature of this kind of test becomes apparent when the context is varied. Suppose the rule has the same logical form but is more abstract "if D, then 3." When subjects are presented with four cards showing "D," "F," "3," and "5" and asked which ones must be turned over to see if the rule is being violated, most chose just "D" or those marked "D" and "3." The correct answer is "D" and "5" (P and not Q). The rule says nothing about whether F and 3 or 5 and D might be associated.

These two problems are formally identical but differ in the context in which they are presented. The evolutionary psychologists Leda Cosmides and John Tooby argue that the different responses reflect evolutionary specialization of the human mind for reasoning about social interactions that involve reciprocity. More specifically, the first example involves reasoning about how to detect cheaters in social exchanges. Cosmides and Tooby have found that in a variety of other pairs of tests, subjects give different responses to purely abstract problems than they do to problems with social content.

A striking pair of similar tests carried out by the psychologists Gerd Gigerenzer and Klaus Hug supports the argument that the human mind is designed to police social exchanges. The tests had the same social context, and the conditional rule was "if an employee gets a pension, that individual must have worked for the firm at least ten years." The two tests differed, however, in the perspectives of the individuals taking the test. One group of subjects was told they were the employer, and the other was told they were the employee.

The four cards were "pension," "no pension," "worked eight years," and "worked ten years." In three separate experiments, 70–80% of the subjects playing the role of employer chose to turn over the cards saying "pension" and "worked eight years" (thus following the rule "if P then not Q"), and 1–5% chose "no pension" and "worked ten years" (applying a different rule, "if not P then Q").

Those subjects cast as employee produced the opposite pattern of responses: only 10–20% chose "pension" and "worked eight years" whereas 55–65% chose "no pension" and "worked ten years." The "employees" seem to have been unable to apply the logic of the rule, but that misses the point. What seems to constitute cheating depends on the subject's perspective. If you are the employer, an employee getting the pension without having worked ten years is cheating. On the other hand, if you are the employee, not getting the pension after ten years of service is cheating.

---

nutritional conditions. Bats that roosted together exchanged blood significantly more often than bats that did not; no animals that roosted together less than 60% of time fed each other. Deprived bats were fed by their roost mates soon after return to the roost, whereas well-fed bats that returned were never fed. There was preferential feeding of recipients in dire need (average of thirteen hours before starvation) by donors in average condition (average of forty hours before starvation). Finally, starved bats that received blood were

significantly more likely to be donors when the predicament was reversed.

Food sharing in vampire bats is a well-documented example of reciprocal altruism in a mammal, dramatic for its occurrence in what one might think was an unlikely place. The example shows that several behaviors that seem to require human cognitive abilities are in fact displayed by animals with very different brains. Vampire bats are able to discriminate the needy from the well-fed, those likely to reciprocate from strangers, and

FIGURE 6.6 Bats are virtually blind and move about by echolocation—listening to the echoes of their own ultrasonic cries. This is the same principle on which sonar works. Note the enormous ears; the part of the brain dedicated to auditory input is also large. This species is not a vampire bat; it catches insects on the wing. Other species eat fruit, and still others catch fish by trolling their sharp toes through the water and gaffing minnows that have come to the surface to feed.

those who have fed them in the past. The presence of memory and the cognitive capacity to engage in complex social relations is therefore not the sole province of humans. Reciprocity in humans, its roots in our primate relatives (Chapter 13), and the role it likely played in our mental evolution requires a separate and more detailed treatment to which we will return in Chapter 14.

PARENT-OFFSPRING CONFLICT

Relationships between parents and their dependent young in the animal world are never completely harmonious, and conflict is usually most intense when the offspring are close to becoming independent. Anyone familiar with breeding domesticated animals has seen the disputes between mothers and their foals, calves, puppies, or kittens over how long they may continue to nurse. And we all know from experience that a primary source of family turmoil during adolescence is differences between actual and expected parental attention and support. Why should such behavioral conflict exist and under what kinds of circumstances should it be more or less evident?

The same Robert Trivers we met earlier in the discussion of reciprocal altruism is responsible for a seminal paper on the evolution of parent-offspring conflict, which was published in 1974. Trivers proposed that conflict in genetic interests is expected between parents and their offspring over the resources or "investment" that parents should allocate to their young. Trivers defined "parental investment" as anything a parent does that increases an offspring's likelihood of survival and reproduction at a cost to the parent's ability to invest in other offspring. For example, when a parent expends energy feeding an offspring and takes risks in protecting it from predators, the ultimate benefits of these behaviors are to the survival and eventual reproduction of

that offspring, and the ultimate costs are to the parent's ability to produce and sustain other offspring, present or future.

The evolution of conflict between parents and offspring over parental investment can be understood by considering how genes for an organism's behavior affect the survival of identical copies in relatives. A parent's genes *for investing in close relatives* have an equal probability ($r = \frac{1}{2}$) of being present in each offspring, so selection should favor genes that *lead the parent to invest equally in each offspring*. An offspring's r to full siblings is also $\frac{1}{2}$, but its degree of relatedness to itself is 1. Consequently, the offspring will value benefits it receives, relative to the costs of benefits bestowed on full siblings, twice as highly as does the parent. We therefore expect parents and offspring to differ in how much parental investment the offspring should receive. Specifically, an offspring will be selected to demand more investment from a parent than the parent is selected to provide.

THE EVOLUTIONARY BASIS OF WEANING CONFLICT

The argument can be made clearer by using Trivers' example of a female mammal nursing her young. The parental investment in this case is milk, and the mother's degree of relatedness to each offspring is $\frac{1}{2}$. From the perspective of the mother's genes, benefits provided to one offspring come at a cost to other offspring, to which her degree of relatedness is the same. She will tend to stop nursing a particular offspring when the benefit B to the recipient becomes less than the cost C to other offspring (i.e., when $B/C < 1$). From the perspective of the recipient offspring's genes, however, the equation is different: he will value benefit to himself (relative to cost to siblings) twice as highly as does his mother, and he will continue to solicit food until $B/C < 0.5$.

A graphical representation of how the *B/C* ratio of a mother's milk is expected to change with the offspring's age is shown in Figure 6.7. The same argument applies, of course, to any other kind of parental care, such as protecting the young from predators, cleaning them, helping them in competition with others (relatives and nonrelatives), and teaching them how to fend for themselves.

Pursuing this reasoning further, there should be a progressive change in parent-offspring interactions because the benefit/cost ratio decreases as the offspring matures. Immediately after birth, when an offspring is small and cannot feed itself, a small amount of the mother's milk greatly benefits its chances of survival, so the *B/C* ratio is very high, and the mother will anticipate and be responsive to its offspring's needs. As the offspring grows in size, the amount of milk (i.e., the cost of production) needed to benefit the offspring by the same amount increases, so the *B/C* ratio decreases. As the *B/C*

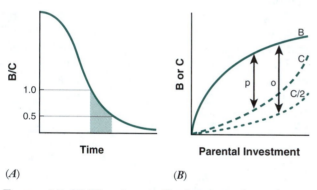

(A)   *(B)*

FIGURE 6.7 (A) Weaning conflict in mammals can be understood in an evolutionary perspective. Just after birth the ratio of the benefit (B) to cost (C) of a mother's milk is high: a little milk makes a big difference as to whether the offspring survives and reproduces (B), whereas the cost (C), in terms of how much that milk detracts from her investment in other offspring, is small. Because the youngster values his siblings relative to himself half as greatly as does his mother, there is a period (shaded) when there is parent-offspring conflict over the delivery of goods and services—when the youngster is eager to have more than the parent is easily willing to provide.

(B) A similar disagreement exists over the *amount* of parental investment provided by the mother. A small parental investment to any one offspring provides a big benefit and entails a small cost, but a large investment carries with it a smaller benefit and a larger cost. From the offspring's perspective, however, the costs will be devalued by a factor of 2. The parent will be inclined to withhold further investment when the difference between benefit and cost is maximal (short arrow labeled *p*), but the offspring will demand more until the difference between benefit and ½ cost is maximal (longer arrow labeled *o*). Remember that in this analysis cost and benefit relate to the lifetime reproductive success of the parent, but the offspring's "interests" in the outcome are influenced by effects on close kin as well as itself.

ratio reaches 1, we expect the mother to curtail nursing by refusing its offspring's demands and decreasing the secretion of milk. At this point and beyond, when the *B/C* falls below 1, the mother should resist nursing. If the offspring can cause the mother to continue nursing, the *B/C* ratio may eventually fall to 0.5, at which point the offspring should stop nursing because further benefits to itself are less than the benefits to its genes in full siblings. Proximate causes of behavioral change include a developing ability to utilize other sources of food.

This sort of change in parent-offspring interactions is seen, in one form or another, in virtually all animals in which the young depend upon their parents for food and protection between birth and adulthood. Parents of birds and mammals are typically very protective and solicitous toward their newborn young, and as the young mature the parents become progressively less responsive. Finally, the young, now grown in size, become so physical and emotional in their demands for investment that the parent's rebuff may not be gentle. In the case of nursing mammals these behaviors are so striking that they have their own descriptive name: "weaning conflict." These interactions have previously been seen as the parents' way of preparing its young to fend for themselves, but this argument does not make much evolutionary sense. Why should young animals resist what is good for them?

By this point you may have recognized ways in which this description of parent-offspring conflict applies to the interactions between human parents and their young. This should not be surprising, for parental investment in humans extends for longer than in any other species and therefore can become complex and intense. Trivers extended his analysis to provide insight into many other aspects of parent-offspring conflict, and a number of these insights follow quite logically from the concept as we have outlined it to this point.

PSYCHOLOGICAL ADAPTATIONS OF OFFSPRING AND PARENTS

Young animals cannot physically coerce their larger parents into extending investment, but psychological manipulations can be used to accomplish the same end. Parents are adapted to respond to the distress cries of offspring, but offspring may feign greater hunger or dependence on the parent than is actually the case, or they may become so loud, disruptive, and interfering in their attention-getting behavior as to "blackmail" the parent into acceding to their demands. A psychological arms race ensues, in which the parent must distinguish between real and exaggerated needs.

Anyone who has watched the interactions between young birds or farm animals and their parents has seen how loud, aggressive, and persistent the young's begging can be and how infantile are their postures and cries. To a young animal the arrival of a new sibling is a sure signal that its parents' investments and attentions will be divided; the behavior often triggered by this event is so intense and characteristic in humans that, like weaning conflict in animals, it has its own name, "regression." Most, if not all, human parents have seen what they consider attention-getting and interfering behavior by their young children; indeed, the descriptive terms themselves imply that the behavior is excessive, a sure sign that there is conflict. But the theory of parent-offspring conflict sees parents as potential psychological manipulators as well. As Trivers notes, the parental argument that curtailment of investment benefits young by preparing them for adult independence could itself be seen as usefully manipulative and self-deceiving.

SOCIAL BEHAVIOR OF OFFSPRING TOWARD NEAR RELATIVES

Why do young siblings so often fight, misbehave, resist control, and in general behave more egoistically than their parents want them to? Why do parents have to threaten, coerce, and preach? The socialization of young animals, especially humans, is often thought of as a learning process that enables them to acquire the behaviors they will need as adults in order to be successful in their family and social groups. If this process is strictly in the best interests of the young, as in the case of weaning conflict, it is a mystery why it is so often disharmonious.

It therefore isn't sufficient to provide answers solely in terms of proximate cause. Johnny snitched his brother's cake because he thought he didn't get a fair share or because his brother did the same thing to him yesterday; or Johnny misbehaved because his parents are not always there and he feels insecure. These kinds of explanation have an immediate validity in the domain of psychology, but an explanation in terms of evolutionary cause can provide a deeper, albeit complementary perspective by addressing why Johnny and his siblings persist in their squabbling, no matter how well-provisioned or secure their lives really are. The theory of parent-offspring conflict provides such an insight.

Just as with weaning conflict, we refer to Johnny's degree of relatedness to himself ($r = 1$) and to his siblings ($r = \frac{1}{2}$). In any perceived conflict over resources, from cake to parental attention, Johnny is therefore inclined to behave in a manner that parents find egoistic. As the parents' r with all their children is $\frac{1}{2}$, they are inclined to view these squabbles not only as disruptive, but they will wish to see the cake and attention apportioned equitably. In other words, in social conflicts with siblings, Johnny's sense of the appropriate B/C ratio (although he will not conceptualize the problem like this!) will differ by a factor of 2 from that of everyone else in the family. The traditional view of developmental psychologists that children must be socialized out of such egoistic behavior is not incorrect; it just doesn't address

the question of why in evolutionary terms the problems of socialization take the form that they do. The case of sibling rivalry is thus an example of how an evolutionary explanation can complement an explanation that is cast solely in terms of proximate cause.

DEPENDENCE ON THE MATING SYSTEM

The examples of parent-offspring conflict we have so far discussed pertain to a monogamous mating system in which all offspring are full siblings and therefore $r = \frac{1}{2}$. Under these conditions conflict is predicted for the period when $1 > B/C > 0.5$. If the mating system is at the other extreme, in which each sibling is fathered by a different male and r between them is therefore $\frac{1}{4}$, more prolonged and intense conflict is expected, namely when $1 > B/C > 0.25$. Indeed, comparisons of the intensity of sibling conflict and parent-offspring conflict in species with known degrees of sibling relatedness have shown that both kinds of conflict increase with decrease in r. In humans, children usually know (or frequently wish to discover) who their genetic parents are. In comparison with families in which there are full siblings, those with half-sibs and stepparents have additional sources of conflict between siblings, between parents and offspring, and between the parents themselves, especially if both bring to their marriage children from previous marriages (Chapter 14).

CONFLICT DURING PREGNANCY

How parents allocate their resources influences their lifetime reproductive success.

> Since parental investment begins before eggs are laid or young are born, and since there appears to be no essential distinction between parent-offspring conflict outside the mother (mediated by behavioral acts) and parent-offspring conflict inside the mother (mediated primarily by chemical acts), I assume that parent-offspring conflict may in theory begin as early as meiosis.

This prescient prediction made in 1974 by Robert Trivers suggested that parent-offspring conflict might begin *in utero* with a fetus attempting to maximize the benefits it receives. Indeed, the evolutionary geneticist David Haig has recently described evidence for parent-offspring conflict during human gestation. The measures and countermeasures seem so clearly antagonistic that he has likened these mother-fetus interactions to a confrontation between two armies. At the very least, they suggest evolutionary compromises reflecting the somewhat different genetic interests of the two parties.

After the embryo implants in the mother's uterine wall, its specialized trophoblast cells invade the uterus, break down the smooth constrictor muscles of the adjacent arteries, and form the "front line" of placental tissue in contact with the mother's circulatory system (Fig. 6.8) As a result of this initiative, the fetus gains control over several important parameters that later affect its supply of nutrients from the mother. First, the uterine arteries cannot respond to substances that are secreted by the mother into her blood to constrict the flow of blood to the uterine wall. She therefore loses this control over the flow of nutrients across the placenta. Second, the fetus can secrete substances directly into the mother's blood that increase the flow of blood and thus the flow of nutrients. Third, the fetus can regulate how much of certain substances in the maternal blood reach the fetus.

The maternal tissues that line the uterus—collectively called the "decidua" because they are shed at birth with the placenta—respond to the fetal cells in a way that looks more like defensive countermeasures than an opened-arm welcome. The stromal cells of the uterine lining secrete macromolecules that form a tough extracellular barrier or capsule around the arteries and in the path of the invading placenta. In their turn, the fetal cells secrete digestive enzymes that break down the barrier, and the uterine cells reply by secreting inhibitors of these enzymes.

Further evidence for *in utero* conflict between the mother and the fetus is seen in the regulation both of glucose concentration in the maternal blood and of maternal blood pressure. During early pregnancy, the mother's blood glucose level between meals falls, but after twelve weeks it stabilizes at a new low level until the baby is delivered. This lowering of glucose, which causes tiredness during early pregnancy, does not seem to be due to the fetus, because its early demands for energy are low and do not increase until later when the maternal supply of glucose has stabilized at the lower level. It appears instead that early in pregnancy the mother resets her blood glucose level to a low value in

Relationship of the chorionic villi to the maternal blood in the uterus

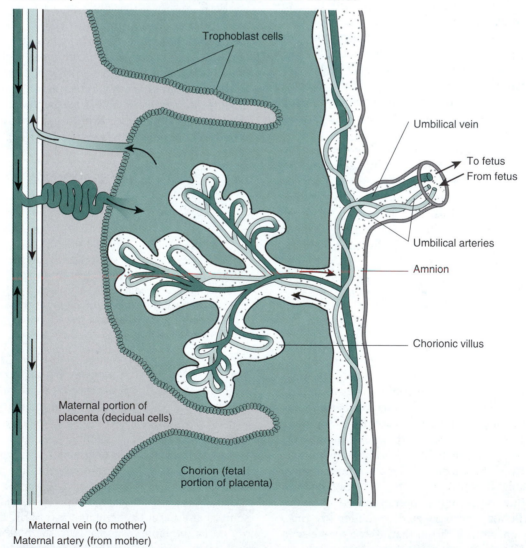

FIGURE 6.8 The relation between the maternal and fetal blood supply. Fetal blood vessels are bathed by blood from the mother's circulatory system.

anticipation of the fetus's demands, so that the fetus cannot remove more than it needs and more than is in the mother's genetic interest.

The fetus, operating from its advantageous placental beachhead, employs countermeasures. During the third trimester of pregnancy, the mother's blood level of insulin (the hormone that promotes removal of glucose from her blood) increases in concentration and at the same time becomes much less effective in removing glucose, especially after meals. This leads to higher glucose levels in her blood, and in extreme cases to gestational diabetes. There is good evidence that two hormones released by the fetus into the mother's blood, placental lactogen and placental growth hormone, interfere with the mechanism by which the mother's insulin lowers her blood glucose. The placental secretion of these hormones cannot be regulated by the mother, and despite their high concentrations neither is essential for a successful outcome of pregnancy. It appears that these hormonal interactions between the fetus and its mother are designed for interactions other than efficient and cooperative communication. As David Haig puts it: "If a message can be conveyed in a whisper [low concentrations of hormones], why shout? Raised voices are frequently a sign of conflict."

REPRODUCTIVE FUTURE OF THE PARENT

In many animals for which the supply of food varies during the rearing of young, the parents produce more offspring than are likely to survive to adulthood if food should become scarce. Under such conditions, some of the eggs or young may be abandoned or cannibalized by the parents and/or the siblings. Examples are found among hawks and owls, whose young grow rapidly in early spring. At this time of year, a late winter storm can make food difficult to find. There may be three young in the nest, one of which is smaller because it hatched last and because its larger siblings frequently monopolize the food as it is brought to the nest by the parents. If one of the young is cannibalized by parents or siblings, it is always the smallest.

We can readily see why and how natural selection has favored this behavior. Conditional infanticide, exercised when food is scarce and the entire brood is threatened, can rescue reproductive success for the current season. From the genetic perspective of the parents, benefits exceed costs ($B > C$) if the sacrifice of one offspring substantially increases the chances that at least some of the young will survive. The argument is equally clear from the genetic perspective of the larger of the nestlings: kin selection will favor siblicide when $B > \frac{1}{2} C$. But even from the perspective of the victim, kin selection should support the behavior if $B > 2C$.

Observations thus demonstrate that parents can assess the reproductive prospects of their offspring and redirect their parental investment so as to maximize the number of grand-offspring they leave, even if it requires reducing the number of young in the immediate future. Once again we must caution you about the simple language we are using. Do not read the word "assess" as implying conscious calculation of probabilities. We simply do not have many common English words that characterize the effects of behaviors without also suggesting human mental processes (Chapter 11).

As an organism ages, its potential for reproduction declines. In time it will not have enough resources or live long enough or be physiologically capable of producing another offspring. (In women this line is crossed at the time of menopause.) If at this juncture the organism has dependent young in its care, selection should favor parents who invest their remaining energies and resources in those last offspring. Such a pattern of behavior has been found: older animals generally feed, protect, and accede to the demands of their offspring more than do younger parents, and parent-offspring conflict is correspondingly less. This pattern of indulging the last child or a grandchild is familiar to humans as "doting."

SEXUAL SELECTION

Males and females are defined on the basis of primary universal differences related to the production of different kinds of gametes: males produce small motile gametes called *sperm*, whereas females produce larger, less-mobile gametes called *eggs*. Eggs are many times larger than sperm because they contain virtually all the cytoplasm that will be present in the zygote as well as nutrients that sustain growth during the early development of the embryo. Eggs can be enormous, particularly in birds. The need to make large gametes containing sufficient reserves to launch development was a by-product of the evolution of multicellular life. To be multicellular is to be larger, and how are gametes to find each other over a distance, particularly if they are big cells? The evolutionary solution was for

FIGURE 6.9 Bull elk following cow. The males of elk, deer, and many other hoofed mammals have large antlers or horns that are the product of sexual selection.

one sex to supply the cytoplasmic reserves in relatively large immobile eggs while the other sex delivered its DNA to the door in small packages (sperm, pollen), frequently motile, and invariably produced in large numbers. This is the quintessential example of disruptive selection, introduced in Chapter 4.

In addition to this primary distinction, males and females also differ in a external genetalia and *secondary sexual characters*. The former are directly related to the transfer and reception of sperm and the nurture of young—for example, the penis of male mammals and the vagina and mammary glands of female mammals. Charles Darwin called attention to an additional array of secondary sexual characters not directly connected with the act of reproduction but used in gaining matings with members of the opposite sex. In his introduction to

FIGURE 6.10 The Hamadryas baboon (*Papio hamadryas*) is an example of a mammal in which there has been considerable sexual selection. The animals feed during the day in groups consisting of one male and several females and their young. Here the large male, with its heavy mane and large canine teeth is threatening another male in order to keep him from the females. Two of the several females in his group are standing behind him, one with an infant, still with its juvenile black coat, clinging to her back. At night the single-male, multi-female family groups congregate in trees or on rocky ledges for protection from predators like leopards. Such a large troop is seen in the background as it disperses for the day to feed. From Wilson, 1975, reprinted with permission of Harvard University Press.

this topic in the 1880 edition of *The Descent of Man and Selection in Relation to Sex* Darwin gave some familiar examples: "... the weapons of offence and the means of defense of the males for fighting with and driving away their rivals—their courage and pugnacity—their various ornaments—their contrivances for producing vocal or instrumental music—and their glands for emitting odours ...." Darwin is referring here to such features as the antlers of male deer and the large canine teeth of male baboons, the bright plumage and characteristic songs of many species of male birds, and the habit of many male mammals of marking their territory with scent glands. When, as in these examples, males and females appear different because of either size or such elaborate secondary characters, they are said to display *sexual dimorphism* (Greek for two forms) (Figs. 6.9–6.12).

Although it is true that males are frequently larger and more combative and possess bigger weapons (antlers, spurs, horns, tusks, canine teeth) or are more colorful and vocal than females, there are many species in which males and females are indistinguishable in behavior and appearance. Furthermore, there are some species in which the usual sexual dimorphism is reversed: the females are larger and more colorful, vocal, and combative than males. What sense can we make of this diversity?

Because these secondary sexual characters are used most conspicuously when males and females either mate or form pair bonds, Darwin suggested that a variant of natural selection, which he termed *sexual selection*, had caused their evolution. Darwin's idea was that although two individuals might differ little in their ability to obtain food and escape predators, they might differ greatly in the ways they acquire mates. One way that differences in mating success arise is by competition among individuals of the same sex (typically males) for access to the other sex. Sexual selection then enhances those characteristics that make individuals successful competitors; for example, the presence of large antlers and great stamina in male deer and elk (Fig. 6.9) and the large size and long canines of male hamadryas baboons (Fig. 6.10). In another form of sexual selection, individuals of one sex (typically females) chose their mates on the basis of characters displayed by the opposite sex. In this variant, selection elaborates those traits that are used in displays. The most familiar example is probably the spectacular ornamental plumage of male birds such as peacocks, pheasants, and birds of paradise that are displayed to females during courtship

(Fig. 6.11). Traits like the peacock's tail seem to have no function other than in courtship displays.

As just described, these two forms of sexual selection represent polar extremes. Thus some traits might be favored both because they are advantageous in com-

FIGURE 6.12 Male red-winged blackbirds are glossy black with bright red and yellow epaulettes. They sing and display (*above*) to defend their territories and attract females. The females (*below*) are brown with streaked undersides, thus camouflaged.

FIGURE 6.11 Male sage grouse (*Centrocercus urophasianus*) displaying before females by spreading its tail and rapidly inflating and deflating air sacks in its neck. The neck sacks are not only visually prominent, they amplify the bird's vocalizations. The sage grouse is an example of a species where males congregate at sites called *leks* and compete with each other through elaborate displays for the attention of the females.

petition among males and also because they serve (indirectly) the genetic interests of the females. For example, large antlers and body size are advantageous to male deer in competition with other males, but in addition, females gain by mating with the winning males because if size of antlers and strength of body are heritable traits, their sons will more likely be successful competitors. Furthermore, as Trivers has suggested, daughters too can benefit. Large bodies and hefty antlers indicate males with superior abilities in procuring food and sequestering the calcium required for growing new antlers every year, two traits that are crucial to a daughter's ability to provide milk for *her* rapidly growing fawns.

There can be an even more direct interaction between male competition and female choice. Male red-winged blackbirds stake out a nesting territory and advertise their presence by singing and with visual displays of their red "epaulets" (Fig. 6.12). They thereby attract females, who in turn chose males that are in possession of the best territories, i.e., those that provide the best cover for nesting sites. At the same time, however, males are competing with other males for possession of the territory and will attempt to drive off rivals that intrude. The capacity to hold prime territory, like the ability to grow big antlers, is likely to reflect a more general genetic makeup that makes such males desirable mates.

The advantages of larger body size and better weaponry in competition are apparent, but it is not obvious why a female should prefer one male over another because it has tail feathers that almost require a valet to carry them about. The male peacock's tail is in fact a handicap outside the mating game, for it makes it more difficult for him to take flight when a fox or a tiger approaches. Why, then, did the peacock's tail evolve?

Darwin's explanation was simply that females are inherently the choosy sex and just prefer ornamented males. He reasoned that because at the microscopic level motile sperm seek out and fertilize the immobile egg, at the macroscopic level it must be males who seek matings wherever they can find them while females do the choosing. This argument is thin. First, there is no reason why adult mating behavior should reflect the behavior of gametes. Second, in many animals the male and female are equally choosy, and in some the females are colorful and combative and court choosy males (see below). Third, and more important, a preference for ornamentation seems to imply an aesthetic choice—a sort of "good taste" for displays that are multicolored, complexly patterned, and symmetrical. It is clear why a female deer might favor a large male with big antlers, but an appeal to aesthetics does not explain how preference for male ornaments could improve a female's reproductive success. Is there not a deeper explanation?

In 1915, the statistician and evolutionary theorist R.A. Fisher provided a possible explanation for how "good taste" for male ornaments could be favored by natural selection. He argued that if a particular male with an unusual variation of song, or color, or pattern of tail feathers happens—for whatever capricious reason—to gain the mating attention of females somewhat more often than other males, it is to the advantage of all females to prefer this "new fashion" because their sons will be more attractive to females as mates during the next generation. As a result, more daughters with their mother's preference for the new trait will be produced, and a runaway selection ensues in which further attractive elaborations reinforce the advantage of female preference for the trait. Selection for male ornamentation and female taste thus proceed hand-in-hand, and, depending upon the evolutionary path initially taken, may produce a complex, colorful, and symmetrical feather display or a song with extravagant or musical qualities.

More recently, however, evolutionists have been attempting to identify utilitarian connections between male ornaments and their reproductive advantage to females. One possibility is that the quality of an animal's display reflects the general quality of its genetic endowment; that is, it is an indication of whether the male has "good genes." This is actually an extension of the idea that large antlers of deer or the capacity of red-winged blackbirds to defend territory signify the presence of "good genes." One way in which such a connection could be established is the "handicap hypothesis" suggested by the biologist Amotz Zahavi. The more conspicuous and encumbering the peacock's plumage—the more of a handicap it presents—the better must be the bird's general genetic endowment that underlies his ability to create and support such an impediment and avoid the increased risk of predation it entails. Theoretical models suggest that such a mechanism could work, but measuring the summed lifetime benefits and costs of a secondary sexual character in different individuals is very difficult.

Another related explanation, offered by William Hamilton and Marlene Zuk, is that parasites have acted as a selective force in establishing a connection between male display and female choice. Parasites, and infectious organisms in general, are a particularly insidious and persistent threat because they are present everywhere, and their short generation times and genetic variability enable them to produce new variants to which their host is not immune. This threat requires constant genetic reshuffling by the hosts in order to mount new defenses against new parasites (Chapter 8), and males may be using more costly (by inviting predation) songs and displays to advertise that their health is not impaired by parasites. Thus females would be selected to mate with such males because their offspring would be more resistant to parasites.

FIGURE 6.13 The long tail feathers of barn swallows are important in mate choice. Experiments have shown that changing the symmetry of the tail by altering the length of one feather can compromise the attractiveness of birds to prospective mates.

Recently symmetry has been suggested as another possible link between male ornamentation, "good genes," and female choice. In most mobile animals the right side of the body is a mirror image of the left, and this symmetry is important in accurately gathering information from paired sensory organs and precisely coordinating movement of paired limbs. The development of complex, macroscopic, paired structures of the same size, shape, and distance from the midline of the body requires tuning of cell migrations, differentiation, and the activation of genes during development (Chapter 10). Moreover, high body symmetry has been found to be associated with higher metabolic efficiency, better immunity to infections, and lowered parasite loads. There is thus evidence for linkage between body symmetry and "good genes." Furthermore, there is increasing evidence from studies of insects, birds and mammals (including humans) that females prefer to mate with males with the most symmetrical features. For example, clipping one of the long tail feathers of a barn swallow (Fig. 6.13) reduces its attractiveness as a mate.

THE ROLE OF PARENTAL INVESTMENT IN SEXUAL SELECTION

The major deficiency in the theory of sexual selection, beginning with Darwin and extending to the middle of this century, was that it did not account for why, in most mating systems, males compete with each other for access to females and females are the choosier sex. Why isn't the reverse found more often?

The first step in answering this question was provided in 1948 in a study of sex differences in mating behavior and reproductive success in the fruit fly *Drosophila melanogaster*. The geneticist A.J. Bateman observed individuals with different genetic markers while the flies were feeding, mating, and laying eggs in closed bottles containing fly food. The pattern that he observed for the fruit fly is typical of many other animals in which the only investment males make in offspring is to contribute sperm during mating. First, male flies attempt to mate with as many females as they can, and females chose the males with which they will mate. Second, there is much greater variation in the reproductive success of males than among females. In Bateman's study the most successful male had three times as many offspring as the most successful female, and whereas only 4% of females had no offspring, 21% of the males fathered none. Finally, whereas the number of offspring a male fly fathered increased in direct proportion to the number of females with which he mated, females gained nothing by mating more than once; in fact, most of them mated only once. This reflects the fact that female *Drosophila* possess sperm storage organs, which enable them to sequester sperm from a single copulation and dispense it over many days.

Why is it in this animal, as in most other sexually reproducing species, that males mate relatively indiscriminately whereas females are more selective? Bateman argued that these differences evolved because of the differences in the metabolic costs of offspring to males and females: one sperm can fertilize an egg, but an egg is likely to be thousands to many millions of times more costly to produce than a sperm. The reproductive success of a male is therefore not limited by the number of sperm it can produce but by the number of females with which it can mate. Under these circumstances natural selection should thus favor males who produce large numbers of sperm and attempt to mate with as many females as possible, irrespective of the quality of the eggs any particular female produces. Conversely, because of the high metabolic cost of eggs, the reproductive success of a female is limited by her ability to find food and convert it into egg yolk. Given the high cost of eggs and the eager, indiscriminate, mating efforts of males, selection should favor females who find food well and choose from among competing males those individuals whose appearance and behavior suggest high quality.

In 1972 Robert Trivers extended Bateman's ideas into a more general conception of how sexual selection is regulated by the relative contributions of each sex to the production of offspring. Earlier we defined *parental investment* as any parental effort that promotes the survival of an offspring at the cost of producing another offspring. Parental investment starts with the making of eggs and sperm but includes all aspects of parental care. Male and female differences in the amount of parental investment are at the center of Trivers' thesis.

MINIMAL PARENTAL INVESTMENT BY MALES IS FREQUENTLY ASSOCIATED WITH POLYGYNOUS MATING SYSTEMS

For each egg that is fertilized, there has to be a sperm. Consequently (for most species, in which the sex ratio is close to 1:1 at birth) the *average* number of offspring produced by females and males is the same. But if the parental investment made by one sex (usually females) is substantially greater than that made by the other, then females become a limiting reproductive resource. In other words, the reproductive success of females is then limited by their capacity for parental investment (providing nutrients for eggs, fetuses, and nurslings), whereas the reproductive success of males is limited only by the number of opportunities they have to find and mate with females. Under these conditions, individuals of the sex that makes the smaller parental investment (typically males) compete with each other for matings. As a result, sexual selection is more intense among males, and its intensity is proportional to the disparity between the sexes in relative parental investment. Moreover, in long-lived species most females that survive to an age where they are capable of reproducing will manage to reproduce. But in the competition among males, only a minority will manage most of the copulations. Therefore, although the average reproductive success of males and females must be equal, there will be greater variation among males.

Finally, in species in which the parental investment of males is much smaller than in females, the mating system is *polygynous* (Gr., many females) in which one male mates with several to many females. This is another manifestation of the greater variation in male reproductive success; because of strong sexual selection among males, a few winners among the competitors are able to control the reproduction of a disproportionate number of females.

As we discussed earlier, greater female parental investment very likely started in the earliest multicellular animals with the divergent evolution of expensive eggs and inexpensive sperm. Once these separate evolutionary paths were taken, individual females would always have more to lose than males by discontinuing investment, so selection for parental investment after fertilization was likely to impact more on them than on males. Additionally, eggs and embryos can be more safely provisioned and nourished by remaining within the body that produced them. With the evolution of mammals, such a self-reinforcing process of increasing female investment has led to large disparities between the sexes in parental investment. A minuscule invest-

FIGURE 6.14 **Elephant seals in the northern hemisphere were driven nearly to extinction by human hunting, but they now breed successfully in a colony on the coast of California where they have been studied extensively by biologists at the University of California at Santa Cruz. In late winter the animals congregate on the shore in large colonies for breeding. The males weigh about 6000 pounds (3 tons!) and are about three times larger than the females. They engage in vicious fights among themselves, and the winners do most of the mating. In one study 4% of the males accounted for 85% of the copulations.**

ment of sperm accomplished by a male in a few moments initiates a costly and risky investment by a female that may take years, from the implantation of the embryo in the uterus, through development, birth, nursing, weaning, and adolescent dependence. In most mammals the mating system is polygynous, male reproductive success is more variable than that of females, and the degree of sexual dimorphism is greater than in any other class of vertebrates (most fish, amphibia, reptiles and birds). Figure 6.14 shows an example of a species with little male parental investment, a polygynous mating system, and considerable sexual selection for male size.

In Chapter 13 we will return to the different breeding systems employed by primates, and in Chapter 14 we will consider the extent to which humans are or are not typical mammals.

LARGE PARENTAL INVESTMENT BY MALES IS OFTEN ASSOCIATED WITH MONOGAMOUS MATING SYSTEMS

Birds occupy a number of points on the continuum of relative parental investment and thus illustrate the consequences for sexual selection and mating systems. In more than 90% of avian species the two sexes contribute approximately equally to the production of young. We might suppose that the large investment of female birds in egg yolk—compared to the tiny male investment in sperm—would make birds polygynous like most mammals. For many species, however, developmental and ecological factors outweigh the relative parental contributions to the fertilized egg (Fig. 6.4). For most of the small familiar songbirds of field and forest, the young hatch in a naked helpless state. Successfully rearing a brood requires that both parents forage for food, and that males make a substantial contribution to feeding the young and protecting them from predators. Additionally, the food of most small terrestrial birds usually consists of small items—insects, seeds, and berries—that are evenly distributed in the environment and can be efficiently exploited and defended by a mated pair. To all appearances, the mating system appears to be monogamous.

But appearances are deceiving. Careful observations of mating behaviors combined with molecular techniques for assigning genetic paternity reveal that in close to 60% of "socially monogamous" species of birds, one or both members are likely to seek extra-pair matings. Among the species in which this behavior occurs, 3–30% of the young are fathered by one or more extra-pair males. In some species the proportion reaches 40–60%.

In most bird species, the females control the success of copulation and sperm transfer, and the evidence suggests that females seek extra-pair fertilizations in order to improve the genetic quality of their offspring. They do not have difficulty in finding willing males, because extra-pair copulations cost males little. For one bird that has been studied in detail, the great reed warbler of northern Europe, females seek extra-pair matings with males that have larger song repertoires than the males with which they are paired. This is a successful tactic, because the offspring sired by these opportunistic males have better survival rates after fledging than young fathered by the resident male.

Although in most small birds the two sexes are similar in size, in many of these species the males sing and display, and there has been sexual selection on males for bright plumage. Female choice in such species is usually more important than in animals with intensely polygynous mating systems. Figure 6.15 shows an extreme example of a species with essentially equal parental investment by the two sexes, little difference in sexual selection, and a monogamous mating system.

In some species of birds, nests are built on the ground, the young hatch from the egg covered with downy feathers, and in temperate latitudes the hatchlings are able to feed themselves immediately. Male parental investment is consequently less. In grouse, sexual selection acting on males has produced ornamentation and noisy displays. These birds have an unusual mating system in which the males gather at display sites called leks, strut about, and winners are chosen by the females (Fig. 6.11).

SPECIES DISPLAYING REVERSED SEX ROLES HAVE POLYANDROUS MATING SYSTEMS

Nature provides a dramatic corroboration of the hypothesis that differences between the sexes in parental investment regulate sexual selection. The confirmation is seen in species in which the usual pattern of parental investment is reversed, and males make a greater parental investment than females. The attention of naturalists has long been drawn to such species of birds, particularly phalaropes, shorebirds that nest in the far north. The females are larger and more colorful than males, arrive earlier in the breeding area than males, and fight with each other to gain and retain males as mates. Moreover, females do not provide any parental care after they lay their eggs. The male broods the eggs and young and protects them from predators.

Ecological factors seem to have caused this reversal of the usual roles of males and females. These shorebirds nest on open ground in northern latitudes where the breeding season is short and the eggs are especially susceptible to predation. The young are able to feed themselves immediately after hatching, so only one parent is needed after the eggs are laid. After laying her eggs the female usually departs and attempts to mate and leave another batch of eggs with a new male. In other words, the mating system is *polyandrous* (Gr., more than one male). On average, successful females mate with more different males than males do with females.

FIGURE 6.15 The extreme southern latitudes pose a severe environmental challenge for ground-nesting penguins. Both the male and female are required to raise a pair of chicks—a single chick in the case of the winter-breeding Emperor penguin; consequently the birds have strong pair bonds, much male parental investment, a monogamous breeding system, and males and females appear identical (i.e., no sexual dimorphism).

Reversal of parental investment and sex roles is also found among some species of fish and frogs, and here too the males' provision of safe places for eggs to develop is a major factor limiting reproduction. In the tropical poison arrow frog, females are larger, more colorful, and more active in courtship than the male, who broods fertilized eggs in his mouth and takes the hatched tadpoles to water. In the sea horse (a fish), the female is active in courtship, and the male takes the fertilized eggs into his ventral brood pouch, kangaroo style, for safe keeping and growth.

To summarize, the reasoning of Bateman and Trivers is confirmed in species exhibiting male-female role reversal. It therefore seems to provide an encompassing understanding for how and why relative parental investment regulates sexual selection, which in turn leads to sex differences in size, morphology, and behavior.

SOME FURTHER BEHAVIORAL CONSEQUENCES

Because females (typically) make the larger parental investment, they have more to lose reproductively should a mating not lead to reproductive success. In species where successful rearing of young requires the continued participation of the male, it is in the interests of the female to mate with males who are likely to provide that additional parental investment. This is the reason for the evolution of courtship periods in which males demonstrate that they have more than "good genes" to offer. An obvious example is the male's possession of territory as a source of food for the young, as described earlier for red-winged blackbirds and reed warblers. Courtship provides an extended opportunity for the exercise of female choice and is an example of an evolutionary outcome that has been driven primarily by the genetic interests of females.

Males, with smaller parental investment than females, can benefit reproductively if they are able to fertilize the eggs of more than one mate, particularly if another male is left to tend the additional set of young. But in species where males make a substantial parental investment, their opportunities for extra-pair matings become more limited. The male who leaves his territory unguarded while in search of reproductive opportunities elsewhere runs the risk of loosing his resource base to another male. He also runs the risk of having his mate inseminated by another male, whose offspring he will then rear. This outcome—known as cuckoldry—is a genetic disaster because his investment is all cost and no benefit. Cuckoldry is a particularly acute problem in species where fertilization is internal and where males must observe their mates constantly to be certain they have not been inseminated by other males. Thus male swallows follow their mates on feeding flights until the female lays her eggs, and male mammals frequently seek copulations with their mates after periods of separation. The human emotion associated with such mate guarding is sexual jealousy, a topic to which we will return in Chapter 14.

MANIPULATING THE SEX RATIO

Why are there approximately equal numbers of males and females at the start of each generation? In 1930, the mathematician Ronald Fisher provided the answer to this question in evolutionary terms. Each fertilization requires the participation of a male and female, so the *average* reproductive success of males will be the same as the average reproductive success of females. This is as true in polygynous species as in monogamous ones so long as sons require the same average parental investment as daughters.

What would happen in a population in which ten times as many sons were produced as daughters? In the following generation, each daughter would have ten times the number of offspring as the sons. Natural selection would then favor individuals that produced more daughters than sons, and this would continue until the sex ratio approached 1:1. In other words, natural selection is expected to favor the sex ratio that gives parents the same return in grand-offspring per unit investment in each sex. In humans, pre-reproductive mortality of males is significantly greater than among females (a consequence of sexual selection), and the human sex ratio at conception and birth is significantly biased toward males.

In 1973 Robert Trivers and Dan Willard extended Fisher's reasoning by proposing that selection should favor parents that produce greater numbers of the sex that under certain conditions are likely to have more offspring than the other. They argued that in highly polygynous mammals, for example, larger males achieve disproportionately more matings than smaller ones, and adult size is closely correlated with birth size. Sons born large—because their mother is well nourished and healthy—are better able to compete for food, with the result that their initial size advantage is actually magnified. Males born smaller and weaker—because when food is scarce their mother is in below-average condition—are likely to remain so and achieve disproportionately fewer matings. The impact of scarce food is less severe on the reproductive prospects of daughters, however. Because the greater parental investment of females makes them a limiting reproductive resource for males, most females who survive to adulthood will manage to breed. Consequently, an adult female of less than average size and vigor is likely to outreproduce a comparably small male.

Trivers and Willard therefore reasoned that when polygynous mammals are able to produce sons that are larger and healthier than the population average, they should favor sons; conversely, if their offspring are likely to be smaller and less healthy than average, daughters are a safer bet.

OPOSSUMS MANIPULATE THE NUMBERS OF SONS AND DAUGHTERS

The Trivers-Willard hypothesis was considered improbable by many, including biologists Steven Austad and Mel Sunquist. So they decided to test the idea by manipulating the food supply in a population of opossums and observing what effect this had on the sex ratio of offspring. Opossums are familiar as the only North American marsupial mammal (Chapter 2). The young are born very immature and are nourished in the mother's pouch, as in kangaroos. Female opossums establish feeding territories, mate with a male, and rear their young alone. During the day a female sleeps in an underground den. The mating system is polygynous: males defend large mating territories that overlap the territories of many females, and they range widely to chase away other males. The larger a male, the more successful he is at dominating other males and mating with females. Body size does not affect a female's ability to mate or produce young, except when she is undernourished.

Austad and Sunquist trapped and radio-collared forty virgin female opossums in order to be able to find their dens and monitor their movements. Twenty females were provisioned with extra food—sardines left at the entrance to their dens—and twenty females were not given extra food. Both sets of females produced the same number of young, but those of the food-supplemented mothers were 50% larger. Along with an increase in size there was a 16% increase in survival of daughters and a nearly 30% increase in the survival of sons.

Most dramatic was the difference in the sex ratio of offspring of the two groups of mothers. Mothers whose food was not supplemented produced equal numbers of sons and daughters, but mothers with the extra rations of sardines produced 40% more sons. This finding led Austad and Sunquist to reexamine data they had gathered over several years on the offspring of aged females (two or more years old), which develop cataracts, lose weight, wean fewer young, and are generally in poorer condition than during their first year. Unlike the younger, well-nourished animals, these females produced 80% more daughters than sons. These experiments confirmed both sides of the Trivers-Willard prediction—the female opossum can bias the sex ratio toward sons or daughters, depending on which sex is

likely to achieve greater reproductive success and thus propagate more of the mother's genes.

RED DEER ALSO MANIPULATE THE SEX RATIO

Another test of the Trivers-Willard hypothesis was provided by a long-term study by Timothy Clutton-Brock and his co-workers on reproduction in a herd of red deer on the isle of Rhum off the coast of Scotland. Socially dominant females are in better physical condition than subordinate females because they are able to gain preferential access to good browse. As a result, dominant females have sons who grow up to be dominant, and subordinate females have sons who grow up to be subordinate. Clutton-Brock and his associates found that dominant mothers produced more sons than daughters, and they had more grandchildren by their sons than by their daughters. Precisely the reverse was found for subordinate mothers. These deer had more daughters than sons, and the daughters experienced greater reproductive success than the sons.

FEMALE DOMINANCE CAN ENHANCE THE REPRODUCTIVE SUCCESS OF DAUGHTERS

One of the many unforeseen discoveries of this research on the sex ratio is the finding that a female's rank in a dominance hierarchy can be more important to her daughter's reproductive success than to the reproductive success of her son. This has been found in two kinds of Old World monkeys: baboons and macaques. In these species daughters remain in their natal group and "inherit" the social rank of their mothers while males emigrate and are unaffected by their mother's rank. In the populations of bonnet macaques studied by the primatologist Joan Silk, dominant females and their allies gain preferential access to food and water and so severely harass the daughters of subordinate females that few of them live. Dominant bonnet macaque females give birth to more daughters than sons, and subordinate females do the reverse. Furthermore, in this social system subordinate females care for their daughters after birth much less than they do sons.

The evidence is now widespread and extensive that Trivers and Willard were right in suggesting that animals might respond to ecological and social conditions by adjusting the sex ratio of their offspring so as to maximize the number of grandchildren. Could this apply to humans, an animal in which the reproductive prospects of offspring can be greatly affected by the social rank and resources of parents? We will return to this question in Chapter 14.

SYNOPSIS

In this chapter we presented a framework for explaining the evolution of social systems. Behavioral interactions between individuals influence the reproductive success of one or both parties; accordingly they can be characterized as selfish, cooperative, spiteful, or altruistic. Behavior that is seemingly altruistic is very common among many species of animals, but it raises a question for evolutionary theory: If individuals who sacrifice their reproductive success for others leave fewer offspring than those who behave selfishly, how can altruistic behavior be supported by natural selection?

There are two ways. First, copies of one's genes are present in near relatives, so it is possible to enhance one's reproductive success through benefits extended to close kin. This is the basis for nepotistic behavior. It has its most elaborate expression in social insects, as we will describe in the next chapter. The changes in gene frequency that result from the effects individuals have on the reproductive success of relatives is *kin selection*.

Second, many social systems include individuals who are not closely related but who interact on a regular basis, and behavioral mechanisms have evolved to facilitate the reproductive success of individuals living in social groups. These behaviors involve reciprocal interactions in which benefits extended are followed by benefits returned. This *reciprocal altruism* is found in long-lived species with sufficient cognitive capacity to recognize individuals and remember and assess the results of previous interactions. We will examine its expression in nonhuman primates in Chapter 13 and in humans in Chapter 14.

The genetic uniqueness of individuals of sexually reproducing species is not only the basis of self-interest, it leads to conflicts of self-interest where one might least expect to find them. Forms of parent-offspring conflict provide examples, some of which can be identified during the gestation of mammals.

Many species are characterized by noticeable differences between the sexes in size, coloration, and behavior. Such sexual dimorphism was described by Darwin, who pointed out that it tended to be greater in species in which males (usually males) compete among themselves for access to females. Darwin coined the term *sexual selection* to designate a kind of natural selection that works on males and females of the same species in quite different ways because the two sexes have different strategies for acquiring mates.

Understanding of sexual selection was extended in the 1970s by Robert Trivers, who pointed out that the controlling factor is the degree of *parental investment* that each sex makes in offspring. Parental investment means anything that a parent does to benefit an offspring that comes at a cost to investing in other offspring, thus possibly compromising the lifetime reproductive success of the parent.

Parental investment is great in female mammals and quite modest in males (although humans are something of an exception to this generalization). The consequent limited lifetime reproductive capacity of females makes them a limiting resource for which males then compete, and the mating system is typically *polygynous*. In many species of birds, the need to feed helpless young requires both parents, and with a greater relative parental investment by the male the mating system is *monogamous* (or nearly so). A powerful demonstration of the validity of these ideas is provided in nature by species in which the degree of parental investment by males and females is reversed, as is sexual selection, and the mating system is *polyandrous*.

In polygynous mating systems, the relative reproductive prospects of male and female offspring can differ, depending on ecological conditions or the dominance status of the mother. There are several examples known where female mammals alter the sex ratio of their offspring so as to maximize the number of grand-offspring.

QUESTIONS FOR THOUGHT AND DISCUSSION

1. Occasionally identical twin sisters marry identical twin brothers. What is the degree of relationship between the offspring of the two couples? What is their nominal relationship: cousins, siblings, or what?

2. In order to attract females as mates, male bower-birds build elaborate displays of shiny and colored objects they have collected, including things discarded by humans. How might this behavior have arisen through natural selection? What would you predict about the degree of sexual dimorphism in plumage coloration in these birds?

3. Is there fundamental conflict or meaningful convergence between the biological concept of sexual selection and the tenets of feminism? Explain. (You may want to revisit this question after reading Chapter 14.)

4. A critic of the theory of kin selection said that it could not explain the evolution of nepotistic behavior in animals because animals know nothing of fractions (by which r, the degree of genetic relatedness is expressed), and it could not apply to humans because fractions are a recent invention in some cultures but are unknown to most. Comment on this criticism.

5. In our society many people adopt and raise children to whom they are not genetically related. Is this evidence against the operation of kin selection during our evolution? Explain.

6. Evolutionary social theory predicts conflict between individuals in inverse relation to their degree of genetic relatedness. Compare the degree of conflict predicted between parents and offspring for a monogamous versus a polyandrous mating system. Consider the behavior of siblings toward each other, the father's and mother's view of how the siblings should behave toward each other, the father's and mother's views of how much each of them should invest in the offspring, and the offsprings' view of how much each of the parents should invest in them.

SUGGESTIONS FOR FURTHER READING

Geary, D. C. (1998). *Male, Female: The Evolution of Human Sex Differences*. Washington, DC: The American Psychological Association. Recent, broad in scope, yet focused on humans, and with many references.

Haig, D. (1993). Genetic conflicts in human pregnancy. *Quarterly Review of Biology, 68*, 495–532. Evolutionary theory provides a fascinating perspective on mother-fetus interactions.

Trivers, R. L. (1971). The evolution of reciprocal altruism. *Quarterly Review of Biology, 46*, 35–57

Trivers, R. L. (1972). Parental investment and sexual selection. In Campbell, B. (ed.), *Sexual Selection and the Descent of Man 1871–1971* (pp. 136–179). Chicago, IL: Aldine.

Trivers, R. L. (1974). Parent-offspring conflict. *American Zoologist, 14*, 249–264. Three truly original and lucid papers that contributed fundamentally to the development of evolutionary social theory.

Trivers, R. L. (1985). *Social Evolution*. Menlo Park, CA: Benjamin/ Cummings. Written as an undergraduate text, this book clearly explains and illustrates the broad biological foundations of evolutionary social theory. Out of print but available in libraries.

PART TWO

Evolution in Action

William D. Hamilton (1936–2000) was one of the most important contributors to Darwinian evolutionary theory of the 20th century. His early work provided the foundation on which evolutionary social theory is constructed (Chapter 6), and it led to an understanding of social evolution in bees, ants, wasps, and termites (Chapter 7). He proposed that a principal reason for sexual reproduction is found in the complexity of host-parasite interactions (Chapter 8). While in Africa looking for evidence of how the virus that causes AIDS in humans may have evolved from a similar virus found in chimpanzees he contracted a fatal case of malaria.

7

The Success of Social Insects

CHAPTER OUTLINE

- *The power of kin selection*
    - *The concept of superorganism*
    - *The natural history of honeybees*
    - *The honeybee's altruistic reproductive behavior has an unusual genetic basis*
    - *Termites are incestuous*
- *Behavior in honeybee societies is both complex and adaptive*
    - *Information is needed within the colony*
    - *Foraging bees communicate their finds to their hive mates*
    - *There is no central authority for assigning tasks in a bee colony*
    - *The behavior of bees appears to be rational*
    - *Colonies of cells and colonies of organisms use similar control principles but different mechanisms*
- *Synopsis*
- *Questions for thought and discussion*
- *Suggestions for further reading*

Why devote a chapter to insects in an account of evolution and behavior that leads to humans and other primates? There are two reasons. First, colonies of bees, ants, and termites are among the largest and most elaborate social structures on earth. Moreover, bees also provided early and important insight into the concept of kin selection.

Second, these creatures are fascinating for the complexity of their behavior. More than a few writers have sought to ennoble humanity by contrasting our free will with the unthinking behavior of social insects. This comparison is so facile, however, that it obscures an even more interesting feature of insect behavior. On looking closely we find that the roles of individual bees and ants are not so rigidly fixed as this contrast suggests. Tasks change, and the work of individuals is so well coordinated that the colony is able to adjust to shifting resources and new environmental challenges. Comparing the processes by which the behavior of social insects and humans are regulated in an adaptive manner reveals the scope and splendor of natural selection.

Photo: Ants, like bees, illustrate the power of kin selection to generate complex social structures. These leaf-cutter ants (*Atta cephalotes*) are farmers. They cut pieces of leaf, carry them to their nests, and grow a fungus on them. They then eat the fungus.

THE POWER OF KIN SELECTION

Bees, ants, wasps (order Hymenoptera), and a more distantly related group, the termites (order Isoptera), live in large colonies in which one or a few members do all of the reproducing and most of the individuals simply assist. In terms of numbers, ants and termites represent particularly successful evolutionary outcomes. In the tropics, where the diversity of species reaches a maximum, there can be 200–300 species of ant in a square mile, and as many as forty-three different kinds of ant have been collected from a single tree. Furthermore, individual colonies of ants and termites can contain millions of individuals, and in tropical rain forests these social insects make up more than three-quarters of the total insect biomass.

Colonies of bees and ants have a "queen" who lays eggs and thousands of workers that tend the eggs and larvae, build and clean the nest, and forage for food. Ants differ from bees in that the colonies are generally larger, the workers do not have wings, and the division of labor among workers is so marked that workers frequently come in two body forms or *castes*, specialized for different functions (Fig. 7.1).

THE CONCEPT OF SUPERORGANISM

A colony of bees or ants is sometimes likened to a superorganism, because the behaviors of the individual insects are so well tuned to collective need that the colony seems to function with the coordination typical of a single animal. The existence of these evolutionarily successful creatures therefore raises a number of important conceptual issues. What is the biological basis for so many individuals foregoing their own reproduction? How is information about the needs of the colony discovered? How do individual insects communicate that information? And how is the behavior of individuals coordinated so that their actions are cooperative and adaptive?

THE NATURAL HISTORY OF HONEYBEES

Although ants are equally interesting, we will use the common honeybee (*Apis mellifera*) to address these questions. Bees are exploited the world over. Honey is a source of food for humans, and in agricultural regions bees are reared commercially so that they are available in large numbers to pollinate cash crops in the spring. Most importantly, bees have been studied extensively, and understanding their way of life is a source of considerable interest.

A colony of bees consists of 20,000 or so worker bees and a queen. The workers, like the queen, are female, but only the queen is capable of laying fertilized eggs. All of the workers are daughters of the queen, and are therefore sisters. There are some males present too, called drones. They are somewhat larger than the workers (Fig. 7.2), but they make no contribution to

FIGURE 7.1 Castes of a species of ant. The queen is at the bottom right, and a male (with wings) just above her. All of the other individuals are nonreproductives: soldiers and workers.

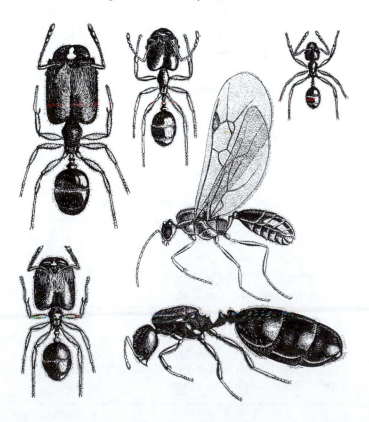

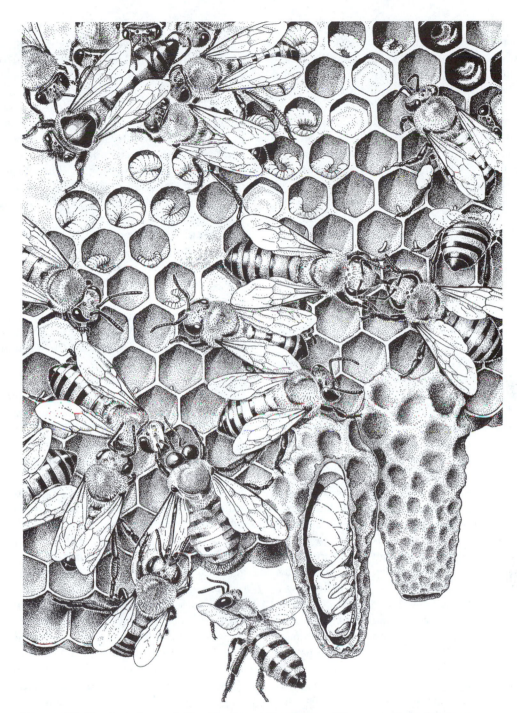

FIGURE 7.2 Portion of a comb from a colony of honeybees. The queen is in the upper left surrounded by several worker attendants. Several of the open cells on the left contain eggs or larvae in different stages of development. The open cells in the upper right contain honey. At the bottom left a drone is being dragged outside by a worker, who has seized a wing. The two large cells at the lower right contain developing queens.

the life of the hive and exist only to fertilize new queens. Their numbers are under the control of the females. When they are needed, the queen lays more drone eggs (you will see presently how she can control which eggs are to become drones and which workers), and when there are more drones present than are needed, the workers drive them from the hive.

Bees form colonies in protected places such as hollows in trees where they secrete combs of wax consisting of precise arrays of hexagonal cells (Fig. 7.2). The cells serve multiple purposes. Some are brood cells, in which the queen lays a single egg. Others are storage cells, containing either pollen that has been collected from flowers, or honey, which the bees make by concentrating the nectar they collect from flowers. Pollen and honey are the food of bees.

The eggs hatch after about three days, and the larvae are fed by nurse bees. About six days later the larvae pupate, still in the comb in their individual wax cells. After another twelve days the adult bees emerge. A worker bee lives for several weeks, except for those present at the end of the season, which live through the winter and continue the life of the colony when the weather warms the following spring. The queen can live for several years.

A worker bee takes on different tasks during its lifetime. The youngest bees remain in the hive and are initially engaged in cleaning the brood chamber (the region of combs used for rearing young) before becoming nurse bees. Nurse bees provide food to the larvae and attend to other tasks such as regulating the temperature in the brood chamber and accepting nectar and pollen from forager bees. Between about ten and twenty days of life the bees develop wax glands in the abdomen and secrete the waxy substance from which they make the combs. The oldest bees in the colony are foragers that leave the hive to find nectar and pollen. There is great flexibility in this sequence of different behaviors, and what an individual bee is doing is not tied in a firm way to its age. For example, if the population of foragers should decline (which can be forced experimentally by closing a hive and moving it during the middle of the day), younger bees can step forward to replace the missing foragers. Even without such drastic intervention, however, the natural needs of the colony require a dynamic shifting of labor.

There is a second mode of reproduction in which the queen and a number of the worker bees depart and form a second colony. This *swarming* happens in early summer as the number of bees in the colony outgrows the available space. Under these conditions, workers feed a few of the developing larvae a different substance, and these individuals develop into queens, with functional ovaries and an unbarbed stinging apparatus. As they near full development, the old queen leaves with about half of the workers to find a new nest site. The first of the new queens to emerge as an adult stings the other developing queens to death, then leaves for a mating flight in which she becomes inseminated by drones, generally from other hives. The sperm that she acquires on that flight are stored internally for the rest of her life and doled out as she needs them to fertilize eggs.

THE HONEYBEE'S ALTRUISTIC REPRODUCTIVE BEHAVIOR HAS AN UNUSUAL GENETIC BASIS

Why do thousands of worker bees forego their personal reproduction and cooperate for the reproductive success of one individual, the queen? Cooperation can even involve sacrifice of life: the end of the stinger of a worker bee is barbed, and it cannot be withdrawn from the victim without pulling the poison gland out of the worker's abdomen, damage that the worker does not survive.

As a prelude to answering this question about the evolution of cooperation, consider the more familiar example of your own body. As we pointed out in Chapter 5 (Cells in an Organism Must Cooperate), your body is a colony of cells, many of which are specialized for different functions such as skin, muscle, liver, and brain. As virtually all of your cells have the genes that were present in the fertilized egg from which you developed, the cells of your body have the same genetic interests as your eggs or sperm. Success of your eggs in getting fertilized, or your sperm in fertilizing the egg of another individual, is success for the genes that are present in all your body's cells. This is the basic reason why organisms can evolve with responsibility for reproduction vested in a small subset of their cells, the germ line. The somatic cells are specialized for roles that enable the reproductive success of the germ line.

But a colony of bees or ants does not consist of genetically identical individuals. Bees and ants are sexually reproducing organisms, and during meiosis and the production of eggs and sperm, recombination produces genetic diversity (Chapters 3–4). The result is that individual, sexually reproducing organisms are genetically unique and therefore have singular reproductive interests. How, then, are we to explain the existence of colonies of bees and ants in which genetically different individuals do not reproduce but instead cooperate with others to ensure the reproduction of a third individual, the queen? The reason that a colony of bees can function as a reproductive unit is an interesting story in genetics, and understanding has undergone significant revision in the last few decades.

Charles Darwin recognized that cooperative behavior of nonreproducing social insects posed an important problem for the theory of natural selection, but knowledge of inheritance in his day was not sufficient for anyone to formulate a plausible explanation. In 1964 the evolutionary geneticist William Hamilton suggested that the explanation could be found in the way sex is determined in these insects. Females develop from fertilized eggs and are therefore diploid (with a set of chromosomes from each parent), whereas males develop from unfertilized eggs and are haploid (with only a single set of chromosomes). This is how the queen can control the production of males; when she does not fertilize an egg with the sperm she has stored, it develops into a drone. Hamilton pointed out that as a consequence of this means of sex determination, the degree of relatedness between sisters should be 0.75 (Box 7.1) rather than the 0.50 that would be expected of full sisters in humans and other monogamous animals. Thus the original explanation for the reproductive altruism of bees and ants was that sisters are more

BOX 7.1
Tracing the Degree of Relatedness of Individual Bees

Recall from our discussion of kin selection in Chapter 6 that the degree of relatedness *r* between any two individuals is the probability that they share genes by virtue of a recent common ancestor. In bees and ants, males develop from unfertilized eggs and are haploid, and this changes the arithmetic. Refer to Figure 7.3 to follow the argument.

(A) To determine the degree of relatedness of sisters, trace the arrows from the dark green to the light green symbol: *r* through their mother (the queen, Q) is $\frac{1}{2} \times \frac{1}{2} = \frac{1}{4}$. That is, the probability that any gene was inherited from the mother is $\frac{1}{2}$, the probability that that gene was passed from the mother to the other daughter is also $\frac{1}{2}$, and the two probabilities are independent. Similarly, the probability that a gene came from the father is $\frac{1}{2}$, but the probability that the sister inherited that gene from her father is $1/n$, where *n* is the number of drones with which the queen mated. If *n* were 1, any gene received from the father by the first sister (dark green symbol) is also present in the second sister, because the drone (their common father) has only one set of genes. On the other hand, because the queen mates with a number of drones (generally about ten), the probability that two sisters share a gene through the same father is diminished by a factor of 1/10. The degree of relatedness of two sisters is the sum of the probabilities through each parent: $r = (\frac{1}{2} \times \frac{1}{2}) + (\frac{1}{2} \times 1/n)$. If $n = 1$, $r = \frac{3}{4}$. If $n = 10$, $r = 0.3$.

(B) A mother is related to her son by $r = \frac{1}{2}$; that is, the probability of any gene being passed to her haploid offspring is $\frac{1}{2}$. This is true for the unfertilized eggs of workers as well as for those of the queen.

(C) A worker's *r* to her sister's son (her nephew) is calculated as for (A), but must extend for an additional generation. Examine the diagram and see that $r = (\frac{1}{2} \times \frac{1}{2} \times \frac{1}{2}) + (\frac{1}{2} \times 1/n \times \frac{1}{2}) = 0.15$ for $n = 10$. This diagram also shows that the queen's *r* to her grandson is $(\frac{1}{2} \times \frac{1}{2}) = \frac{1}{4}$.

(D) The degree of relatedness of a male to his daughter is 1; that is, because he is haploid, all of his genes are present in every one of his female offspring. He is not related to any of the queen's sons, because they develop from her unfertilized eggs. His average degree of relatedness to offspring of the queen is thus 0.5 (or would be if she were not carrying sperm from other males).

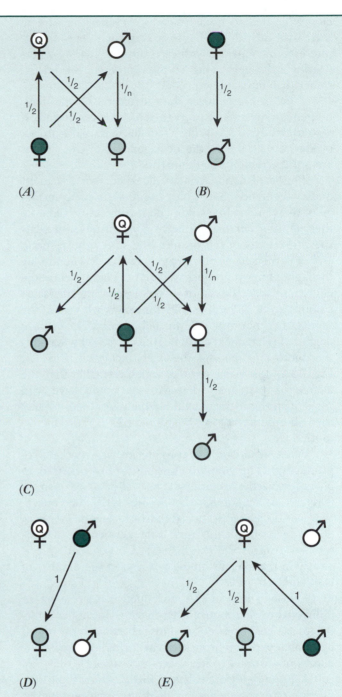

(A)   *(B)*

(C)

(D)   *(E)*

FIGURE 7.3 Tracing degrees of relatedness among bees: (A) worker females to each other; (B) workers to their sons; (C) workers to brothers and to the sons of the other workers; (D) males (drones) to their offspring; and (E) drones to their siblings. See the text for an explanation of each calculation.

(E) The degree of relatedness of a male to his siblings is 0.5. Note again, a male has no father, but he does have a grandfather by way of his mother, the queen.

The consequences of this mode of inheritance for the colonial life of bees are described in the main text.

closely related to each other (0.75) than they are to their own offspring (0.5), so it is in their reproductive interests to cooperate with each other in rearing sisters (through the egg-laying of their mother, the queen) rather than in trying to produce their own daughters.

Why are the females the "altruists"? Drones have no genetic incentive to forego reproduction. In fact, it is to their advantage if more females than males are reproduced, because in the next generation their genes are only carried by daughters (Box 7.1).

This explanation has an elegance that led to its quick acceptance, but it has a serious problem in that all of the workers in a colony do not share the same father. During her mating flight a virgin queen accepts sperm from about ten drones. She stores those sperm and uses them to fertilize eggs more-or-less randomly throughout her reproductive life. As a consequence, the degree of relatedness of two workers in the same colony can be as high as 0.75 for sisters who share the same father, but only 0.25 for half-sisters. The average degree of relatedness r will be $(0.25 + 0.5/n)$ where n is the number of matings the queen achieved (Box 7.1, Fig. 7.3A). If $n = 10$, $r = 0.3$, which is smaller than the degree of relatedness that workers would have with their offspring if they could lay their own eggs. What, then, keeps the colony functioning as a reproductive unit?

What would be the consequences of workers reproducing? The simplest possibility for workers is to lay unfertilized eggs. A female bee—worker or queen—is related to her son by $r = 0.5$ (Fig. 7.3B). The queen, however, is related to her grandsons—any drones produced by unfertilized eggs from her worker daughters—by only 0.25 (Fig. 7.3C). Consequently, it is clearly in the queen's interest to restrict the number of eggs laid by workers.

Although it is in the reproductive interests of an individual worker to produce sons, it is not in her interests to see her sisters do so. This is because if the queen has mated with ten drones, the average degree of relatedness of a worker to her sister's sons is only 0.15 (Fig. 7.3C). Because a worker has so many sisters, their production of drones could readily dilute any reproductive success she might achieve by laying eggs herself. It therefore becomes a much better tactic for her to vest her reproductive success in the queen's sons (her brothers), to all of whom she is related by $r = 0.25$ (Fig. 7.3C). Note that because drones develop from unfertilized eggs, the degree of relatedness of a worker with her brothers is independent of the number of matings the queen achieved.

Can the workers reproduce? A small number of workers (about 1%) have functional ovaries, and about 10% of the unfertilized (drone) eggs in a colony are laid by workers, but only about 0.1% of the drones in a colony have developed from eggs laid by workers.

Something is restricting the production of drones by workers, and both the workers and the queen play a role. The workers actively destroy unfertilized eggs laid by other workers (as we just saw, it is in their genetic interests to police each other), and the queen appears to mark the eggs she has deposited with a chemical signal (a *pheromone*), allowing the workers to identify which drone eggs should not be destroyed. Another queen pheromone, which is distributed around the colony by workers, prevents swarming by inhibiting the workers from rearing additional queens. Whether it also inhibits the worker's own reproduction is not known.

The result is a balance in which the disparate, selfish interests of genetically different workers are largely quelled by other workers, to their own genetic advantage. The result is an evolutionary compromise in which the workers seek reproductive success through the rearing of brothers rather than a mixture of sons

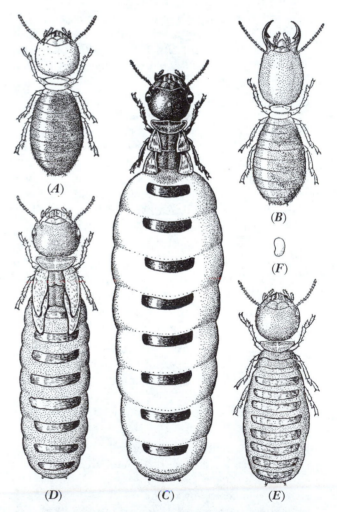

FIGURE 7.4 Castes of termites: (A) worker, (B) soldier, (C) queen, (D) and (E) secondary and tertiary queens; (F) is an egg. Compared to the nonreproductives, the queens are massive egg-laying machines. The primary queens can live for many years.

and nephews. *Just as the evolution of nepotistic behavior of sexually reproducing vertebrates can be understood in terms of kin selection and degrees of relatedness of near relatives, the social structure of bee colonies, and the "altruistic behavior" of workers is based on the genetics of honeybees. Although for bees the details are more complex, kin selection is also responsible.*

TERMITES ARE INCESTUOUS

Massive insect social systems organized around a few reproductives and much larger numbers of sterile individuals have evolved in a second group, the termites. This group of insects is related to roaches and is more primitive than the Hymenoptera. For example, termites grow through successive molts and do not have separate larval, pupal, and adult stages. Termite social systems are therefore probably older than those of bees and ants. As with ants, however, the nonreproductive termites can form both worker and soldier castes (Fig. 7.4). Moreover, termite colonies rival those of ants in size. The termite mounds of a fungus-growing African species house millions of individuals and can be nearly thirty feet tall (Fig. 7.5). All termites eat woody material, although only about ten percent of termite species

are economic pests. Many termites are dependent on intestinal protozoa or bacteria for digestion of cellulose (the polysaccharide of which wood is composed), and most species must remain within damp chambers of their own construction in order to keep from desiccating. Ants are a major predator of termites.

The genetic basis for termite social structure is very different from that of ants and bees. The nonreproductive castes consist of both males and females, and both are diploid. What, then, accounts for the presence of such large-scale reproductive altruism in this group of insects? The likely explanation is thought to involve inbreeding in the following way.

New colonies are periodically formed from the mating of winged males and females that have emerged from two different parent colonies. As the primary and, at first, the only reproductives, the king and queen produce all of the new colony members. Secondary reproductives are formed later, and they remain within the colony and contribute more offspring. This intensive inbreeding leads to an increase in homozygosity. Consequently, when winged individuals are produced and leave the colony, the chromosomes of each homologous pair are homozygous at many loci.

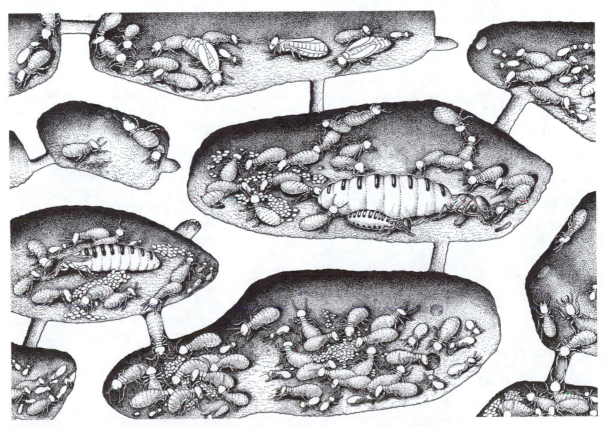

FIGURE 7.5A Internal structure of a small part of a termite colony. The primary queen is in the central compartment, and there is a secondary queen in the compartment below and to the left.

FIGURE 7.5B A large termite mound in East Africa. The animals sitting on the top are baboons.

When two individuals from different colonies mate and start a new colony, all of their offspring are genetically very much alike, just as were the offspring of the two long-inbred strains of tobacco in Figure 4.6. This is because each parent, being homozygous at many loci, produces gametes that are genetically very similar. The offspring are therefore heterozygous for the parental chromosomes, but each offspring has a very high degree of relatedness to all its siblings. Sexual reproduction by offspring leads to recombination, with the result that secondary reproductives are more related to

FIGURE 7.6 The social system of this hairless subterranean mammal is similar to that of termites, with much inbreeding and therefore high genetic relatedness among colony members. This makes it possible for reproduction to be vested in one female.

their brothers and sisters than their offspring. There is thus a genetic incentive to forego reproduction and assist the king and queen in making more siblings. When the founders of new colonies are the result of cycles of inbreeding within the old colony, however, the conditions are set for the process to repeat.

Termites thus illustrate how natural selection has found more than one way to form colonies of organisms in which reproduction is vested in one or a few individuals that are supported by numerous nonreproductives. There is in fact a mammal, the naked mole rat of east Africa, where a single female produces young and is attended by a colony of near relatives (Fig. 7.6).

BEHAVIOR IN HONEYBEE SOCIETIES IS BOTH COMPLEX AND ADAPTIVE

INFORMATION IS NEEDED WITHIN THE COLONY

The colonies of bees and ants consist of thousands of females, each of which must achieve reproductive success by working to support the colony. We have seen how genetic relatedness lies behind the extraordinary social structure, but in order for the colony to thrive, the activities of the individual worker bees must be coordinated. The problem has a superficial likeness to the running of an automobile plant. If everyone in the plant should start making fenders, the production of finished cars would soon come to a halt. Similarly, if all the bees in a colony should devote themselves to secreting wax and building comb, there would be no honey to store and nothing to feed the larvae. We saw earlier that bees perform different tasks as they grow older, but what a bee is doing at any particular time is not tied rigidly to its chronological age. The colony maintains a flexibility that enables it to respond to changes in weather and to exploit the ever-changing sources of nectar and pollen in the surrounding area.

If someone in an automobile assembly plant does not order parts in timely fashion, the plant will be forced to shut down. Similarly, bees must bring supplies into the hive—nectar, pollen, and water. Therefore, there must be an allocation of labor among the forager bees such that these items are procured as required. During a period of cool, wet weather the bees are unable to forage, and the stores of pollen in the hive may become very low. As soon as the weather permits, however, the bees replenish their supplies by devoting extra effort to collecting pollen.

Bees regulate the temperature of the hive in the region of the brood cells with remarkable precision. The temperature is held at $34.5° \pm 1.5°$ C, and on a typical day it varies within only $1°$ C. In hot summer weather when the nurse bees are unable to keep the temperature down by stirring the air with their wings, they hang drops of water and lower the temperature by evaporative cooling. But for this they need an extra supply of water, which requires water transport by a subset of the foragers.

In the automobile plant, suppliers may change their prices or be unable to deliver when required, so buyers must be ready to seek alternative sources. In the hive, the collection of nectar requires even more active shifts of effort. Patches of flowers can bloom and wither on a daily basis as the season progresses, so new sources must be discovered promptly and exploited efficiently as the blossoms open. This, too, necessitates a constant shifting of labor among the foragers.

Success leads to growth. If automobile sales are brisk and demand continuous, there may be an economic advantage to building a second plant elsewhere. In a colony of bees, rapid growth typically exhausts the available space in the hive, and there is a reproductive advantage to swarming, a process that increases the population of near relatives.

Both automobile plants and bee colonies require information in order that appropriate decisions are made. In the case of an automobile plant, a small group of individuals (top management) gathers information and hands down decisions that are intended to maximize company profits. To the degree that they are successful in furthering that goal, the decisions are intelligent. Most individuals in the company accept these decisions and do as they are instructed. Such hierarchy is a common feature of human social structures; military establishments are prime examples.

The honeybee colony works on a very different model. There is no centralized control; the queen has no imposing authority. She does little more than lay eggs. Furthermore, neither she nor any other member of the colony has all the information required for an intelligent managerial decision. The colony nevertheless responds to changes in the outside world in a reproductively adaptive fashion, *as though* the response were intelligent. How does this occur?

The answers illustrate the enormous power of natural selection. The honeybee colony is a social system in which no individual has access to more than a small fraction of the total information that is being exploited by the entire hive. Nevertheless, 20,000 individual insects behave in a coordinated fashion, achieving success for the entire colony, providing numerous examples of interlocking behavioral adaptations.

FORAGING BEES COMMUNICATE THEIR FINDS TO THEIR HIVE MATES

One of the most fascinating features of bees is the way in which foragers communicate with their sisters in the hive about sources of food and water that they have lo-

cated. When a forager has found a particularly fine source of nectar—one with a high concentration of sugar that can be easily exploited—she returns to the hive and performs a "dance" on the surface of the combs (Fig. 7.7).

The dance consists of "waggle-run" in which the bee moves forward a couple of centimeters while shaking her abdomen from side to side, then circles about and repeats the waggle-run again, remaining near the hive entrance in a restricted region referred to as the "dance floor." The hive is dark, and the dancer also

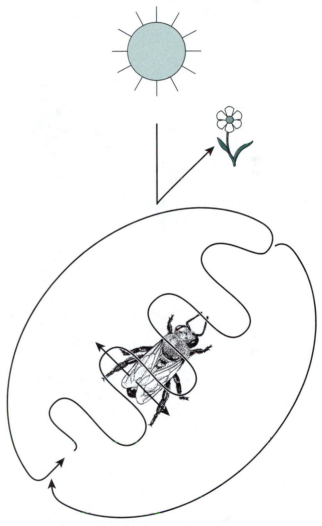

FIGURE 7.7 The waggle dance of honeybees is performed in the dark on the vertical surface of the comb by returning foragers who have discovered a particularly profitable source of nectar or pollen. The bee advances a short distance emitting a low-frequency sound and waggling her abdomen from side to side. The frequency of repetition of the waggle-run indicates distance to the food, the total number of repetitions indicates the quality of the source, and the angle between the waggle-run and the vertical indicates the angle a recruit must keep with the sun as she flies out to find the food source.

Box 7.2
Animal Communication

Communication is the exchange of information, and many animals are capable of communication. We usually communicate with each other by spoken and written language, but in suitable context a wink, a blush, or a smile can convey volumes. Although human language amplifies enormously our capacity to exchange information, it does not define communication.

This point is illustrated by Figure 7.8, which shows profiles of dogs' faces in a 3 × 3 matrix. Moving down the first column, these faces display increasing fear: the lips are closed, the hair lies flat, and the ears are laid back. Moving from left to the right across the top row, the faces display increasing rage; the lips are curled exposing the teeth, the hackles rise, and the ears become erect. Fear and rage are not mutually exclusive, and both can be seen in the lower right, which could be a threatened female with young pups. Anyone who knows dogs will know what these signals mean and would be reluctant to extend a hand to pat the animals that appear angry. Bared teeth and raised hackles are common among carnivorous mammals. Animal communication thus has an objective reality that can sometimes be interpreted even by other species.

We study the communication systems of animals by observing the effects the signals have on the behavior of other individuals. The dances of bees were decoded by observing their effects on the bees that surround the dancers and then by conducting experiments to verify the interpretations. For example, bees were trained to visit artificial feeding sites at various directions and distances from the hive and the orientation and tempo of their dances were carefully measured. These experiments revealed what information is communicated by the dances.

That bees communicate with each other does not mean they are consciously aware of new sources of nectar or form mental images of flowers of the sort we experience. The bees convey information to each other, but as observers we know nothing about the mental processes that are occurring in the animals' brains. We can observe the effect the information has on the behavior of the recipient, and this can tell us the nature of the information being conveyed. But what, if anything, an animal "thinks" is not revealed by watching its behavior. The only possible exceptions involve our closest living relatives, the great apes, to which we will return in Chapter 11.

FIGURE 7.8 A familiar example of animal communication. These profiles signal increasing fear from top to bottom and increasing rage from left to right.

emits a soft, low buzzing sound, which is important in attracting other bees and bringing them into physical contact with her.

Although she moves around on the comb while dancing, the orientation of the waggle-run with respect to the vertical remains the same, and that angle tells other workers who contact her during her dance the direction they must fly to reach the source of nectar or pollen. The angle between the waggle-run and the vertical is the angle that a recruit must maintain between her outward flight path and the sun. (Bees that are crawling on a vertical surface can sense their orientation to the vertical by the weight of the head on sensory hairs between the back of the head and the front of the thorax.)

Note that interpreting the dance requires a transposition of coordinates; the angular orientation of the dance on a vertical surface corresponds to the horizontal angle between the sun and the direction to the food source.

The frequency with which the waggle-run is repeated conveys information about the distance to the food source; the closer the flowers, the more often the waggle-run is repeated (Fig. 7.9). The dance therefore conveys specific information about both direction and distance to the food source.

A dancing bee makes other information available to her hive mates. Generally she will have picked up the odor of the flowers she has just visited, and this can be detected by chemoreceptors on the antennae of

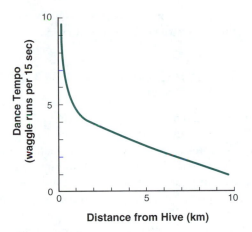

FIGURE 7.9 The relation between the frequency with which the waggle-runs are repeated and the distance to the target (nectar, pollen, water, new hive site).

other bees. This is not information coded in the dance, but it is useful to newly recruited foragers as they approach the area where the flowers are located. Once they have found the flowers the first time, they are able to locate them on subsequent trips using visual cues.

This same dance is used in other contexts besides foraging. When the colony swarms, the departing bees hang in a cluster in a bush or tree until the scouts have located a new nest site. They report their findings by dancing, but the waggle-runs are performed on the surface of the swarm, on the bodies of the other bees. The dances are followed by recruits that then visit the potential site. On their return they may also dance.

A final point about the dances: they are sometimes referred to as the "dance language" of bees. This is not only because they convey a fairly sophisticated array of information, they do it in symbolic fashion. Although bees sometimes dance on a horizontal surface, in which case the orientation of the waggle-run indicates directly the route to the food source, they usually dance on a vertical surface of a hanging comb inside the hive. Under these conditions the dance is not simply an enactment of the outward flight path, for the directional information is coded, with "up" representing the position of the sun as seen on the flight from the hive. Moreover, the interpretation of the dance by other bees depends on the context. Nevertheless, the communication system bears no real comparison with the rich syntax of human language.

THERE IS NO CENTRAL AUTHORITY FOR ASSIGNING TASKS IN A BEE COLONY

Consider first the control of queen production and swarming. Earlier we mentioned that the queen produces a chemical signal, a *pheromone*, that is distributed

about the colony by worker bees when they come in contact with her. That pheromone inhibits workers from feeding larvae the substance that causes them to develop into new queens. When the hive grows to a size at which the concentration of the queen's signal is too low, the colony prepares to swarm. Nurse bees rear new queens, but before the new queens emerge, the old queen leaves with about half of the workers. Although it is unknown what triggers the old queen's departure, the falling level of her pheromone has set in motion a series of behavioral responses among the worker bees that culminate in a division of the colony. The "decision" to swarm is therefore a collective response of the colony to overcrowding, initiated by a chemical signal.

A forager only dances when she has located a particularly profitable source of food. As she dances, however, her assessment of its value—the concentration of sugar in the nectar—is also conveyed to the hive. This final piece of information, like distance and direction to the food, is coded in the dance, but unlike distance and direction *it does not become available to any single bee*. The higher the quality of the food source, the *longer* she dances, and for a particularly good source she may perform a hundred waggle-runs. But an unemployed forager who has just been attracted to the dance floor does not survey all of the dances that are being performed. She follows a single dancer, probably at random and for only a few waggle-runs, before she leaves the hive and follows the dance instructions. Each new recruit therefore has no information about the relative quality of the source that she is attempting to locate. The information about quality is exploited by the hive as a whole, however, because the better the food source is, the longer the dancer performs, and the more recruits will be attracted to her waggle-runs. In similar fashion, the dances performed during swarming also contain information about the quality of the prospective nest site— size of the cavity and nature of the entrance hole.

An individual bee "knows" only about the single source of food she is visiting (or has just been induced to visit), but the population of foragers will be attracted to different sources in rough proportion to the array of individual 'decisions' the dancers have made about their quality. Furthermore, the system prevents the population of foragers from being drawn to a single, advantageous resource that can fail abruptly. The colony hedges its bets, so to speak, and is able to abandon declining resources quickly and shift foraging efforts to more favorable sources as they become available.

How does this happen? If a resource fails, bees will stop dancing about it, and foragers will abandon it. Conversely, if another patch of flowers should become profitable, it will elicit dancing, and new recruits will be attracted to it from the population of unemployed foragers. The shift of the population of foragers from one source of food to another is rapid and can be completed in several hours.

How are new sources discovered? At any time about 1% of the foragers (about a hundred bees) are functioning as scouts. They do not follow the dances of other bees, but instead they explore the vicinity of the hive in a radius of several miles in order to locate new sources of food.

The collection of nectar can be influenced by the internal state of the colony as well as by the available supply in the surrounds. The time required for a returning forager to find a recipient for her load reflects the balance between nectar foragers and nectar processors; a long time signifies there are too few processors. If a forager returns from a rich source and is able to disgorge her load within about twenty seconds to a bee who is making and storing honey, she is likely to perform a waggle dance. On the other hand, if she is unable to find a recipient for her load within about fifty seconds she performs a "tremble dance" in which she walks about the hive making trembling movements, both side-to-side as well as fore-and-aft, and continually facing in different directions. The effect of these tremble dances is to increase the rate of nectar handling by bees inside the hive while decreasing the number of waggle dances. Another behavior of foragers returning from a profitable source is to move about the hive shaking other bees. This signal serves to increase the number of foragers.

When returning foragers experience difficulty in finding bees to unload their nectar, it can mean that the hive is running short of water. This can have two causes. Water is needed to prepare food for larvae as well as for evaporative cooling when the brood chamber overheats. Both of these needs are detected by bees working in the brood chamber, either assessing the water content of larval food in their mouths or with their temperature receptors. The information is conveyed to foragers by the nurse bees decreasing their receptivity for nectar. This signal diverts foragers from the collection of nectar to the collection of water from nearby streams or ponds. A small fraction of the foragers (about 1%) seem to be dedicated to collecting water.

The regulation of pollen collection is less well understood, but here too the nurse bees are involved, because they work directly with the pollen stores. They seem to be able to inhibit pollen foragers by feeding them a secretion from their hypopharyngeal glands.

Comb production is expensive; it requires several grams of carbohydrate (sugar, honey) in order to manufacture a gram of wax. The bees therefore do not invest in more comb than is needed. The amount of unfilled comb is seldom more than 20 percent of the total that the bees have made. The combined presence of few free cells and a brisk influx of nectar to the hive trigger production of new comb. Both conditions are required to induce the bees to secrete wax.

THE BEHAVIOR OF BEES APPEARS TO BE RATIONAL

All of these behavioral strategies of the colony are rational in the sense that they are decisions that you or I would likely make if we were colony administrators. We reach rational decisions by reasoning about cause and effect, and as colony administrators we would likely apply practical, utilitarian criteria: Does redirecting workers to collect water serve the purposes of the hive? Specifically, is lowering the temperature of the brood chamber necessary to insure the survival of eggs or larvae or pupae, thus the continued prosperity of the colony? If we thought the future of the colony were at stake, we would likely decide to change the behavior of the bees in the same ways that now occur without the involvement of an administrator.

The bees' behaviors are adaptive: they foster reproductive success of the colony, and they are the products of natural selection. The bees do not use reason, but we nevertheless see their behavior as exquisitely reasonable. There is thus a similarity, which is not accidental, between the effectiveness of adaptations and the utilitarian criteria humans use in much of their everyday reasoning.

COLONIES OF CELLS AND COLONIES OF ORGANISMS USE SIMILAR CONTROL PRINCIPLES BUT DIFFERENT MECHANISMS

The mammalian body regulates its salt and water balance with receptors that monitor the concentration of salts in the blood and communicate with the kidney via a pituitary hormone. The body regulates gas exchange by monitoring CO_2 in the blood and modulating the rate of breathing. And it regulates the energy supply, in part by eating and in part by mobilizing stored reserves. In short, bodies maintain a state of internal homeostasis in which needs (demand) are adjusted to the external environment (supply). This is accomplished by processes of which we are largely unconscious, as they involve the autonomic nervous system and hormones circulating in the blood. By these means, cells, tissues, and organs with different functions are kept working in a coordinated fashion so that the organism continues to thrive.

As we have just seen, a colony of social insects faces the same general challenge. Individual animals must perform tasks in a coordinated yet ever-changing manner in order that the entire colony can exist. In each case—mammalian body or colony of ants or bees—internal and external conditions must be constantly monitored in order to determine what corrective actions need to be taken.

Evolution has necessarily arrived at somewhat different outcomes in the two cases. Integrated control systems are possible for the intimate collection of cells that form a single mammalian body. A subset of cells—neurons and endocrine system—specialize in monitoring and control. A much more distributed system, however, is necessary for an assembly of individual organisms in a colony of social insects, a superorganism. The solution for the insects is probabilistic and also quite different from a social group of humans with a leader. No one individual has access to all of the necessary information, but the behaviors of the individual insects are such that statistically, their summed responses are adaptive for the colony as a whole.

SYNOPSIS

The insects are evolutionarily the most successful multicellular animals, in terms of both numbers of individuals or numbers of species. Social species of insects are striking for their intricate societies.

Darwin saw that colonies of bees and ants consisting of a relatively few individuals capable of reproduction supported by much larger numbers of sterile workers posed a challenge to his theory of natural selection. The dilemma: How could selection either favor individuals that do not reproduce or cause the evolution of morphological and behavioral adaptations in such sterile individuals? An explanation couched in genetic terms began to emerge only about forty years ago. It is now clear that these insect societies with their sterile castes are nature's most dramatic manifestation of the power of *kin selection*. In the bees and ants (order Hymenoptera) these social systems are based on a form of sex determination in which males develop from unfertilized eggs, and are thus haploid. In termites (order Isoptera) it is probably based on extensive inbreeding.

The social insects also display extraordinary behavior involving division of labor in the colony, communication among individuals, the capacity of individuals to learn and remember, and flexibility of the colony in meeting short-term environmental challenges. There is a constant flow of information among individuals about the needs of the colony and the availability of resources in the immediate environment. This information is used to direct foragers to new resources and to different kinds of resources. The mechanisms, however, differ from those found in human social groups, where individuals either take orders from a central authority or share in the decisions by a process of mutual understanding and consensus. In a colony of honeybees, by contrast, no one individual has access to all the information available to the colony, and no one bee makes decisions for the group. Behavior is redirected in adaptively important ways by statistical processes. Falling concentration of a chemical signals the size of the colony; longer waggle-dances signal high-quality food sources and thus recruit more foragers; and longer unloading times experienced by returning foragers signal a surfeit of supply and redirect foraging efforts to other resources. Behavior of groups of related individuals is therefore so complex and adaptive that it appears rational to a human observer. It illustrates the power of natural selection in ways that will be useful to have in mind when we return to behavior in Chapter 11.

QUESTIONS FOR THOUGHT AND DISCUSSION

1. Why are bee colonies called "superorganisms"?
2. Suppose—perhaps in another galaxy—evolution were to produce a life form with cognitive abilities (intelligence) comparable to humans but with sex determination similar to bees or termites. What kind of social systems might emerge? How might they be similar or different from large corporate structures or feudal systems that exist (or have existed) on earth? Consider division of labor, authority, and the utilization of information.

SUGGESTIONS FOR FURTHER READING

Seeley, T. D. (1995). *The Wisdom of the Hive*. Cambridge, MA: Harvard University Press. The details of how a colony of bees adjusts to changes in their world is revealed when all of the bees are individually marked. Natural history at its best.

Wilson, E. O. (1972). *The Insect Societies*. Cambridge, MA: Belknap Press of Harvard University Press. One of the world's authorities on ants describes the diversity and complexity of social insects.

8

Parasites, Hosts, and the Evolutionary Warfare of Infectious Disease

CHAPTER OUTLINE

- *Uninvited guests*
    Why consider pathogens in an evolutionary context?
- *Evolutionary views of infectious diseases*
    Pathogens' strategies: the evolution of virulence of infectious diseases
- *Ecological factors in the emergence and spread of disease*
    Influences of humans on natural selection among diarrheal pathogens
    Pathogens in hospitals
    Ecology and emerging diseases
- *Evolutionary epidemiology*
- *Our immune system: A microcosm of biological evolution*
    HIV and AIDS
- *Other ways hosts fight pathogens*
- *The Red Queen*
- *Synopsis*
- *Questions for thought and discussion*
- *Suggestions for further reading*

Great fleas have little fleas upon their backs to bite 'em,

And little fleas have lesser fleas, and so ad infinitum.

Augustus de Morgan (1806–1871)
A Budget of Paradoxes

UNINVITED GUESTS

Neither Augustus de Morgan, a well-known mathematician and puzzle enthusiast of his time, nor Jonathon Swift, from whom he got the idea for this rhyme, were biologists. However, de Morgan's poem is biologically perceptive. A variety of parasites inhabit virtually all cellular and multicellular organisms, including parasites themselves. The sequence of ever-smaller parasites within parasites ends only when we arrive at the sequences of DNA or RNA that parasitize individual cells. There are thus more different species of parasites than there are species of free-living organisms. Humans are no exception: we are host to many species of mites and lice that live in our skin, dozens of species of worms that can live in our gastrointestinal

Photo: Whenever people are crowded together with neither a safe supply of drinking water nor facilities for the proper disposal of human waste, conditions favor rapid reproduction and evolutionary change in microbial pathogens. Refugee camps are places where human behavior creates favorable ecological conditions for the spread of cholera, typhoid fever, and dysentery. Smoke from cooking fires hangs over this camp of Rwandan refugees.

Box 8.1
What is a Virus?

A virus is smaller than a bacterium, and the simplest viruses consist of some genes surrounded by a capsule of protein. A large virus can have several hundred genes, but there are only nine genes in the human immunodeficiency virus (HIV), the virus that is responsible for AIDS (acquired immune deficiency syndrome). Some viruses (of which HIV is an example) also contain a few enzymes that are necessary for their replication, and the protein capsule is surrounded by an outer envelope of lipid.

A virus is not able to obtain energy for itself, as it does not have the necessary enzymes. Viruses therefore depend on living cells for an environment in which they can replicate, and they exist outside of living cells as dormant bits of genetic information wrapped in a protective protein coat. In this state they do not have a life of their own, and unless they come in contact with an appropriate host cell, they do not have a future either. But viruses also illustrate dramatically how efficient and self-serving DNA and RNA can be in replicating. They exhibit parasitism in its most elemental form.

To replicate, viral genes must enter a living cell. Once inside, the virus commandeers the metabolic machinery of the cell to make copies of itself. There are many kinds of viruses. Some infect plants, others animals, and some, called *bacteriophage*, are parasites of bacteria. Some use DNA for their genetic material, others employ RNA. In some the nucleic acids are organized as a double-stranded helix, in others as a single strand. Viral genes code for a variety of proteins, most importantly the enzymes needed for their replication and the coat proteins that protect the genes as the virus moves between hosts. Viruses can override the host cell's mitotic cycle, so they can replicate much faster than the host cell.

A common outcome of a viral infection is that the host cell becomes filled with newly synthesized virus particles and bursts. The lesions of the skin associated with chicken pox, smallpox, or cold sores are places where large numbers of infected epithelial cells have ruptured and released viruses. Alternatively, some viruses bud off from the host cell without killing it, wrapping themselves in host cell membrane as they depart (Fig. 8.1).

Viruses can have specialized ways of getting into cells. Some bacteriophage inject their nucleic acid into the host bacterium, leaving their protein capsule outside. The RNA viruses that cause diseases such as influenza and AIDS induce the host cell to take them up by endocytosis (Figs. 3.2 and 8.7), but once inside they escape destruction by the degradative enzymes of the lysosomes and use their own proteins to initiate replication.

DNA viruses take over the protein-making machinery of the cell for themselves. Their DNA serves as a template for making mRNA, and the cell obliges by making viral protein. The RNA viruses are more varied. In some the RNA can serve as its own messenger. In others, the *retroviruses*, an enzyme called *reverse transcriptase* transcribes the RNA back into DNA, which can then be incorporated into the host's chromosome, where it is replicated and transcribed as part of the host's genome. The virus that causes AIDS operates this way, and we will see later in the chapter why this process poses such a difficult medical problem. Some of the varied ways that viruses treat their host cells are depicted in Figure 8.1.

---

tract and even invade our tissues, and hundreds, if not thousands, of different viruses, bacteria, yeasts, and fungi that thrive inside and all over us.

The word "parasite" derives from Greek and one meaning is "he who eats at the table of another." In everyday language the word is still used to characterize people who do not reciprocate. In biology the term refers to an organism that lives in, on, or with another organism, or to a virus (Box 8.1), obtaining benefits for itself but at a cost to its host. By this definition everything from a twenty-foot tapeworm in our intestine to

a submicroscopic virus in the cells of our nasal passages is a parasite. If the invader is a virus or a single-celled organism (bacterium, yeast, fungus, protozoan) and ordinarily causes the host to become ill, the parasite is referred to as a *pathogen* (disease generating). Organisms of different species that coexist intimately in a more even relationship in which both benefit are called *mutualists*, or more generally, *symbionts*.

Why are we and all other multicellular organisms such inviting havens for so many uninvited guests? It is because in creating a favorable physical environment for

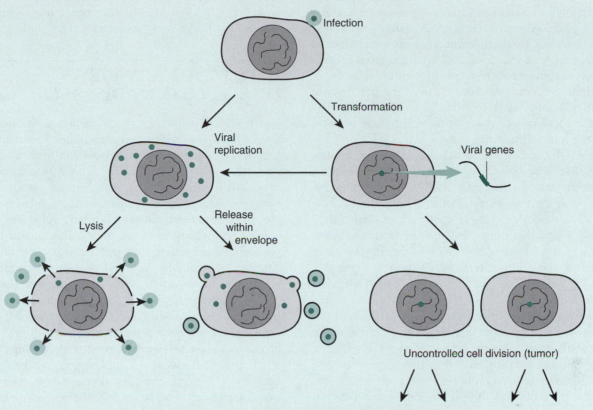

FIGURE 8.1 Some of the possible fates of cells that are infected with different kinds of viruses.

(*Left*) The virus may use the protein synthesis machinery of the cell to manufacture more of itself. Depending on the kind of virus, this may eventually cause the cell to rupture (called *lysis*), thereby liberating many newly synthesized viruses and killing the cell. Or the viruses may leave the cell without killing it, in some forms of virus, coated with lipid membrane derived from the membrane of the cell.

(*Right*) Some kinds of viruses insert their genes into the DNA of the host cell's chromosomes, a process called *transformation* because it literally transforms the genetic endowment of the cell. A number of outcomes are then possible, again depending on the nature of the virus. Viral genes may remain quiet for a number of cell generations, hiding, so to speak, but replicating along with the host's genes. They may eventually turn on, be transcribed and translated, and escape as viruses to infect other cells. Or they may replicate by causing the host cell to divide in an uncontrolled fashion, leading to a tumor. Such *oncogenes* are one cause of human cancers.

Some viruses carry their genes as DNA, others use RNA. The process of transformation by the RNA virus responsible for AIDS is described further in Figure 8.7.

its own cells, an organism invariably creates an agreeable home for other species. Consider the features that our remote ancestors evolved as they became multicellular and later, in colonizing the land. A covering of skin—impermeable to water, ions, and solar radiation—creates an isolated internal environment in which our other cells can thrive. This internal environment is also carefully regulated: the concentrations of ions, nutrients, gasses, and waste products are monitored and maintained within narrow bounds. And in birds and mammals, the internal temperature of the body also is

held nearly constant. Such a Garden of Eden is an invitation to other organisms that can penetrate the host's defenses, so it is little wonder that so many kinds of parasites and so many kinds of parasitism exist.

WHY CONSIDER PATHOGENS IN AN EVOLUTIONARY CONTEXT?

We introduce the subject of pathogens and their evolution for several reasons. First, infectious diseases continue to be the major source of human suffering and mortality worldwide. In spite of the great advances made by medical science in treating and preventing infectious diseases, new strains of old pathogens—such as the bacteria that cause tuberculosis—continue to reemerge in both the developed and developing world in forms that are more resistant to antimicrobial drugs than their predecessors. Second, over the past twenty years the pathogens responsible for a number of diseases have had an increasing impact on human populations: the human immunodeficiency virus (HIV), which causes AIDS (acquired immune deficiency syndrome); viruses that cause hemorrhagic fevers (such as dengue); and the bacteria that cause Lyme disease, Legionnaire's disease, and toxic shock syndrome. Third, the life histories of some pathogens—how, where, and when they multiply within their hosts and how they move between hosts, which in some cases involves moving between different species, illustrate some of the most complex suites of adaptations that evolution has produced. Fourth, reducing the threat of infectious diseases to humans is made easier by an evolutionary analysis of the life history of pathogens. Understanding their ecology also reveals how humans can alter natural selection among pathogens, thus changing the severity of disease. The same understanding of evolution that so successfully accounts for morphological and behavioral adaptations also applies to how and why different genetic strains or species of pathogens are favored by natural selection, what can be done to redirect such evolutionary change, and how humans react to pathogens in the ways they do. And finally, the warfare between diseases and hosts offers a window on a larger evolutionary issue: the ubiquitous role of other organisms as part of the environment in which natural selection operates.

We start with human diseases. The main reason why infectious diseases continue to be such a grave threat to human health becomes clear when we compare the rate at which prokaryotes and viruses can evolve new strategies for invading a large, multicellular host and the rate at which the host can evolve defensive strategies to thwart them. A bacterium living under optimal conditions can reproduce every twenty minutes, whereas the generation time for humans is unlikely to be much less than twenty years. Because of this difference, viruses and prokaryotes can evolve many times more rapidly than humans and other multicellular organisms. In response to this formidable challenge, vertebrates are equipped with a flexible mechanism to recognize invading pathogens and target them for destruction. This defense is provided by the *immune system*, and the feature of its operation that makes it so flexible and efficient starts with its use of special mechanisms to increase genetic diversity among particular classes of white blood cells. Natural selection then improves the genetic information used by these cells as

they recognize and disable invading pathogens. There is, in short, a form of evolution occurring within lineages of your cells every time you catch a cold.

EVOLUTIONARY VIEWS OF INFECTIOUS DISEASES

The traditional view of why the virulence of infectious diseases changes over time is exemplified by the following sorts of statements: "Given enough time, a state of peaceful coexistence eventually becomes established between any host and parasite." Or, "disease usually represents the inconclusive negotiations for symbiosis . . . a biological misinterpretation of borders." This sort of argument seems to make sense, for a pathogen that multiplies rapidly and kills its host will die with its victim. The alternative—to survive in a sort of peaceful coexistence within the host—would seem to hold more promise. And this alternative is what we seem to see when a pathogen is transmitted to an unresistant population: an epidemic ensues, but eventually it subsides. It is as if the pathogen and its host were "learning to get along."

Is this really what is happening? Most likely not, because evolutionary theory suggests that the traditional argument is based on the implicit but faulty assumption that natural selection always favors organisms that cooperate with one another so as to ensure their mutual survival. This logic in turn suggests that the balance of nature is the result of cooperative interactions among organisms, and that new epidemics represent a temporary disruption in a network of mutually beneficial, or at least neutral, interactions.

We saw in Chapter 6 that cooperation between individuals can be favored by natural selection, but only insofar as it enhances the reproductive success of both participants. There are in fact some cases of cooperative interactions involving organisms of different species, but it is hard to make the case that pathogen-host relationships provide such an example. Although it is easy to see why a pathogen might spare its host, the host has nothing to gain and possibly everything to lose by allowing the invader to take up residence in the first place. In short, there is nothing good to say about hosting pathogens. Indeed, the way in which our immune system works suggests that it was designed for war with pathogens, not for peaceful coexistence.

Both pathogens and their hosts vary. There is variation among individuals, and both individuals and populations change with time. Some strains or species of pathogens may be more successful than others, but their success depends on what hosts are available. Similarly, some individual hosts may be more vulnerable than others, either because of their genetic endowment, age, environment, history of prior infections, or some combination of these factors. What we call the balance of nature is usually not the outcome of cooperation for mutual survival, even among individuals of the same species; rather, it represents a standoff between one competitor and another, between predator and prey, and, in the case of infectious diseases, between pathogens and their hosts. Who has the advantage very much depends on circumstances and therefore changes with time and place.

PATHOGENS' STRATEGIES: THE EVOLUTION OF VIRULENCE OF INFECTIOUS DISEASES

If mutualism is not the expected outcome in the coevolution of pathogen and host, what reasoning should we apply in order to understand why some infectious diseases cause minor discomfort whereas others are life-threatening? Or why an outbreak of disease may be serious at one place or time but have a less serious impact under other circumstances?

One clue emerges if we consider the various ways in which infectious diseases are transmitted. For example, rhinoviruses, which cause the common cold, produce the conditions that insure their transmission from one host to another. They reproduce within cells of the nasal passages, but instead of killing the host cells, they cause them to secrete fluids. The newly made viruses are shed into this moist and mildly irritating environment, inducing the host to sneeze and expelling an aerosol suspension of millions of viruses. If other potential hosts are in the vicinity, which is frequent in a social species like ourselves, they inadvertently inhale the viruses. If the virus is not transmitted in this manner, it has other possibilities. The nasal secretions easily get on our hands and are then transmitted to others directly (by touching) or indirectly (by touching the same objects). The rhinovirus is an example of a pathogen transmitted directly from one host to another.

The organism that causes malaria, however, presents a more complex life cycle with a very different mode of transmission. This protozoan invades red blood cells, divides rapidly, and eventually causes the red blood cells to burst, thereby triggering bouts of chills and fever. The newly released organisms quickly invade other red blood cells. When there are many of these organisms present, some mature into gametes that can be picked up by mosquitoes, where they complete their life cycle. The mosquito also transports them to other hosts. The life cycle of the protozoan is adapted to utilize both of these hosts in different ways (Box 8.2), an elegant example of the power of natural selection. From the perspective of human medicine, however, the mosquito is simply a *vector* that transports the infection from one person to another.

These two pathogens illustrate alternative forms of transmission of infectious diseases: direct in the case of

the rhinovirus, and borne by a vector in the case of the *Plasmodium* that causes malaria. Are these different modes of transmission of infectious agents simply an interesting curiosity, or is there some wider significance? Figure 8.3 compares the severity of infectious diseases according to whether they are transmitted directly or through insect vectors. In this figure, severity of infection is measured by the percentage of untreated infections that lead to death. In the left panel you can see that about 77% of the directly transmitted pathogens cause death in less than 0.1% of the untreated infections, and fewer than 7% of these pathogens cause death rates greater than 10%. In general, then, the infections caused by most directly transmitted pathogens are relatively mild.

In the right panel you can see that the group of vector-borne diseases appears to be relatively more severe. Only about a third of the pathogens lead to death rates as low as 0.1%, whereas 28% kill their hosts in 1–10% of untreated infections, and nearly 20% are lethal in more than 10% of the infections.

Paul Ewald, an evolutionary biologist, has suggested that the level of virulence to which pathogens evolve—the extent to which they are able to overcome the host's defenses—is heavily influenced by how they are transmitted from one host to another. His reasoning is based on evolutionary concepts and is illustrated by the examples of rhinoviruses and the malaria-causing *Plasmodium*. Successful transmission of a pathogen like the rhinovirus is clearly easier if the host remains mobile. If the virus were to become more virulent—by invading and dividing rapidly in other tissues and thus destroying them—it would immobilize its host, likely decrease the host's contact with other people, and thereby decrease its own chances of transmission.

Each year between 1 and 3 million humans die of malarial infections and millions more are chronically debilitated. Its impact on human health makes it one of the most serious human diseases. Malaria is most prevalent in tropical climates where there are ample populations of mosquitoes. Where there is no netting to exclude mosquitoes nor insecticides to kill them, sick hosts are inviting targets, less able to fend off mosquitoes than people who are well. When the *Plasmodium* infects a person, the protozoans are like passengers on a crippled ship. But it matters little if they sink their ship, because there are plenty of lifeboats available in the form of mosquitoes to get them to another vessel.

There is more to the story, however, as some diseases are spread by direct contact and are nevertheless very virulent. Smallpox, now believed to have been eradicated, is such an example. When it reached the New World with Europeans, its effects on the native population were devastating. Smallpox virus is different from the pathogens that cause common colds, however,

BOX 8.2
The Traveling Protozoan that Causes Malaria

The single-celled organism that causes malaria is a sporozoan, a kind of protozoan belonging to the genus *Plasmodium*. Its life cycle takes it back and forth between a vertebrate (a bird or a mammal, depending on the species of *Plasmodium*) and an arthropod, such as the mosquito *Anopheles*. This anatomically simple organism, consisting of a single cell, displays an amazing range of adaptations that enable it to exploit the tissues of two very different hosts (Fig. 8.2).

While residing in a vertebrate, *Plasmodium* infects red blood cells and reproduces by mitotic divisions (1 in Fig. 8.2). The near-synchronous bursting of large numbers of red blood cells at intervals of three or four days is associated with the severe chills and fever that characterize malaria. The numerous sporozoans that are released (2) are able to infect other red blood cells. Eventually some of the pathogens (which are haploid at this stage of the life cycle) develop into proto-gametes, in which form they can be transmitted to another host.

The vertebrate host provides a massive reservoir of nutrients, but to get from one vertebrate host to another, the *Plasmodium* utilizes an alternate host, a mosquito, as a vector. Some of the gametes of the pathogen are drawn into the mosquito's gut when it takes a meal of blood (3),

and there the gametes fuse to form a zygote (4). This diploid cell worms its way into the wall of the mosquito's gut and forms a little cyst or spore (5), within which it undergoes meiosis. The haploid cells then reproduce further by mitotic divisions. The form of the *Plasmodium* at this stage of its life cycle is quite distinct from the form that reproduces in the red blood cells of the vertebrate host.

Eventually the cyst breaks, and the *Plasmodium* migrates to the salivary gland of the mosquito (6). When the mosquito bites a vertebrate for its next meal of blood, it injects an anticoagulant from its salivary gland, thus enabling the *Plasmodium* to reenter another vertebrate host (7). The *Plasmodium* converts to the form that is able to exploit red blood cells, and the life cycle is complete.

In most of the organisms with which you are familiar, haploid cells exist briefly and only in the service of reproduction. During most of the life cycle of *Plasmodium*, however, the cells are haploid, and the diploid phase is condensed to a short period within the mosquito. Nevertheless, the importance of genetic recombination associated with sexual reproduction is not negated by this twist, and it doubtless contributes to the lack of success so far in developing a vaccine against *Plasmodium*.

There is some evidence that the *Plasmodium's* presence in the mosquito is not benign. There appears to be some damage to the salivary gland that requires the mosquito to make more and shorter feeding stops. Although this may endanger the mosquito, it likely increases the number of vertebrate hosts that will be infected.

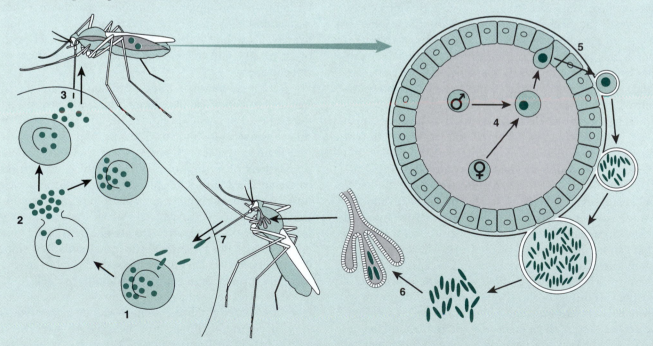

FIGURE 8.2 Life cycle of the single-celled organism, *Plasmodium*, that causes malaria. See the text for a description of what is happening at each of the numbers on the diagram.

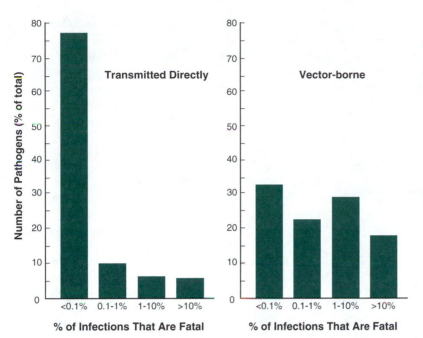

FIGURE 8.3 Diseases for which the pathogen is transmitted by a vector are significantly more likely to cause death than diseases that are spread by direct contact.

in its ability to survive outside a host. Whereas the common pathogens of the respiratory tract that kill less than 0.1% of those who become infected ordinarily survive no more than a few days outside of a host, the smallpox virus can lie in wait for a decade or more. The bacterium responsible for tuberculosis is another example of a virulent pathogen that is transmitted directly from host to host, and it too can survive for many months outside of the human body. This capacity to survive in a dormant state for a long time between infections removes the need for rapid transmission to another host. Being able to rely on the mobility of susceptible hosts has allowed these agents to become very virulent.

These several examples of human pathogens illustrate an important correlation between the virulence of an infectious pathogen and its mode of transmission. Natural selection, acting on the pathogen—not on the pathogen and its host together in a symbiotic adjustment—tunes its mode and rate of reproduction as well as its virulence to the level that enables transmission to a new host. Killing the host without a means of transmission to another host is an evolutionary dead end for the pathogen. As we shall see in the next section, supporting evidence for this relationship is supplied by examples where humans have unwittingly created new ways for pathogens to be transmitted. As a result, some diseases have become more virulent than they previously were. On the other hand, the virulence of other diseases has decreased as transmission has become more difficult. In these cases, selection has occurred among related organisms that express different degrees of virulence, depending on the opportunities for transmission available to them.

ECOLOGICAL FACTORS IN THE EMERGENCE AND SPREAD OF DISEASE

Besides creating conditions that alter natural selection on pathogens, humans have also changed their own environments in ways that can either promote or decrease their exposure to pathogens.

INFLUENCES OF HUMANS ON NATURAL SELECTION AMONG DIARRHEAL PATHOGENS

If asked what the principal cause of human death has been throughout recorded history, many people in developed countries would guess war or starvation. They would be wrong. Four to twenty million people, most of them children, die each year of bacteria-caused diarrheal diseases, including dysentery, cholera, typhoid fever, and those caused by virulent strains of the common human intestinal bacterium, *Escherichia coli*. Human behavior can alter the frequency and even the virulence of these diseases. Sometimes the outcome is intended and reduces the impact of these diseases on human health; other times it is unintended and has just the opposite effect.

Intestinal pathogens that cause the host to lose massive amounts of fluid (diarrhea) are simply seeking another host to infect. This is the quickest way out of the host's body and is the characteristic mode of transmission of these water-borne bacteria. Where drinking water is drawn from wells or streams, casual disposal of human wastes frequently leads to contamination of the water supply by these organisms. Humans can thereby provide *cultural vectors* of disease. Let's examine a couple of examples in more detail.

There are several species of the dysentery-causing bacterium *Shigella*, and they differ in both virulence and their dependence on water as the means of transmission between hosts, as shown in the following table.

Organism (species of *Shigella*)	Mortality (deaths per 100 untreated infections)	Transmission (% of outbreaks that are waterborne)
S. dysenteriae, type 1	7.5	80.0
S. flexneri	1.35	48.3
S. sonnei	0.45	27.8

The species that is most dependent on water in getting from one host to another is the species that causes the most serious infection in humans. The same evolutionary logic that accounted for the relation between transmission by vectors and virulence can be applied here: if water offers a relatively sure means of transmission between hosts, there can be selection for species or strains with high virulence.

History provides a test of this hypothesis. In the United States, major improvements in water purification were made during the first half of the twentieth century, and the most deadly dysentery bacterium, *Shigella dysenteriae* type 1, was replaced by the less virulent *S. flexneri*, which was in turn displaced by the still less severe *S. sonnei*. The correlation is in fact even more significant, for the same sequence has been seen elsewhere around the world when public water supplies were improved. For example, in Britain the improvement in water supplies and the evolutionary changes in *Shigella* preceded those in the United States by about thirty years; in much of China they occurred later.

Cholera provides another example. The bacterium *Vibrio cholerae* occurs many places in nature, but two forms produce toxins and infect the gastrointestinal tract of humans. The toxin alters the permeability of the epithelial lining of the intestines, causes massive loss of body fluids, and flushes out the pathogens along with competing microorganisms. The toxin is thus the cholera organism's ticket to its next host as well as lessening competition for those cholera bacteria that remain within the infected host.

There are two genetic types of this pathogen that differ in their production of toxin and thus their virulence. The "classical" type produces more toxin and more severe symptoms, but throughout much of the world it has been replaced by a milder "el tor" type as water supplies have been improved. In Bangladesh, however, the water supplies are not yet reliable, and 1980 saw an outbreak of the classical type.

The competition between classical and el tor types of *Vibrio cholerae* also illustrates how relative selective advantage depends on conditions. The massive production of toxin by the classical type facilitates transmission where the water supply is easily contaminated. The el tor type puts fewer of its metabolic resources into making toxin, but because it resists desiccation, it is able to survive for a longer time outside a host. The purification of water supplies thus does not simply kill pathogens. It also causes evolutionary changes in populations of pathogens in the larger environment.

When people are crowded together under dirty conditions such as refugee camps, the opportunities for diarrheal pathogens are amplified and outbreaks of diseases like cholera are particularly severe. Consider what happens when a person living in unsanitary conditions is immobilized with severe diarrhea. Bacteria are released into their clothing and bedding, which may then be washed without disinfection in the nearest stream, thereby contaminating sources of drinking water. Moreover, it becomes increasingly difficult for those who are tending the sick to keep from being contaminated and contaminating others through direct contact. As transmission becomes independent of host mobility, selection for virulent strains of the pathogen occurs.

The conditions under which humans live—the technology that is available to them, their customs, and their behavior—can thus facilitate the movement of these pathogens from one host to another. This result is a clear example of human behavior providing a cultural vector for a disease.

PATHOGENS IN HOSPITALS

"If you want to get really sick, go to a hospital." Variations on this ironic humor reflect the reputation that hospitals have, more in the past than now, as places where we are more likely to acquire a serious infection than in the outside world. Most people attribute this risk to some combination of three reasons: hospital patients are in a weakened condition, either when admitted to a hospital or after they have undergone surgery, so they are more susceptible to infections; there are more pathogens in hospitals because they are brought there by their seriously ill hosts; and pathogens in hospitals are more infectious because they have become resistant to the antibiotics frequently used there.

Not surprisingly, all of these factors are known to contribute to the increased likelihood of acquiring serious infections that kill an estimated 20,000 people a year in U.S. hospitals. But something else may also occur: taking care of people in hospitals provides cultural vectors of disease. Hospital attendants may transmit pathogens from one patient to the next if they do not disinfect their hands or instruments properly. For example, hospital nurseries offer ideal environments for the selection of virulent strains of diarrheal pathogens. There are several reasons for this. First, newborn humans are much more susceptible to infections than adults: they lack the immunity that has been acquired by adults; they lack the benign bacteria of the gut that compete with the more virulent species; and they have not yet developed the high stomach acidity that kills many bacteria before they are able to enter the intestines. Furthermore, many studies have shown that contaminated hands are a major cause of hospital-acquired infections, and that hand-contacted objects in rooms housing infants with diarrheal diseases become heavily contaminated within a day. Doctors or nurses touch many babies many times a day, especially those with the most severe diarrhea. Moreover, the hands of attendants are more easily contaminated by these infants. Clearly, bacteria that multiply rapidly and are expelled in a transmissible medium (diarrhea) will be the first to be transmitted to the next host.

Transmission of pathogens can occur anywhere attendants care for one individual after another over a short period of time—nursing homes, kennels, stables. There is much evidence that such places have been the site of origin of outbreaks of infectious diseases, not only in the past but more recently as well. By providing cultural vectors of disease, we thereby create conditions that favor the spread of disease, and where ready transmission fosters selection for more virulent strains or species of pathogen or strains that have evolved resistance to antibiotics, the problem is compounded. The converse is also true. By suitable behavior we can also make it more difficult for pathogens to use us as unwitting vectors in their transmission to new hosts.

ECOLOGY AND EMERGING DISEASES

Expanding population has put people in environments that were sparsely populated in the past. Changes in technology have altered the landscape. Forests are cut down for farms, swamps and marshes are drained, rivers are dammed, water is either made safe for drinking or polluted, and suburbia invades former forest or farmland. These changes are usually undertaken for someone's immediate economic gain, but they can carry with them unanticipated consequences for health as the ecology of pathogens or their vectors are altered in unexpected ways.

Dengue fever, which is caused by a mosquito-borne RNA virus and results in fever, rash, and muscle pain, has been recognized in human populations in Asia and the Americas for some two hundred years. It has increased in incidence and in geographic range over the last twenty years, largely because human populations and the pathogen's vector, mosquitoes, have increased in numbers in close proximity to each other. Human populations have both increased and occupied new areas. Mosquito populations have also changed. In some places they have been controlled, but elsewhere increases are traceable to new breeding sites provided by poor urban sanitary conditions, to relaxation of mosquito control programs, and to increase in the resistance of mosquitoes to insecticides.

Lyme disease, a bacterial infection that causes arthritis-like symptoms and nerve damage if it is left unchecked, gets its name from a cluster of cases near Lyme, Connecticut. However, a similar outbreak was described in Europe more than a hundred years ago, so the disease is not new. The bacterium that causes the disease is transmitted by ticks to humans from two "reservoir hosts," mice and white-tailed deer. The incidence and geographic range of Lyme disease has increased in New England, and the reason is thought to be because these reservoir hosts have increased in numbers with the loss of farms and consequent reforestation of farmland areas.

Legionnaire's disease, a pneumonia caused by the bacterium *Legionella pneumoniae*, and toxic shock syndrome (TSS), a vaginal infection caused by the bacterium *Staphylococcus aureus* when it multiplies within tampons, are examples of how human technology has created new favorable environments for pathogens. *Legionella* bacteria normally inhabit aquatic environments, either as free-living organisms or as parasites of protozoa; they only cause disease in humans when inhaled into the lungs in suspension in water droplets. The air-conditioning systems of large public buildings such as hotels provide both the aquatic environment required for growth of *Legionella* and an aerosol suspension of bacteria in air that is recycled through the building and breathed by humans.

The *Staphylococcus aureus* that causes toxic shock syndrome in women is common, for example, in our nasal passages; from there it is probably hand-transmitted to super-absorbent tampons inserted into the vagina, where it is nourished and thrives. In this special environment it also produces a toxin, something it does not do in most locations.

Efforts to control the spread of disease are frequently conscious and planned. The provision of safe drinking water in much of the world during the last century is a public health measure of enormous importance and may have saved more lives than all the activities of physicians combined. But an insufficient understanding of the life histories of pathogens and of the force of natural selection has frequently led to human behavior that increases the impact of infectious diseases. Massive and indiscriminate use of antibiotics has spurred the evolution of antibiotic-resistant strains of microorganisms, which are now posing new and difficult medical problems.

This history illustrates the importance of viewing pathogens from a Darwinian perspective. Epidemics are not momentary lapses of nature in which disease-causing organisms have temporarily disrupted a peaceful coexistence with their host. Nor are they phenomena that beg for a supernatural explanation. Pathogens, like all organisms, are vehicles for self-replicating nucleic acids and are driven by natural selection to outreproduce each other by whatever means they find available. If they should wipe out their host and be unable to adjust to another, they join the long roll of organisms that have evolved and then become extinct.

These principles should inform our actions. A deeper understanding of our environment and the organisms with which we interact can alert us to possible consequences of technological change. Even after a disease emerges, ecological knowledge can sharpen our insight about possible causes.

EVOLUTIONARY EPIDEMIOLOGY

We have been arguing that pathogens and their hosts should be seen in the context of their evolutionary and ecological relationships. If there is a reliable way for a pathogen to escape a doomed host and find a new one, or if a pathogen evolves a means of extending its survival outside its host, selection will favor increased virulence. Generally this process occurs through changes in selection for different preexisting and competing strains or species, but it is possible for pathogens to generate novel mutations that are then selected. This latter process makes bacterial strains resistant to antibiotics.

Hosts, however, also have the potential to evolve resistance. Those who survive epidemics are likely to be the individuals with the best defenses—either because they are young, healthy, and well nourished, or have an array of alleles that enables them to resist the pathogen particularly effectively, or some combination of these characteristics. Defenses that vary because of the host's genotype are of course subject to natural selection.

But prior exposure to a pathogen can also improve an individual's defenses. As we will explain later, the immune system has a "memory" for previous infections and is able to mobilize an effective response if the same foreign agent is encountered again. When a population of hosts lives for years with a virulent pathogen, many vulnerable individuals are killed, often before they reproduce. Many other individuals recover from the disease, but in the process they develop immunity that protects them from future infections. When the pathogen encounters a population of hosts that has never experienced the disease, however, no one has immunity, and the ravages of the disease can be particularly severe. For example, in the sixteenth through the eighteenth centuries, many more native North Americans died from smallpox introduced by Europeans than as a result of armed conflict or any other cause.

The hosts' relationship with frequently encountered pathogens does not necessarily remain stable. The population of either host or pathogen may change with time, and another epidemic can occur later. For example, the virus that causes influenza evolves so rapidly that new vaccines have to be developed each year.

OUR IMMUNE SYSTEM: A MICROCOSM OF BIOLOGICAL EVOLUTION

The immune system of vertebrates provides a remarkable array of defenses against rapidly reproducing pathogens. Two categories of *lymphocytes*, a kind of white blood cell, are critical: B cells, which are formed in the bone marrow, and T cells, which mature in the thymus. In addition there are several other places in the body where these lymphocytes congregate as they circulate through the body in the blood: lymph nodes, spleen, tonsils, and the appendix. In the overview of the immune system that follows, we will focus on the behavior of the B cells.

Whenever large foreign molecules appear in our bodies, the cells of the immune system recognize and attempt to destroy them. Foreign molecules that activate the immune system are called *antigens* (Fig. 8.4). Virtually any large molecule such as a protein or a polysaccharide (a polymer of sugars) can be an antigen. Many antigens are present on the surfaces of invading pathogens.

The B cells produce proteins called *antibodies*, which circulate in the blood and bind specific antigens, thus marking or "flagging" them for destruction. Bacterial toxins (for example, produced by the organisms that cause tetanus or the food poisoning known as botulism) act as antigens. Toxin molecules are bound to antibodies and are then destroyed by proteolytic (protein-destroying) enzymes in the blood that only become active in the presence of antigen-antibody complexes. When the surface of a cell or virus has been marked with antibodies, it is recognized and engulfed by white blood cells called *macrophages* (meaning "eaters of large things") (Fig. 8.4).

The binding between an antibody and an antigen is similar in principle to the way a substrate is recognized by an enzyme (Chapter 3): there is a fit between the three-dimensional shapes of antigen and antibody. The degree of specificity and the tightness of the association between antigen and antibody varies, however. For reasons that will become clear shortly, the stronger the binding, the more effective is the immune response.

Since both proteins and other macromolecules can serve as antigens, there are a great many unique antigens that the immune system is likely to encounter. In

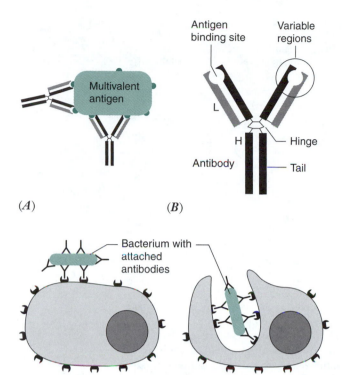

Multivalent antigen

Antigen binding site — Variable regions

L

H — Hinge

Antibody — Tail

(A)   (B)

Bacterium with attached antibodies

Macrophage with receptors for the tails of antibodies recognizes and engulfs an antibody-coated bacterium

(C)

FIGURE 8.4 **(A) An antigen can have more than one molecular site that will be recognized by an antibody and is thus *multivalent*. Antibodies are Y-shaped proteins that have two identical antigen-binding sites at the ends of the arms of the Y. The strength of binding of an antibody to an antigen is much stronger if both sites are used, and a flexible "hinge" at the fork in the Y extends the reach of the antibody.**

(B) The antibody molecule is formed from two identical heavy chains making up the stem of the Y and extending into each arm (each with about 440 amino acids) and two identical light chains confined to the arms (each with about 220 amino acids). The antigen-binding sites engage both the heavy and light chains. These ends are very variable in their amino acid sequences, which makes for the great diversity of antibodies.

(C) A bacterial cell that has been decorated with antibodies can be recognized by a white blood cell (e.g., a macrophage). The macrophage then engulfs and destroys the invader. The macrophage has *recognition sites* on its surface that will bind to the tail of antibody molecules when the antibodies are in turn bound to antigens. This is only one of many ways the immune system deals with antigens.

fact, each pathogen, or even a single kind of protein molecule, may present a variety of antigenic sites, so over your lifetime you must be able to make specific antibodies against hundreds of thousands, maybe millions, of different possible invaders. Our immune sys-

tem can actually manufacture hundreds of millions of different antibodies. But to make a protein requires a gene, and we are not born with genes for many millions of proteins. A mammal has only about 100,000 genes for all of the enzymes and other proteins that it uses during its entire life! But not only does the immune system have the ability to recognize many antigens and to make a virtually limitless number of different antibodies, most remarkably it also has the capacity to improve the strength and specificity of the initial binding between antigen and antibody.

What, then, is the source of the genetic information for each antibody, and how is that information made more effective with time? The answer is that new genetic information arises in your immune system by some tricks of recombination not found in other cells, and this novel information is then expanded by mutation and sifted by natural selection. These latter processes take place repeatedly whenever you have an infection, and they are rapid, ordinarily happening over several days.

B cells both recognize antigens and manufacture antibodies. Most of the antibodies they make are secreted into the blood, but they insert some into their cell membranes, and these function as molecular receptors. The immune response begins when one of these cell-surface antibodies binds to an invading antigen. Because there are millions of B cells in the lymphatic tissue and blood, each making and bearing a *single kind* of antibody that is different from the antibodies of virtually all of the other B cells, at least a few of these lymphocytes are likely to recognize a novel antigen.

The binding of antigen to the surface of a B cell acts as a signal, and with the participation of T-helper cells, the lymphocyte is stimulated to divide by mitosis. The antigen-antibody reaction thus leads to one or more lineages of cells containing the genetic code for making successful antibodies. Cells arising from a parental cell by mitosis are called a *clone*. The more B cells there are producing active antibody, the more effectively the foreign antigens are neutralized. And because binding of antigen selectively triggers division of only those B cells with effective antibodies, those B cells increase by *clonal selection* (Fig. 8.5)

But B cells, like all of our somatic cells, are descended from the same initial zygote by mitosis, so how can there be millions of genetically distinct kinds, each with a gene for a different kind of antibody? How can we make more different antibodies than we have inherited genes to code for them? The basic mechanisms by which such cell-based antibody diversity is generated are easily understood, although the many details can occupy an entire course in immunology.

An antibody is a Y-shaped protein consisting of subunits called "heavy" and "light" chains (Fig. 8.4B). The two identical heavy chains form the stem of the Y,

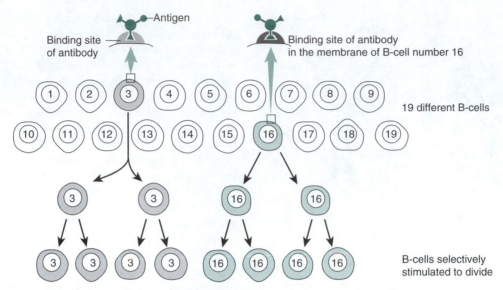

FIGURE 8.5 Clonal selection. Among the many B cells, each with its own *specific anti-body* serving as a *membrane receptor*, only a very few will recognize (bind to) a novel antigen. As suggested by the diagram, a novel antigen is likely to be recognized by more than one antibody, and as shown in Figure 8.4A, an antigen is likely to bear more than one antigenic site. Different antibodies may recognize the same antigen because they bind different sites (as suggested here) or because they bind the same site, but with different specificity and strength.

The binding of antigen to antibody membrane receptors stimulates the B cells to divide, producing daughter cells expressing the same antibody. These cells there-fore outreproduce the other B cells, creating clones, each of which is able to synthe-size and secrete an antibody specific to the antigen. As described in the text, the antibodies can be further refined during this process.

and one extends into each arm of the Y. The two identi-cal light chains are located in the arms of the Y. Each of these chains contains regions of amino acids that are relatively constant as well as regions that are variable. The variable regions are located at the ends of the arms of the Y where the two antigen binding sites are located.

Differences between the constant regions are re-sponsible for five different classes of heavy chains and two classes of light chains, all differing in their amino acid sequences. Different combinations of these classes produce ten general classes of antibodies, each with somewhat different functions. The enormous variety of antibodies within each general class, however, is mainly provided by differences between the variable regions of the two chains.

Diversity of antibodies is created by a kind of "ge-netic differentiation" of B cells that is unique to lym-phocytes in the immune system. Whereas all other somatic cells become different (that is, differentiate into various tissues) during development by selective activation of genes that they all possess (Chapter 10), B cells become different in the antibodies they produce by rearranging, deleting, and mutating their genetic material. The rearranging is a form of genetic recom-bination analogous to the recombination that occurs

by crossing over during meiosis, but in B cells it is not associated with cell division and is confined to the genes for antibodies. For example, prior to making mRNA for the heavy chain, four different segments of DNA must be linked together, a C segment (for the constant region) and V, J, and D segments for the vari-able region. Each of these V, J, and D segment genes are present in multiple, somewhat different copies, and during the maturation of each B cell, one of each of these codes is randomly cut out and stitched together to form the heavy chain gene. The other possible codes for V, J, and D segments are not used for the an-tibody unique to that B cell (Fig. 8.6). For the human heavy chain, there are six codes for the J segment, twenty-seven for D, and fifty-one for V, which makes possible $51 \times 27 \times 6 = 8{,}262$ different heavy chains. There are 320 possible light chains, and therefore $8{,}262 \times 320 = 2.64 \times 10^6$ different antibodies. Actu-ally, the number is even larger, because the cutting and rejoining of DNA that is required to assemble the genes from their separate segments is sloppy. Dele-tions of nucleotide bases from the ends of the seg-ments are common, and insertions occur as well. These changes frequently shift reading frames (Chap-ter 3), a mutation that leads to more changes in the

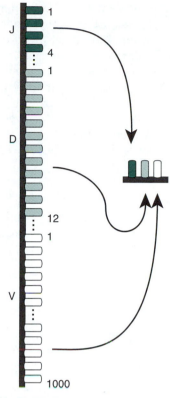

The Origin of Antibody Diversity

FIGURE 8.6 The enormous diversity of antibodies arises during development of B cell lymphocytes. The immature cells contain pools of different coding segments for the heavy and light chains of antibody molecules. Each cell makes functional genes by combining single segments from each of these pools. This diagram illustrates how just the variable portion of the gene for the heavy chain of mouse antibodies is assembled. It is spliced together from three coding segments, *J*, *D*, and *V*. There are four *different* copies of the *J* segment present in the genome, twelve of the *D* segment, and about a thousand of the *V* segment. But in assembling the gene, a B cell uses only one copy of each kind of segment. Consequently, in mice there are 4 × 12 × 1000 different possible combinations of segments for this region of the gene. Light chains also have a similarly variable region of their gene.

amino acid sequence of a protein than any other. The result is an enormous increase in antibody diversity.

The lymphocytes congregate in lymph nodes, and here they are presented with foreign antigens circulating in the blood. With more than 100 million different kinds of antibodies, it is not surprising that one or a few of them will bind to a novel antigen. Once this happens, as we described above, a B cell making a binding antibody is stimulated to divide.

To summarize, this sequence of events produces evolutionary change in the population of lymphocytes in several steps. The maturation of lymphocytes produces great *genetic diversity*, which is reflected phenotyp-

ically as variation in the array of antibodies. The selective interaction of antibodies with environmental antigens in turn leads to *differential reproduction*—clonal selection—of a small subset of the lymphocytes. Here, then, are the minimal requirements for natural selection as described in Chapters 1 and 5: a population of replicating entities that differ from one another in a heritable property that influences their reproductive success.

But this mechanism for the differential reproduction of cell lines within the body is just the first step in "evolving" new antibodies to fight an infection. When these selectively activated cells begin to divide, the rate of point mutations (Chapter 3) during replication of the variable regions of their antibody genes is allowed to increase a millionfold. As a result, roughly half the daughter cells of any division have slightly modified versions of the antigen-binding site. Since the mutations are random, most of the modified antibodies will bind the antigen less effectively. A few, however, will bind antigen more tightly. The tighter the binding, the stronger is the signal to divide. The result is that those B cells that produce improved antibodies are the ones that will reproduce the fastest and give rise to the most successful clones. The immune response therefore strengthens with time, and within a week or ten days your cold is gone. Natural selection has been used by your immune system, itself a product of natural selection, to improve the precision of the tools it uses to excise invaders.

The genes for antibodies that have been effective in countering an antigen do not disappear when the invading antigen has been removed because some B cells with the gene for the new antibody persist for months or years. The immune system therefore has a "memory." If the antigen is reintroduced later, the immune system is then quickly mobilized to destroy it. This response is the basis for vaccinating against diseases. A vaccine consists of antigens such as a virus that has been chemically crippled so that it cannot cause the disease but still has antigens that elicit antibodies to the undamaged infectious form. When the vaccine is injected into the body, the immune system makes antibodies and stores the antibody genes in a clone of B cells that preserve the memory of the antigen and are ready to pounce should a "live" influenza or polio virus be encountered.

The biological evolution of antibodies in our immune system is speeded up in two ways: by shortening the division time of B cells and by increasing the rate at which genetic variation and mutations are introduced into the selection process. The division time of B cells is shorter than for many mammalian cells, eight hours, but it is still much longer than that of many bacteria or viruses. By far the most important factor in speeding up antibody evolution is the B cells' ability to increase selectively the mutation rates of their antibody genes a millionfold. This high mutation rate of antibody genes makes it possible for the immune system to compete with the rapid evolutionary potential of bacteria and viruses.

Perhaps you are imagining that if a pathogen could evolve an "antigen-evolving" mechanism like the antibody-evolving mechanism of the immune system, it could generate new antigen "disguises" for the pathogen too rapidly for the immune system to keep up. One of the malaria-causing protozoans does precisely this during its life cycle, and as a result it is able to win a few more rounds of hide-and-seek with the immune system than most other pathogens. Consequently, it has been very difficult to produce a vaccine for malaria. A similar thrust and parry goes on between the immune system and the virus responsible for AIDS, an evolutionary dual that can last for years before the immune system is finally overcome.

HIV AND AIDS

The human immunodeficiency virus, or HIV, illustrates a number of the ideas that we have been discussing in this chapter: emergence as an epidemic, rapid evolution of the pathogen, the role of human behavior as a cultural vector, and the response of the immune system. Between 40 and 50 million people are estimated to be infected worldwide, and the disease is continuing to spread. AIDS is so menacing because at this writing there is no vaccine or cure, and most of the people who are infected cannot afford the drugs that slow its course.

HIV is present in the blood, semen, and vaginal secretions of infected individuals, and it is transmitted during either hetero- or homosexual intercourse. Transfusions of HIV-contaminated blood and the sharing of unsterilized, HIV-contaminated hypodermic needles can also result in transmission of the virus.

HIV is an example of a *retrovirus* (Fig. 8.7). The genes are carried as RNA, and one of the viral enzymes is *reverse transcriptase*. Copies of the reverse transcriptase molecule are packaged with the virus, and the enzyme is able to go to work as soon as it enters the cytoplasm of a host cell. Reverse transcriptase is a DNA polymerase, but it has the ability to use either RNA or DNA as a template. It first makes a complementary strand of newly synthesized DNA, using the viral RNA as a template. The DNA strand is then used as a template to make a section of DNA double helix, which is in turn inserted into the host chromosome. The viral genes can hide for years in an inactive state in the host's genome, but eventually new virus is made as its genes are transcribed (making multiple copies of viral RNA) and translated by the host's protein-synthesizing machinery (making additional viral protein).

The cells infected by the HIV are a class of T cell lymphocytes that are critical for the proper functioning of antibody-producing B cells. When these T cells are

killed faster than new ones arise, the immune system becomes so depleted that other normally mild pathogens—yeast, fungi, bacteria, protozoa, and other viruses—are able to multiply and cause the symptoms collectively described as acquired immune deficiency syndrome or AIDS. Clearly, a pathogen that can score such a preemptive strike—by destroying the cells that are supposed to destroy it—has evolved a uniquely deadly strategy. Moreover, the fact that the viral genes can hide in the host's chromosomes, as well as their ca-

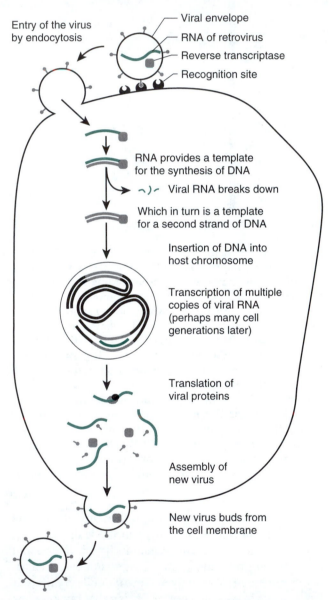

FIGURE 8.7 HIV is an example of a *retrovirus*. The genes consist of RNA, from which DNA is made (thus the "retro"), which is in turn inserted into the host DNA. After a variable time, new virus is then made by conventional transcription and translation.

In HIV infections the hosts are a class of T cell lymphocytes that are critical for the proper functioning of the immune system.

pacity to mutate during viral replication, make HIV a difficult target for both the immune system and medical science.

From where did this virus come and how did it become so virulent? Related viruses are present in other primates (Fig. 8.8), and the present evidence indicates that they can sometimes move between species. This is most likely to happen when individuals of one species kill and eat another. Human butchering of chimpanzees for food has recently been implicated. The AIDS epidemic began about twenty years ago, but it is possible that the virus has been present in humans for much longer than that. What, then, happened to trigger the epidemic?

There is indirect evidence that a cultural change may have been involved. In central and east Africa during the 1960s and 1970s large numbers of men seeking industrial jobs migrated from rural to urban areas where, without their families, they had unprotected sex with prostitutes. In some areas the prostitutes, two-thirds of whom became carriers of HIV, averaged a thousand sexual contacts per year, and their male clients averaged about thirty regular partners per year.

This cultural change could have initiated an epidemic in a couple of ways. If the virus had previously been present in a small and relatively isolated population of rural Africans who had developed immunity over time, its introduction into a larger population with little or no resistance could have led to an epidemic. In this view, it is the absence of resistance in the host population that is the critical result of evolutionary and social history.

But there is another, albeit not mutually exclusive possibility. The virus does not survive for long outside a host. This suggests that it is a pathogen whose degree of virulence may increase if there are more frequent opportunities for transmission. Among the modes of transmission of HIV, sexual intercourse has the greatest potential for increase in frequency. If mating is monogamous, the virus has few opportunities for transmission. Mating between partners may be frequent, but if the virus tends to kill both parties, evolutionary reasoning suggests there will be selection for less virulent strains as well as host resistance. If mating is promiscuous and frequent, however, there will be numerous opportunities for the virus to find another host, and more virulent forms can be selected.

There is good evidence that the virulence of HIV strains has been increased in this way. In populations in which promiscuity is high, the strains have become more virulent, and in populations in which promiscuity and unprotected sex has decreased, so has the virulence

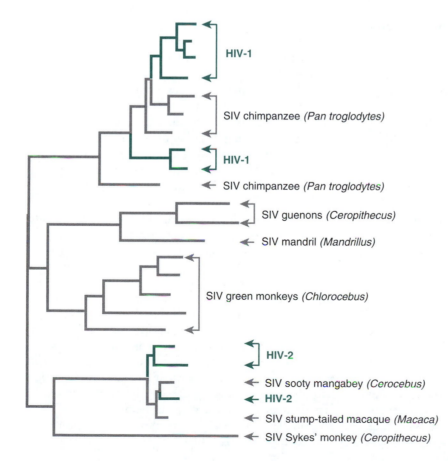

FIGURE 8.8 An evolutionary tree of HIV (the virus causing AIDS), based on genetic differences between the various strains. SIV (simian immunodeficiency virus) is the form of the virus found in apes and monkeys. It appears, however, that the virus has jumped between species on more than one occasion. How this happens is uncertain. It is very unlikely that the virus survives in mosquitoes and other insects that feed on mammalian blood. It is quite possible that the virus entered humans as a result of butchering chimpanzees for food. If an individual who cleaned an infected animal had a small scratch, there would be a significant possibility of transmission across species. For further details see the reference to Hahn and colleagues in the general bibliography.

of the HIV strain These evolutionary outcomes are consistent with the inability of HIV to survive for long outside of a host. Its capacity to mutate is why it poses such a challenge to the immune system.

OTHER WAYS HOSTS FIGHT PATHOGENS

Evolutionary thinking is not only useful in helping us to understand how and why pathogens spread and reemerge in more threatening forms, it also helps us understand the mechanisms that have evolved for resisting infections. Paul Ewald, Randolph Nesse, and George C. Williams have argued that many of the *symptoms* of infections—our physiological responses to disease—are not just annoying effects of the pathogen but are evolved defense mechanisms of the host.

Fever is a common response to infections, and we commonly try to suppress it with such drugs as aspirin or acetaminophen. But experiments with animals have shown that blocking fevers may jeopardize and delay their recovery from infections. Similarly, controlled experiments with human volunteers who were deliberately given colds have shown that those individuals who took a placebo (pills not containing a drug) produced more antibodies, had less congestion, and were infectious for a shorter time than those who took pills containing acetaminophen or aspirin.

What does fever do? By letting the body temperature rise, the host is trying to alter its internal environment in a way that will give an advantage to its immune system, but the mechanism is not understood. Clearly the struggle can get out of control, for too high a fever for too long can itself be dangerous.

Another likely adaptive response is the way in which our bodies withhold iron from bacterial pathogens. This element is crucial to their high rates of reproduction, and during an infection the concentration of iron in our blood is reduced, we absorb less through our intestines, and our appetite for iron-rich food declines. Even when we do not have an infection, iron is closely guarded: a special protein, *transferrin*, binds it tightly and releases it only to cells with the proper transferrin-recognizing protein on their membranes. The diarrhea-causing bacteria can increase their production of toxins when iron is scarce, thus making the host sicker. But by inducing diarrhea, the pathogen enhances the possibilities for its transmission. This is yet another example of the ongoing warfare between host and pathogen. The host's efforts to slow the growth of bacteria by limiting the availability of an essential mineral (iron) makes the environment less desirable for the bacteria. The response of the bacteria is to seek a new host by the only mechanism in their arsenal, secreting toxin and inducing fluid loss from the host.

There are some first-line defense mechanisms that try to keep pathogens at the gate. Saliva, which contains digestive enzymes, is constantly secreted in the mouth and either kills pathogens or flushes them into the stomach where the acid environment and more concentrated digestive enzymes destroy them. Similarly, tears, which also contain bactericidal compounds, constantly wash the eyes, and more are secreted when foreign objects come to rest on the cornea. Large particles that enter the trachea or lungs are expelled by coughing, while smaller particles, such as dust, smoke particles, or pathogens, are trapped in mucus and propelled upward by beating cilia until their accumulation triggers coughing. Suppression of coughing and sneezing by medication can restrict the expulsion of such foreign matter.

Our digestive system uses expulsion to get rid of pathogens and poisons. The bacteria and fungi that infect food frequently produce distinctive odors, and our sense of smell provides an early warning system. If ingested, the toxins that are made by bacteria may induce us to vomit. Vomiting is acknowledged to be an important defensive mechanism, and unlike diarrhea it is usually not suppressed with medication.

THE RED QUEEN

The warfare between pathogens and their hosts has even broader significance than we have suggested so far. The host is the environment in which a parasite exists for most of its life, so this is the principal environment in which its evolution occurs and to which it must adapt. But by the same reasoning, the parasite or pathogen is part of the host's environment, and therefore each is responding to the other. Toxins produced by bacteria induce physiological responses in the host that increase the probability of escaping and infecting another host. Having an immune system, in all its complexity, or running a fever are adaptive responses to the challenge presented by pathogens.

These kinds of evolutionary ploy and counter-ploy are present everywhere. Organisms are invariably critical parts of each other's environment. The leaves of plants are eaten by hoards of animals from insects to mammals. Plants are the energy resource that has permitted the evolution of all animals, but from the genetic perspective of a plant, it is not on earth to provide fodder for organisms that cannot photosynthesize for themselves. Rather, it is here to make more of its genes and, as a means to that end, more of itself. So plants have ways of regenerating their green parts from less tasty tissues, and they synthesize chemical compounds that are toxic to animals that eat them.

Animals, in turn, evolve chemical receptors and detoxifying enzymes that help them avoid being harmed by these substances; our own taste perception of bitter is just such a defense mechanism. Some animals take advantage of these noxious compounds. For example, monarch butterflies sequester milkweed plant poisons in their bodies, thus making birds that eat them sick. These and other butterflies advertise their distastefulness with brightly colored wing patterns, which birds quickly learn to associate with a bad meal, and thus avoid.

Humans use plant-produced spices—many of which prevent or retard the growth of fungi and bacteria—to preserve food. The evolutionary biologist Paul Sherman and his students have shown that there is a latitudinal gradient in the traditional use of spices, being greatest in warm climates, where food spoils quickly.

Predator and prey also have an ongoing evolutionary relationship, but the relations are not cooperative. The lion is dependent on the antelope or the zebra for food, thus for its reproductive success. The antelope's or the zebra's reproductive success, however, depends on not being eaten by the lion. Each has evolved to deal with the other according to its own genetic interests.

Flowers and their animal pollinators provide examples of coevolution with benefit to both parties. The plants utilize a bird or insect to transfer pollen, and they frequently try to make this process efficient by attracting (with color or fragrance) and feeding (with nectar, a sugar solution) insects or birds that will quickly learn the signals and visit other plants of the same species. As the insects and birds find the flowers useful sources of food, they have evolved an impetus to learn the appropriate sensory signals.

We have also seen a preview (Chapter 6) of how important the behavior of animals in social groups can be. Other animals of the same species can therefore be a critical part of the environment, and the social interactions of an individual can determine whether or not it reproduces successfully.

These diverse examples illustrate a few of the many ways in which an organism's environment is composed, not just of the physical world, but of other organisms. Moreover, as the discussion of pathogens illustrates vividly, other organisms can present a serious evolutionary challenge when they do not hold still genetically. Evolutionary biologists call this the "Red Queen Effect" in reference to the character in Lewis Carroll's *Through the Looking Glass* (Fig. 8.9), who informed Alice "Now, *here*, you see, it takes all the running *you* can do, to keep in the same place." Because the biological environment is itself being constantly changed by natural selection, organisms must continually respond in kind simply to maintain the *status quo*—that is, to stay in place *relative* to other organisms. Sexual reproduction, by increasing the genetic variation of progeny, represents one answer to this challenge.

FIGURE 8.9 Alice and the Red Queen, a metaphor for the evolutionary arms race between hosts and pathogens. From the 1872 illustration by Sir John Tenniel in Lewis Carroll's *Through the Looking Glass*.

SYNOPSIS

An evolutionary perspective on infectious diseases shows that host-pathogen interactions are like two adversaries endlessly at war. No holds are barred, no weapons are banned, and if the parties continue to coexist, it is not because they are cooperating or because a truce has been declared. It is because pathogens find ways to hide from their host's continuous expeditions of search and destroy, and pathogens must adjust their invasions and conquests so as not to jeopardize their chances of transmission to the next host. Understanding this evolutionary relationship reveals how human activities and man-made environments can either assist or hinder the organisms responsible for infectious diseases, and it can help guide health policy and environmental practices toward avoiding and controlling the ravages of these organisms.

The capacity of viruses and prokaryotes to mutate and reproduce at much higher rates than multicellular organisms illustrates the power of evolution on a time scale that is familiar to everyday human experience. Viruses, consisting of only a few genes in a protective protein wrapper, can appropriate the biochemical machinery of an infected cell and put it to their own use. Some can even integrate their genes into the chromosomes of their hosts.

The immune system of vertebrates provides a rapid defensive response to viruses, bacteria, and other cellular invaders. *Antigens* are foreign molecules that stimulate certain white blood cells of the host to make *antibodies*, proteins that bind to the antigens, causing them to precipitate or otherwise marking them for destruction. The response of the immune system of vertebrate animals employs genetic mutation and recombination in novel ways to generate a diversity of antibodies far greater than the number of genes present in the fertilized egg.

The ubiquitous presence of parasites illustrates a larger point: the environment of every organism includes other organisms. Furthermore, other organisms make the environment unpredictable, because they, too, evolve. One of the likely reasons why sexual reproduction is so common is that genetic recombination increases the possibilities for dealing with rapidly changing, unpredictable environments.

QUESTIONS FOR THOUGHT AND DISCUSSION

1. What are the pros and cons of considering viruses alive?

2. Can a genetic disease also be an adaptation? What criteria would you use to decide? (This question also draws on concepts in Chapters 3 and 5.)

3. What would you expect to happen to the virulence of Lyme disease in humans if the pathogen that causes it (a spirochete bacterium) were to become sexually transmissible between humans?

4. What insights from evolutionary biology might be usefully considered in formulating policy in medical decisions, public health, and agriculture? Consider local, national, and international settings.

SUGGESTIONS FOR FURTHER READING

Ewald, P. W. (1994). *Evolution of Infectious Diseases*. New York, NY: Oxford University Press. Pathogens and hosts have conflicting evolutionary interests, and how a pathogen is transmitted is the major determinant of its virulence.

Janeway, C. A.; Travers, P.; Hunt, S.; and Walport, M. (1997). *Immunobiology: The Immune System in Health and Disease*. New York, NY: Garland Press. An authoritative reference on the immune system.

Nesse, R. M., and Williams, G. C. (1995). *Why We Get Sick: The New Science of Darwinian Medicine*. New York, NY: Random House. A physician and a biologist use evolutionary theory to broaden our understanding of health and disease.

PART THREE

The Biology of Behavior

White-fronted bee eaters nest in burrows they dig in clay banks along rivers in eastern Africa. The female bird uses information about the reproductive status of her relatives and the dominance status of potential mates to decide whether (a) to remain in or leave her natal area, (b) breed or not breed for the season, (c) help or not help her parents rear siblings, (d) parasitize other birds by laying eggs in their nests, or (e) pair and raise young with a male. The ability of these birds, and of animals in general, to make adaptive choices based on information from their physical and social environments depends on the functioning of nerve circuits in their brains (Chapter 9). These nerve circuits develop by complex interactions between the organism's genes, cells, tissues and learning (sensory) experiences (Chapter 10). If the developmental program for the construction of these nerve circuits has been influenced by natural selection, the behavioral decisions of these animals should tend to maximize inclusive fitness under the conditions that elicit them (Chapter 11). Most of the time they do.

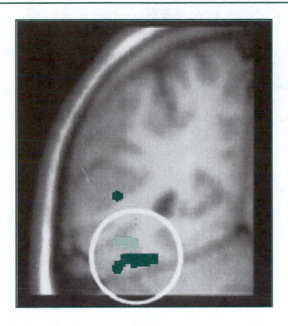

9

Nervous Systems Generate Behavior

Photo: Using a technique known as functional magnetic resonance imaging, small regions of active neurons can be identified in human subjects. This individual was looking alternatively at letter-strings and pictures of faces, one fading away to be replaced by the other every several seconds. These two forms of visual stimulation are processed in distinct but nearby locations in the cingulate gyrus of the cerebral cortex represented here by two shades of green (see Fig. 9.16D, page 209). In the brains of monkeys, cortical neurons have been found that respond only to faces. Apes and humans are able to "read" a great deal of information in the faces of other individuals, and it is not surprising that there are regions of the cortex that reflect that evolutionary adaptation.

THE FIRST APPEARANCE OF NEURONS OPENED NEW EVOLUTIONARY POSSIBILITIES

The sun is setting on the Serengeti Plain. A zebra sees or smells a slowly approaching lion, and suddenly predator and prey explode into action. Or the place is center court at Wimbledon. A tennis player waits to receive a serve, crouching slightly, swaying expectantly on the balls of her feet, every needed nerve and muscle in a state of high alert, set to meet a projectile that will travel across the net in a small fraction of a second.

The active life of animals is possible because of the presence of cells that are capable of rapid communication. With a nervous system, one part of the body can communicate with another in times measured in thousandths of a second. Nervous systems process sensory information, and they control the contraction of muscles, enabling an animal to move about and interact with its environment in ways impossible for plants and other stationary organisms.

With a nervous system, lions and tennis players not only can organize complicated motor acts like running down a zebra or meeting an on-rushing ball, they can plan ahead. Based on its hunting experience, the lion can decide which zebra is the easiest prey, and it can

change its target as the pursuit develops. The tennis player can anticipate her opponent's likely play and attempt a return that will exploit any apparent weakness. Nervous systems are therefore more than devices for detecting and processing the sensory information of the moment. They can also use information stored from previous experience, combine it with current sensory information, and select a behavioral response.

This comparison of lion and tennis player seems to imply some conscious understanding of the problem of catching a zebra, but we have to be careful not to attribute our own mental processes to other animals. Conversely, however, the description of the tennis player may imply more conscious thought than she gives the problem. As she moves to meet the ball, she is not thinking about where she is putting her feet or what she is doing with her racket arm. She is an accomplished player, and she is now running on "automatic pilot." In fact, if she has to think about the mechanics of her movements, she is not likely to make a successful return. Similarly, in driving an automobile you are able (up to a point!) to shift gears, stop for traffic lights, and avoid other cars while pondering other matters. You are not conscious of much of the behavior that your nervous system generates, and there are entire aspects of neural function of which you remain entirely unaware except when there is a problem. For example, to use some terminology borrowed from desktop computing, your nervous system is always "running in the background"—controlling your body temperature, regulating the balance of salt and water in your blood, monitoring the CO_2 in your blood, adjusting your rate of breathing, and causing the slow contraction of smooth muscles in your blood vessels and digestive system. None of this ordinarily receives any conscious attention, unless you have a fever, are very thirsty, have just run a mile, or are suffering from an upset stomach.

Much mammalian behavior is influenced by internal states of the nervous system. Hunger, thirst, fear, anger, and libido are the classic familiars. We can also recognize in ourselves "feelings" that motivate us—hate, envy, pride, guilt, curiosity, ambition, and needs for friendship and love.

This chapter about the nervous system has three overarching messages. The first is that nervous systems employ processes that can be understood in terms of physics, chemistry, and cell biology. Consequently, brains can be studied by the methods of science. The second is an extension of the first. Higher mental functions—the cognitive processing of sensory information, thoughts and imaginings, feelings and emotions, in short, what we refer to as the mind—are due to the activities of large numbers of interconnected *neurons* (nerve cells). Psychology is the study of the emergent properties of these ensembles of cells.

Finally, the behavior of animals, no less than their morphology, has been shaped by natural selection. Nervous systems, like other features of the body, can therefore only be fully understood if they are recognized as products of evolution. In order to understand how behavior can evolve, however, we first need to have a basic understanding of how nervous systems operate.

NERVOUS SYSTEMS ARE CONSTRUCTED OF TWO KINDS OF CELLS

Like the entomologist in pursuit of brightly colored butterflies, my attention hunted in the gray matter [of the brain], cells with elegant forms, the mysterious butterflies of the soul, the beatings of whose wings may some day—who knows—clarify the secret of mental life.

Santiago Ramón y Cajal
(on receiving the Nobel Prize in 1906)

NEURONS ARE THE PRINCIPAL ACTORS

During the nineteenth century, improvements in optical microscopes and the techniques for using them made it possible to examine biological tissues. Many of the chemical stains that were first used by anatomists showed that brains seemed to consist of a large number of cell nuclei gathered together in vaguely discernible layers, but they revealed little else. In the late nineteenth century an Italian, Camillo

Golgi, discovered that when pieces of brain are treated with solutions of silver salts, many nerve cells (*neurons*) stain black. Individual cells become visible in their entirety, unobscured by the many other cells that surround them. This technique, which is still used today, was applied with enormous effectiveness by the Spanish scientist Santiago Ramón y Cajal, with whom Golgi shared a Nobel Prize in 1906. Cajal was able to see that nervous systems are composed of discrete cells and not an interconnected network of tiny tubes. This finding aligned the study of the nervous system with other tissues of the body, broadening the concept that cells are the universal building blocks of living organisms.

Examples of Golgi-stained neurons are shown in Figure 9.1. The cell body bears numerous thin projections called *neurites*, giving the many different kinds of nerve cells their characteristic *shapes*. Cells with particular shapes occupy characteristic *positions* with respect to other cells. Cajal appreciated the importance of this distinctive set of anatomical relationships. He postulated that when a cell was stimulated at one end, it conveyed the stimulus to the other end, where it then influenced the activity of other cells. This view of the functional role of neurons, although much simplified, is basically correct.

Figure 9.2 shows the parts of a neuron in more detail. The cell body or *soma* contains the nucleus, and as in other cells the soma is the major site of protein synthesis. Neurons actively synthesize proteins, for they have many slender, branching processes to maintain. The *dendrites* receive input over their surfaces from other neurons. One process, longer than all the

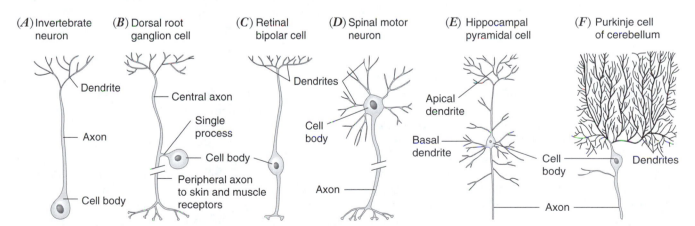

FIGURE 9.1 Some representative neurons, based on Golgi-stained preparations and illustrating the variety of shapes of nerve cells. B and D are sensory and motor neurons whose axons leave the central nervous system. The rest are *interneurons* contained within the central nervous system.

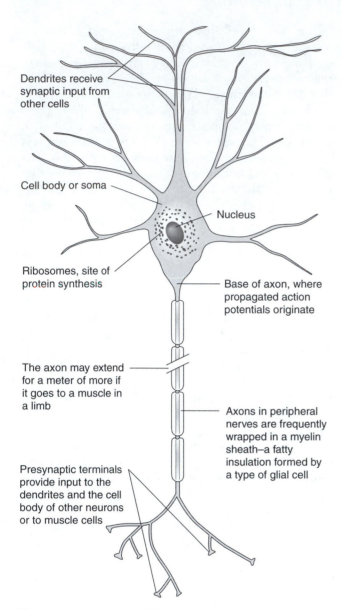

Dendrites receive synaptic input from other cells

Cell body or soma

Nucleus

Ribosomes, site of protein synthesis

Base of axon, where propagated action potentials originate

The axon may extend for a meter of more if it goes to a muscle in a limb

Axons in peripheral nerves are frequently wrapped in a myelin sheath–a fatty insulation formed by a type of glial cell

Presynaptic terminals provide input to the dendrites and the cell body of other neurons or to muscle cells

FIGURE 9.2 Generalized diagram of a neuron.

others, is the *axon*. The axon usually carries information away from the cell body and toward other neurons or muscles. At its destination an axon characteristically branches into a *terminal arborization* and makes multiple contacts with other cells. Notice how the words to describe the shapes of nerve cells conjure images of bushes and trees: dendritic branches, terminal arborization.

Several of the neurons in Figure 9.1 are *interneurons* and are contained entirely within the *central nervous system* (CNS). In a vertebrate, the CNS consists of the brain and spinal cord, whereas the peripheral nervous system consists of nerves entering and leaving the CNS. A neuron that activates a muscle (rather than another nerve cell) is called a *motor neuron*. Its cell body

typically lies within the spinal cord, but the axon extends beyond the bounds of the CNS and into the periphery where it excites a muscle, causing the muscle to contract. A *sensory neuron*, by contrast, carries information from the periphery toward the CNS. In a large animal, axons of sensory and motor nerves can be several meters long.

A *nerve* is a bundle of axons from different cells, and each axon in a nerve is called a *nerve fiber*. Peripheral nerves can contain sensory fibers, motor fibers, or in some cases, a mixture of axons from both sensory and motor cells. Pressure on your sciatic nerve (a major bundle of axons running from the spinal cord in the small of the back to the leg) can therefore activate both sensory and motor axons, causing muscle contraction as well as pain. Conversely, nerve damage can produce loss of both.

In vertebrates, many axons are wrapped in a lipid sheath known as *myelin* that increases the speed with which information can be carried along an axon. *Multiple sclerosis* is a devastating disease in which myelin sheaths degenerate, and normal neural function of sensory and motor nerves is no longer possible.

Within the CNS, bundles of axons are called *tracts*. White matter consists of nerve tracts whereas gray matter is largely cell bodies and dendrites. It is the presence of myelin sheaths on the axons that makes white matter appear different from gray matter. Within the gray matter and between layers of cell bodies there are dense felt-works of dendrites and axon terminals called *neuropil* where the connections between cells can only be resolved with the high magnification of an electron microscope.

NEUROGLIA PLAY SUPPORTING ROLES

More than 90% of the cells in your nervous system are not neurons. They are cells called *neuroglia*, or simply *glia*. The name of these cells is from a Greek root meaning glue, and refers to the fact that glia occupy the bulk of the space around the cell bodies and dendrites in the CNS. Glia differ from neurons in that they do not generate signals. They also differ from almost all neurons in being capable of cell division in adult animals. In humans, almost all brain tumors are the result of uncontrolled proliferation of glial cells, not neurons.

Basically, glia do housekeeping. Some are phagocytes, engulfing dead or foreign material. Others have many processes connecting capillaries with neurons and are likely involved in the movement of nutrients. Some glia take up the K$^+$ ions that neurons lose when they are active. Glial cells can also recycle transmitter molecules that neurons release at the junctions between cells (see p. 202). Still other glial cells form the myelin sheaths around axons.

THE SPECIAL PROPERTIES OF NERVE CELLS EVOLVED FROM BASIC FEATURES OF CELL MEMBRANES

ANIMAL CELLS EXTRUDE SODIUM IONS IN ORDER TO KEEP FROM SWELLING AND BURSTING

The signals that neurons employ are based on movements of ions through the cell membrane. In order to understand these processes and how they arose in evolution it is useful to consider the osmotic dilemma that is faced by all cells in an animal's body.

In many respects the membranes surrounding neurons are typical of all animal cells. The membrane consists of a layer of lipid two molecules thick in which are embedded protein molecules with a number of different functions. Among these proteins are *ion channels*. This fragile surface separates the contents of each cell from its environment, and it controls the rates at which molecules and ions are able to enter and leave the cell through the channel proteins (Fig. 9.3). The proportions of different ions within cells are different from those in the fluids that bathe them (blood and lymph), and the cell membranes are specialized to maintain these proportions. Similarly, the whole organism must maintain a constant internal environment—in the chemical composition of its fluids and temperature—so that all cells can be bathed in the same surround. Here we will focus on the first of these processes, because the special properties of nerve cells have evolved from basic molecular mechanisms that regulate the concentrations of ions in all cells.

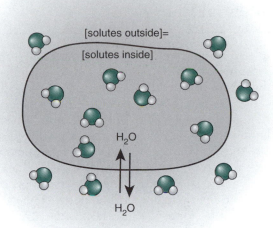

[solutes outside]= [solutes inside]

H_2O

H_2O

(A)

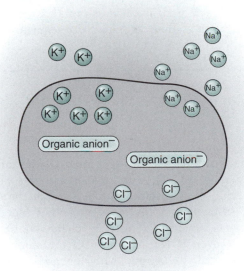

K^+ K^+

K^+ K^+

K^+ K^+ K^+

Organic anion⁻

Organic anion⁻

Na^+

Na^+ Na^+

Na^+

Na^+

Na^+

Na^+

Na^+

Cl^- Cl^-

Cl^- Cl^-

Cl^- Cl^-

(B)

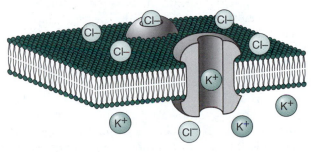

FIGURE 9.3 Neurons, like other cells, are surrounded by a membrane consisting of a bimolecular layer of phospholipid molecules with their polar head groups facing outward and their hydrocarbon tails oriented within the membrane. This diagram shows two *ion channels*, proteins that form small pores through the membrane, allowing water and ions to enter or leave the cell. As described in the text, these channels can be very selective in the ions that they allow through as well as the circumstances in which they are open.

FIGURE 9.4 (A) Because the total concentration of solutes inside an animal cell is the same as in the surrounding aqueous medium, the diffusion of water through pores in the membrane is the same from out-to-in as it is from in-to-out. This is a necessary condition for cells to maintain a constant volume; if it is violated, the cells either swell or shrink.

(B) Although the total concentration of the solution inside the cell is the same as it is outside, the solutions are different. Externally the cell is bathed in a salt solution similar to dilute sea water (mostly Na^+ and Cl^-). Internally, however, the major ions are potassium (K^+) and a variety of organic anions, including proteins.

BOX 9.1
Concerning Different Forms of Potential Energy

Molecules tend to move from regions of high to low concentration. How does this happen? Imagine two solutions of some common substance like sugar at different concentrations, separated by a membrane through which molecules of the dissolved substance (the solute) can pass (Fig. 9.5). Because there are more solute molecules in the left compartment than in the right, random molecular motions will lead to more collisions with the membrane from the left side than from the right, and in time there will be a net movement of solute from the left compartment to the right. The presence of the membrane is not an essential part of the argument. In its absence there will be a net movement of solute from left to right through the plane represented by the membrane. Conclusion: substances *diffuse* from regions of high to low concentration.

Osmosis is just the movement of the solvent (water) through a membrane in the direction that reduces the concentration gradient of solute. Consequently, if cells

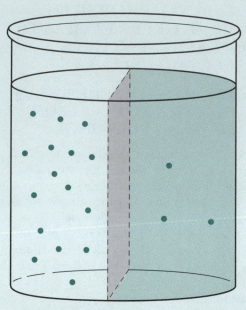

FIGURE 9.5 Diffusion from a region of high to low concentration results from the *random movement* of solute molecules. As there are more solute molecules on the left, the *probability* that there will be movement through the central plane from left to right is greater than the probability that molecules will move from right to left. The final condition is that the concentration is uniform on both sides of the chamber.

Blood and lymph consist primarily of a solution of NaCl (about 0.1 M), with smaller amounts of other ions such as Ca^{++}, some proteins, and in the case of blood, suspended erythrocytes and other blood cells. The principal cation (positively charged ion) within cells, however, is K^+, and most of the anions (negatively charged) are organic, with only a relatively small concentration of Cl^- (Fig. 9.4). The total *concentration* of dissolved material inside the cell is the same as outside, however, for if it were not, water would enter or leave the cell by osmosis, and the volume of the cell would change. Maintenance of a nearly constant cell volume is one of the regulatory challenges of animal cells, because if they take up too much water, they will swell and burst.

How are different concentrations of the principal ions maintained on the two sides of the membrane while keeping the *total* concentration of solute the same? Most of the organic anions within the cell such as proteins are trapped inside with no way to get through the membrane, but Na^+, K^+, and Cl^-

ions can leak through membrane pores. Because there is a concentration gradient from outside to in, Na^+ ions slowly enter the cell. If this process were allowed to go unchecked, Cl^- ions would follow, the total concentration of salt inside the cell would increase, water would enter, and the cell would swell and burst.

This fatal outcome is prevented by the presence of an ion pump that ejects Na^+ as fast as it diffuses in. The inward diffusion of sodium is akin to water leaking into a boat, so removing Na^+ ions by pumping requires the cell to expend energy (Box 9.1). The pump consists of a protein within the cell membrane, and it utilizes ATP that is generated in cellular respiration.

What about K^+ ions? The movements of K^+ and Na^+ ions differ in two ways. First, because of the properties of their respective channels, K^+ is able to move through the membrane at a higher *rate*. In other words, the membrane is more permeable to K^+ than to Na^+. Second, unlike Na^+, K^+ is *free to diffuse*. By this we mean its *final distribution* is largely *passive* and is not de-are placed in distilled water, they rapidly swell by osmosis and burst, and if placed in concentrated brine, they shrink. The osmotic equilibrium of living cells requires that the total *concentration* of solutes inside the cell be the same as outside, although the solutes inside need not be the same as outside.

All of this may be familiar, but let's push the argument a bit further. If diffusion spontaneously eliminates concentration differences, the creation of concentration differences must require the input of energy. If you want to recover salt and produce drinking water from seawater, what must you do? One approach is to boil the seawater and recapture the water vapors by condensing them on a cold object. There are other procedures that would work, but all require the expenditure of energy.

Another way to view the matter is to recognize concentration differences as a form of potential energy, like a bucket of water that has been carried to the tenth floor, or a cube of sugar, or a battery. A weight loses gravitational energy when it is lowered, a cube of sugar yields energy stored in chemical bonds when it is burned, and the battery employs chemical reactions to produce a separation of electrical charges at its terminals. Falling water can turn a turbine and generate electricity; separated electrical charges can be used to start engines and light lamps; and the

burning (oxidation) of sugar can be done by cells to make ATP (Chapter 3), which in turn can supply the energy required by muscles to carry water to the tenth floor. Various forms of energy are thus interconvertible.

Left alone (without an input of energy), systems tend to relax to states of lower potential energy. Osmosis, like the diffusion of solute molecules, is a movement of the system toward a state of lower potential energy. Sometimes the spontaneous change in energy state may not be evident if there is an energy barrier that must be surmounted to start the process. The bucket of water on the tenth floor is very unlikely to leap to the window and fall by itself, and a cube of sugar will oxidize rapidly only if it is heated in a flame or brought into contact with water and appropriate enzymes for metabolizing sugar.

Back, now, to neurons. The separation of charge across the membrane (expressed in volts) and the concentration differences of K^+ and Cl^- ions between the inside and outside of the cell are forms of potential energy. There are ways to relate these two forms of energy quantitatively—to derive an equation for the membrane voltage in terms of the concentration gradients of diffusable ions. By utilizing that equation, the theory of membrane voltages has been verified under different experimental conditions.

termined by the presence of an ion pump, as is the case for Na^+.

K^+ is present inside the cell because it is attracted to the negatively charged organic ions that are trapped there. Similarly, Cl^- is in high concentration outside the cell as a counter ion to Na^+. Thus, the distributions of K^+ and Cl^- are passive (or nearly so) because the cell does not have to do additional work to maintain the differences in concentrations of these two ions on either side of the cell membrane.

THE ASYMMETRIC DISTRIBUTION OF IONS GENERATES A VOLTAGE ACROSS THE CELL MEMBRANE

If you think about Figure 9.4B and the argument that has been presented to this point, you may be puzzled. If Na^+ ions have to be pumped out of the cell to keep a higher concentration outside than inside, why do passive movements of K^+ and Cl^- ions lead to asymmetric

distributions of these ions? Why do these ions not diffuse down their concentration gradients?

Consider what happens as soon as a small amount of K^+ diffuses down its concentration gradient and leaves the cell. It carries a small amount of positive charge with it, leaving the inside of the cell slightly negative. Likewise, as a small amount of Cl^- enters the cell, it brings negative charge with it, also making the inside of the cell slightly negative. These movements of *very small numbers of ions* thereby generate a voltage across the membrane, with the inside of the cell negative to the outside by 0.06–0.09 volts (60–90 mV). This small voltage gradient opposes the further diffusion of both K^+ and Cl^-.

The passive forces controlling the movements of K^+, Cl^-, and Na^+ ions are summarized in Figure 9.6 with only slight simplification. The concentration gradient of K^+ is oriented from out to in, but it is opposed by an electrical gradient oriented in the opposite direction. Consequently K^+ flux is in a steady state, with no *net* movement through the membrane in either direc-

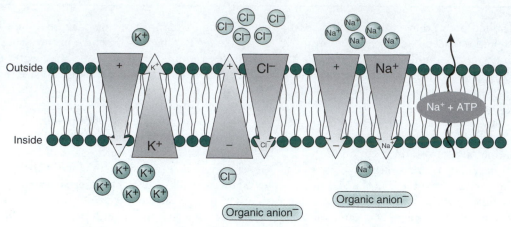

FIGURE 9.6 The asymmetric forces on the major ions bathing cell membranes. Each ion has a tendency to diffuse down its concentration gradient, but the movements of very small numbers of K⁺ and Cl⁻ ions through the membrane cause a separation of electrical charge and produce a voltage across the membrane. This voltage opposes further diffusion. The electrical and chemical gradients for K⁺ and Cl⁻ ions are therefore oriented in opposite directions, and there is no net movement of these ions through the membrane of a resting cell. Sodium, on the other hand, has a steep electrochemical gradient directed from out-to-in. The slow, inward leakage of Na⁺ is countered by the sodium pump, which uses metabolic energy stored in ATP to extrude sodium. (The pump moves K⁺ in as Na⁺ is extruded, but to simplify the account we are ignoring this detail.)

tion. A similar argument applies to Cl⁻, but the orientations of the concentration and electrical gradients are opposite from what they are for potassium. Sodium ions, however, do not distribute passively. Both the electrical and the concentration gradient are directed inward, and accumulation of sodium is prevented by the sodium pump.

In summary, organic anions (e.g., proteins) are trapped within the cell, sodium is pumped out of the cell as it enters, and potassium and chloride ions diffuse passively through open channels. This results in a separation of charge of nearly 0.1 volt across the membrane. Every animal cell therefore has such a *membrane voltage* (or *membrane potential*) with the inside of the cell negative to the outside. In nerve cells that are not engaged in signaling, this voltage is called the *resting potential*. When neurons are actively signaling (i.e., not at rest), however, additional kinds of ion channels open and close, allowing brief currents of ions to pass through the membrane and producing transient changes in membrane voltage. *Nerve signals are generated by regulating the opening and closing of these membrane channels.*

AXONS CARRY SIGNALS OVER LARGE DISTANCES

THE NERVE IMPULSE IS CAUSED BY A BRIEF MOVEMENT OF SODIUM IONS THROUGH THE AXON MEMBRANE

All cells in the body must remain in osmotic equilibrium, and all do it by restricting the accumulation of sodium ions. Nerve and muscle cells employ a special trick for rapid signaling in which they allow a brief inward movement of Na⁺ ions. Their membranes contain an ion channel specific for sodium ions. (This protein is not the same as the sodium pump.) When the nerve cell is at rest (i.e., not signaling), this channel is closed and sodium is unable to get through. The channel protein can change shape, however, causing the pore in the channel to open and allowing Na⁺ ions to pass through. Because Na⁺ ions are more concentrated outside the cell than inside, and because the inside of the cell is negative with respect to the outside, when the channel opens Na⁺ ions start diffusing in. The channel remains open for only about a thousandth of a second (1 msec), and not many Na⁺ ions

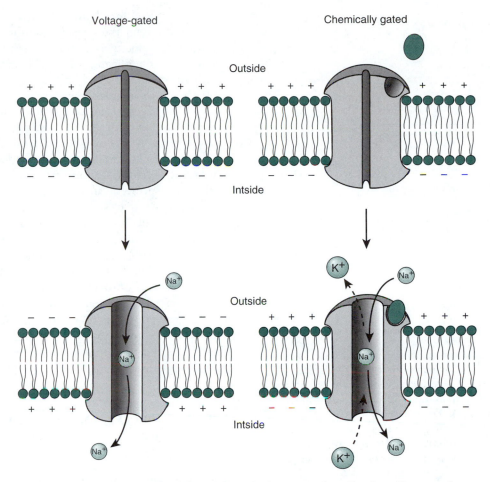

Voltage-gated   **Chemically gated**

Outside

Intside

Outside

Intside

FIGURE 9.7 **Two examples of gated channels that are ordinarily closed but can be opened to allow ions through. The channel on the left is** *voltage-gated*: **a change in membrane potential induces conformational changes in the channel protein and increases the size of the central pore. Voltage-gated sodium channels underlie the action potentials that propagate along axons.**

The channel on the right is *chemically gated* **and opens when a specific small molecule binds to a site on the end of the channel projecting outside the cell. Channels like this occur on postsynaptic cells in conjunction with synapses. This example allows both sodium and potassium ions to pass and would be located at an excitatory synapse.**

have time to enter, but while the channel is open the membrane voltage reverses and becomes positive inside.

What causes the channel to open? The channel protein has an ionic group that makes it sensitive to the voltage difference across the membrane. If the inside of the cell becomes less negative than the resting membrane voltage by several mV, the ionic group moves, the channel protein changes its conformation, and the pore opens (Fig. 9.7). Channels that can open and close are like gates, and these channels are said to be *voltage-gated* because a separation of charge across the membrane causes them to open.

What causes the channels to close? They close automatically, because the voltage-sensitive gates behave like doors on springs. Even if the inside of the cell is held at a positive potential to make them open (which can be done experimentally by passing an electric current across the cell membrane), the channels close. As the channels close, the membrane starts to repolarize to the *resting voltage*, which is largely determined by the difference between concentrations of potassium ions inside and outside the cell. The speed of repolarization is hastened as voltage-sensitive K^+ channels also open.

This transient reversal of membrane voltage is a nerve impulse or *action potential*. The top panel in Figure 9.8 is a graph of membrane voltage during a single nerve impulse. The lower panel shows the underlying movements of Na^+ and K^+ ions through the membrane.

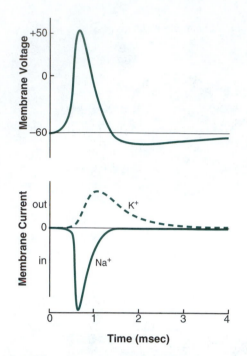

FIGURE 9.8 The action potential in a nerve axon lasts for several thousandths of a second (msec). During this time the voltage across the membrane reverses from the resting value of −60 mV to about +50mV (inside voltage minus outside voltage). These changes are caused by the successive opening of voltage-gated sodium and potassium channels. The currents (carried by ions) through these channels are shown in the lower graph. The sodium channels close automatically (*inactivate*); the potassium channels close as the membrane repolarizes to the resting potential. The small increase in negativity of membrane voltage at the end of the action potential is caused by the membrane's transient, high-permeability to potassium while the voltage-gated potassium channels are still open.

ACTION POTENTIALS PROPAGATE ALONG THE LENGTH OF THE AXON

The entire purpose of axons and action potentials is to move information quickly from one part of the body to another. The action potential lasts for a few msec. It starts locally, at one end of the fiber, but as Na-channels open at that site, other channels nearby are activated by the change in voltage, and the action potential begins to propagate along the length of the axon (Fig. 9.9). The speed of *conduction* is not the speed at which an electric current travels in a wire.

Time is required for channels to open and charges to redistribute along the surface membrane of the cell. Conduction velocity is faster in larger fibers and in fibers that have sheaths of myelin around them. In myelinated fibers, the only places that sodium can get through the axon membrane are at tiny gaps in the sheath that are present every 1–2 mm. The sodium channels are concentrated at these gaps, and the nerve impulse jumps from one gap to the next.

Large myelinated fibers can conduct impulses at 100 m/sec, which is about ten times faster than a world-class sprinter can run. The smallest unmyelinated fibers conduct at less than 1 m/sec. If you have stubbed your toe in the dark or struck your thumb with a hammer, you have experienced different conduction velocities of nerve fibers carrying somewhat different information. The first sensation comes from receptors reporting aspects of touch and pressure. Realizing what you have just done to yourself, you still have an instant to anticipate the agonizing pain that you will experience when reports traveling on small unmyelinated fibers have had time to reach your brain and be consciously registered.

THE STRENGTH OF A SIGNAL IS CODED BY THE FREQUENCY OF ACTION POTENTIALS

The code that is available to axons is limited because it consists of identical action potentials. This is not a very rich code with which to convey information, and it is used to communicate only the strength of the signal. For a weak signal such as a dim light, a soft sound, or a gentle touch, only a few nerve impulses are sent along a sensory axon. For stronger signals—brighter lights, louder sounds, a stubbed toe—the sodium channels open repeatedly, and the *frequency* of nerve impulses increases (Fig. 9.10). The strength of excitation is therefore not conveyed instantaneously, but is coded in the number of action potentials in a given time.

AN AXON CONVEYS QUALITATIVE INFORMATION BY THE ANATOMICAL CONNECTIONS IT MAKES

The frequency code is common to all nerves, sensory and motor. If all that a nerve fiber can convey to its target cells is the strength with which it has been excited, how is qualitative meaning conveyed? How is the brain to know whether the report it is receiving from the senses means a bolt of lightning or a clap of thunder? That confusion is never pre-

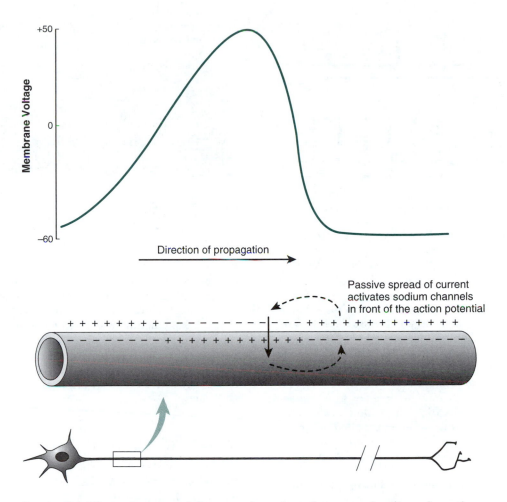

Direction of propagation

Passive spread of current
activates sodium channels
in front of the action potential

FIGURE 9.9 **The action potential occurs where the voltage-gated sodium channels have opened. The diagram at the bottom shows an influx of sodium ions (in the region around the short vertical arrow) that causes the inside of the axon to become positive with respect to the outside. This creates a voltage gradient along the *length* of the axon, so that there are passive currents carried by potassium ions (dashed arrows) in front of the action potential. These currents depolarize the membrane in front of the action potential, opening more sodium channels, and causing the action potential to propagate from left to right.**

The graph at the top shows the membrane voltage as a function of distance along the axon at one instant in time. The axon is not drawn to scale. It is a small fraction of a mm in diameter, but several cm of its length are depolarized at any instant.

sent because every nerve fiber is committed to a very specific task that is determined by the cells to which it connects. Each fiber in the optic nerve reports from a small region of the retina, corresponding to a restricted portion of visual space, and connects to cells in a correspondingly localized region in a part of the CNS that is dedicated to processing visual information. Similarly, a fiber in the auditory nerve is excited by a certain range of frequencies of sound pressure detected in the ear, and

it delivers its message to an auditory center in the CNS. Claps of thunder are not confused with bolts of lightning because the sensory input paths for sound and light are anatomically separate. This association of function with distinct anatomical paths is referred to as the principle of *labeled lines*. The same principle applies to motor nerves. Your toes do not wiggle when you reach for your fork because different nerve cells are engaged in moving the muscles of your hands and feet.

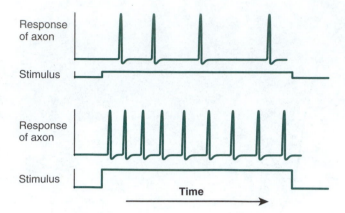

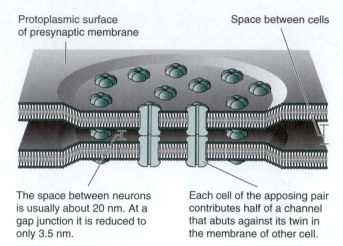

Protoplasmic surface of presynaptic membrane

Space between cells

The space between neurons is usually about 20 nm. At a gap junction it is reduced to only 3.5 nm.

Each cell of the apposing pair contributes half of a channel that abuts against its twin in the membrane of other cell.

FIGURE 9.11 A *gap junction* is a structure where the membranes of two cells are closely apposed and connected through a patch of channels called *connexons*. The channels allow ions to pass from one cell to the other. A gap junction therefore couples two cells electrically and is a form of synapse that is found where the postsynaptic cell must be rapidly excited by the presynaptic neuron. Gap junctions are also found in embryonic tissue where they may facilitate exchange of small molecules.

FIGURE 9.10 As the strength of stimulation increases, an axon generates action potentials at a higher frequency (i.e., more impulses per unit time). In this diagram the stimulus is represented in a general way. If the axon is in a sensory nerve, the stimulus could be light, sound, pressure, or whatever the appropriate form of excitation is for that particular nerve ending. If the axon is part of an interneuron in the CNS, the stimulus would come from synapses impinging on the cell from other neurons.

Notice that the time axis is compressed compared with Figure 9.8. That is necessary in order to see that a stimulus that lasts much longer than a few msec generates a *train* of action potentials occurring one after another. Action potentials are sometimes called *spikes* because of their appearance when a recording from an axon is made on a time scale of seconds rather than msecs.

This diagram also shows that trains of spikes characteristically slow in frequency, a process called *adaptation*. Adaptation occurs even when the intensity of the stimulus remains constant. (Refer to Box 5.3 on page 118 to review the multiple meanings of the word "adaptation.")

NERVE CELLS COMMUNICATE WITH EACH OTHER AT JUNCTIONS CALLED SYNAPSES

THE PROBLEM OF GETTING INFORMATION FROM ONE CELL TO ANOTHER

So far we have been considering the messages that travel *along* axons, but what happens at the specialized junction called *synapses* where one cell ends and another begins? *Gap junctions* illustrate how specialized pores are necessary for currents generated in one cell to spread to an adjacent cell (Fig. 9.11). These *electrical synapses* account for only a small fraction of all the synaptic junctions in the nervous system. For example, they are present in circuits such as escape reflexes where speed of transmission is important.

SYNAPTIC TRANSMISSION ORDINARILY INVOLVES THE SECRETION OF CHEMICAL MESSENGERS

In most synapses, however, arrival of a nerve impulse in the axon terminals of the presynaptic cell causes the secretion of a chemical from the nerve endings, which in turn affects the flow of ions through the membranes of the postsynaptic cell.

The axon of the presynaptic neuron branches and ends in small swellings in close contact with a dendrite or the soma of the postsynaptic cell. A presynaptic fiber may contact one or many postsynaptic cells, and a postsynaptic cell can receive input from many presynaptic cells.

Figure 9.12A shows a diagram of a single synaptic ending. On the presynaptic side there are numerous small hollow spheres, each bounded by a membrane. Each of these *synaptic vesicles* contains several hundred molecules of a *neurotransmitter*, the substance used to communicate with the postsynaptic cell. There are many different neurotransmitters, but only one or two are produced at any synapse.

When a nerve impulse arrives at the presynaptic ending, a small number of Ca^{++} ions enter the axon terminal and cause some of the synaptic vesicles to fuse with the cell membrane, thus emptying their contents into the narrow cleft separating the two cells (Fig. 9.12B).

The transmitter molecules diffuse across the narrow synaptic cleft and bind to *receptor proteins* embedded in the membrane of the postsynaptic cell. Many of these receptors are ion channels (Fig. 9.12B). Because they are opened or closed by the binding of transmitter molecules, they are *chemically gated* channels, in contrast to the electrically gated channels of the axon.

A common channel permits the ready passage of Na^+ and K^+ ions when it opens. This influx of positive ions causes a small *excitatory postsynaptic potential (EPSP)* (Fig. 9.12C). Unlike the "all or none" character of the action potential in an axon, which rises to its full amplitude when the nerve is excited, postsynaptic potentials are *graded* in size. The more channels that are activated, the more positive charge enters the cell, and the greater the depolarization. The number of open channels depends in turn on the number of vesicles that have released their contents into the synaptic cleft. A single action potential in the presynaptic fiber causes only a few of the vesicles to discharge, but if additional action potentials follow quickly after the first, more vesicles are recruited. A presynaptic axon that is signal-

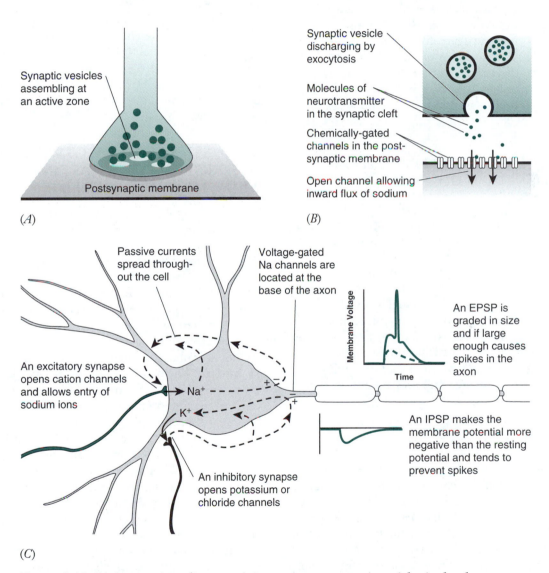

(A)

(B)

(C)

FIGURE 9.12 **(A) A synaptic ending containing numerous synaptic vesicles is closely apposed to the membrane of a postsynaptic cell.**

(B) Enlarged view showing a vesicle discharging its contents of several hundred transmitter molecules into the synaptic cleft. Two of the molecules have bound to receptors on the postsynaptic membrane. These receptors are chemically gated cation channels that pass an inward current of sodium ions when they are open.

(C) Synapses can be either excitatory or inhibitory. An EPSP depolarizes the membrane from the resting voltage. An IPSP holds the membrane potential close to the resting value, frequently making it more negative.

ing strongly with a volley of nerve impulses therefore produces a larger EPSP in the postsynaptic cell than when it fires only one or a few impulses.

SYNAPTIC POTENTIALS SUM ON THE CELL BODY OF THE POSTSYNAPTIC CELL

The cell membrane covering the dendrites and cell body of a neuron does not contain voltage-gated sodium channels. Consequently, action potentials only arise at the base of the axon. Postsynaptic potentials, however, arise at many local sites on the dendrites and soma where synapses are located. These postsynaptic potentials do not propagate like an action potential. They spread over the surface of the cell, decaying with distance from their site of origin, but summing with each other. If the depolarization at the base of the axon is sufficient, the postsynaptic cell fires impulses in its axon (Fig. 9.12C).

PROPAGATED ACTION POTENTIALS ARE NOT NECESSARY IN MUCH OF THE CNS

Many nerve cells in the CNS do not generate action potentials. The propagated impulses of axons exist only to transmit information rapidly over relatively long distances via peripheral nerves or fiber tracts in the brain. Within the gray matter, however, layers of cell bodies alternate with *neuropil*—dense tangles of dendrites, short axons, and a profusion of synaptic contacts. In many places the distances across cells, from synaptic input to output, are so short that postsynaptic potentials alone are sufficient to modulate the release of transmitter, and propagated impulses are not necessary.

SYNAPSES CAN BE INHIBITORY AS WELL AS EXCITATORY

Consider again the hunting lion and the tennis player with which we began this chapter. A sudden change in direction of the fleeing zebra or the abrupt appearance of a young animal that can be more easily brought down is followed by an equally rapid change in the lion's behavior. Similarly, the tennis player starts to meet the ball, but at the last instant decides the serve is likely a fault and lets it pass. In each case the action of one set of muscles abruptly replaces another. The ability of nervous systems to process information is enormously enriched by the capacity to inhibit some cells while exciting others. In the examples of the lion and the tennis player, actions change when the motor cells to some muscles are inhibited and others excited, but the interplay of excitation and inhibition is found in every task that nervous systems perform.

An *inhibitory postsynaptic potential* (IPSP) occurs when the channels in the postsynaptic membrane allow the passage of either K^+ or Cl^- but not Na^+ ions. K^+ and Cl^- are the ions to which the membrane is ordinarily most permeable, and enhancing their permeabilities holds the membrane voltage close to the resting value or even causes it to become more negative. This counters the effects of any increase in permeability to Na^+ ions occurring at nearby excitatory synapses (Fig. 9.12C). The relative positions of excitatory and inhibitory synapses on the postsynaptic cell, along with the number of transmitter molecules released at each site (Box 9.2), determines the output of the postsynaptic cell: whether it will fire action potentials or depolarize enough to activate the synapses on the cells to which it connects. Each cell thus combines or *integrates* the various synaptic inputs it receives at every instant.

SOME SYNAPTIC EFFECTS ARE BRIEF; OTHERS ARE LONG-LASTING AND CAN AFFECT GENES

The postsynaptic effects of transmitters we have considered so far are relatively brief. One or a few impulses in the presynaptic nerve causes a postsynaptic potential that lasts for several thousandths of a second before the transmitter is removed or inactivated. Channels in the postsynaptic membrane then close, the synaptic currents decay, and the voltage across the postsynaptic membrane returns to the resting value.

Another kind of synaptic event modulates the membrane voltage in the postsynaptic cell for seconds or minutes. The neurotransmitter reacts with a receptor protein, but the receptor is not a channel. Instead, the activated receptor initiates a cascade of chemical reactions within the cell that include the formation of a diffusable *second messenger*. This molecule brings about longer-lasting effects by chemically modifying channels that are localized elsewhere on the postsynaptic membrane. Second messengers can modulate the sensitivity of the postsynaptic cell to other synaptic input, for example, increasing its sensitivity by closing a particular class of potassium channels.

Interestingly, the final targets of second messengers are not limited to ion channels. Moreover, second messenger cascades are activated by hormones as well as neurotransmitters. This sort of signaling is used in regulating many cell functions, including the oxidation of sugars and the activation of genes. In principle, then, this is a mechanism by which the nervous system can influence gene expression. We will illustrate these points with a specific example in a later section on cellular changes that take place during learning.

Box 9.2
Neurotransmitters, Disease, and Drugs of Abuse

Most neurotransmitters are small molecules, frequently amino acids or their derivatives, but some are peptides (chains of amino acids). A few are very common and are used in many places in the nervous system, but there are likely dozens of different transmitters, many of which are not yet identified. Whether a transmitter is excitatory or inhibitory depends on the postsynaptic receptor with which it reacts.

There are cells located in small subcortical structures that send processes to many parts of the brain and are involved in broad classes of behaviors or general moods and feelings. These cells use a few clinically important transmitters—*dopamine*, *norepinephrine*, and *serotonin*. When there is a defect in the formation or use of these transmitters, the behavioral manifestations can be severe.

Parkinson's disease, with its characteristic tremors and stiffness of movement, is caused by a loss of cells that synthesize and use dopamine. These cells account for 80% of the dopamine in the brain. Schizophrenia, which in its psychotic phase is characterized by disturbances of thought, delusions of control by outside agents, and hallucinations (e.g., hearing voices), is often treated with a variety of *antipsychotic* drugs. These drugs (with such trade names as Thorazine and Haldol) have the common property of blocking receptors for dopamine. One of the contributing factors in schizophrenia therefore appears to be an overabundance of dopamine.

Other mental disorders affect mood: depression, sometimes coupled to a manic phase, is a familiar example. These states can be treated with drugs that affect the metabolism of norepinephrine and serotonin. As more has been learned about the biochemical pathways for synthesizing, storing, secreting, recovering, and disabling these transmitters, and about their mechanism of action, many possible targets for antidepressant drugs have been identified. For example, after a neurotransmitter has been secreted into the synaptic cleft, its action must be stopped. At some synapses, the transmitter is broken down with enzymes; at others, the transmitter is taken back into the presynaptic cell. Drugs are now available that can inhibit both of these processes As an illustration, the drug Prozac, which is taken by individuals who have mild clinical symptoms, is an antidepressant that works by inhibiting serotonin uptake.

The involvement of dopamine, norepinephrine, and serotonin in psychiatric conditions tells us that the various neurons using these transmitters are engaged in important (and complex!) interactions that modulate the mental state of people who are mentally well. For example, in monkeys, serotonin levels rise and fall in individuals as their social dominance changes. Similarly, among college men who belong to athletic teams and fraternities, serotonin levels are highest in the leaders.

Unfortunately there is an illicit market for drugs that alter one's senses of self and reality. These agents have their effects by changing the balance of important transmitter molecules. For example, cocaine inhibits the uptake of serotonin. LSD binds to serotonin receptors and produces hallucinations. And an addictive drug (phencyclidine) with the misleading street name "angel dust" binds to an important class of glutamate receptors that are involved in learning, and it produces psychoses similar to schizophrenia.

TARGET CELLS CAN BE LOCATED AT A DISTANCE

When a presynaptic cell speaks, only a few postsynaptic sites listen. Both the identity of the listener and the speed of its response depend on the intimacy of the synaptic contact. The membranes of the pre- and postsynaptic cells are separated by a small fraction of a μm, so the transmitter not only diffuses across the gap in less than a msec, it cannot diffuse far before it is either destroyed or recycled. Consequently, even within the profusion of neurites and synapses that make up the gray matter of the brain, the activity of one neuron can have very local and selective effects on the activity of another.

Nerve cells can also communicate with very distant targets by using the circulation of the blood for a delivery system. This process is known as *neurosecretion*, and the secretory product is called a *neurohormone* rather than a synaptic transmitter. The target cell is frequently not another neuron but an organ like the kidney or the uterus. Such secretory cells are found in the hypothalamus, a phylogenetically old part of the brain involved in regulating the internal environment of the body. For example, salt concentration and blood pressure are monitored by the brain. When the body is losing water faster than it is being replaced by drinking, the hypothalamus tells the kidney to make less urine. Neurosecretory cells in the hypothalamus synthesize the peptide hormone vasopressin in their cell bodies and transport it down their axons to terminals located on a capillary bed in the posterior pituitary (Fig. 9.13). When the hormone is needed, the cells fire action po-

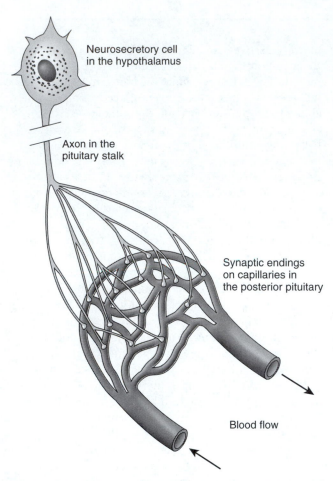

Neurosecretory cell in the hypothalamus

Axon in the pituitary stalk

Synaptic endings on capillaries in the posterior pituitary

Blood flow

FIGURE 9.13 Neurosecretory cells in the hypothalamus secrete peptide hormones. Compared to ordinary neurons, the target cells lie a great distance away in organs such as the kidney or the uterus. The synaptic terminals discharge the peptides into capillaries in the posterior lobe of the pituitary gland, and the hormones are carried to their target organs by the blood.

tentials that cause the hormone to be discharged into the blood, much like vesicle release during synaptic transmission. Binding of vasopressin to receptors in the kidney causes the kidney tubules to resorb some of the water that has been filtered from the blood. Oxytocin, another peptide hormone secreted in the posterior pituitary, causes contractions of smooth muscle in the uterus and ejection of milk from the mammary glands.

CREATING BEHAVIOR WITH NEURONS

So far we have seen that although neurons differ greatly in shape, they communicate with each other in a limited number of ways. If communication must be rapid and the receiving cell is far away, self-propagating action potentials are sent along the axon. Commu-

nication between cells, however, usually involves chemical messengers that alter the properties of channels in the postsynaptic cell. Action potentials in axons are remarkably similar everywhere, but synaptic transmission is more flexible. There are many different kinds of ion channels, and they differ in how they are regulated (voltage, binding of a transmitter, mechanical stretch, covalent attachment of a chemical group mediated by an enzyme), in the ions they allow to pass (K^+, Na^+, Ca^{++}, Cl^-), and in their specificity for transmitter molecules. The interaction among ion currents arising at different places on the membrane of a postsynaptic cell can also vary, depending on the distribution of synapses the cell receives from presynaptic neurons.

Behavior results from patterns of activity of large numbers of neurons working together. These patterns are generated by the right *combinations* of neurons, each signaling in proper *sequence*. As there are about 10^{12} neurons in a human brain, each with hundreds of synapses, understanding our behavior in terms of neuronal activities is an immensely difficult challenge. Fortunately there are some additional general concepts that will help us to make out the forest, even if many of the trees are still obscure.

The importance of *patterns of connections* between neurons was illustrated by the way in which sensory information is conveyed—how action potentials in the optic nerve are interpreted as visual signals because the cells that provide the information are located in the retina and the axons terminate in a part of the brain dedicated to visual processing. A great deal of evidence shows that *many neural activities can be associated with identifiable locations* in the brain. As a corollary, some information is *processed serially*, for example in getting from the eye to the appropriate part of the brain.

During the last 200 years the idea that different parts of the brain are specialized to perform different functions has had a seesaw history. One reason for the confusion is that loss of particular functions caused by local injury can sometimes be overcome when other areas of the brain take over the task. Furthermore, much cognition requires *parallel processing* of information in different parts of the brain. For example, when you see and recognize a friend, different networks of cells are engaged with various aspects of the sensory information. You infer the presence of a human face, associate it with a particular individual, and draw on stored information about that person's behavior in previous contexts. The recognition may, in turn, trigger an emotional response, depending on whether or not you want to see the person then. To a large extent, processing must therefore be distributed among many neurons. Furthermore, remembering your friend requires that neurons make enduring changes in the brain, for how else could memories have a physical basis?

In following sections, we will use the cortex of the brain to illustrate how some functions are localized. The properties and arrangement of neurons in the visual cortex also show how sensory systems have evolved to analyze particular kinds of information and how that analysis affects our perceptions. This discussion will also provide background for understanding the sorts of changes that occur during learning and how experience changes the synaptic architecture of neurons during development (Chapters 10 and 11).

CORTICAL MAPS

One of the features of mammalian evolution has been the enlargement of the cerebral cortex. As our behavioral uniqueness is largely associated with the cortex, this is where we will now focus. What goes on in the cortex? How is it organized? And how do we know these things?

Understanding the functions of different parts of the brain is based on a variety of observations. Some of the first were made by nineteenth century physicians who were able to correlate behavioral deficits of stroke victims with local tissue damage discovered when the brain was examined after death. (A stroke is caused by interruption of the blood supply in an area of the brain.) In the twentieth century the invention of electronic amplifiers and other measuring equipment enabled rapid progress in understanding how neurons work and opened many new possibilities for studying the nervous system. For example, in animals under anesthesia it is possible to record electrical activity from parts of the brain while stimulating the sense organs, or alternatively, to stimulate the spinal cord or brain locally and study how muscles are affected. In an extension of these techniques, electrodes, anchored in the bones of the skull with the animal under anesthesia, can stimulate or record from local parts of the brain when the animals are awake and alert.

One picture that emerges from such studies is summarized in Figure 9.14, a side view of a human cortex. The cortex is a sheet of tissue several mm thick that nearly covers deeper and evolutionarily older neural structures. The surface area is large, but furrows and crevices enable it to fit within the skull. Based on some of the major grooves (each called a *sulcus*, pl. *sulci*) and the positions of the overlying bones of the skull, several different zones are recognized (*frontal*, *parietal*, *occipital*, and *temporal lobes*), each performing somewhat different functions.

Information from the senses—eyes, ears, and receptors in the skin for touch, pressure, and temperature—is projected in *cortical maps*. The *somatosensory projection area*, located immediately behind the central sulcus in parietal cortex, illustrates this concept most simply (Fig. 9.15). The receptor neurons of the body surface report to cortical cells in this region. The body surface is represented here as a map; touch of the lips is registered by cells in one part of this region, a pinch of the toe by other cells. The map, however, is distorted. Those regions of the body that are richly innervated and are thus very sensitive, like the lips and fingertips, are well represented in the map with many cells. Other regions, such as the middle of the back, project to fewer cortical neurons and therefore occupy a much smaller area of the map.

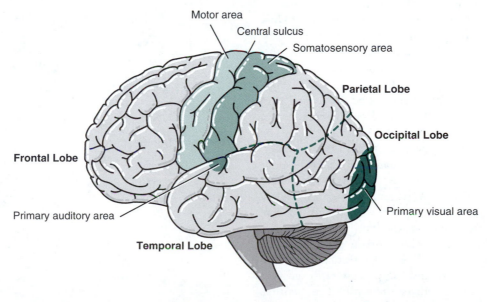

FIGURE 9.14 View of the left side of the cortex showing several anatomical regions described in the text.

FIGURE 9.15 (A) The cortical somatosensory projection is represented by a homunculus whose distortion represents the relative numbers of cortical neurons devoted to analyzing sensory information from different regions of the body surface. The shaded area represents the back side of a vertical slice through the brain made just behind the central sulcus. Only the top left side of the brain is included in the picture, so only the somatosensory map for the right side of the body is present. It is formed by cells near the surface of the cortex.

(B) The analagous motor map for the other side of the body.

(C) Relative representations of the somatosensory maps of several species of mammal.

Information from the eyes goes to an area in the occipital cortex, and from the ears to a region of temporal cortex (Fig. 9.14). The map in the visual cortex corresponds to a spatial map of the retina. The auditory map in the temporal lobe is different. Different tones, i.e., frequencies of sound, stimulate the cochlea in the ear at different positions along its coiled length. Correspondingly, the auditory cortex contains a *tonotopic* map in which different frequencies of sound pressure are represented at different places in the auditory cortex.

Information from different sense organs does not remain separate throughout the cortex. For example, you are able to understand both spoken and written language, and there are places where this equivalent information from eyes and ears is interpreted and related. Such regions are referred to as *association cortex*. Some of the parietal, temporal, and occipital lobes function as association cortex.

During the 1950s, Wilder Penfield, a Canadian neurosurgeon, made some dramatic observations on patients he was operating on for severe epilepsy. During epileptic seizures, regions of the brain cease to function in coordinated fashion. Neurons become uncontrollably active, and the disturbance spreads to nearby regions, causing convulsions and loss of consciousness. Severe cases that cannot be controlled with drugs are treated surgically, sometimes by removing a small part of the brain where the trouble arises. The surgeon must find the right tissue to remove, however, and at the same time try not to damage healthy tissue.

Penfield was operating on patients with severe temporal lobe epilepsy. A patient was appropriately anesthetized but conscious, with the surface of one side of the brain exposed. In order to identify the properties of tissue prior to cutting, the surgeon then applied a mild stimulating electric current at different points on the surface of the brain, and the patient reported verbally the effect it produced. Penfield found that stimulation of the temporal lobe could elicit vivid memories of events from the distant past, memories that had

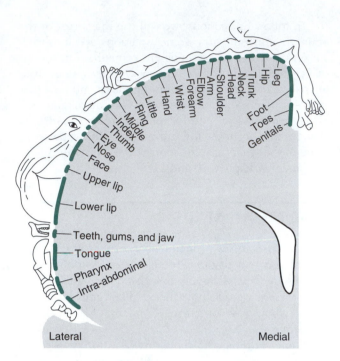

(A) **Sensory homunculus**

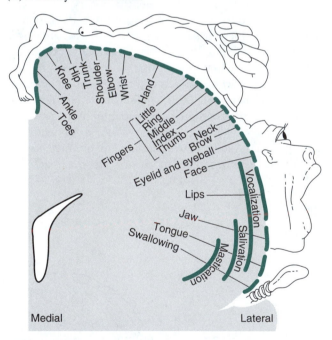

(B) **Motor homunculus**

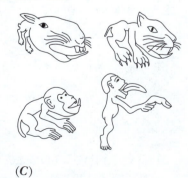

(C)

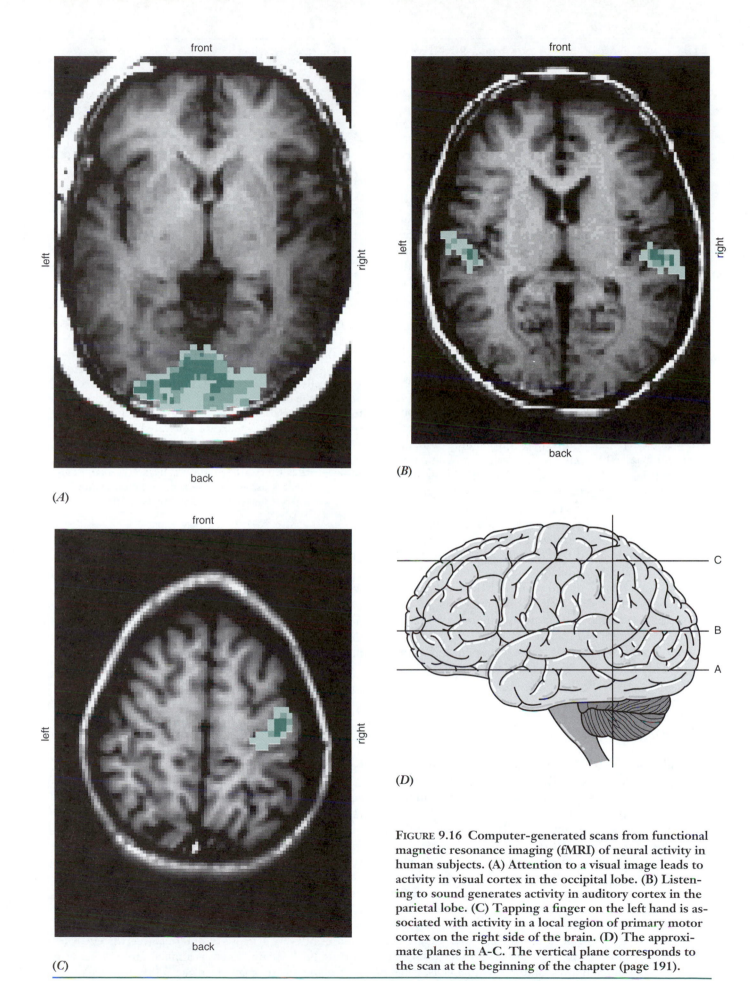

(A)

(B)

(C)

(D)

FIGURE 9.16 Computer-generated scans from functional magnetic resonance imaging (fMRI) of neural activity in human subjects. (A) Attention to a visual image leads to activity in visual cortex in the occipital lobe. (B) Listening to sound generates activity in auditory cortex in the parietal lobe. (C) Tapping a finger on the left hand is associated with activity in a local region of primary motor cortex on the right side of the brain. (D) The approximate planes in A–C. The vertical plane corresponds to the scan at the beginning of the chapter (page 191).

nothing to do with the patient's present state. Based in part on these observations, the temporal lobe and underlying hippocampus are recognized as important in the formation and recall of memories.

Lying in front of the central sulcus is a *motor area* in which the body is also represented as a map. Motor commands to muscles throughout the body receive a final organization here. Much of the frontal cortex, however, is association cortex, dealing with the planning and organization of behavior that requires recall of earlier sensory information. There are still other association areas in the limbic system, a ring of evolutionarily old cortical tissue out of sight in the view in Figure 9.14. This neural tissue is closely associated with the hypothalamus and the autonomic nervous system and deals with emotions and judgments about the consequences of behavior.

Recently a technique known as *functional magnetic resonance imaging* (fMRI) has made it possible to detect local regions of brain tissue where oxygenated blood is being delivered to active neurons. The procedure is non-invasive and can be performed on people. By scanning with detectors in different positions around the head, the location of active neurons can be resolved to within several mm (Fig. 9.16). These observations provide exciting confirmation that different perceptual, motor, and cognitive tasks employ distinct regions of the brain.

This summary of the principal features of the cortex suggests that some functions are more discretely compartmentalized than they really are. Virtually all behavior is complex, utilizing multiple sensory cues, drawing on many previous experiences, choosing among different motor acts. Different parts of the brain are specialized for particular functions, but in order for an animal to initiate behavior or respond to sensory input, neural activity must be organized and integrated by the brain working more as a whole.

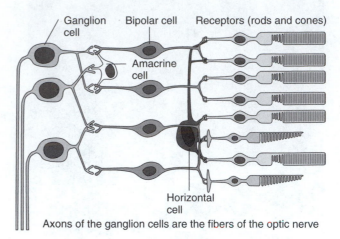

FIGURE 9.17 The retina contains several kinds of neurons in addition to the photoreceptor cells (the rods and cones). Information flows from photoreceptor to bipolars to ganglion cells, whose axons go to the brain. Horizontal cells and amacrine cells make lateral connections between the other cell classes. The retinal neurons interact with each other almost exclusively with graded synaptic potentials. Spikes are generated in the axons of the ganglion cells, however, as information leaves the retina. There are about a hundred times as many photoreceptors as there are ganglion cells, so the map of visual space that is present in the rods and cones is much modified when it leaves the retina for the brain.

glion cells generate action potentials. There are about 100 million photoreceptors but only 1 million ganglion cells. Therefore *the representation* of the optical image of the world that is cast on the retina—colors, movements, spatial relationships of objects, in short, everything that you see—*converges* on the ganglion cells before it is sent to the brain as trains of nerve impulses.

PROCESSING VISUAL INFORMATION

We will now take a closer look at one of the primary sensory projection areas where studies of single cells tell us much about the principles that are at work in the early stages of perception.

The retina is a thin layer of cells at the back of the eye. It contains *photoreceptor cells* called *rods* and *cones*, as well as several kinds of interneurons (Fig. 9.17). Light is absorbed in the rods and cones, and ensuing chemical reactions cause changes in sodium channels, membrane voltage, and synaptic activity. Rods and cones synapse on *bipolar cells*, and the bipolar cells in turn synapse on *ganglion cells*. The axons of the ganglion cells form the only connections that the eye makes with the brain, and unlike bipolar and horizontal cells, gan-

THE RETINAL GANGLION CELLS HAVE CONCENTRIC RECEPTIVE FIELDS

The information that reaches each ganglion cell has been shaped by passage through at least two synapses. Throughout most of the retina each ganglion cell receives input from many receptor cells. More importantly, the outputs of bipolar and ganglion cells are influenced by events that occur elsewhere on the retina. This is because *horizontal cells* and *amacrine cells* make lateral synaptic connections with other retinal cells. Some of these synapses are excitatory and others are inhibitory.

This anatomy sounds complicated, but we do not have to trace every connection and examine the responses of every synapse to find out what the eye is reporting to the brain. There is a much more direct way. By placing a very small electrode on the surface of the retina it is possible to record the traffic of nerve impulses generated by individual ganglion cells and thereby discover how the retina is packaging visual information. This is not an experiment that can be done on people, but it can be performed on animals such as frogs, cats, or monkeys.

Each photoreceptor cell is excited by a tiny fraction of the light in the retinal image, much as a silver grain on a sheet of film registers only the light that reaches it. The little bit of visual space that a photoreceptor "sees" is known as its *receptive field*. The concept of receptive field can be applied to bipolar cells, ganglion cells, and cells upstream in the central nervous system, as well as cells in other sensory modalities. The receptive field of a ganglion cell is that part of the retina where stimulation of photoreceptors influences the output of the ganglion cell. It can be measured by exploring the retina with small spots of light while recording nerve impulses in the axon of the ganglion cell. The receptive fields of ganglion cells typically include many receptors. Because ganglion cells are at least two synapses removed from the receptors, their responses are influenced by both excitatory and inhibitory synapses.

When this experiment was first done with mammals, the result seemed perplexing (Fig. 9.18A). In the dark, ganglion cells generate action potentials at a low frequency. Flashing a small spot of light in the receptive field caused a brief burst of nerve impulses, but a larger spot of light, rather than causing a faster burst of action potentials, had little or no effect. The mystery was soon resolved. When the stimulus is an annulus (ring-shaped), the dark discharge is actually inhibited. The receptive fields of ganglion cells thus have two concentric regions: a central area and an antagonistic surround. For example, in Figure 9.18A, the center is excitatory and the surround is inhibitory, and the cell fires a burst of nerve impulses when the light comes on in the center of the field. This inhibition from the periphery of the receptive field is mediated by horizontal cells. As you can see from Figure 9.18B, however, the receptive fields of other ganglion cells are organized in the opposite sense. The cell is excited by light in the periphery of the receptive field and inhibited when the stimulus is confined to the center. Notice how such cells fire a burst of nerve impulses when a spot of light in the center of the field goes off (Fig. 9.18B).

The receptive fields of ganglion cells overlap extensively (Fig. 9.19). This means that a particular receptor can contribute to the receptive fields of many ganglion cells. The lateral connections made by horizontal cells are important in creating the antagonistic surrounds, and horizontal cells, too, can influence the receptive fields of many ganglion cells.

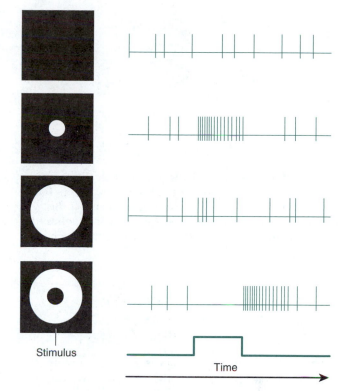

(*A*) Is an "on-center" unit.

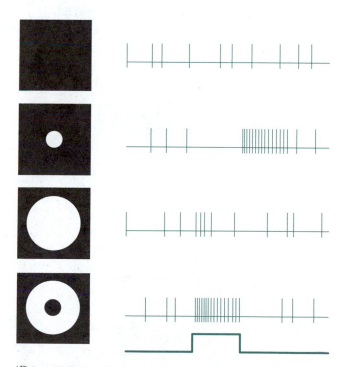

(*B*) Is an "off-center" unit.

FIGURE 9.18 Pattern of action potentials from the axons of two ganglion cells. (A) is an "on-center" unit. (B) is an "off-center" unit. The shape of the stimulus is shown on the left; its duration is shown by the colored line under each set of records.

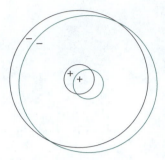

FIGURE 9.19 Receptive fields of ganglion cells overlap extensively throughout most of the retina. Except for a small region of the retina where visual acuity is the greatest, the center of a receptive field corresponds to an angle of visual space about the size of the moon.

VISUAL INFORMATION PASSES THROUGH ONE SYNAPTIC RELAY STATION BEFORE REACHING THE CORTEX

The path to the visual cortex is shown in Figure 9.20. Some of the axons cross the midplane of the body (forming the *optic chiasm*), and all synapse in a sensory relay structure, the lateral geniculate nucleus (LGN) of the thalamus. The axons of LGN neurons then synapse in turn with cells in the cortex. Some features of this pathway will be important for what follows. First, notice the pattern of crossing of the optic nerve fibers. Information from the left visual field is imaged on the right side of each retina and ultimately projects to the

right side of the visual cortex. Similarly, the right visual field is imaged on the left side of each retina and projects to the left cortex. Each side of the cortex therefore analyzes sensory information from the opposite side of the body, but it is getting that information from both eyes. The information from the two eyes is not mingled in the LGN, however. It is still traveling on separate axons when it enters the visual cortex, and only there does it start to converge.

LINES AND EDGES ARE EMPHASIZED EARLY IN CORTICAL PROCESSING

Just as with ganglion cells in the retina, the receptive fields of neurons in the brains of anesthetized cats and monkeys can be measured with tiny electrodes. The

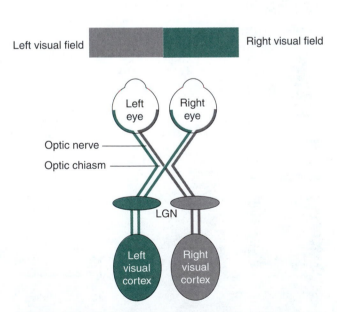

Left visual field   Right visual field

FIGURE 9.20 The path of visual information from eye to brain. Information from the left visual field of both eyes projects to the right visual cortex, and the right visual fields project to the left cortex. The information from each eye remains separate until it reaches the cortex.

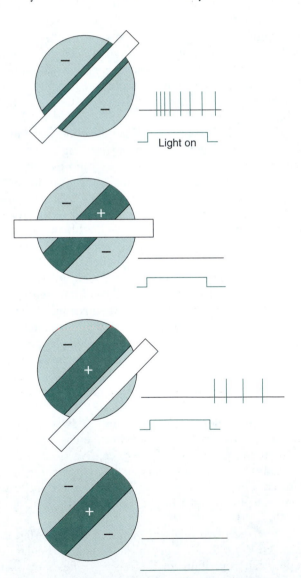

FIGURE 9.21 The response properties of a cortical neuron involved in an early stage of visual processing. The receptive field has excitatory and inhibitory regions, but the optimal stimulus is a properly oriented bar or slit of light.

LGN neurons hold no obvious surprises, for their receptive fields are similar to those of retinal ganglion cells. The visual cortex, however, contains a variety of cells with very different properties. The simplest kind is illustrated in Figure 9.21. The best stimulus is a bar or slit of light, but its orientation is critical. If it lies along the axis of either the excitatory or inhibitory regions, its effect is optimal. On the other hand, if it lies at an angle to the axis of the receptive field, it has no effect.

These cortical cells receive convergent input from neurons with concentric receptive fields (Fig. 9.22). In this model, each of the presynaptic cells from the LGN produces a small postsynaptic potential in the cortical neuron. The cortical neuron only fires, however, when several postsynaptic potentials occur simultaneously and sum with each other. This only happens when the bar or slit of light is oriented across the centers of the receptive fields of several of the presynaptic neurons.

These cortical neurons differ in the optimal orientations of their receptive fields, and other cortical neurons have more complex receptive fields. For example, some are sensitive to movement of oriented edges, even movement in one direction only. This single example, however, should serve to illustrate the principle of how the interactions of presynaptic inputs can mold and shape the responses of postsynaptic cells.

THE MAP OF VISUAL SPACE IS ORGANIZED IN COLUMNS OF CELLS EMPHASIZING DIFFERENT FEATURES OF THE RETINAL IMAGE

The two-dimensional optical image of the world that is focused on the retina is mapped onto the visual cortex. Adjacent objects in visual space stimulate receptors that are near neighbors in the retina, and these in turn project to cortical cells that are also close together. There are many more neurons in the visual cortex than there are axons in the optic nerve, so the anatomical convergence of presynaptic neurons on retinal ganglion cells is reversed at the cortex. This allows adjacent columns of cortical neurons to represent different features of each local part of the retinal image.

Figure 9.23 illustrates this principle with a much-simplified diagram. The surface of the cortex containing these cells is a few mm thick. Imagine a block of this tissue with the surface of the cortex located at the top. If you looked at a vertical face of the block with a suitable microscope, you would resolve half a dozen layers differing in the relative numbers of cell bodies and neuropil. The axons from the LGN terminate in the cortex in one of the middle layers, and most of the synaptic connections are made with cells that branch toward or away from the surface.

A tiny electrode that penetrates vertically from the surface samples cortical neurons that share the same

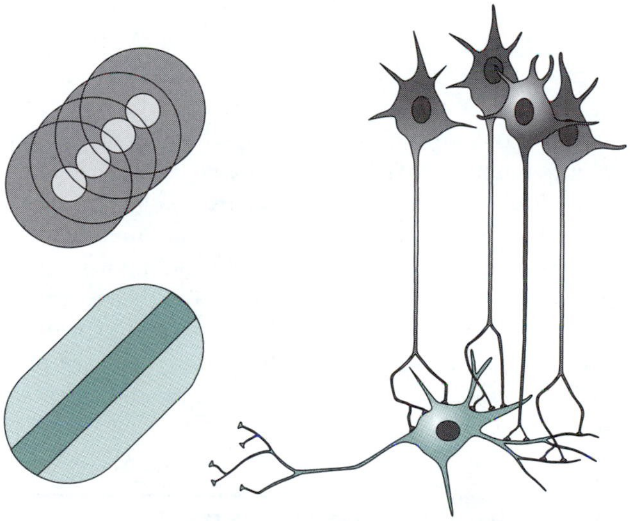

FIGURE 9.22 A model suggesting how convergent input of presynaptic neurons with concentric receptive fields can give rise to cortical neurons preferentially sensitive to oriented edges. The receptive fields of four presynaptic neurons (*left, top*) lie in a row. The postsynaptic cell is optimally sensitive when all of the presynaptic cells are active. Its receptive field (*left, below*) is therefore an edge or slit lying along the centers of the receptive fields of the presynaptic neurons.

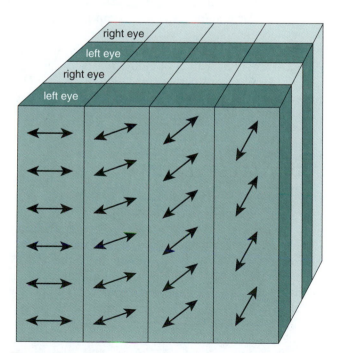

FIGURE 9.23 The visual cortex is comprised of columns of cells with similar orientation sensitivity and ocular dominance. The columns are much less regular than this diagram suggests, however.

sensitivity to angle of orientation of edge or bar. In an actual experiment it is virtually impossible to move the electrode so precisely, so as it goes deeper it encounters sudden shifts of about 10° in the orientation sensitivity of cells that it samples. Cells with similar orientation sensitivity therefore occur in slender vertical *columns*. The columns are about half a mm wide, and there is no visible boundary between them. They can only be detected by virtue of the physiological properties of their neurons.

If an electrode is in a layer above or below the site of entry of LGN axons, the cells are not only sensitive to the angular orientation of the light stimulus, they are also driven preferentially by one eye or the other. Here is the first place where input from the right and left eyes converge, but it is not balanced; most of the cells are unequally stimulated by the two eyes. Each eye maps equivalent information about the image to the same local point on the cortex, and a cell that is driven by both eyes has the same orientation sensitivity for each eye. Like orientation sensitivity, the cortex is thus also functionally organized into *ocular dominance columns*. As the tip of an electrode moves laterally about a mm it encounters cells that are driven preferentially by the other eye.

Figure 9.23 is a very idealized view of these columns. They are drawn as a precise array at right angles to each other, but in reality they both are quite irregular and it is not easy to visualize any relation between them. It is possible to see them under the microscope, but this involves very special methods. For example, ocular dominance columns can be made visible by injecting one eye with a radioactive substance that is taken up by cells and transported across synapses to the cortex. The location of radioactive cells can then be made visible by placing a thin photographic emulsion over a thin slice of cortical tissue and letting the radioactivity expose it.

RESPONSES OF NEURONS UNDERLIE PERCEPTIONS

This brief journey through the first stages of visual processing illustrates an important general principle. Serial processing is able to extract and emphasize particular features of a stimulus. Moreover, these are the features that prove to be important in higher-order perceptual processing. Again, a very simple illustration provides a context for thinking about more complex examples where less is known about the participating neurons. The visual system does a large amount of analysis by detecting edges and emphasizing contrast across boundaries. The simple cortical cells we described above are working at an early stage in this process. Figure 9.24 is an array of uniform gray bars of

FIGURE 9.24 **The visual system amplifies contrasts at edges and borders. Each of these stripes is uniformly gray, but they appear to be darker near the edges that are next to a lighter stripe and lighter on the opposite side, where they are adjacent to a darker stripe. This perceptual phenomenon makes objects more distinct against their background. The brain has many adaptations for filtering sensory information, of which this is a very simple example.**

increasing darkness. But notice how they do not appear to be uniform. Each seems darker on the side that is next to a lighter bar and lighter on the side that is adjacent to a darker bar. Edges are made more prominent by enhancing the contrast across the boundary.

These examples illustrate how our internal representation of patterns of light and shadow, shape and color, emphasize particular features of the environment. Here, in the activity of a few cells, we see a model for how evolution has filtered and tuned the way we perceive the world. More is probably known about the roles of individual cells in the visual system than elsewhere in the CNS, but this knowledge provides a framework for thinking about how more complex processing occurs, such as in understanding and speaking language.

UNDERSTANDING AND COMMUNICATING LANGUAGE

Some victims of strokes subsequently have difficulty speaking. About the time of the American Civil War, Pierre Broca, a French physician, reported a patient who had lost the ability speak or write in organized sentences. (Such a language impairment is called an

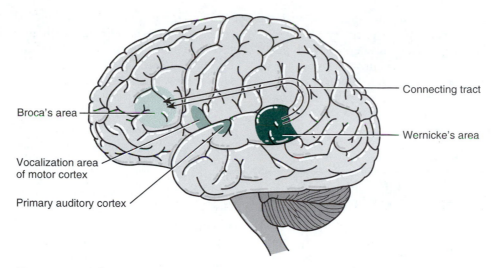

FIGURE 9.25 **The positions of some cortical areas involved in the interpretation and generation of language.**

aphasia.) After the patient's death his brain was examined, and damage was found in a region of frontal cortex anterior to the temporal lobe (Fig. 9.25). This is now known as Broca's area, and it is association cortex involved in organizing and generating grammatical language.

Fifteen years later, Carl Wernicke, a German scientist, described patients with another aphasia. They have difficulty in understanding language, and their speech, although grammatical, contained incorrect words and distorted sounds. Their brain lesion was in the temporal lobe, in association cortex approximately between the sensory projections from the visual and auditory systems. This is therefore a region of the brain where visual and auditory input can be integrated, and it plays an important role in interpreting language (Fig. 9.25). These two regions of the brain are connected, and both are involved in the normal understanding and generation of speech. Dyslexia, a congenital impairment of reading, seems to include deficits in the recognition of sequences of letters and their interpretation as language, and individuals with this condition use alternative neural paths. Figure 9.26 is an fMRI image showing neural activity during speech.

These observations not only tell us more about how the brain is organized to perform various functions, they say something important about the origins of language. The human capacity for language has been thought by some scientists to defy explanation by known natural processes. But as we have learned more

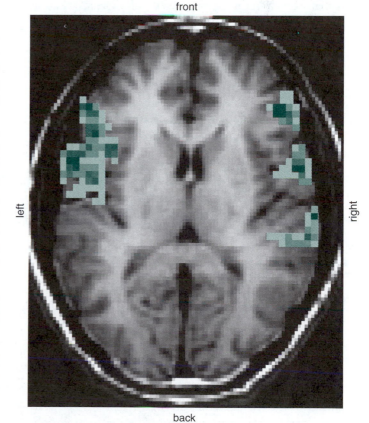

FIGURE 9.26 **An fMRI image made while the subject was talking. The plane of the image passes through Broca's area (approximately the same as B in Figure 9.16D). Note the asymmetrical activation of neurons on the two sides of the brain.**

about how the brain works we find that the capacity for language is not a generalized property of a large brain. Rather, language is comprehended and produced by specialized groups of neurons whose injury or death results in characteristic behavioral deficits. There is only one way that nature has of producing such specialized structures: evolutionary change through natural selection.

THE RIGHT AND LEFT HALVES OF THE BRAIN ARE DIFFERENT

When illustrating the cortical somatosensory projection area in Figure 9.15 we pointed out that sensory information from the right side of the body projects to the left cortex and vice versa. Similarly, information from the right visual field is represented in the left visual cortex and vice versa (Fig. 9.20). In an equivalent way, voluntary control of muscles on one side of the body arises in neurons in the opposite side of the cortex (Fig. 9.16C).

A point we did not make about language, however, is that in most individuals Broca's and Wernicke's areas are in the left cortex. Speech is typically controlled from the left side of the brain in both right- and left-handed people. About 4% of right-handed and about 30% of left-handed individuals are exceptions, and about half of the left-handed exceptions actually show no dominance of either side of the cortex. What is the homologous tissue in the other side of the brain normally doing? It is involved in language, but its role is closer to the emotions, as displayed by inflection and gesturing.

The condition of some patients with particularly severe epilepsy that spreads throughout the brain can be improved by surgically cutting the *corpus callosum*, the massive fiber tract that connects the two halves of the cortex. Such individuals then have a *split brain*, with the two hemispheres unable to communicate with each other. Most of such people behave completely normally after the operation, and their perceptual and linguistic abilities have been carefully studied by Roger Sperry, Michael Gazzaniga, and others. The following observations illustrate dramatically the *lateralization* of the cortex.

A young woman with a severed corpus callosum sits in front of a screen onto which is projected a picture of a familiar object. The woman is looking straight ahead at a small black target, and the picture is positioned so that its image falls entirely on either the right or left side of each retina. The picture remains on the screen for only about a tenth of a second, however. This is enough time for the object to be recognized, but not enough time for the viewer to make an involuntary eye movement and bring the image onto another part of the retina.

The top diagram in Figure 9.27 shows the path of information in an individual with an intact corpus callosum. It moves from one side of each retina through one of the lateral geniculate nuclei to one side of the cortex. From there it passes to the other hemisphere by way of the corpus callosum.

When the young woman with a severed corpus callosum is shown a cup in her right visual field, the image registers in the left cortex (middle diagram, Fig. 9.27), and when asked what she had seen, she reports verbally that it was a cup.

Then she is shown a picture of a spoon in her left visual field (bottom diagram, Fig. 9.27). This time when asked what it was she disclaims having seen anything. The experimenter then asks her to reach under the screen with her *left* hand and by touch alone select the object whose picture had just appeared—and which she had just denied seeing. She feels among several possible objects and holds up the spoon (which is still out of her view). She is now asked what she is holding, and she says that it is a pencil.

What is going on? When the picture registered in the left cortex, she was able to report verbally that it was a cup, because the left hemisphere is the one that can communicate through language. When the picture of the spoon registered in the right hemisphere, however, the left hemisphere had no clue. In fact it denied any knowledge of the spoon because it hadn't seen anything! The right hemisphere knew, however, but the only way it had to communicate was via the left hand. When the left hand reported tactile stimuli that matched the visual identification of spoon that was stored in the right hemisphere, the right hemisphere had the left hand hold up the spoon (still out of sight). But when asked to identify the object, the left hemisphere was still mystified—thus the answer, "a pencil."

It may seem surprising that individuals with a split brain can lead as normal lives as they usually do, but you can see from the description of this experiment that rather special conditions are required to insure that the two halves of the cortex receive different information. Nevertheless, some individuals have found that the two halves of their brain can sometimes go separate ways. One hand may reach for an article of clothing while the other either tries to stop it or picks up something else. These reports confirm in an interesting way that much that goes on in the brain is inhibitory. The simplest behaviors such as walking involve alternative excitation and inhibition of muscles, and in organizing more complex behavior the brain is constantly selecting among various possibilities. What finally happens, at the expense of what might have happened, results from the interplay of excitatory and inhibitory activity of neurons.

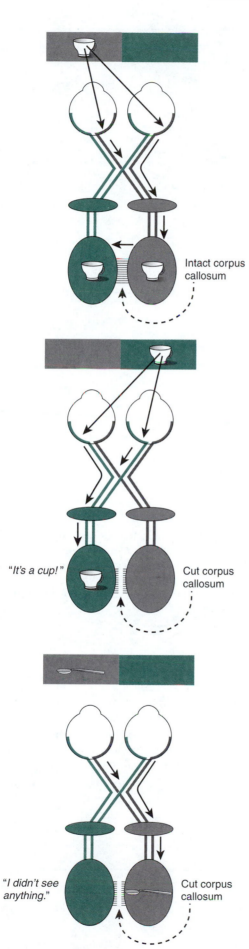

FIGURE 9.27 **What happens when the two halves of the cortex are surgically separated? See the text for a description.**

NEURONS CHANGE DURING LEARNING

LEARNING AND MEMORY ARE TRADITIONALLY STUDIED BY PSYCHOLOGISTS

Until very recently there have been no methods for detecting cellular changes associated with learning and memory, and virtually everything known about these processes has been based on studies of behavior and cognitive psychology. These studies nevertheless reveal important features of learning and memory.

Some memories are of motor skills that have been learned by repetition and do not involve much thought, whereas others require conscious participation in both learning and recall. Learning to ride a bicycle is not the same as learning about biology; these processes involve muscles to very different extents, and they engage different parts of the brain.

Events of which you are conscious are initially registered in short-term memory before they are consolidated in long-term memory. Short-term memory has a limited capacity. In humans tested with words or objects, fewer than a dozen items can be remembered, and depending on circumstances, this memory may last only seconds.

New information is interpreted in relation to old information, and long-term storage is easier if the new information relates sensibly to older memories. If new information has no meaningful context in which it can be placed, it is more difficult to remember without practice or repetition. Memories are continually changed, and education is thus cumulative.

Amnesia, a temporary loss of memory, can result from head injury. The most recent memories are the last to be recovered, and in severe cases this takes weeks or months. That memories eventually return indicates that they had been stored in long-term form at the time of the trauma. In addition to memories, there is therefore also a separate search process for scanning the memory bank, and it can be separately damaged. One does not have to be hit on the head, however, to be frustrated by unsuccessful recall.

In previous sections we have emphasized that parts of the brain are specialized for different functions. Forty or fifty years ago most psychologists and neuroscientists were largely convinced that there is no part of the brain specifically involved in learning and memory. This view was based on the finding that rats could learn

Box 9.3
Changes in Neurons that Underlie Learning

Invertebrates are very useful for studying neural processes. Their nervous systems are relatively small and have many cells that can be individually recognized. In some species the cells are also large and can be readily penetrated with tiny electrodes. Furthermore, these animals are capable of simple forms of learning. The cellular basis of learning and memory has been extensively studied in a large, soft-bodied sea slug *Aplysia* (a nudibranch mollusk) by Eric Kandel and his colleagues at Columbia University. Their experiments illustrate some of the sorts of plastic changes in neurons that likely occur in all nervous systems.

The simplest sort of learning is *habituation*, in which an animal ceases to respond to repeated stimuli that prove to have no consequences. Gently touching the skin of *Aplysia's* siphon causes the animal to retract its gill. This protective reflex is mediated by a very simple circuit involving a couple dozen sensory neurons, some interneurons, and half a dozen motor cells (Fig. 9.28). If the touch is repeated, withdrawal of the gill becomes progressively weaker. Electrical recording from the motor neurons shows that the excitatory postsynaptic potentials become smaller, and this is because the sensory neurons discharge fewer vesicles with each stimulation (the colored presynaptic endings in Figure 9.28). If the animal is given several training sessions, the effects of habituation can last for several weeks. Under these experimental conditions the number of presynaptic contacts that the sensory neurons make with motor cells actually decreases.

Figure 9.29 shows another kind of experiment. A weak electric shock delivered to the tail at the same time the siphon is touched increases the sensitivity of the gill-withdrawal reflex. This *sensitization* is mediated by facilitating interneurons that make synapses with the same presynaptic terminals of the sensory neurons that were the site of habituation in the previous experiment. The synaptic endings of the sensory neurons are therefore both presynaptic with respect to the motor neurons and postsynaptic with respect to the facilitating interneurons.

The transmitter released by the facilitating interneurons is serotonin (Fig. 9.30). The receptor on the postsynaptic membrane (the synaptic terminals of the sensory neurons) is not a channel but lies at the

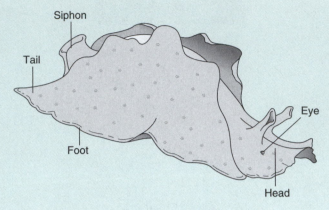

(A)

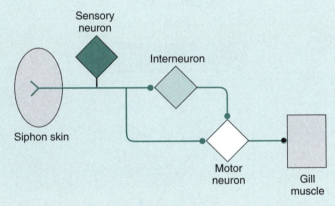

(B)

FIGURE 9.28 (A) The sea slug *Aplysia*. (B) A simplified diagram of the neural circuit involved in the gill withdrawal reflex of *Aplysia*. Repeated gentle touch of the siphon skin causes the response to habituate due to both short- and long-term changes in the synaptic terminals that are dark green.

head of a cascade of chemical reactions inside the cell leading to the production of a second messenger. The second messenger (which in this cell is cyclic AMP) has two effects. The first is to activate an enzyme that attaches phosphate groups to the intracellular (cytoplasmic) ends of potassium channels, thereby blocking the movement of K^+ ions through the channels' pores. Recall from our earlier discussion of action potentials that a voltage-gated potassium channel is involved in hastening the end of an action potential. When Na^+ rushes in, the cell depolarizes, and the K^+ channel opens, helping to drive the membrane potential back to

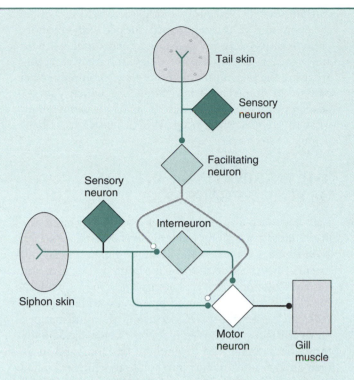

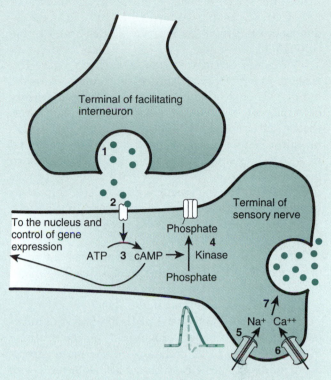

FIGURE 9.29 Sensitization of the gill withdrawal reflex to touch occurs via facilitating interneurons. The terminals of the sensory neurons from the siphon skin are the targets: transmitter release is increased and more terminals form. See the text and Figure 9.30 for how this happens.

the resting value. In these *Aplysia* cells, blocking this K$^+$ channel makes the individual action potentials in the terminals of the sensory neurons last several msec longer than normal. This longer depolarization causes more Ca^{++} to enter the cell, more synaptic vesicles are released, and a larger postsynaptic potential arises in the motor neurons. The effect persists because the phosphate groups remain attached to the K$^+$ channels for a long time.

The other effect of the second messenger is more far-reaching. cAMP diffuses back to the cell body where it leads to the attachment of a phosphate group to a protein that is involved in the control of transcription. Genes are thereby activated, new proteins are synthesized, and most interestingly, the sensory neurons make new synaptic endings!

These experiments are important in showing how the functional properties of neurons can be modified by sensory input, so as either to decrease or increase the likelihood of a behavior. The effects range from transient to long-lasting. They do not involve special

FIGURE 9.30 Changes in a synapse resulting from simple learning. (1) Arrival of an impulse in the facilitating interneuron causes release of transmitter (serotonin). (2) Serotonin binds to a receptor on the postsynaptic cell— the axon terminal of a sensory neuron. (3) This activates a second messenger cascade leading to the formation of cyclic AMP. (4) The presence of cAMP in turn activates an enzyme (kinase) that attaches phosphate groups to K$^+$ channels, thereby inactivating them.

(5) The subsequent arrival of an impulse in the axon of the sensory nerve opens Na$^+$ and (6) Ca^{++} channels. (7) Elevated Ca^{++} causes discharge of the transmitter secreted by the sensory axon. (Inset) With the K$^+$ channels blocked, the action potential lasts longer (solid compared with dashed curve), more Ca^{++} enters, and more transmitter is released.

Prolonged sensitization causes the cAMP to reach the nucleus where another protein kinase phosphorylates a transcription factor, leading to the activation of genes.

neurons that exist just for learning and memory, and they produce structural changes. Some changes, like the attachment of phosphate groups to ion channels, last for hours or days, and the altered structure is a molecule. Other changes involve protein synthesis and are more persistent. New synapses are formed, and the neurons themselves are altered structurally.

mazes after the cortex had been extensively cut, or even removed. There thus seemed to be an enormous redundancy of memory circuits and plasticity for learning throughout the brain.

To some extent this view is correct. There is a large amount of parallel processing going on in the brain, particularly when there are multiple sensory cues that can be used in the learning task. Moreover, the capacity of sensory input to make lasting changes in motor output is a very general property of neural tissue, even as simple as an isolated ganglion from an insect nerve cord. As we saw earlier, however, the temporal lobe of the cortex is particularly important in eliciting memories. Both the temporal lobe and the underlying hippocampus are involved in both consolidation and recall.

Perhaps the most fundamental finding about memory is that it survives deep anesthesia or chilling of the brain. This means that memories are not stored as ongoing activity of neurons. If the dynamic activity of neurons can be slowed or stopped without loss of memory, the plastic changes that occur during learning must reflect *structural* alterations in the brain. Moreover, inhibiting protein synthesis also interferes with learning. These changes are now beginning to be explored in single neurons (Box 9.3).

THE PERCEPTION OF SELF

Each of us has a unique sense of being, of who we are, our likes and dislikes, our past experiences, our plans and ambitions, and how we fit into our immediate social environment consisting of friends and relatives. The most sweeping challenge in neuroscience is to understand these emergent properties of the human brain—the most complex structure in the known universe. "*Cogito, ergo sum*" ("I think, therefore I am") said the French philosopher and mathematician René Descartes (1596–1650), but he had the proposition just backwards.

From this brief description of the problem you can infer, correctly, that much of the brain is likely to be engaged in the perception of self. For example, imagine that you are greeted by a voice on the street. You turn, and a view of a face confirms the preliminary identification that you made based on the sound of the voice. Memory is stirred: the time and place of your last encounter (was it at Mary's party?), the nature of your relationship (perhaps you are in love with her, or maybe he still owes you money). Mental images flash into your consciousness (how smashing she looked at the party, some unpleasant consequences of being delayed now). Your feelings—and that is how we often refer to this inner state—are not to be compared with a computer working out its next move in a game of chess. Unlike the computer, your thoughts are charged with emotion, perhaps pleasure at this unexpected encounter, perhaps annoyance at the unwelcome delay. In either case, you are using the present moment to relate a complex social past to an immediate social future for which you are now going to make a series of small decisions about what to say and what to do. What, physically, is happening in your brain?

THE FASCINATING CASE OF PHINEAS GAGE

Let's approach an answer to that question through one of the most astonishing medical cases ever recorded. In 1848 Phineas Gage was an up-and-coming, 25-year-old foreman for the Rutland and Burlington Railroad, and his construction crew was preparing a new roadbed. Blasting powder was poured into holes drilled in solid rock, plugs of sand were added, and the sand tamped down with an iron rod. At one point Gage accidentally thrust his tamping rod (over three feet long and more than an inch in diameter) into a hole to which sand had not yet been placed. The rod struck a spark that ignited the powder, and the rod rocketed upward, pointed end first, entered the side of Gage's face, exited the top of his skull, and landed about a hundred feet away. Gage may have lost consciousness, but only for a few moments. Not only did he survive the accident, he lived for another thirteen years!

Two features of this case give it greater importance than the astounding fact that Gage was not instantly killed. First, he was attended by a young physician, John Harlow, who documented a profound change in Gage's personality. Before the accident Gage was judged to be astute, energetic, likeable, and effective—certain to rise in the company. Following the accident, however, he became (in Harlow's words) "fitful, irreverent, indulging at times in the grossest profanity which was not previously his custom, manifesting but little deference for his fellows, impatient of restraint or advice when it conflicts with his desires, at times pertinaciously obstinate, yet capricious and vacillating, devising many plans of future operation, which are no sooner arranged than they are abandoned. . . . A child in his intellectual capacity and manifestations, he has the animal passions of a strong man." Needless to say, Gage never held a responsible job for the remainder of his life.

The second feature of this case that has made it medically important is that following Gage's death in 1861, Harlow persuaded the family to have the body exhumed and the skull (along with the tamping rod) saved in a medical museum at Harvard University. In the early 1990s Hanna Demasio and colleagues were thus able to use modern, computer-based imaging technology to reconstruct the path of the rod and identify the ventromedial (lower, near the midline) region of the prefrontal cortex as the part of the brain that had been most severely damaged. Together with other evidence from the effects of strokes and tumors, the story of Phineas Gage has given us a picture of a brain region that is devoted to a particular kind of reasoning. It is reasoning about oneself in the context of social relations, and it is reasoning with important emotional content. Individuals with damage in this part of the brain may be able to analyze and discuss a social situation yet be unable to decide what to do. This contrasts with reasoning that is centered in more dorsal and lateral regions of prefrontal cortex where damage produces additional deficits that are more general (keeping attention focussed) and involve objects, words, or numbers rather than social relations. As we will see in later chapters, this physical evidence for a cortical module directing social behavior is particularly interesting when placed in an evolutionary context.

EMOTIONS, IMAGES, AND INTERCONNECTIONS

Our thoughts are invariably colored by emotion, ghosts summoned from the limbic system and hypothalamus, evolutionarily ancient parts of the brain that equip us with regulatory circuits and motivations that have proved useful in our evolutionary past. Like the headlights, brakes, and air bags of automobiles, they have become standard equipment. As we saw when discussing the hypothalamus and neurosecretion, some of this apparatus is beyond ordinary conscious control. But emotions are powerful motivators that work together with the frontal cortex. To say that we are rational decision makers is an abstraction more at home in theoretical microeconomics than in psychology. Both memories and decisions about future behavior are invariably tinged by emotions, and there are corresponding neural connections between frontal cortex and the limbic system. Our emotions are hierarchical; some are very general and others are linked to past experiences in very personal and specific ways.

The linkage of conscious will and the frontal cortex with deeper emotional centers is illustrated by a very familiar example. When we are posing for a picture and told "now smile," the result is generally disappointing. The smile looks forced and unnatural, and we know it while we are trying to smile. On the other hand, if the photographer says something amusing, our smile is both spontaneous and more genuine. The reason is that we have conscious control over only part of the musculature that is used in smiling, whereas deeper cortical regions activate all of it. A smile that is produced with no emotional content is thus unconvincing.

When we plan or remember, we create images "in our mind's eye," although the images may involve sound, taste, or smell as well as vision. We can picture a face, a room, or a sunset, we can "hear" sirens or symphonies, but we do not confuse these mental constructs with current sensory input. Whether we are summoning memories or creating a possible future with these images, we are quite clear that we are imagining and not sensing events then occurring around us. Interestingly, however, evidence from PET scans suggests that we use some of the same cortical areas for both processes. We can detach our attention from the surrounding world and draw upon internal resources to manufacture images in regions of cortex where sensory information is also projected. Primary sensory cortex not only forwards information to association cortex, it can receive information sent back from higher centers.

A recurrent theme in science fiction is the idea of a disembodied brain dwelling in a transparent barrel of nutrient broth and thinking great thoughts. It is an amusing scene to contemplate, but it could not be a human brain. Quite aside from the technical problems posed by sustaining the brain in broth, it is unrealistic for another, more fundamental reason. The brain requires ongoing sensory input. It constantly monitors the surround through eyes, ears, and nose, as well as through receptors in the body surface that assess skin temperature, touch, and pressure. Moreover, the brain does not rely exclusively on motor commands to the body muscles to know where it has moved the limbs or which way it has turned the head. It receives a steady stream of information about the degree of contraction of the muscles and the positions of the joints, none of which enters conscious awareness without express command from the cortex. Moreover, the brain is engaged in chemical monitoring and regulation of internal temperature as well as sodium, oxygen, carbon dioxide, and sugar in the blood. The basic point is that the "output" of the brain—its normal function—requires all of this input. This is why one of the most serious and debilitating things that can happen to a prisoner is to be kept in isolation without sensory input. The brain is an evolved control system, and it operates at many levels, but all of them require input.

DREAMS

The creation of sensory images—most importantly visual images—is particularly interesting during dreams. Everyone dreams, and in many cultures, including our own, dreams have been given important religious significance as omens or as messages from a deity. In medicine, the physician Sigmund Freud founded a school of therapy based on the interpretation of dreams. There are therefore ample reasons for wanting to understand what the brain is doing when it dreams.

During a phase of sleep characterized by rapid eye movements, neural signals arising in the brainstem ascend through the thalamus to the cortex. While one is in this state, inhibitory discharge of acetylcholine-secreting neurons quenches virtually all sensory input to the cortex. The cortex, now uncoupled from sources of information about the external world and the positions of the body, is nevertheless not quiescent. It conjures images in meaningless fashion, influenced by those events of the day that are not yet secured in the vaults of long-term memory, as well as by emotions. The images can be frightening or erotic and are generally chaotic and senseless, fantasy that would be improbable or impossible while awake. A few strong sensory stimuli may leak through the barricades of inhibition and become incorporated into the dream as the cortex tries to give them meaning. Disconnected from immediate external reality, however, the cortex fabricates significance, and the alarm clock or a loud noise outside the window becomes a fire alarm or a gun threatening the dreamer's safety. The cortex may believe that the body is running for its life, but it is not receiving confirmation from the limbs. In the dream the body is wading through glue and getting nowhere, while the approaching terror is gaining. And so forth.

Bouts of dreaming are cyclic, because in time the neurons responsible for isolating an active cortex are in turn inhibited by other neurons in the brainstem. Moreover, dreams are not readily consolidated into memory and are difficult to recall more than a few minutes after awaking. Their function—if any—is not clear. Although they tap into emotions that are important tools in our psychological survival kit, there is growing reason to doubt Freud's belief that they represent censored (repressed) ideas that the cortex is attempting to keep out of consciousness and which can only be understood (interpreted) when described to a trained psychoanalyst. Rather, they appear to be the mental meandering of a temporarily disconnected cortex, more akin to the uncensored visions of awake but psychotic individuals who are not processing sensory information in a normal manner. It is probably this mysterious and unsettling nature of dreams that has made it so easy for shamans, priests, and others of like authority to endow them with spiritual significance.

THE MATTER OF "FREE WILL"

In a universe where every event has a physical cause (or causes), every state is determined by the conditions that immediately preceded it, and so on backwards in time. In what sense, then, can a brain have open choices before it? Is not the universe deterministic? How can we square this idea with the commonsense experience that we seem free to choose alternatives hundreds of times a day?

This is not the paradox that it seems to be. Consider a deck of cards that has been thoroughly shuffled. The sequence of cards has been determined by the sequence of physical forces that occurred during the shuffling, but the nature of shuffling is such that it is impossible for a person to achieve the same outcome twice. The application of forces varies every time, and when finished, the cards are in different sequences. Suppose you are asked to name the top card; you have 1 chance in 52 of being right and 1 chance in 4 of naming the suit correctly. There will be some subtle cause for your choice, but it will not affect your chances of being right.

Now consider the game of bridge. The deck is shuffled, the cards are dealt, and their distribution around the table, we say, is due to chance. What we mean is that we are unable to predict the distribution because we lack detailed information about what occurred during shuffling. Bridge, however, is not a game of chance. As play progresses, bits of information become available to the players. The bidding gives a sense of who has some of the high cards of particular suits, and as each trick is played, four cards are turned face up. Good players are able to assimilate this information and use it to make educated guesses as to who has which of the cards that are still concealed in the other players' hands. Constrained only by the rules of the game (such as following suit whenever possible), each player is free to decide at every turn which one of the remaining cards should be played. How players decide depends on how well they have assimilated the information that has gradually become available since the cards were dealt.

This description of the game of bridge illustrates two points. First, many events that occur in the world have a probabilistic nature. In other words, when repeated, the outcomes vary. The occasional mutation that occurs during the replication of DNA, the recombination of genes during meiosis, the number of synaptic vesicles discharged as an action potential arrives at a nerve terminal, and next year's rainfall in your hometown are all examples. This does not mean that these events have not been determined by material causes; it means only that the causes are sufficiently complex that the outcome varies in a manner that can be described by the mathematics of statistics but cannot be predicted in any particular instance.

Second, the brain is an information-processing machine, and that is the linkage between behavioral cause and effect. The brain adjusts the amount of antidiuretic hormone delivered from the pituitary to the kidney based on its assessment of the concentration of salt in the blood. Or the bridge player decides whether to take the trick with the king of hearts depending on his assessment of whether the person to his left is likely to hold the ace of hearts. In each case information is utilized, but the second case is more complex because the brain is not simply scaling output (secretion of a hormone) to input (a sensory measure of sodium concentration). In playing bridge, the brain has enough information about the game to understand that it lacks all the information it needs in order to optimize its next move. It thus consciously weighs the relative likelihood of different possibilities using the incomplete information that it possesses. That is how natural selection has "designed" it to function.

In summary, free will does not imply that decisions are independent of materialistic causation. Nor does it suggest the alternative extreme that all possible behavioral outcomes are equally probable. Our choices are constrained by both personal and evolutionary history as well as the information that is available to us in the time available for us to choose a behavior. But within these bounds we use information to weigh consequences, and if our brains are functioning as evolution crafted them, we *do* decide, sometimes well, sometimes poorly, and seldom without the interplay of reason and emotion.

SYNOPSIS

Just as knowledge of genes is required to understand how organisms grow and develop and how evolution proceeds, so a basic knowledge of nerve cells is necessary for understanding how brains process information and generate behavior. New methods for visualizing active parts of the brain, recording the activity of individual cells, and measuring the movements of ions through single membrane channels have recently added considerably to our knowledge of how nervous systems work.

Nerve cells, *neurons*, signal by two basic processes that are based on properties found in all cells. In order to keep from filling with water, all animal cells *pump sodium ions* from the inside to the outside of the cell. In neurons, a brief opening of *voltage-gated sodium channels* sweeps from one end of an axon to the other, causing an electrical signal that can propagate the length of the body in a small fraction of a second.

Neurons communicate with each other at *synapses*. The presynaptic cells secrete small molecules, *neuro-*

transmitters, that bind to protein receptors on the postsynaptic cells, causing ion channels to open. Depending on which ions are able to pass through the activated channels, the postsynaptic cells can be either excited or inhibited. Some of these molecular signals trigger the synthesis of a "second messenger" within the postsynaptic cell, with longer-lasting effects on the target, including the activation of genes.

When large numbers of neurons assemble together they have emergent properties that result from the activities of networks of cells. Coordinating brain and kidney to maintain a constant salt concentration in the blood or recognizing a face in a crowd are the work of *systems* of neurons that have to be understood in terms of their patterns of connection.

Different components of behavior or of information processing are associated with different parts of the brain, but the separation is not complete. Parallel processing and the integration of information from multiple sources require that many parts of the brain function simultaneously.

As is often the case with evolutionary outcomes, the mammalian brain has a quirky design that reflects its history. The *cortex*, whose expansion is one of the main features of mammalian evolution, is layered over phylogenetically older tissue that controls the internal physiological environment and the emotions. The cortex processes and integrates complex sensory information, plans behavior, and in humans it understands and generates language. Awareness of self is a further manifestation of this complexity.

QUESTIONS FOR THOUGHT AND DISCUSSION

1. What evolutionary innovations at the cellular level enable the active life of multicellular animals? Consider the acquisition of information about the environment and its processing and movement within the organism.

2. Some economic theorists view individuals as "rational actors." To what extent is this characterization justified by current knowledge of how and why the brain and mind function as they do?

3. Discuss the role of the emotions in deception and self-deception. Is there a place for the process of evolutionary adaptation in this relationship? Explain.

4. What sorts of processes occurring in the brain are related to your sense of self? Do you think any nonhuman animals have a sense of self? What evidence would you want to have in order to answer that question definitively?

SUGGESTIONS FOR FURTHER READING

Damasio, A. B. (1994). *Descartes' Error: Emotion, Reason, and the Human Brain*. New York, NY: Avon Books. A discussion of the important interactions between rational thought and the emotions, written by a neurologist with a gift for communicating to a general audience.

Kandel, E. R.; Schwartz, J. H.; and Jessell, T. M. (2000). *Principles of Neural Science*, 4th ed. New York, NY: McGraw Hill. Detailed information on the structure and function of the brain, but not for the faint-hearted.

Pinker, S. (1997). *How the Mind Works*. New York, NY: W.W. Norton & Co. A lucid account by a cognitive scientist with a deep interest in evolutionary theory.

Springer, S. P.; and Deutsch, G. (1997). *Left Brain, Right Brain: Perspectives on Cognitive Science*. New York, NY: Freeman. Includes many interesting details about the lateralization of function.

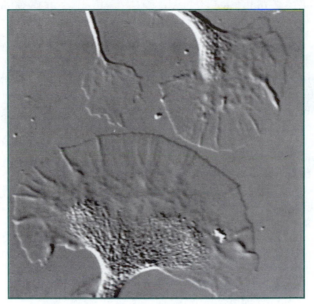

10

Individual Organisms Are the Product of Development

THE UNDERSTANDING OF DEVELOPMENT IS ENTERING A NEW PHASE

The emergence of a fully formed organism from a single cell is one of the most striking phenomena displayed by nature. Over a period of days, weeks, or months development creates new orders of complexity,

Photo: During development, growing axons of nerve cells must find their way for considerable distances, influenced by molecular signals they encounter as they extend. In vertebrates, many more neurons start the journey than survive in the adult animal. These are the growth cones at the tips of axons of *Aplesia* neurons growing in tissue culture.

producing an exquisitely crafted plant or animal. It is easier to describe the process of development than it is to account for its inner workings, but that is because techniques for manipulating the underlying molecular events have only recently become available.

Like evolution, development involves irreversible change. As with time itself, neither process runs backwards. Although evolutionary futures are unknown, developmental outcomes are normally foreseeable, at least within predictable boundaries. The fertilized eggs of humans develop into people, not fruit flies or oak trees, and for each species the final variation among individuals has characteristic limits that are infrequently crossed. But in both development and evolution, organisms interact with their environment in intricate and inseparable ways.

Once set in motion, development ordinarily proceeds to a product that is dependent on genes, but genes alone do not account for development. The fates of individual cells and the activities of their genes are constantly influenced by events that are taking place outside the cell. Some of these influences arise within the embryo itself, but others come from the external world. Development can now be explored in terms of molecular processes: how the activities of genes change with time, how these activities vary among the cells of the embryo, how genes are controlled by external signals, and how cells influence each other. The goal is to see how molecular processes underlie the changing shapes and behaviors of cells as they mold a new organism. The study of *developmental biology* is one of the most active and exciting areas of contemporary biological research.

This molecular approach to understanding development is powerful and exciting, but as the growing organism becomes more complex, additional perspectives are useful. The development of behavior is the ultimate example. Influences of early learning and their effects on subsequent conduct are ordinarily cast in the language of psychology. This is because the environmental influences and their effects on behavioral development are measurable features of entire organisms, and the behavioral sciences provide appropriate techniques, descriptions, and interpretations. But for every behavior described by psychology, there are causal processes going on in nerve cells and networks of nerve cells. The construction and alteration of these networks involve sensory stimulation, cell growth, and the selective activation of genes. For every change in behavior there are underlying changes in the nervous system.

This rather obvious point has not always been appreciated. In some scholarly traditions, "biology" has been assumed to end at birth, and the behavioral development of humans is considered entirely a cultural phenomenon for which a biological perspective is both unnecessary and unwelcome. But developmental processes do not end at hatching or birth. The birth of a mammal is indeed an important transition, for the newborn abruptly becomes independent of the mother's circulatory system for oxygen and nourishment. Attaining sexual maturity is another important time of developmental change, but development, including that of the brain, is a continuous process that eventually merges with changes that are ordinarily thought of as aging.

In this chapter we describe some basic features of development and how recent knowledge of genes provides new understanding of this intricate process. You will see the inseparable interplay of genes and environment: how genes are influenced by chemical signals arising within the organism (the *internal environment*) as well as by sensory cues originating in the outside world (the *external environment*). These themes are illustrated with examples showing the emergence of behavior in the life of individual animals.

EGG AND SPERM ARE ELABORATELY SPECIALIZED CELLS

Development begins with *fertilization*, the union of a small motile sperm cell with a larger egg. Sperm were first observed under early microscopes in 1678, but their role was not understood for 200 years. For a time they were even thought to be parasites. In 1876, however, their true nature was clarified definitively when sperm nuclei of sea urchins were observed fusing with the nuclei of sea urchin eggs.

Like the sperm of many other animals, those of mammals are little more than traveling haploid nuclei with a long flagellum for propulsion, some mitochondria to supply energy, and a few enzymes to assist them in penetrating the thick envelope of glycoprotein (the *zona pellucida*) that surrounds the egg (Fig. 10.1). Sperm do not wander aimlessly; they are guided to their destination by species-specific chemical signals secreted by the eggs. This is obviously essential in an aquatic environment like a coral reef where many species of fish and invertebrates are living together and external fertilization is common. But even in mammals, where fertilization is internal, these signals are important. Of the nearly 300 million sperm in the ejaculate of a human male, only a few hundred arrive in the upper end of the oviduct in the vicinity of the several eggs that are produced at each ovulation.

When a sperm cell encounters an egg it secretes enzymes that soften a path through the zona pellucida (Fig. 10.1). The sperm extrudes a structure called the *acrosomal process* with protein receptors that mediate the

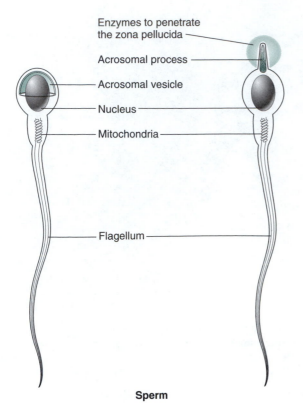

Enzymes to penetrate the zona pellucida

Acrosomal process

Acrosomal vesicle

Nucleus

Mitochondria

Flagellum

Sperm

In transit   During fertilization

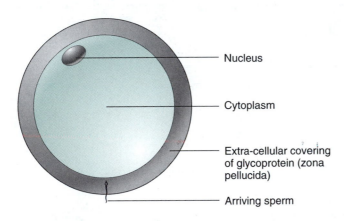

Nucleus

Cytoplasm

Extra-cellular covering of glycoprotein (zona pellucida)

Arriving sperm

Egg

FIGURE 10.1 The human egg is small compared with the eggs of many animals: about 0.1 mm (100 µm) in diameter. Nevertheless, it is muh larger than a sperm, as suggested by the sperm shown penetrating the zona pellucida at the bottom of the egg. Sperm cells are shown above greatly enlarged.

egg's further activation of the sperm's flagellum. Following fusion of their cell membranes, the nucleus of the sperm cell enters the cytoplasm of the egg. In most mammals, including humans, only then does the egg nucleus complete the second meiotic division. At each

of the two meiotic divisions, one of the daughter nuclei (called a *polar body*) is discarded, so at the end of the second meiotic division there is only one haploid set of maternal chromosomes. In mammals these line up with the set provided by the sperm and the first mitotic division of the fertilized egg (*zygote*) begins.

In contrast to the sperm, the egg is enormous (Fig. 10.1). The cytoplasm contains many ribosomes and much mRNA that does not become translated until after fertilization. All of the mitochondria of the zygote are supplied by the egg. The egg not only has chemical signals to attract sperm and communicate with the acrosomal process, but as soon as one sperm nucleus has entered the egg cytoplasm, the egg changes its surface properties and prevents additional sperm from getting through.

SOME BASIC FEATURES OF DEVELOPMENT

CELLS INCREASE IN NUMBER BY MITOSIS

After fertilization, the zygote undergoes a series of mitotic divisions (*cleavage*), partitioning the large amount of egg cytoplasm among a number of daughter cells. In humans, as in other mammals, the first four cell divisions occur over several days, during which time the ball of newly formed cells is slowly swept toward the uterus by cilia lining the walls of the oviduct. After four cell divisions only a couple of the sixteen cells are destined to become part of the new organism; the remainder will form the embryo's contribution to the placenta. After a few more cell divisions the embryo consists of a partially hollow sphere, and in mammals this *blastocyst* is then capable of implanting in the wall of the uterus and developing further (Fig. 10.2). Some mammalian embryos never implant and are lost from the reproductive tract. If implantation is successful, mitotic divisions continue throughout much of development, but at a somewhat lower rate.

CELLS BEGIN TO MOVE WITH RESPECT TO ONE ANOTHER VERY EARLY IN DEVELOPMENT

Following the first several cell divisions and the formation of the blastula there is a coordinated movement of regions of cells, a process called *gastrulation* (Figs. 10.3, 10.4). This rearrangement of cells produces three layers: *ectoderm* destined to become the epidermis and the nervous system; *endoderm*, which will line the digestive tract; and *mesoderm*, which gives rise to blood, bone, connective tissue, muscles, gonads, and kidneys.

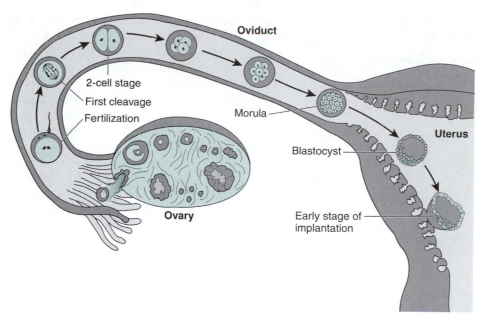

FIGURE 10.2 After fertilization in the upper oviduct, the human zygote starts to divide. The embryo moves down the oviduct, and at about 5 days it reaches the uterus, "hatches" from the zona pellucida, and implants in the uterine wall.

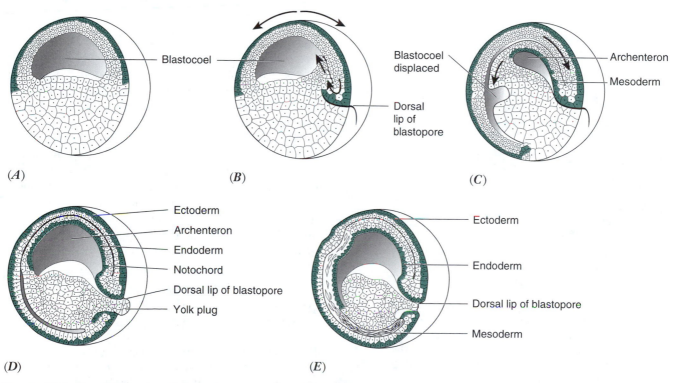

FIGURE 10.3 Gastrulation in a frog embryo. Superficial cells (colored) move in the directions shown by the arrows, ingressing at the blastopore. These movements create three layers, ectoderm on the outside, endoderm lining the internal cavity, and mesoderm sandwiched between, each giving rise to different tissues as the embryo develops. The cell movements are easier to visualize in these spherical frog embryos than they are in mammals.

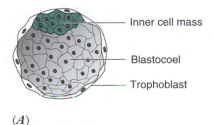

(A)

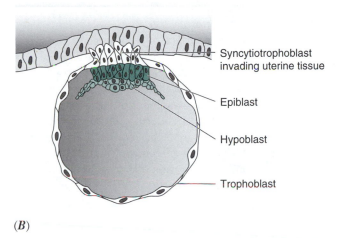

(B)

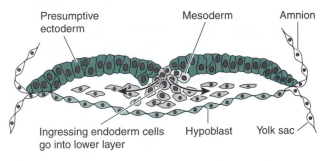

(C)

FIGURE 10.4 (A) Mammalian blastocyst just prior to the start of gastrulation. Only some of the cells of the inner cell mass are destined to become part of the fetus. Monozygotic ("identical") twins occur at this or earlier stages when the inner cell mass splits into two separate clusters of cells.

(B) A later stage during gastrulation. Epiblast and hypoblast have arisen from the inner cell mass. Only some of the cells of the epiblast will contribute to the fetus. Note that some of the cells above the epiblast are invading the uterine lining in the first stages of making the placenta.

(C) A cross section showing the migration of cells during gastrulation. Cells that migrate inward will form mesoderm and endoderm. The yolk sack reflects the reptilian heritage of mammals. Reptiles, like birds, have large eggs with much yolk to nourish the growing embryo.

Important cell movements occur throughout development, sometimes as the movement of sheets of cells as in gastrulation, sometimes as the migration or growth of individual cells. We will consider how movements of cells are guided when we describe the growth of nerve cells in a following section.

DIFFERENTIATION IS THE COMMITMENT OF CELLS TO SPECIALIZED ROLES

As cells move and the embryo begins to take new shape, individual cells start to perform specialized tasks. The process by which lineages of cells become specialized as nerves, muscle cells, liver cells, blood cells, and other tissues is called *differentiation*. An underlying cause of differentiation is the selective activation of specific genes, a process that was described in Chapter 3. Differentiation is ordinarily an irreversible process; once a cell line is committed past a certain point it ordinarily cannot retrace its steps (Fig. 10.5).

There is much variation as to when cell lines become committed, both within and among species. For example, even before fertilization the cytoplasm of the egg of a frog is regionally specialized. Consequently, the cells that result from the first several cleavages are not identical, and those that are destined to form the germ line are identifiable at this very early stage. In mammals, on the other hand, little has been determined by the blastula stage.

FIGURE 10.5 In this metaphor for differentiation, the fate of a cell is like a boxcar rolling into a railroad switching yard. How and where it ends is the result of its passage through a series of decision points. Time extends away from you into the branching tree of track, and once a car has rolled over a switch its fate is determined and it is not free to come back and choose again. The photograph was used by the English embryologist Joseph Needham to make this point in 1936.

A CELL'S NEAR NEIGHBORS CAN DETERMINE ITS FUTURE

Cells generate signals that influence the fate of other cells. Some of these signals are molecules that are secreted into the fluid-filled spaces between cells, move by diffusion, and bind to receptors on neighboring cells. Other signal molecules protrude from the cell membrane, so two cells must come into contact for the signal to be received.

A cell may provide a particular signal only at a certain time during development. Similarly, a cell may have the right receptors for the signal and the ability to respond only at a restricted time. Both providing specific signals and having the competence to respond to those signals are therefore stages on the paths to differentiation.

Cells whose fate has been set are said to be *determined*. Cells become determined before they differentiate into recognizable types such as nerve cells or liver cells. The determination of one group of cells by signals from another adjacent group is called *induction*.

Induction is illustrated by the following events occurring early in development. As gastrulation proceeds, a region of ectoderm thickens to become the *neural plate*, and its edges roll together to form the *neural tube*. These are the cells that will become the central nervous system. Even before they differentiate, however, the ectodermal cells of the prospective neural plate have become determined What makes them different from other ectodermal cells is that they have received a molecular signal from the underlying mesoderm. This is demonstrated by a classical experiment on frog embryos in which a bit of tissue from the surface of an early embryo—tissue located at the dorsal lip of the blastopore and destined to become mesoderm after gastrulation—was removed and transplanted onto the surface of another early embryo at a site that would ordinarily become ventral belly skin (Fig. 10.6). Cells at the dorsal lip of the blastopore are determined, and after transplanting they therefore formed another site of invagination and moved within the embryo to become mesoderm. Moreover, they provided an inductive signal to the overlying ectoderm so that after gastrulation the recipient embryo had two neural tubes, one in the normal place and a second over the transplanted mesoderm (Fig. 10.6).

FIGURE 10.6 **An experiment illustrating neural induction. (A) In early gastrulation, a piece of tissue that will become mesoderm after gastrulation is transplanted from one frog embryo to another and placed in a site that would ordinarily become ventral epidermis. (B) Involution (the movement of cells from the outer surface of the blastula to form an inner layer of cells that takes place during gastrulation) occurs not only at the normal site on the opposite side of the embryo, but under the transplanted tissue as well. The transplanted tissue is thereby carried within the gastrulating embryo and induces a second longitudinal body axis with a notochord and neural tube. (C) Later in development. This spectacular experiment done by a student led to a Nobel prize.**

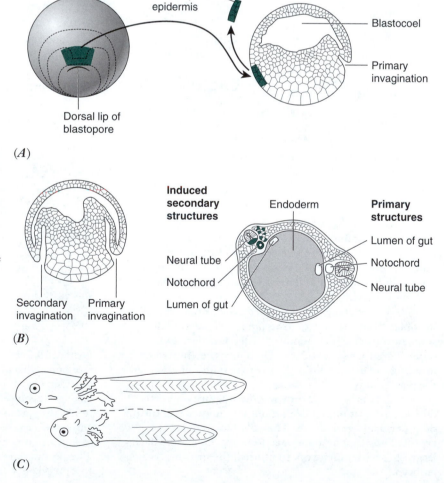

DEVELOPMENT INVOLVES THE INTERPLAY OF GENETIC AND EPIGENETIC PROCESSES

The example of neural induction illustrates an important principle. Although every cell in the embryo has an identical set of genes, cells differentiate along different tracks because of influences that arise outside the cell. The molecular signal from mesoderm that caused ectodermal cells to develop as nerve cells is an example of an *epigenetic* effect (*epi*, on top of, or in addition to). It is epigenetic because it arose outside the genome of the ectoderm. It is true that the molecular signals that caused differentiation were coded in the genome of the embryo, but from the perspective of the cells they affected, they were an external influence.

The cell's decision to activate a gene—to switch tracks (Fig. 10.5)—must occur at the right time in order for development to proceed normally. And as we have just suggested, that decision is largely determined by the cell's environment, by the signals the cell is receiving. But whenever a cell responds to these signals by switching one or more of its genes either on or off, it is likely to send signals of its own to its neighbors. Development therefore consists of a meshwork of interactions between cells in which the results of one set of events sets the stage for succeeding processes.

A cell's changing capacity to produce or receive specific signals is frequently stated another way: There are *critical* or *sensitive periods* during development at which particular epigenetic events must occur for development to proceed normally. We will see later in mammals that similarly there are critical periods for sensory experience occurring after birth.

CELLS NEED TO "KNOW" WHERE THEY ARE IN THE EMBRYO

As the embryo grows, it begins to take shape. The head end becomes different from the tail, limbs form from little buds of protruding tissue, and internal organs such as liver or brain begin to develop from characteristically different tissues of differentiated cells. One of the centrally important conceptual problems is how a cell "knows" where it is in the embryo and what is its appropriate role. Consider, for example, the development of an arm. First, arms are not the same as legs, and they develop at different places on the body. Second, an arm has a proximal end at the shoulder and a distal end with a hand. Moreover, an arm does not have radial symmetry like a rod: the top is different from the bottom and the front is different from the back. Cells in the developing limb therefore have different roles to play, depending on which limb they are in, where in the limb they are located, and to which tissue they are contributing. In order to form a properly shaped arm that bends at the right places, responds to motor nerves when it is commanded, and reports tactile information to the right places in the central nervous system, each cell must grow and develop in ways appropriate to its changing position relative to other cells. The example of a developing limb illustrates a very general problem: how are positions of cells specified, and how is that information utilized by other cells?

We saw earlier how transplanting cells between early embryos revealed the role of mesoderm in neural induction. We now explore several more examples in which experimental manipulations and a growing knowledge of genes are enriching our understanding of development. Each example illustrates one or more of the developmental concepts introduced in this section.

THE EMERGENCE OF FORM: GENES THAT CONTROL OTHER GENES

In Chapter 3 we described how genes are under the control of proteins that interact with DNA, either to activate or repress transcription of genetic information into mRNA. Furthermore, because these control molecules—*transcription factors*—are proteins, they are themselves coded for by genes. One gene can therefore regulate the activities of other genes in cascades that can become quite complex. Such processes hold considerable interest not only for understanding how development occurs but how developmental pathways can evolve.

Even the most cursory survey of the animal kingdom reveals a diversity of body plans; worms, insects, and vertebrates all have characteristic features shared within the group but different from other groups. Insects are small, with three pairs of legs, frequently wings, and a hard, tough covering on the body that doubles as skeletal support. Vertebrates, on the other hand, have two pairs of limbs and an internal skeleton of bone. Although both insects and vertebrates form from single fertilized eggs, the embryos look very different through most of the course of development. Furthermore, the architectural differences between the adults are traceable to the earliest fossils of the two groups and reflect early divisions in the evolution of multicellular life.

HEADS OR TAILS?

Until very recently no one suspected that these very different body plans of vertebrates and invertebrates could share similar genes that control their development. In fact, major differences in body plan seem to raise important issues for understanding macroevolution: did natural selection, working on small heritable differences in phenotype, generate this diversity of body plan and developmental pathway? Recent discoveries support the view that the problem of macroevolution does not have to be framed in quite those terms.

The fruit fly *Drosophila* plays a major role in this story, just as it did in the early part of this century when the formal rules of genetics were being demonstrated. The body of an insect like a fruit fly is segmented along the longitudinal axis (Fig. 10.7). The segmentation is relatively uniform when the animal is a wormlike larva, but it becomes more specialized when the fly metamorphoses into an adult. The head of the

adult develops from three segments; the thorax has three segments (each with a pair of legs); and the abdomen has nine. During the development of a fly, the tissues in each segment differentiate as the fly takes on its adult form.

Some genetic mutants of *Drosophila* (mutants that were actually discovered before 1900) alter this normal developmental outcome in bizarre and fascinating ways. These are called *homeotic* mutants (from homeo, meaning like) because when present they cause one segment of the fly's body to look like another. One such mutant (called *antennapedia*) causes a pair of legs to appear on the head where sensory antennae should be. Another mutation in another gene (called *bithorax*) causes a second pair of wings to grow on the third thoracic segment where there is normally a pair of small

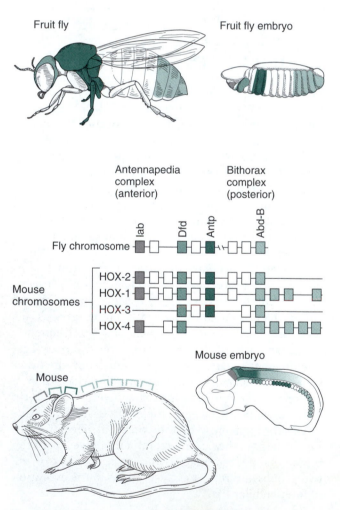

FIGURE 10.7 In insects and mammals very similar genes—and in the same order—control the front-to-rear development of the body. These homeobox genes are located on one chromosome in the fly and on four different chromosomes in mammals.

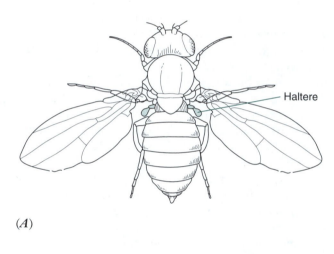

(A)

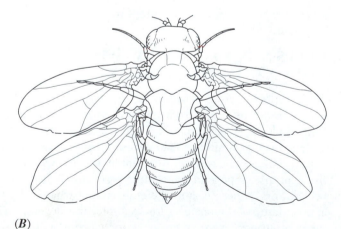

(B)

FIGURE 10.8 (A) Normal (wild type) fruit fly *Drosophila*. Note the single pair of wings, characteristic of flies. Instead of a second pair of wings, the third thoracic segment bears a pair of small, club-shaped balancing organs called halteres.

(B) The mutant *bithorax* has a duplicate copy of the thoracic segment that carries the wings.

balancing organs known as *halteres* (Fig. 10.8). There are eight of these genes in the fruit fly, and although most of their mutations cause lethal developmental defects, it is clear that as a group the genes are important in specifying which segments are to become head, thorax, or abdomen along the body axis of the fly from front to back. To do this, they become active during development in overlapping domains of cells from the anterior to the posterior of the fly.

Not only do these genes function in linear sequence in the tissues of the fly, they are linked in the same sequence on their chromosome. Moreover, all contain regions of DNA that has been conserved in evolutionary time. These similar regions of DNA that are shared by the homeotic genes are called *homeoboxes*; they code for similar segments of protein.

What are the gene products of the homeotic genes and how do they orchestrate this feature of development? The proteins for which these genes code are regulatory molecules that control the activities of still other genes. Furthermore, as each of these homeotic genes is present in all cells, their activation must be controlled along the anterior-posterior axis of the embryo in order for the normal segmentation of the body to develop. For example, the mutation in *antennapedia* that causes legs to appear on the head is due to activation in the head of a gene that is normally active only in the thorax. Although much remains to be discovered about these and similar genes, it is clear that they function as developmental switches. Not only can homeotic genes switch development along a particular pathway in a local region of the embryo, mutations in these genes can produce quite substantial changes in phenotype. Recognition that selection can work on developmental switches at various stages during development is an important conceptual refinement in thinking about the evolution of different body plans.

THE FLY IN THE MIRROR

Homeotic genes in the fruit fly might be little more than an esoteric sideshow in the cellular mechanisms of development if they were confined to insects, but homologous genes with very similar function occur in a diversity of animals including mice and people. In mammals these genes are called *Hox* genes (standing for homeobox). They fall into several groups located on different chromosomes, but like the corresponding genes in *Drosophila*, the linear order on the chromosome is reflected in a linear sequence of activation along the body axis of the developing embryo (Fig. 10.7), and if expression of one of these genes is experimentally induced in mouse cells where it would normally be silent, morphological deformities appear. Conversely, preventing the activation of a Hox gene

shows that these genes are necessary for normal morphological development.

Interestingly, there is more extensive homology between homeobox domains of corresponding mouse and fruit fly genes than within the entire Hox complex of mice. For example, the homeobox component of the fly *antennapedia* protein differs from its counterpart protein in the mouse at only four of sixty-one amino acids. This comparison implies that these gene families existed at a time perhaps 700 million years ago when there was a common ancestor to both vertebrates and invertebrates, and that there has been very strong selection to maintain the primary structure of the proteins for which these genes code (Box 10.1).

THE DEVELOPING NERVOUS SYSTEM

KNOWING ONE'S PLACE: CONNECTING THE EYE TO THE BRAIN

In Chapter 9 you saw how sensory systems project as maps in the central nervous system. Cells from the same small region of the retina project to the same place in the visual cortex, thereby creating a two-dimensional representation of visual space. During development, the axons from the retina must grow to a relay station in the thalamus, and axons of thalamic neurons must in turn grow and find synaptic sites in the visual cortex. How do the axons find the right path to their destination? How do the retinal neurons recognize the correct cells with which to synapse in the thalamus, and how do the thalamic neurons in turn recognize their appropriate cortical targets? Are the relations among the central neurons specified before the axons arrive from the eye and thalamus, or is order imposed by the presynaptic cells? These questions are examples of a larger problem that we introduced in discussing a developing arm: how do cells "know" where they are in the embryo and the role they are destined to play in development?

For technical reasons, some of the seminal experiments on early development of the visual system have been done on amphibia (frogs, salamanders) and fish rather than mammals. The eggs of these animals are fertilized and develop externally; consequently their embryos are much easier to manipulate than are those of mammals. Furthermore, frogs and goldfish can regenerate cut nerves much more effectively than can mice or other mammals.

In frogs and fish the axons from the retina synapse with a part of the brain called the *optic tectum*. Although this structure is not homologous to the visual cortex of mammals, the retina projects onto the optic tectum as a

Box 10.1
Evolution and the Development of Eyes

> To suppose that the eye, with all its inimitable contrivances for adjusting the focus to different distances, for admitting different amounts of light, and for the correction of spherical and chromatic aberration, could have been formed by natural selection, seems, I freely confess, absurd in the highest possible degree.
>
> Charles Darwin
> *On the Origin of Species*, 1859, pg 186

These words have frequently been quoted by creationists to support their beliefs in nonscientific alternatives to evolution. But was Darwin actually saying that natural selection was an inadequate process to produce an eye? Not at all. Darwin was acutely aware of the difficulties that many people would have in conceiving how natural selection could generate structures with the seeming perfection of the human eye, and he followed this passage with a general account of how the eye could have arisen through natural selection. His words continue—and illustrate how a writer's meaning can be twisted when removed from its original context:

> Yet reason tells me, that if numerous gradations from a perfect and complex eye to one very imperfect and simple, each grade being useful to its possessor, can be shown to exist; if further, the eye does vary ever so slightly, and the variations be inherited, which is certainly the case; and if any variation or modification of the organ be ever useful to an animal under changing conditions of life, then the difficulty of believing that a perfect and complex eye could be formed by natural selection, though insuperable by our imagination, can hardly be considered real . . . .

The great diversity of eyes was illustrated in a much simplified form in Figure 5.12 (page 121) when we introduced the concept of adaptive landscapes. Darwin was of course familiar with the presence of various eyes, although much has been learned about their diversity in the intervening years.

Nature has generated two basically different optical designs for complex eyes (Fig. 10.9). Optically, our eyes are similar to a camera. A lens in front projects a two-dimensional image of the world through a cavity onto a concave sheet of light-sensitive cells (the *retina*) in the back of the eye. Like a camera, the walls of the eye are black to absorb stray light, and like the diaphragm on a camera, the pupil can open to increase

sensitivity when there is little light available. Eyes with these optical features are found in all vertebrates. Optically similar eyes are also found in large visually active invertebrates like squid and octopus (Fig. 5.11). Octopus and mammal eyes differ so profoundly in the cellular organization of the retina, however, that their similar optical design is a good example of convergent evolution.

An alternative optical design occurs in crabs, lobsters, insects, and other arthropods. Here a large number of light-sensitive elements are assembled on the convex surface of the body to form a *compound eye*. Each one of these many tiny tubular structures (singular, *ommatidium*, plural *ommatidia)* collects light from a small solid angle of visual space in front of the eye, and the image of the world available to the brain is a mosaic of hundreds or thousands of reports of different intensities from the individual ommatidia.

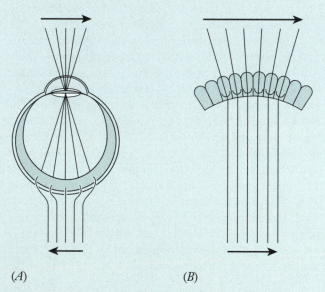

(A)   *(B)*

FIGURE 10.9 (A) Optically, the familiar eyes of vertebrates are elaborations of the simple eyes of flatworms (Fig. 5.12, page 121). They are larger, with many more cells in a light-sensitive layer called the *retina*, and they are much improved by the addition of a lens to focus the light. The axons that go to the brain leave through the back of the eye instead of the front. (B) Insects, crabs, and other arthropods have *compound eyes* composed of hundreds of small, cylindrical elements (called *ommatidia*) consisting of a small lens, eight photoreceptor cells, and a sleeve of opaque pigment cells. Each ommatidium looks at a small solid angle about a degree across, and the central nervous system assembles a mosaic map of visual space from the separate reports of the many ommatidia. Compare with Figure 5.12.

Note that each of these major designs can be improved by small changes. The "camera" eye of a vertebrate will be more sensitive the larger it is, ultimately limited by the size of its owner's head. That is because a larger lens is able to collect more light. Second, the ability to resolve sharp images can be improved by increasing the refractive properties of the lens and the number of light-sensitive cells per unit area of retina. Increasing the number of light-sensitive cells on the retina is equivalent to using a fine-grained film in a camera. The more light-sensitive elements on the retina or film, the sharper the image.

Likewise, the effectiveness of a compound eye can be improved by the addition of individual tubular elements. Increasing their diameters allows them to gather more light and increases their sensitivity, and decreasing the angular separation between their axes improves the resolution of images.

All of these improvements can be made by natural selection operating on small quantitative differences. It has been calculated that with only a 0.005% increase in these characters per generation—which is actually a minute change compared with what is possible in natural selection for quantitative characters—elaborate eyes could evolve in 400,000 generations. This number may seem large, but it is readily accommodated within the time evolution has been taking place. Even with a generation time of twenty years (long even for a large mammal), 400,000 generations occur in a mere 8 million years, a longer span of time than has passed since the divergence of humans and chimpanzees from a common ancestor (Chapter 12).

So although Darwin's insight about what is possible for natural selection to accomplish is supported by current conservative estimates of rates of evolutionary change, recent work with genes has uncovered a whole new aspect to the evolution of eyes. Like the evolutionary divergence of body plans, "camera" and compound eyes have traditionally been deemed independent evolutionary inventions, even though both structures share a number of very ancient molecular constituents.

The key discovery was a gene in the fruit fly *Drosophila* whose mutant allele is known as *eyeless* because its phenotype is a reduced or absent eye. This is a control gene that acts early in development. The mutant allele is not particularly interesting; its existence shows how important the normal allele is for development of an eye. However, when the *normal* allele of the gene is made active in tissues where it is usually inactive (an experimental trick that can be accomplished by molecular geneticists), eyes develop in odd places. Fig-

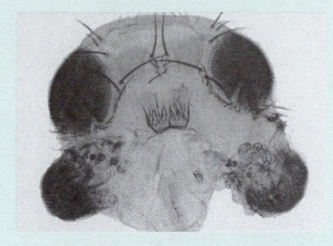

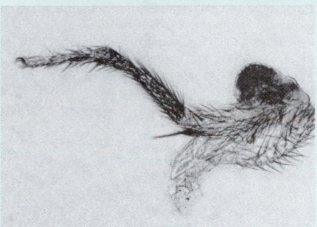

FIGURE 10.10 Extra eyes caused to form on the head or leg of a fruit fly by activating a control gene in tissues where it is ordinarily not expressed.

ure 10.10 shows such eyes that have developed where they normally would not.

As with the homeotic genes of the fruit fly, this gene has a homolog in mice and humans, and through some complex technical manipulations it is possible to insert the mouse gene into a fruit fly and make it active in developing tissues where an eye would not normally be. The result is an extra eye—not, however, a "camera" eye of a mouse but a compound eye of a fly. The specific, developmental mechanisms producing the eye are therefore those of the host fly, but they can be stirred to activity by a control gene from an animal in another phylum! Clearly, therefore, the switching mechanism itself is widely shared across phyla. Findings such as this are providing unexpected and important insights into both development and evolution.

map, just as it does in mammals (Fig. 10.11). Roger Sperry (who received a Nobel Prize for this and related work) removed the eyes of frogs, rotated the eyes 180 degrees, and replaced them. The optic nerve was cut when the eyes were removed, but the frogs not only recovered from this operation, the severed axons of the retinal ganglion cells (which are the fibers of the optic nerve) regenerated, growing back to the optic tectum and forming new synapses. Furthermore, the axons sought and connected to the very tectal cells with which they had synapsed prior to the operation. Consequently, when the frog was presented with a small object to eat, it struck at the food in the wrong direction.

Figure 10.12 shows why the frog's strike was misdirected. An insect that is in front of and above the frog is imaged on the lower-rear quadrant of the retina. The little arrows on the retina are for orientation, and in Figure 10.11 you can see where this region of the right eye maps on the left tectum. After the eye has been rotated, the cells in the lower-rear quadrant of the retina are not in their original position: they belong in the original upper-front quadrant, and when their axons regenerated they made synaptic connections in the tec-

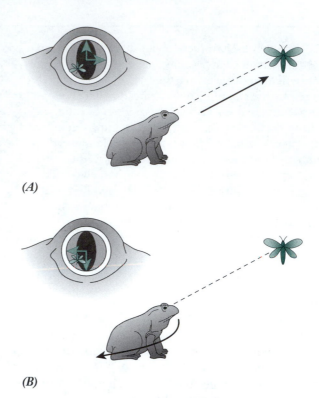

(A)

(B)

FIGURE 10.12 (A) A control frog sees an insect in its visual field (*dashed line*) and strikes (*solid arrow*). The insect is above and to the right of the frog, and its image is cast on the posterior-ventral quadrant of the retina.

(B) After rotating the eye 180 degrees and waiting for the nerve fibers of the optic nerve to regenerate, the frog's strikes at prey are misdirected. That is because neurons in the (now) posterior-ventral quadrant of the retina have located and connected with their original postsynaptic tectal neurons. Visual information is therefore interpreted incorrectly.

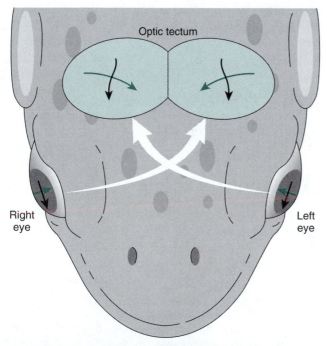

FIGURE 10.11 The retinal ganglion cells of frogs connect in orderly fashion with neurons in the optic tectum of the brain, projecting a map of the retina. Neurons from the left eye synapse with cells in the right tectum and the right eye projects to the left tectum. The pairs of arrows on the retina run from back to front and from ventral to dorsal; similar arrows show the corresponding gradients of cells in the tectum. The tectum can be mapped by recording electrical activity from cells while stimulating different points on the retina.

tum appropriate for their original position. The frog therefore misinterprets the position of the insect and snaps in the opposite direction.

Frogs with rotated eyes never learn to adjust their behavior. There is therefore a map of visual space laid down in the optic tectum that cannot be altered by experience. But did this map exist before the first axons from the retina reached the brain during development? Or was the map imposed by the first set of synaptic connections that was made? The answer to that question was found by Marcus Jacobson, who rotated eyes early in larval development and tested the animals later to see whether the tectal projection developed normally, as in Figure 10.11. He found that the tectal map becomes specified during a *critical period* prior to the arrival of the first retinal axons. During about a day the anterior-posterior axis of the map becomes specified, then the dorso-ventral axis (arrows in Fig. 10.11).

Axons from the retina arriving in the brain must thus find specific cells with which to connect. This is true both during larval development and in adult frogs that have had their optic nerves cut surgically. The tectal neurons are therefore labeled in some fashion that allows them to be recognized by the growing tips of the advancing axons. In mammals the number of genes ($\approx 10^5$) is far smaller than the number of neurons in the brain ($\approx 10^{12}$), and as a general rule it is doubtful that in any animal all individual nerve cells are uniquely tagged with their own molecular label. It seems likely that there are more general mechanisms for specifying the relationships between cells in developing embryos.

One model for how an axon could find its proper place on the surface of the frog's optic tectum could be a *concentration gradient* of a specific marker protein present on the surfaces of the tectal cells. A change in concentration of the marker along the long axis of the tectum and a second gradient of another protein at right angles could provide the necessary information for a retinal axon to locate its target neuron. Evidence for such gradients exists, because it is possible to make antibodies (Chapter 8) to cell surface proteins, bind the antibodies to a fluorescent dye, expose developing neural tissue to a solution of these fluorescent antibodies, and observe a gradient of dye by examining the tissue under a microscope.

During development there is a dynamic relationship between the optic nerve fibers and the tectal neurons with which they synapse. That is because the retina grows by adding cells all around its edges, whereas the tectum grows along the medial and posterior borders. The general shape of the map is unchanging, but cells must exchange connections as retina and tectum enlarge in this asymmetric fashion. We shall return to the question of how cells consolidate and validate their synaptic connections in following sections.

We introduced the idea that cells must signal their position relative to other cells when we pointed out the consequences of structural asymmetry in developing limbs. The problem of specifying position thus has many manifestations of which the optic tectum provides a model example. Many kinds of experiments have led to the concept of *morphogenetic fields* in which gradients, sometimes of diffusible molecules, sometimes of cell surface markers, are important in specifying the approximate positions and the future roles of cells.

HOW GROWING AXONS FIND THEIR WAY

When axons from the frog's retina connect with their proper tectal cells they are making the last of a number of "decisions." The cell bodies of many neurons are located far from the ends of their axons, and when the axons grow during the development of the nervous system they must find their way through a three-dimensional maze of other cells. There may be scores or hundreds of possible wrong turns along the way. How do axons navigate without becoming lost?

Much of the information that we have on this problem comes from studies of invertebrates, particularly insects and a small nematode worm. Once again, technical considerations influence the choice of experimental material. A feature of these invertebrates is that they have relatively small nervous systems. Moreover, it is frequently possible to recognize specific nerve cells in individual animals and to perform experiments to study the cell lineage and behavior of particular neurons. This approach is made even more powerful when it is also possible to use genetic mutations that affect the growth of neurons or their capacity to form synapses.

At the tip of every nerve fiber is a fan-shaped expansion called the *growth cone* (Fig. 10.13) from which extend slender spikes of cytoplasm. The spikes contain the protein *actin*, which is also found in muscle as well as elsewhere in the cytoskeleton (Chapter 3). The growth cone probes constantly, contracting and expanding, and there is directional movement of the cell when spikes adhere to the external substrate and a force pulls the rest of the cone forward. Growing axons thus contain receptors that make adhesive contact with the *extracellular matrix*, which consists in part of proteins with attached sugar residues. These molecules (*laminin* and *fibronectin* to name two) are important in holding cells together, but they also provide a substrate to

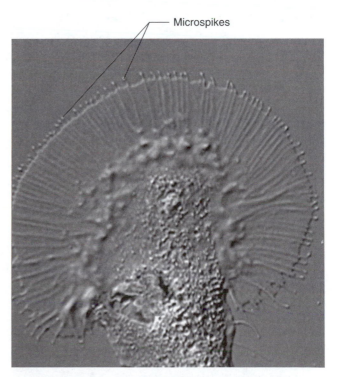

Microspikes

FIGURE 10.13 The growth cone at the tip of a growing axon is very motile and contains fibrous proteins that are constantly changing length.

which growing axons can stick. Gradients of adhesion as well as gradients of diffusible molecules like the protein *nerve growth factor* therefore provide some important general cues to guide and even to stimulate growth of nerve cells.

The path of growth, however, can be very specific. Some insect axons follow the course of earlier axons (called *pioneer fibers*) that grow before the embryo becomes so crowded with cells. But later cells have to recognize the pioneers. Furthermore, axons frequently follow prescribed paths among other fibers, ignoring many possible places to turn or branch before coming to a decision point where they change direction, fork, or make a synapse with another cell. These patterns are highly regular; in insects, where specific nerve cells can be readily recognized, the paths taken by individual axons are identical from embryo to embryo. Such observations mean that there are forms of labeling that contain more information than gradients of adhesivity or of diffusible molecules in the extracellular space.

Specificity of labeling probably involves more than spatial patterns of labeling molecules distributed among the cells of the embryo. Growing cells change their sensitivity to particular labels with time, and specificity can therefore be enhanced when a cell is receptive to a sequence of signals. Understanding the details of how the movements of cells are guided during development is a very active field of experimental investigation.

THE ROLE OF CELL DEATH

An interesting and important feature of development is that more cells are produced than are used in the finished organism. A familiar and dramatic example is the tadpole's tail, which simply disappears as the tadpole changes into an adult frog. A more subtle example is the loss of webbing between the toes of early vertebrate embryos. The webbing persists in ducks and seals, but is lost in chickens and humans, species that do not have webbed feet as adults.

Many more neurons form than are finally incorporated into the adult nervous system. For example, about twice as many motor neurons form in the spinal cord than are ultimately used to stimulate muscle. Survival of a motor neuron depends on making synaptic contact with a limb muscle; those cells that make synapses receive a signal from the muscle that is necessary for the neuron's continued life. If growing motor nerves are prevented from reaching muscles, they die. If extra limb buds are grafted onto an embryo, providing more than the usual number of muscle cells, more motor nerves make synapses and survive.

The formation of synaptic contacts does not necessarily insure the survival of a neuron. Both on muscle cells and in the central nervous system there is a competition between presynaptic nerve cells for final sites on postsynaptic cells. Activity in the presynaptic fiber is an important factor: those neurons that display the least activity typically lose their influence on the postsynaptic cell and retract their synapses. In limb muscle the losing neuron then dies. In the central nervous system the result may be more complex, as we will describe in the following section.

That cell death is part of the program of development is illustrated very clearly by the precision of loss of cells in the nematode worm *Caenorhabditis elegans*. This organism is studied intensively because it is only a millimeter in length and it is possible to know the origin and fate of every cell in its body. During its maturation it has 1090 cells, exactly 131 of which die. Furthermore, there are mutants in which nerve cells that would ordinarily die persist.

Programmed cell death is not limited to development. Although there are few if any new nerve cells born during the adult life of a mammal, other tissues replenish themselves on schedule. Skin cells and the epithelial lining of the stomach and intestines, blood cells, and cells of the uterine wall all die on a regular basis and are replaced by new cells of the same type. These processes are typically distinct from cell death that is caused by injury: when an injured cell loses its capacity to regulate its internal ionic environment, it swells and ruptures. When cell death is programmed, however, the cells look different. The cells shrink, the surface forms transient blebs of cytoplasm, the DNA and protein in the nucleus condense into shapeless bodies, and the nucleus then disintegrates. In some cases the dead cells persist and perform a final function. For example, the outer layers of the skin consist of dead cells filled with a protein (*keratin*). They finally wear off due to gentle abrasion of the surface of the skin accompanied by the formation of new cells below. Some forms of cancers are now attributed to a loss of control over programmed cell death.

THE NEED FOR SENSORY EXPERIENCE IN TUNING SYNAPTIC CONNECTIONS

As a nerve cell differentiates it assumes a characteristic shape, grows to a specified location, and forms appropriate synaptic connections with other cells. At a molecular level it expresses various ion channels in different parts of its cell surface as well as the enzymes that it needs to synthesize and recycle the neurotransmitters that it will use in communicating with other cells. But the relationships that it develops with other cells in the synaptic architecture of the nervous system are also influenced by use.

The importance of neural activity in altering synaptic structure was described in two earlier contexts. In Chapter 9 we discussed how synaptic structure and the efficacy of synaptic transmission change during learning. And a few paragraphs earlier in this chapter we described how the competition between motor neurons is important in determining which of the developing nerve cells will make permanent synapses and which will die. We turn now to an example in the developing visual system of mammals, where the competition between neurons for synaptic control of cortical cells is adjusted by sensory experience during a *critical period* after birth.

These important observations were made about 35 years ago by David Hubel and Torsten Weisel (another Nobel Prize here), who were then studying the receptive fields of neurons in the visual cortex of cats and monkeys (Chapter 9). In order to test the effects of visual deprivation, they sewed shut one eyelid of a week-old kitten, just prior to the time the eyes would open naturally. This was not done out of idle curiosity. At the time there were medical reports of children who had been born with cataracts but who had not had their opaque lenses removed until about age 8. When a cataract is removed from the eye of an elderly person, their vision improves. These children, however, never had useful vision, even after the operation. There was clearly some defect in development of their visual system that occurred during the first few years of life when they had been unable to see clearly. Hubel and Wiesel's experiments on kittens were motivated by this history and a consequent wish to learn more about the development of the eye. Their findings led to an important change in medical practice.

The kitten's eye was opened about ten weeks later, and the responses of cells of the visual cortex were studied. In a cat that has had normal visual experience from birth, most cortical cells are stimulated to some extent by both eyes, and many cells are about equally influenced by cells from either eye. But when a kitten's eye is prevented from seeing normally during this early ten-week period, the eye loses the ability to influence cortical neurons.

Cats treated in this manner are permanently blind in the deprived eye. If the "good" eye of such an adult cat is temporarily covered with an opaque contact lens, the cat blunders about, walks into objects, falls over edges. The defect is not due to the absence of light during the first several weeks after birth; if instead of sewing the lids closed the eye is covered with a piece of translucent plastic like a piece of ping-pong ball, the kitten still becomes permanently blind in that eye. The eye must therefore see images in order for its vision to develop normally. A corresponding loss of vision cannot be produced in adult cats.

Kittens are susceptible to visual deprivation starting at about four weeks, peaking a few weeks later, and tapering off until about four months of age. Similar effects are obtained with young monkeys, but their *critical period* starts at birth, peaks at two weeks, and wanes slowly

thereafter. As with cats, when one eye is deprived during the sensitive period, the loss of vision is irreversible.

Does the effect of deprivation simply mean that the visual system must develop its appropriate synaptic connections through experience gained during the first few weeks of life? This interpretation would be consistent with the traditional view of the newborn brain as a *tabula rasa*, a blank slate, an unprogrammed memory chip ready to receive information. Study of newborn monkeys, however, shows that this interpretation cannot be right. First, a newborn monkey is visually alert and curious about its surroundings. This is not the case for kittens, whose eyes do not even open for ten days, nor is it true of human infants, whose visual system is not fully developed at birth. Second, cells in the visual cortex of newborn monkeys respond to oriented edges, just as the cells of adults (Chapter 9). Clearly the visual system is ready for use *before* the monkey leaves its mother's womb.

Why, then, does early visual deprivation of one eye cause the animal to become blind in that eye? Muscles weaken if they are not used. Is this effect on the visual system an example of deterioration brought about by disuse? But if that is the explanation, why is the effect seen only during a few weeks after birth? This reasoning suggests an alternative: a developmental process is being interrupted.

Important clues were discovered by a pair of additional experiments. If *both* eyes are covered during the critical period, the animal does not become completely blind as one might expect if failure to use the eye were the explanation for the deficit. The cortex of such animals is not normal, but there are cells that are driven by one eye or the other.

In the second experiment, some of the muscles that move the eye in the socket were cut. This operation makes it impossible for the animal to bring both eyes to bear on an object simultaneously. (Once again, there was a medical reason for doing the experiment. Some people have impaired vision because they are unable to focus both eyes together; in common language they are said to have a "squint" or are "cross-eyed." Some of these individuals lose function in one eye, so there was ample reason to try to understand the defect better.) When the ocular muscles of a kitten or monkey were cut during the critical period for sensory deprivation, both eyes continued to drive cortical cells, but there were very few cells that had input from *both* eyes.

These two experimental observations made it clear that some final feature of synaptic wiring of the cortex requires interaction between the two eyes after birth. Not only does each eye have to see images, they have to see the same images at the same time. Only then do axons from the two eyes simultaneously report equivalent information. When these streams of information converge in the cortex there thus appears to be a competition for synaptic space. If axons reporting from the right and left eyes provide similar input—fire action potentials at about the same rate at the same time—their input is balanced. But if one eye is covered, it reports little and loses in the competition for postsynaptic space. If both eyes are seeing images, but the images on equivalent parts of the retina are not in register, there is equal competition, but the pathways do not manage to converge on the same postsynaptic neurons in the visual cortex. Cortical cells of adults will be driven by one eye or the other, but very few will receive input from both eyes.

The changes in the arriving axon terminals that occur during these events are diagrammed in Figure 10.14. Recall from Chapter 9 that the cortex consists of several layers of cell bodies and neuropil. All the afferent axons enter layer 4. At this point in the afferent path, information from the two eyes is still separate. In

the *adult* monkey the axon terminals branch into discrete patches about 0.5 mm wide, and patches from the right and left eyes alternate. In this layer of the cortex the adult ocular dominance columns are quite distinct. The postsynaptic cortical cells (not shown in this diagram) send branches into other layers of the cortex and connect with still other cells. Neurons with input from both eyes are found in these other layers.

At birth the arriving axons do not terminate in layer 4 in discrete patches as they do in the adult. As shown in the upper left of Figure 10.14, the terminal branches of axons carrying information from the two eyes overlap extensively. During the critical period of postnatal development, when covering one eye has such a pronounced and permanent effect, the terminal branches of these axons retract into the separate areas that characterize the adult state.

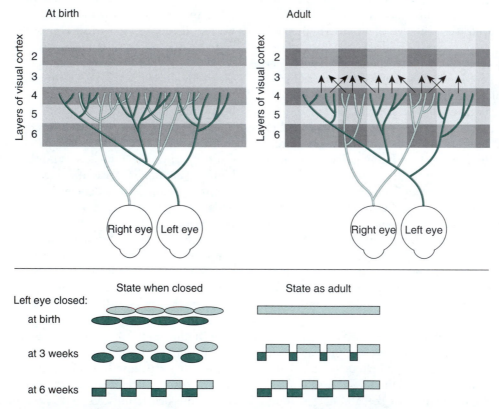

FIGURE 10.14 **The upper diagrams represent axons carrying visual information as they arrive at layer 4 of the visual cortex. In the *adult*, axons synapse in discrete patches, right and left eye alternating. The short arrows running upwards into layer 3 represent the flow of postsynaptic information carried by cortical cells. Outside of layer 4 lateral branches of these postsynaptic units make cells sensitive to input from both eyes.**

The diagram at the left shows that at birth the arriving axon terminals are not neatly sorted into separate patches. In monkeys the retraction of branches in layer 4 occurs during the first several weeks after birth.

The lower figure shows how the fields of these axon terminals decrease their overlap during the first several weeks of postnatal development. It also shows the effect of covering one eye at different times after birth. A more detailed description of these diagrams is found in the text.

The lower diagram in Figure 10.14 depicts changes in the extent of branching of the axon terminals and how it is altered by covering one eye. If the left eye of a monkey is covered at birth, the axons bringing information from the right eye expand their number of terminals and take over virtually all of the synaptic opportunities. If visual deprivation of the left eye does not start until three weeks after birth, the effects are less severe. The competition between axons is now not completely one-sided, and narrow ocular dominance columns driven by the left eye are present. At six weeks, covering the left eye has very little effect, and nearly balanced ocular dominance columns form. The critical period is largely over.

The development of the visual cortex of mammals provides a clear example of how sensory input is necessary in order to adjust the final array of synaptic connections in a part of the brain. The sensory experience is an epigenetic influence, it occurs at a critical period of development, and it illustrates how signals from outside the organism can help to shape the final stages of assembly of the nervous system. It provides a model, a paradigm, to have in mind when we return to the development of behavior in the next chapter.

HOW SEX IS DETERMINED IN MAMMALS

THE INTERPLAY OF GENETIC AND EPIGENETIC PROCESSES IN SPECIFYING SEX

The early embryonic gonads of mammals are undetermined: they can become either ovary or testes. The choice of developmental path that is taken depends on a gene. In mammals, one of the pairs of chromosomes is different from all the others in that one member of the pair is much smaller than the other. These are the sex chromosomes; the larger is called the X chromosome and the smaller is the Y chromosome. Females have two X chromosomes; males have an X and a Y. Sperm, which are haploid, have either an X or a Y chromosome, and whether the zygote develops into a male or a female therefore depends on which of these two chromosomes the sperm is carrying (Fig. 10.15). This is ironic considering the number of women who have been blamed through history for an "inability" to bear sons. Nevertheless, there is evidence that in some mammals females have significant control over the proportions of male and female embryos that survive (Chapter 6).

What is it about the presence of a Y chromosome that causes the fetus to develop as a male rather than a female? At about forty days of gestation (in humans), a gene on the Y chromosome becomes active. It is a control gene, and its product is a transcription factor that throws a developmental switch in the tissues of the as-yet undetermined gonads. As a result, the gonads start to differentiate into testes. In the absence of this gene on the Y chromosome, the gonads become ovaries.

In genetic females, embryonic tissues called the *Müllerian ducts* subsequently differentiate into oviducts and uterus. But when testes start to form instead of ovaries, they secrete two molecular signals into the surrounding tissues. The first is a protein (another switch signal) that causes the Müllerian ducts to degenerate. The second is the male steroid hormone *testosterone*, which reacts with receptors in the cells of another nearby tissue called the *Wolffian ducts*. The effect of this signal is to activate still other genes and start the Wolf-

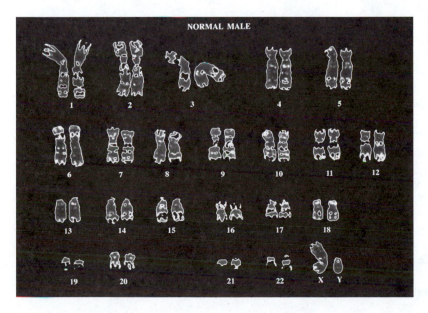

FIGURE 10.15 The 23 pairs of human chromosomes. The X and Y chromosomes are at the lower right.

BOX 10.2
How Is Sex Determined in Other Animals?

Among the vertebrates, birds and mammals have sex chromosomes, so the sex of offspring has a direct genetic basis. (Birds differ from mammals, however, in that the Y chromosome is inherited from the mother.) Fish and reptiles, however, lack sex chromosomes, and the activation of testosterone synthesis and the genes for testosterone receptors, both of which are necessary for the development of maleness, are triggered by environmental cues. For example, in many reptiles, the temperature at which the eggs are incubated determines the sex of the hatchling. Reptiles are not known for parental care, and in some species the eggs are simply buried in sand or deposited in nests of decaying vegetation. In many turtles males then develop if the temperature is relatively low, whereas in many lizards and alligators males result from incubation at the high end of the temperature range. In still other species, including crocodiles, males develop in a mid-range of temperatures and females at the extremes. Although temperature varies continuously, it does not produce a gradation in anatomical sexual features in these embryos. Its effect is that of a switch, directing the developmental outcome to either males or females. The temperature experienced during development can, however, modulate adult behaviors such as aggression that is characteristically associated with males.

An even more fascinating example of sexual outcome is provided by the *social* environment. Certain species of fish that live on coral reefs can change their sex, even as adults. Their gonads maintain the capacity to form either eggs or sperm. An individual fish may experience greater reproductive success as a female while small, but beyond a certain size it may do better as a male, able to defend a territory and fertilize the eggs of a number of females. In these species, social stimuli activating the brain appear to modulate the internal hormonal environment of the animal, and when a female transforms to a male there are dramatic changes in both behavior and coloration of the body (Fig. 10.16).

As birds and mammals arose from reptilian lineages, the appearance of sex chromosomes in these groups is an evolutionary innovation. Both birds and mammals maintain a constant body temperature, and in mammals, where gestation takes place within the body of the female, it would not be possible to regulate the sex of offspring by means of temperature.

The appearance of sex chromosomes in birds and mammals is not, however, a unique evolutionary result, for sex chromosomes are found in insects such as the fruit fly *Drosophila*. The Y chromosome is not involved in sex determination, but it is necessary for males to make functional sperm. Sex is determined by the number of X chromosomes relative to the autosomes; thus a fly with one X chromosome is male and with two is female. Furthermore, circulating hormones are not involved in sex determination of insects, and if an X chromosome is lost from one of the two daughter cells that results from the first mitotic division of an XX fertilized egg, the resulting adult fly is female on one side and male on the other!

fian ducts differentiating into the accessory sexual characters of males: the tissues associated with the delivery of sperm from the testes (epididymis, vas deferens, and seminal vesicle), and following conversion of testosterone to a chemically related hormone (dihydrotestosterone), the penis and scrotum.

The activation of genes in the cells of the Wolffian duct by an external signal, testosterone, originating elsewhere in the embryo is another example of an *epigenetic* effect. Although the environment is the internal chemical environment of the embryo, from the standpoint of the cells of the Wolffian ducts this is the milieu in which they exist.

Whether Wolffian or Müllerian ducts develop is thus dependent on earlier events in another tissue, the gonad. This is another example how differentiation depends on sequences of epigenetic effects that are in turn set up by preceding events elsewhere in the embryo. Furthermore, the delivery of testosterone to the Wolffian ducts occurs at a *critical period* in order for development to proceed normally.

What happens when the normal sequence of events is altered? In order for testosterone to be detected by the cells of the Wolffian duct, the cells must have testosterone receptors. These receptors are a protein, and there is a mutation in the gene coding for this protein that makes the receptors nonfunctional. In genetic males (i.e., individuals with a Y chromosome) who have a faulty gene for this testosterone receptor, the phenotypic consequences are profound. The testes develop, but their testosterone has no effect on the Wolffian ducts, so the accessory sexual structures of males fail to develop. These individuals have the appearance and even the libidinal interests of

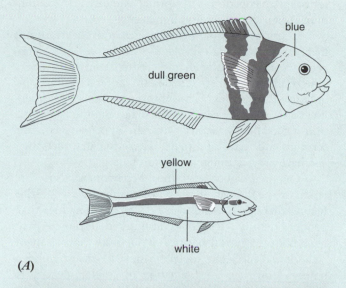

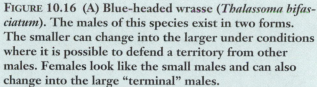

FIGURE 10.16 (A) Blue-headed wrasse (*Thalassoma bifasciatum*). The males of this species exist in two forms. The smaller can change into the larger under conditions where it is possible to defend a territory from other males. Females look like the small males and can also change into the large "terminal" males.

(B) The female stoplight parrotfish (*Sparisoma viride*, right) is black and white on the dorsal side and red below. The male (left) is blue with a yellow spot at the base of the tail, and is about 50% larger than the female on the right. Females turn into males when they become large. When the fish are small, they are able to maximize their reproductive success by being females. As large males, however, they can defend a territory, exclude small males, and fertilize the eggs of multiple females. Thus, above a certain size, lifetime reproductive success can be maximized by changing from female to male.

(A)

(B)

reproductively capable women, but their gonads do not become ovaries. Moreover, because the testes produce the protein that inhibits the differentiation of the Müllerian ducts, there are no oviducts or uterus, and the vagina ends blindly. Rudimentary testes are present in adults, but they do not descend.

Many other alterations of the developmental pathway are known. There is a mutation in a gene affecting steroid metabolism in the adrenal glands that leads in genetic females to a surge of hormone with testosterone activity early in development. This hormone can be sufficient to activate the development of the Wolffian ducts, and the individuals with this mutation can be mistaken for boys at birth. As another example, genetic males who are unable to convert testosterone to dihydrotestosterone have functional testes, but the penis and scrotum fail to develop.

These individuals have a vagina, closed at the upper end, and an enlarged clitoris. They are frequently raised as girls. Box 10.2 provides a glimpse of the many ways sex is determined in other animals.

ARE THE BRAINS OF MALES AND FEMALES DIFFERENT?

The influence of testosterone at a critical period early in development is not confined to the formation of accessory sexual structures associated with the reproductive tract: hormones also alter the course of development of parts of the brain. Brains, therefore, are sexually dimorphic. For example, during the reproductive cycle of female mammals, there is a periodic increase in estrogen formed in the ovary in association with the maturing egg. This estrogen stimulates neurons in the brain—specifically, in the preoptic nucleus of the hypothalamus—and this in turn activates synapses and leads to neurosecretion of other hormones from the pituitary. Only in females, however, are the neurons in the preoptic nucleus sensitive to these spurts of estrogen.

A female does not develop this sensitivity if as an embryo it is exposed to testosterone at an early critical period in development. Rodents are a convenient experimental animal for examining this process because the critical period for the action of testosterone on the development of the brain extends for several days after birth. Female rats injected with a single dose of testosterone at four days of age fail to secrete luteinizing hormone (one of the pituitary hormones involved in reproduction) cyclically when they reach sexual maturity. Conversely, male rats that are castrated a day after birth (thus deprived of their source of testosterone) develop a female-like sensitivity to estrogen in the hypothalamus. If tested as adults by implanting ovaries, these males secrete luteinizing hormone cyclically.

In rats, the critical period for the action of testosterone ends several days after birth. If males are castrated when they are a week or ten days old, the critical period has passed, and the adult hypothalamus does not show the sensitivity to estrogen that is characteristic of females.

The different roles of males and females in reproduction are reflected in their behavior, and behavioral differences are necessarily part of the mating act. When male rodents mount, the females adopt a receptive posture (called *lordosis*) in which the rump is raised and the rest of the back is lowered. Male rats castrated immediately after birth display lordosis when other males attempt to mount them. There is little tendency to lordosis, however, if castration occurs ten days after birth, following the critical period for the effect of testosterone on the developing brain.

Sexual dimorphism of behavior is not limited to the act of mating. As we described in Chapter 6 when discussing sexual selection, animals frequently have species-typical courtship rituals, sometimes involving elaborate visual displays, vocalizations, and male-male competition, the latter frequently of an aggressive nature. In mammals, these phenomena are clearly related to the early effects of testosterone and are revealed by observations of development. In rodents, where litters of several young are usual, a female embryo that implants between two brothers can be influenced by their testosterone. As an adult it will be more aggressive than other females and less attractive to males. The *freemartin* results from a related phenomenon. It is a sterile female calf that has been masculinized in the uterus by the testosterone of a male twin. Natural variation in the amount of testosterone seen by developing

fetuses may contribute to normal phenotypic variation in many species, including humans.

A curious example of the effect of exposure of female embryos to high levels of testosterone in the placenta occurs normally in the spotted hyena of Africa. The result is that adult females are larger and more aggressive than males. The clitoris is large and when erect looks like the penis of the male, and the labia fuse into a fat-filled structure with the appearance of the scrotum and testes of a male. These females nevertheless have ovaries and give birth to young.

The presence of sex-typical behaviors is clearly related to the human concept of gender identity—the feeling that one is either male or female. The idea that gender identity is solely culturally determined may have political appeal because it suggests that we need only devise the right social formula to ensure gender equity. However, the formation of gender identity is more complex than this. Whereas the emerging gender identity of young children is typically reinforced by the behavior of adults, there is no reason to believe that the development of the primate brain differs from that of other mammals in its sensitivity to early hormonal influences. In fact, when female rhesus monkeys are exposed to testosterone at an early critical period their juvenile behavior is masculinized: there is more rough play, including aggressive encounters with young males, and less "mothering" behavior of the sort seen in young females.

Several small clusters of cells in the hypothalamus of mammals are sexually dimorphic. One of the better studied examples is larger in males than females, but the function of none of these nuclei is understood. One is reported to be larger in homosexual men, another smaller, but the significance of these observations is not known. It is not unreasonable to hypothesize, however, that one or more of these structures is related to gender identity or sexual orientation and influenced by early developmental events in the uterus.

There is some evidence that male and female brains may differ in other ways. In primates, the lateralization of the brain discussed in Chapter 9 appears to start earlier, to be more extensive in males than in females, and to be influenced in the expected fashion by early exposure to testosterone. To put it another way, association between the two halves of the female brain remains somewhat more flexible. There may also be more fibers connecting the two halves of the brain in women than in men. Some tests find that women are statistically more adept verbally and better at judging the emotional state of others from facial expressions, whereas statistically men are somewhat better at visualizing in three dimensions and in certain kinds of mathematical reasoning. These cognitive differences are modest, there is enormous overlap between the sexes, and all of these higher-order abilities can be changed by experience. At most, these cognitive tests suggest small biases in the development of the nervous system. They provide no rationale for discriminatory educational or employment policies based on sex.

THE EFFECTS OF HORMONES DURING ADOLESCENCE

The effects of early exposure to hormones that we have just described produce permanent developmental effects in the reproductive tract and in the nervous system, although some of the latter can likely be tuned by subsequent behavioral experiences. In mammals there is a second surge of hormones that takes place at sexual maturity—*puberty*. In humans, females start to deposit more body fat than males, whereas males show a proportional increase in muscle mass and changes in the larynx that cause the voice to deepen. Estrogen from the ovaries triggers the enlargement of the female breasts, and the functional integration of ovaries and uterus with the hypothalamus and pituitary leads to the first menstrual period.

In the brain, however, late hormonal influences are believed to activate circuits that were formed earlier. This is in contrast to the hormone-induced changes in the brain that took place early in development and that were instrumental in organizing synaptic connections. In an adult male, testosterone can elicit aggression, but it does so by stimulating preexisting circuits, not by laying down new ones.

SYNOPSIS

Each new generation of organisms is an array of phenotypes different from each other and different from all that have gone before. Each round of sexual reproduction offers new variants on which natural selection can act. But within each species, the variation has relatively narrow bounds. Sexually reproducing organisms are constrained to make copies that are sufficiently similar that individuals can find and attract mates and, in many species, live with others in a wider social structure.

Development is the process by which phenotypes are generated from single fertilized eggs. Genes do not provide a blueprint, for a blueprint is a detailed picture that shows every part in its precise relation to every other. Blueprints are instructions for how to assemble the whole from a set of preexisting parts, but development does not begin with all the parts. It starts with just one cell.

Richard Dawkins has suggested that "recipe" is a better metaphor for the role of genes in development. This comes closer, for a recipe simply specifies the ingredients, their proportions, the order in which they are to be mixed, and a few rules for cooking. If recipe implies the presence of a cook, an outside agent following directions, this analogy is also incomplete, for it does not convey the idea that development is a complex process of *self-assembly*.

During development many additional cells are made (mitosis), and through the selective activation of subsets of genes they synthesize different proteins, take on different shapes, and perform increasingly specialized functions. For many cells, this process of *differentiation* is an irreversible commitment.

The instructions for activating genes, directing the growth of cells, and regulating cell death can come from other genes. Genes can control other genes, singly or in groups, when the gene product is a transcription factor. Many signals, however, come from other cells and activate transcription factors within the target cells. Other signals provide cues to guide growth of cells such as neurons. Signaling between cells is therefore an essential feature of development, and it starts even before sperm meets egg. Sometimes the signals are borne on the cell membrane, sometimes they are released to diffuse in extracellular space or to be carried to their target cells by the blood. Other signals are laid down in the extracellular matrix. Receiving a signal requires that the target cell have a receptor molecule, sometimes in the cell membrane, sometimes within the cell.

Specific signals are sent and target cells are receptive at *critical periods* of development. Development consists of a series of local events, each with consequences for further changes both in the immediate vicinity and in other parts of the embryo. Where and for how long a gene will be active is not information embedded in the DNA; it is information that arises as a result of the history of changes taking place in the developing embryo. Development is thus an interplay of *genetic* and *epigenetic* events.

Epigenetic influences diversify as the embryo develops. Gradients of cell surface molecules and the concept of *morphogenetic fields* illustrate how cells locate themselves in relation to other cells as the embryo takes on increasingly complex form. Epigenetic influences eventually include events arising after birth and outside the organism as the neuronal architecture of the brain is consolidated and tuned. Reliance on the correct epigenetic cues at the appropriate times is an integral feature of development and can be compromised by mutations that block molecular signals or by extreme environmental perturbations. Development thus consists of an inseparable play and counterplay of genetic and environmental influences.

QUESTIONS FOR THOUGHT AND DISCUSSION

1. How is the control of gene expression described in Chapter 3 important in development?
2. How does increasing knowledge about the process of development contribute to further understanding of emergent properties of the brain?
3. A geneticist and professor of medical science has written that "Western culture is committed to the idea that there are only two sexes...," pointing out that perhaps as many as 4% of births are intersexuals, with varying expression of testes, ovaries, and associated sexual characters. Furthermore, both medical treatment and legal tradition tend to force such individuals into either male or female anatomy as well as social roles. In some states adults are free to choose the sex with which they wish to identify; in others they may be bound to the identity defined by their chromosomes.

   How does a broad knowledge of biology bear on the view that it is a cultural contrivance to think that mammals come in only two sexes?

   What are the possible meanings of "natural" when considering those individual mammals that are not clearly either male or female? Consider this question from the perspective of the individual phenotype and from the perspective of a population of genotypes. (Put another way: What is the natural course of development? Are mutations natural?)

   Finally, what important ethical and political question(s) are raised by this issue? (Consider the early age at which medical decisions about sex are generally made, the difficulties of rearing children,

   single-sex athletic competition, and any other matters you think important.) What approach(es) to policy would you suggest on the basis of these considerations?

4. Recent advances in molecular genetics will provide new reproductive possibilities. For example, if a husband is sterile because he does not produce sperm, his wife might carry to term an embryo created by replacing an egg nucleus with a diploid nucleus from one of his somatic cells. Such an offspring would be a genetic clone of the husband. How much like the husband would you expect this individual to be? What sorts of influences could make him different? What does knowledge of monozygotic (identical) twins add to this discussion? How might the couple have a daughter by this technique?

SUGGESTIONS FOR FURTHER READING

Gilbert, S. F. (1997). *Developmental Biology*, 5th ed. Sunderland, MA: Sinauer Associates. A detailed and authoritative book on the cellular and molecular processes of development.

Hubel, D. (1988). *Eye, Brain, and Vision, A Scientific American Book*. New York, NY: distributed by W. H. Freeman Co. Written for a general readership, this book includes a description of the role of sensory input in the development of the visual cortex of the mammalian brain.

Silver, L. M. (1997). *Remaking Eden: How Genetic Engineering and Cloning Will Transform the American Family*. New York, NY: Avon Books, Inc. A lively and perceptive discussion of the inevitable impact of genetic technology on human reproduction and related social policies.

11

Behavior as Phenotype

Photo: Cuneiform writing from the 6th century B.C., Temple of Darius, Persepolis, Iran. The capacity to communicate by spoken language is an evolved feature of humans. Language is processed in specific parts of the brain and is most easily learned during a critical period of early childhood. The ability to represent objects with abstract symbols was exploited by humans beginning (at least) with the introduction of agriculture. As cultural systems were invented for representing the sounds of spoken language visually, the flow of information across generations was vastly expanded. Both the specific sounds of spoken language and their symbolic representation in writing are arbitrary and vary from one culture to another. The neural circuits engaged in deciphering written language thus evolved for other purposes and have been put to this use only in the last several thousand years and not in all cultures. Interestingly, the capacity to learn arbitrary visual symbols for objects is latent in the brains of apes.

BEHAVIOR AND EVOLUTION

The yellow bill of a herring gull has a bright red spot near the tip. When a newly hatched gull is hungry, it pecks at the red spot on its parent's bill, and this "knock at the door" causes the adult bird to feed the youngster by regurgitating food from its crop. The color of the spot is an important feature of the adult bill, for experiments show that this is the stimulus that induces the chick to peck, and it is the pecking that in turn spurs the parent to provide food. The chick's hunger is communicated to the parent, and the parent responds appropriately.

Behavior is frequently as diagnostic as anatomy in characterizing a species of animal. The way individuals behave and the circumstances that elicit certain behaviors are often predictable and associated with particular kinds of animals. Behavior such as the communication between a gull chick and its parent is therefore referred to as *species-specific* or *species-typical*. This means that the behavior is usually seen in all members of the species of the same sex and age. Of course there are many behaviors that are shared by numerous species of animals. Many animals feed their young by regurgitation, but the sensory cues that trigger feeding vary among species. When your dog tries to lick your face in greeting, he is using the same signal that a puppy employs for begging food. What makes a behavior species-specific is thus to be found in the details. Close observation of many individuals under natural conditions is as necessary as meticulous anatomical study in order to understand not only how an animal behaves, but what details of the environment are important to it.

Like an anatomical trait, a species-specific behavior can be viewed as a defining characteristic of the animal, for like anatomy, behavior has been shaped by evolutionary history. Behavior is important evolutionarily because small differences in behavior can have large effects on reproductive success. Is it then useful or appropriate to think of behaviors as adaptations?

Behind every adaptation there must be one or more genes, so to answer this question we need to consider the relationship between genes and behavior. Behavior is different from anatomy in a very important way. Anatomy is *structure*, a complex assembly of mole-cules, cells, and organs that is put together during development. Behavior, on the other hand, is *activity* of the organism, so what links genes to behavior?

The connection is important to understand. Behavioral activities are directed by nervous systems, and nervous systems are structures. Genes provide information needed for the construction and operation of nervous systems, and selection has produced nervous systems that are sensitive to particular kinds of sensory information, that analyze information in particular ways, that generate predictable motor outputs when activated by particular stimuli, and that are fashioned to learn and remember particular classes of information. So our answer to the question is a qualified "yes": behaviors can be thought of as adaptations where the structure of the brain underlying the behavior has been molded by natural selection.

The evolution of behavior raises another issue. The behavior of many animals is very flexible, and we need to understand how this flexibility is related to species-specific behavior, to genes, and to adaptation. We saw in Chapter 10 that the development of the brain, like the construction of the rest of a new organism, involves an ongoing interaction between genes and several environments: the cytoplasm of the immediate cell, the genes and cytoplasm of neighboring cells, the hormones of the embryo, the maternal environment at the placenta (for a mammal), and after birth or hatching, the sensory environment of the newborn. A brain and the behavior it produces are therefore the products of an intimate sequence of gene/environment interactions. Exposing the relations between genes, evolution, development, and behavior is the principal task of this chapter.

BEHAVIOR AND DEVELOPMENT

The complex sequence of gene/environment interactions that occurs during development does not stop at birth or hatching and includes effects of sensory input on the maturing, postnatal brain. Some of this "external programming" during maturation is familiar to us in another context, and we think of it as learning. There is, however, a seamless connection between many processes that we commonly refer to as learning and earlier developmental events.

The development of every species is guided along channels, and that is what makes each species different. But what an animal can readily learn is also characteristic of its species. As the psychologist Martin Seligman has put it, animals are evolutionarily prepared, unprepared, or even counter-prepared to learn specific behaviors. These capacities to learn have resulted from natural selection. As described above, they have their bases in the structure of the brain, which determines

Box 11.1
The Misleading Concept of Instinct

The word *instinct* suggests a behavior that is preprogrammed, hard-wired, under tight genetic control, and invariant. All of these terms contrast instinct with behavior that is learned, environmentally determined, open-ended, and variable. This rigid dichotomy is part of our cultural heritage, but it is fundamentally false. It has its roots in the belief that humans are uniquely endowed with reason and therefore are not part of the natural order. In this basically theological view, which dates at least from medieval times, some concept other than reason was needed to refer to animal behavior that appeared to be appropriate, purposeful, or even intelligent. Instinct was thus associated with nonhuman behavior. During the nineteenth century, however, the dichotomous animal/human concept of instinct was replaced as instinct came to mean the alternative to learned behavior.

Behavior cannot be pigeon-holed in either way. The word instinct nevertheless continues to evoke these old confusions. We call bad behavior in others "brutish," ascribe it to "base instincts," and characterize its perpetrators as "animals." When we see our pets learn or we think we see signs of reasoning, we raise their status to "almost human."

The basic difficulty with these simplistic categories of behavior is that they are not based on suitably careful observations of nature. First, some capacity for learning is a property of virtually all neural tissue. Even an isolated ganglion of an insect can display simple learning. Second, traditional ideas of instinct ignore the inseparable relations between genes and environment. And finally, because humans are the product of evolution, it is possible to identify some common threads in the behavior of both humans and other animals, clues that help us to understand the evolution of our own mental abilities. Later in the chapter we will explore this proposition in more detail.

---

the kinds of sensory stimuli to which the brain will respond and the conditions under which that sensory input will be processed and remembered.

Even within a species, ourselves included, behaviors vary in the amount of external programming or learning that is required for their expression. Moreover, until the development of a behavior has been studied in detail, it is not possible to predict what experience the animal must have in order for the behavior to develop normally. The word *instinct* therefore does not carry a precise meaning and is difficult to use unambiguously (Box 11.1). Species-specific (or species-typical) behavior is a more useful concept because it makes no assumptions about the developmental events required for the behavior to occur. It also suggests possible relationships among behavior, natural selection, and evolutionary history that are appropriate topics for study and analysis. These ideas will become clearer if we consider some specific examples.

A BEHAVIOR THAT APPEARS WITHOUT LEARNING MAY NEVERTHELESS BE CONTINGENT AND FLEXIBLE

One of the arguments used to suggest that biological and evolutionary reasoning has no relevance to analyzing human cultural phenomena is the mistaken belief

that "what is natural must be universal and invariant." So in this fallacious thinking, if a single cultural exception to a general rule can be identified, the validity of the generalization is promptly questioned. We will analyze some specific examples in the final chapter, but at this point we explore the idea in relation to the larger concept of instinct.

The acorn woodpecker is a familiar bird to residents of California and the Southwest. It is named for its favorite food and its interesting way of stuffing single acorns into small holes that it has drilled in the trunks of trees or the wooden eaves of buildings, much as a cork is pushed into the top of a wine bottle (Fig. 11.1). One population of these birds lives in a relatively arid part of southern Arizona in which the acorn supply varies from year to year, depending on the rainfall. In good years, in which there is ample rain and a sufficient supply of acorns, the birds nest communally and do not migrate south in the winter. In bad years they pair to breed and move further south in the winter. These alternatives involve two core features of birds: migration and reproductive behavior.

Peter Stacy and Carl Bock have shown that these alternative behaviors are not learned by individual birds; they are simply triggered by environmental conditions. Here, then, is a classical "instinct" in the sense of requiring no learning, but it is flexible, contingent, and tuned to available resources.

FIGURE 11.1 The acorn woodpecker (*Melanerpes formicivorus*) stores acorns by stuffing them into holes in the trunks of trees.

Animals of many kinds are frequently confronted with environmental challenges, and there are numerous examples of adaptive solutions that do not require the sort of external programming we refer to as learning. Spiders spin their first web with no practice, but it must necessarily fit the space that is available, taking advantage of suitable attachment points. Learning becomes particularly useful to an animal, however, when the challenges are numerous and unpredictable. Learning provides a way of extending the capacity of the brain to deal with the most unpredictable contingencies. But evolutionary history has influenced how the external environment modifies and adjusts the ability of each individual brain to generate behavior.

LEARNING MAY BE NECESSARY TO COMPLETE THE DEVELOPMENTAL PROGRAM

Ducks and geese nest on the ground and lay large clutches of eggs. When the young hatch, they are covered with downy feathers and are able to run about and feed themselves within minutes. They are quite vulnerable to accidents and predation, however, and follow the mother duck or goose until they have grown larger. This "following" behavior is part of their developmental program, but at hatching there is an important piece of information that is missing: the identity of mother. This information is supplied by the sight of a large moving object close to the nest, which in nature is invariably the parent. The developing brain is programmed to process this visual information in a very specific way. Thus the ducklings or goslings become *imprinted* on the parent and follow her when she walks away from the nest to take them to the relative safety of a pond and to feed. Through this experience they have *learned* the identity of mother and protector. Furthermore, the birds are only susceptible to this experience shortly after hatching. Imprinting is therefore an example of a change that takes place during a *critical period* in development, a concept we met previously in Chapter 10.

Another characteristic behavior of the young of many ground-nesting birds is that they take cover and crouch at the sight of moving objects overhead. This behavior protects the young from birds of prey. At first the crouching behavior of chicks is evoked by virtually any movement overhead, but in time the young birds *habituate* to innocuous objects like falling leaves and the flight of adults of harmless species. (When an animal is presented with repeated stimuli that do not have any negative consequences, it may gradually cease to respond. This simple form of learning is called *habituation*.) The young birds continue to respond, however, to the short-necked, long-tailed silhouettes of hawks that appear overhead less frequently (Fig. 11.2).

In this example, the birds hatch from the egg programmed to escape a particular class of potential danger by crouching and hiding, but experience shapes the behavior. The behavior is clearly adaptive, but it requires some learning to become optimally tuned.

The acquisition of song by birds like white-crowned sparrows and chaffinches provides a third example of the interplay of internal and external programming. These birds ordinarily hear the songs of other males of their species when they are young, and when they start to sing, they hear themselves. When the birds are reared in a controlled environment, it is possible to disrupt this process by several means. The birds may hear only tape recordings when they are young, either vocalizations of other kinds of birds or artificial songs pieced together from phrases of their own species' song. Alternatively, they may hear no songs during early life, or be deafened so they are unable to hear themselves sing.

Experiments such as these reveal two complementary features of how these birds come to sing their characteristic song (Fig. 11.3). First, the song that an adult bird normally sings is influenced by what it heard during an early sensitive period in its life. Therefore, in some species there are regional "di-

(A)   (B)   (C)

FIGURE 11.2 Developmental refinement of the crouching response of young chicks to the silhouettes of other birds, some of which are predators. (A) Chicks originally crouch in response to any movement overhead. (B) In time they habituate to long-necked silhouettes of commonly seen birds. (C) They continue to crouch when short-necked birds of prey appear overhead.

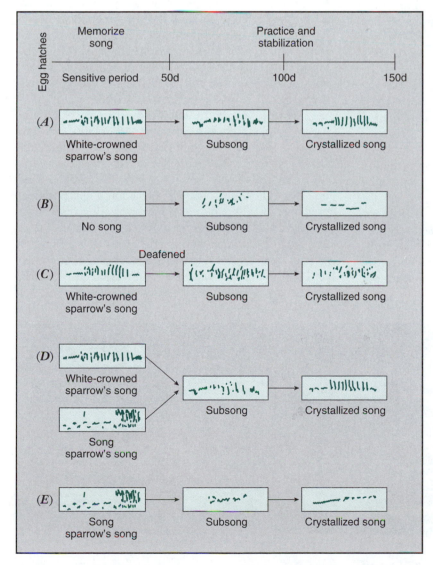

FIGURE 11.3 Sonograms illustrating the development of song in white-crowned sparrows.

(A) The bird hears its species song during an early critical period. It later sings a "subsong," which in time matures into a characteristic song of the species.

(B) If the bird fails to hear a song during the critical period, the subsong never develops into the normal final song.

(C) If the bird is deafened after it hears the correct song during the early critical period, the subsong also fails to mature properly.

(D) The young bird discriminates between its species song and that of another species during the early critical period.

(E) In fact, in these experiments the white-crowned sparrow appears to be incapable of learning the song of another species. In other work, however, what the bird can learn is shown to be influenced by social contact with other birds, and under the right conditions, white-crowned sparrows can learn to sing the song of another species.

alects" in which the species-specific song varies in a recognizable fashion in different geographical areas. Second, there are nevertheless strict bounds on how much the external model can vary and still be mastered. It is as though the brain contains an internal template of the song that limits the amount of variation that can be learned and against which the bird measures its own performance. If the external model is either absent or departs too far from this template, the final song is a jumble of phrases drawn from the bird's limited vocal capacities. Development after hatching thus involves an interaction between an innate program for learning and the experience that the bird acquires through hearing and practice. In nature, exposure to the song of an adult takes place during a sensitive period for learning.

THE DEVELOPMENTAL PROGRAM CAN BE FOOLED BY UNUSUAL SENSORY CUES

In the example of imprinting of goslings we saw that sensory information completed the developmental program in which the young discover what object they should follow as mother. An important advance in understanding this process came from the discovery that a variety of large moving objects in the vicinity of the nest shortly after hatching could substitute for the mother goose: a child's balloon, even the ethologist Konrad Lorenz, who discovered this phenomenon (Fig. 11.4). Thereafter the goslings follow that surrogate as though it were really an adult goose. This kind of experiment tells us the sorts of sensory information that the animal uses to complete the developmental program.

Eberhard Curio carried out some remarkable experiments on the acquisition of mobbing behavior and alarm calling in the European blackbird that also illustrate how nature and nurture are so intertwined as to be inseparable parts of a single process. Wild-caught adult blackbirds were placed in wire cages (Fig. 11.5), and a rotatable, chambered box was placed between them. Each bird could see only one chamber of the box at a time, but it could see and hear the blackbird in the other cage. Each blackbird was first shown a stuffed Hawaiian honey creeper, a harmless species that neither had seen before and did not resemble any of the predators with which they were familiar. Neither blackbird seemed alarmed.

Next, one blackbird was shown a stuffed owl, a predator known to both of them, while at the same time the other was shown another honey creeper. The blackbird that could see the owl gave an alarm call and tried to chase the owl away. The blackbird in the other cage, who could not see the owl, watched and listened to this behavior and then tried to chase away the honey creeper.

Finally, both blackbirds were shown honey creepers. The behavior of the one that had learned from its previous experience to mob honey creepers now caused the other to treat honey creepers as though they were potential predators. In additional experiments, Curio also showed that young blackbirds would learn to mob honey creepers or even laundry detergent bottles by watching their parents. In the acquisition of this behavior there is thus a potential for transmission both within and across generations, which is how culture is transmitted in human behavior.

Alarm calling and mobbing by these blackbirds in nature is likely elicited when a nest is attacked. But Curio's experiments suggest that when blackbirds are recruited to a mob they remember their adversary and in the future will readily mob other individuals of that kind on sight. They thus obtain information about potential predators by observing the behavior of other blackbirds, which means that engaging in defensive measures does not require a prior attack on their own nest. But it is clear that in this learning they are attending to a limited set of all the information potentially available. The fact

FIGURE 11.4 These young geese were imprinted on the ethologist Konrad Lorenz, and they follow him as they would a parent.

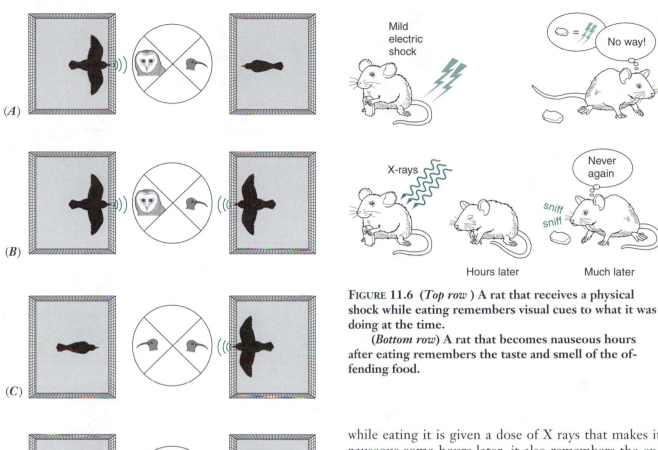

FIGURE 11.6 (*Top row*) A rat that receives a physical shock while eating remembers visual cues to what it was doing at the time.

(*Bottom row*) A rat that becomes nauseous hours after eating remembers the taste and smell of the offending food.

FIGURE 11.5 In A the blackbird on the left tries to mob a stuffed owl that it can see through the screening of its cage. The blackbird on the right can hear the ruckus, but it sees only a small, innocuous, stuffed, nectar-feeding bird called a honeycreeper.

In B, however, it has begun to act as if the honeycreeper is a dangerous predator.

In C, the first bird now hears blackbird alarm calls but it sees a honeycreeper, and in D, it behaves as if honeycreepers are dangerous.

that detergent bottles are harmless does not influence their behavior if other blackbirds are attacking.

ANIMALS ARE EVOLUTIONARILY "PREPARED" TO LEARN PARTICULAR ASSOCIATIONS

If a rat is given a mild but unpleasant electric shock while eating a novel morsel, it remembers the experience and in the future avoids food that *looks* like what it was consuming when shocked. Alternatively, if while eating it is given a dose of X rays that makes it nauseous some hours later, it also remembers the experience, but in the future it avoids foods with the same *taste* or *smell* (Fig. 11.6). This phenomenon (known as the Garcia effect for its discoverer) is a curiosity in learning theory because of the several hours that can intervene between stimulus and response, but when viewed from an evolutionary perspective it makes perfect sense.

Why should the association between food and delayed nausea involve taste and smell rather than vision? The answer is so obvious that it has probably already occurred to you. Many plants manufacture poisons to inhibit browsers, and if food is putrefying, it is likely to contain bacterial toxins. Taste and smell are the sensory channels that provide information about the chemical composition of food, and if a meal makes you sick later, it is well to recall chemical cues that you may have ignored while eating. (If you were a microorganism consuming rotting food, how would you prevent a larger scavenger from ingesting you and your meal?) This phenomenon is not restricted to rats; many people have taste aversions based on single unpleasant experiences with food.

Compared to mammals, most birds have a poorly developed sense of smell, and they can employ visual cues to avoid distasteful or dangerous food. This capability is exploited by certain butterflies that mimic the color patterns of another species that accumulates plant toxins that are distasteful to birds.

EVEN VERY SMALL NERVOUS SYSTEMS CAN LEARN

The behavior of insects is frequently contrasted with that of vertebrates as the epitome of "blind instinct." In Chapter 7 we described the social structure of bees and ants and explained how information about the environment is utilized in a functional and adaptive way by a colony of bees. At that time we referred to the fact that bees learn and remember sources of nectar and pollen. As their brains are tiny compared to those of even the smallest vertebrate, this ability deserves some further consideration.

Honeybees make their living by collecting nectar and pollen from many different kinds of flowers. Flowers bloom at different times during the summer and they vary in color, shape, and odor. Moreover, they often grow in scattered patches. They thus represent a part of the environment that is of great importance to bees but that is also very unpredictable.

Bees will alight on small, brightly colored, flower-like patterns and probe with their mouth parts for nectar. Consequently, foraging bees can be readily attracted to artificial flowers as simple as patches of colored paper. The attractiveness to bees of a potential refueling stop can be greatly enhanced by the presence of a floral odor such as oil of peppermint. If the bees' visits are sufficiently rewarded with sugar solution, they will remember the source and return for another drink.

Experiments show that odor cues are more quickly associated with food than color, and color more readily than shape. Once bees are trained to a specific target, the relative importance of these cues can be tested by seeing which of two sources is preferentially visited if they differ in one or two features (Fig. 11.7). The order of importance of cues is the same as the order of ease with which they are learned: first odor, then color, and finally shape. Moreover, colors at the blue and violet end of the spectrum are learned more effectively than reds and yellows.

The learning of bees, like that of rats, is thus evolutionarily directed. The sensory cues that are most reliably associated with possible sources of food are stimuli to which bees most readily attend. Because the availability of nectar and pollen varies from day to day or even with time of day, the capacity to find, remember, and relearn the locations and features of many different flowers is a critical ability for bees to possess. These behavioral capacities are thus part of the complex of adaptations that comprise the evolutionary "design" of the honey bee's brain.

Bees are also able to use separate kinds of learned visual cues about their home area to solve novel navigational problems. In the experiments illustrated in Figure 11.8, James Gould trained bees to a feeding site and then captured individuals in a darkened container as they were leaving the hive for a new load of nectar

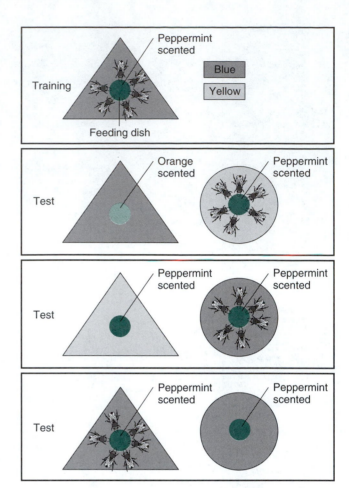

FIGURE 11.7 Bees were trained to associate food with a blue triangle scented with oil of peppermint and then tested with various combinations of peppermint or orange scent, blue or yellow color, and triangle or disc. Odor, color, and shape cues are all remembered, but the memories differ in their effectiveness in eliciting subsequent feeding behavior. All of a bee's memories are not equivalent.

and pollen. The captured bees were then transported in this container to a clearing in a nearby woods and released. From this place the bees could see neither the hive nor the feeding site.

Gould reasoned that the response of the bees to this novel situation would reveal how they navigate. Recall (Chapter 7) that bees orient their dances with respect to the position of the sun so that new recruits to a feeding site can orient their outward flight path. If the bees in Gould's experiment used only the sun as a reference point, they should fly from the release point in the same compass direction they would have flown if they were leaving the hive for the feeding site (Fig.11.8, route A). Alternatively, if the bees found their way from hive to feeding site by remembering a specific sequence of landmarks, they would have to search for the hive and then proceed to the feeding site. Finally, if the bees had

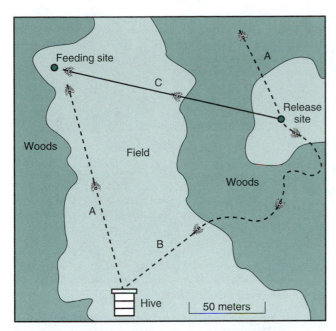

FIGURE 11.8 **Several possible outcomes when bees were released at a novel site. They flew directly to the feeding site (route C).**

a more general representation of their home area, made by remembering and relating various visual cues from their previous travels, and if they could recognize these cues from different directions, they might be able to fly directly back to the feeding site (Fig. 11.8, route C).

The route taken by Gould's bees was in fact route C, showing their ability to solve this novel problem in navigation. It is novel, because in the absence of human intervention, bees never find themselves mysteriously and blindly transported from one spot to another. Many other animals can move efficiently through familiar territory. For example, dogs and cats are also able to use current sensory information and travel by the shortest route to a remembered destination. But to find this and other complex learning in animals with brains no bigger than a couple of cubic millimeters is a reminder of both the power of adaptation and the latent capacities of even very small nervous systems.

SOME UBIQUITOUS HUMAN BEHAVIOR REQUIRES NO EXTERNAL PROGRAMMING

We think of our behavior as flexible, malleable, and determined by our experiences. Of course that is to a large extent correct, but can we identify any important features of our behavior that develop with little or no external programming or that occur cross-culturally with such frequency that they are properly considered species-specific?

Several human behaviors used in nonverbal communication qualify as species-specific. You may not have been consciously aware of them, precisely because no external programming is needed for their development, but they certainly are familiar once pointed out. Eyebrow flashing, in which the brows flick upwards for about a third of a second, appears to be a universal sign of friendliness in people (Fig. 11.9). In addition to greeting, it is also seen in agreeing, encouraging, and flirting. The ethologist Irenäus Eibl-Eibesfeldt has photographed people in many cultures as they engage in their daily activities. (He uses a camera with a right-angle prism so that he does not face his subject and appears to be pointing the camera in a different direction.) He has found that the frequency of eyebrow flashing varies among different cultures. In the Pacific islands it is frequently seen when strangers meet, although in Europe and the United States it appears when the parties know each other, and in Japan it is observed only among children.

These brief upward flashes of the eyebrows are different from a more sustained arching of the brows, which has an equally widespread but different meaning. This facial expression has strong negative connotations, conveying indignation and rejection. The familiar expression "that will raise a few eyebrows" reflects astonishment and concern about another person's behavior.

Smiling is another form of communication that takes place without words and is understood cross-culturally. Even congenitally blind babies smile, so it is not learned by mimicking another person, either consciously or unconsciously. Smiling can signify mild amusement, and in social encounters it is a sign of pleasure or reassurance. Thus a nervous subordinate can be put at ease with a smile. But smiling can also be consciously employed to deceive. Success is not assured, however, because we are also evolutionarily prepared to look for signs of deceit in our social interactions.

PEOPLE LEARN THEIR FIRST LANGUAGE DURING A SENSITIVE PERIOD OF DEVELOPMENT

One of the more humbling experiences you can have is to study a foreign language in school, travel to a country where it is spoken, and be confronted by a seven-year-old native who speaks it fluently and thinks your accent and grammatical mistakes are funny. Your reward for all of those hours of study is a perplexed look or a giggle.

What is the difference between learning a language as an adult and learning one as a very young child? If you learned a second language as an adolescent or adult, you probably decided consciously when and how much you were going to study and then launched into sessions

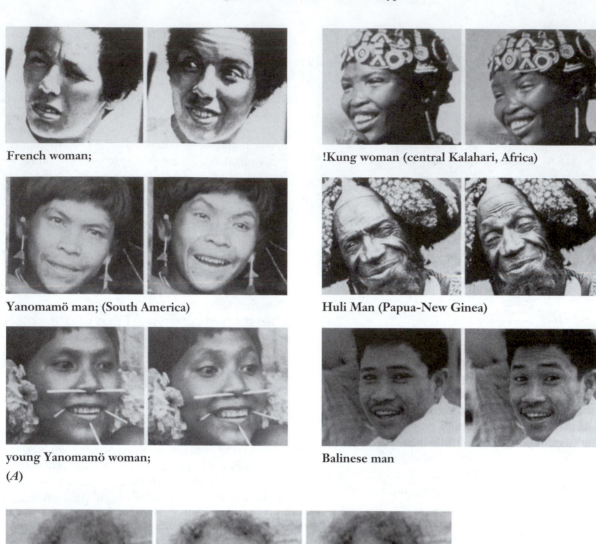

French woman;

!Kung woman (central Kalahari, Africa)

Yanomamö man; (South America)

Huli Man (Papua-New Ginea)

young Yanomamö woman;

Balinese man

(A)

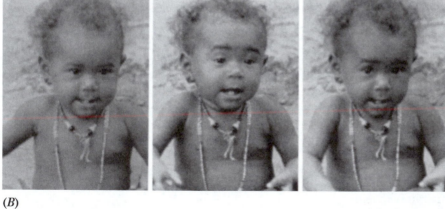

(B)

FIGURE 11.9 (A) Eyebrow flashing.
    (B) Infant girl (16 mo, Trobriand Islands) indicating her readiness for contact
by raising her eyebrows.

of memorizing and practicing. On the other hand, you likely remember nothing about how you learned your first language. That is because you were very young, you did not "decide" to learn, and you were not aware that you were doing something "difficult." Between the ages of two and five, when many simple behaviors such as eating with a spoon or drinking from a cup require obvious practice and encouragement, all humans in all cultures automatically acquire the language (or languages!) they hear spoken around them.

LANGUAGE IS AN EVOLVED FEATURE OF THE HUMAN PHENOTYPE

All human groups have language. Furthermore, there is no evidence that any human culture in recorded history lacked language or spoke a substantially more primitive language than those that exist today. Spoken language leaves no fossils, so nothing is known of its early evolution.

Some have argued that language was a conscious human invention, like a new tool, that spread rapidly because everyone found it useful and could learn it easily as a result of a generalized high intelligence that did not evolve to support language *per se*. After all, no one would argue from the universal use of fire or the wearing of ornaments that humans are specifically endowed with programs to learn these behaviors, so why postulate one for language?

The universality of human language, the spontaneity of its learning during a critical period of development, and the evidence that parts of the brain are specialized for its recognition and generation (Chapter 9) all indicate that the use of language is an evolved feature of humans. Steven Pinker, an evolutionary psychologist and linguist, has drawn attention to several natural experiments that further support this interpretation. By natural experiment we mean a condition that was not planned or caused by the observer but which can nevertheless be analyzed and from which useful conclusions can be drawn.

Linguistic communication can occur by visual as well as auditory signals. Sign language is a rich and effective mode of communication among the deaf, and like spoken language, sign language is learned most easily when children are young. Just as the spoken language of people who have had their hearing restored after early deafness is not as rich or clear as that of individuals who heard normally during the sensitive period for language acquisition, people deaf from birth who were introduced to sign language only later in life are not as fluent as those who learned it early.

In 1979 the Sandinista government of Honduras founded schools in which deaf children, most of whom were eight years of age or older, were brought together for instruction in lip reading and speech. Outside the classroom, however, they communicated with each other by simple signing that was largely devoid of grammar. Something fascinating happened, however, when children younger than five arrived in the school community and started using the established signing of the older children: their signs became more compact and standardized, and most remarkable, they introduced grammatical constructions where there had been none before. In other words, a language with grammatical structure recognizable in other spoken languages began to appear quite naturally.

This is not the only situation in which young children have imposed grammatical order in the process of acquiring language. Pidgin refers to a polyglot of words drawn from several languages and has been used for basic communication by people who have been confined together but who share no language in common. The children growing up in this linguistic environment, however, impose grammatical order, creating a language from that mixture of words, the technical name for which is a creole.

A second kind of natural experiment shows that linguistic abilities can be dissociated from virtually all other aspects of cognition. Local injuries to the brain can produce general (Chapter 9) or very specific deficits in the use or comprehension of language. For example, past tense endings of verbs or plural endings of nouns may be omitted, although such persons may be able to understand and describe their handicap. Another group of language disorders with the catchall label "specific language impairment" runs in particular families, suggesting the presence of genetic mutation and a failure of normal development. These individuals display specific deficits in using and understanding grammar, but their other mental capacities are normal. Conversely, many impairments of other mental capacities such as schizophrenia or even stages of Alzheimer's disease do not affect a person's ability to understand language or speak without grammatical errors. What these people say may be illogical, fantasy, or unconnected to previous conversation, but their language competence is not diminished.

All of these observations support the view that evolution has shaped parts of the brain as an organ for generating and understanding grammatical language. They are not so readily compatible with the idea that human language is a conscious invention like the computer "languages" BASIC, Pascal, or Java. Of course what *specific* language people speak depends on what they heard during their early years. We are describing a generalized capacity for language that is shared among all people, a capacity that is one of the defining characteristics of our species.

SOME IMPORTANT SENSORY PROGRAMMING IS VERY GENERAL IN NATURE

The examples that we have discussed so far all involve either little or no external programming or learning that addresses rather narrowly defined behaviors. The longer an animal lives, the more different kinds of environmental challenges it is likely to encounter. As we shall discuss later in this chapter, this is especially true for social mammals. But long life and unpredictable challenges are best dealt with by a nervous system that can respond flexibly, can store records of its experiences, and can draw on those memories as the immediate situation requires. This ability does not free primate or even human behavior from its evolutionary heritage, but it changes a lot of the learning that occurs during the later stages of development.

One of the most characteristic features of young mammals belonging to social species is the amount of time they spend interacting with each other, frequently playing and often trying to induce an older individual to join. This is particularly familiar behavior in puppies, because for household pets their human owners are *de facto* members of the dog's social group and are often pestered until they participate.

When the experimental psychologist Harry Harlow studied the role of social interaction in young monkeys, he found that isolation produced serious and lasting effects on their behavior. If a young rhesus monkey had no living companions but a cloth doll, it spent much time hugging the doll. The monkeys were given ample food with a baby's bottle attached to the doll, and they grew physically. Emotionally and behaviorally, however, they became seriously defective, for when they grew up and were placed with other monkeys they were incapable of relating to peers in normal fashion. Sometimes they were withdrawn and seemed almost autistic (Fig. 11.10); sometimes they were too aggressive. They also proved to be sexually inept; the males had great difficulty just getting into the right position to copulate with an estrous female, and the females did not wish to mate at all. If forcefully inseminated by a normal male, these socially deprived females became incompetent mothers, refusing to nurse their young and otherwise mistreating them. The degree of behavioral abnormality depended on the length of time the young had been isolated: a few months produced a reversible depression; a year did irreversible damage.

Additional experiments showed that the most critical social input in early development came from interacting with other young monkeys. Monkeys with only the cloth doll for a mother but other young monkeys for companions developed nearly normally. On the other hand, monkeys with access to their mother but none to age peers became very dependent on the mother and displayed too much aggression toward other individuals later in life.

These experiments provide a glimpse of the importance of social experience: an extended sensitive period of development is required for the behavior of adults to be normal. During this period the monkeys are learning general rules about interacting with others, and it is clear that if this experience does not occur at the right time, the effects can be both catastrophic and permanent. Unlike the previous examples of external programming, the details of social interaction are haphazard. Play with peers of course improves coordination and agility, but the social skill that emerges is a general guide for how to behave in a social environment where details are unpredictable but where mastery is crucial for reproductive success (Box 11.2). In the need for appropriate timing and in the lasting changes that are produced in the nervous system, however, the example of social experience provides another illustration of an important principle of developmental biology that runs through this chapter. As the rest of this account unfolds, the likely relevance of Harlow's studies for human development should become even clearer.

HOW MUCH DO ANIMALS KNOW ABOUT THE WORLD?

WHAT DO WE MEAN BY THE PURPOSE OF BEHAVIOR?

Much of the behavior of animals appears to us to be purposeful in that it fulfills desirable or necessary goals. A pelican dives into the water and surfaces with a fish in its beak, a prairie dog utters an alarm cry at the sight of a falcon and then takes refuge in its burrow, and a male grouse struts and drums in front of potential mates. Feeding, escaping from a predator, and securing a mate all require that an animal expend energy, and these activities may even entail danger. Animal behavior is therefore not a random activity; it has aims and goals. What gives these examples of behavior purpose is that, directly or indirectly, they all contribute to reproductive success. This is not a new idea; we met it when we considered how information is processed by a colony of

FIGURE 11.10 (*Above*) A young socially-deprived monkey (*right*) is approached by a normally-reared age peer. (*Below*) It cringes in fear, unable to participate in play.

Box 11.2
Getting Ahead and Getting Along in Toddlers

People compete for resources of many kinds: not just material resources like food and shelter, but friendship, attention, influence, love, and, of course, sex. The competition is unequal, so people find themselves in hierarchies in which some individuals are more successful than others in one or more of these spheres. Hierarchies are present even in young children who play together.

The developmental psychologist Patricia Hawley has looked at the social development of young children from an evolutionary perspective. Among primates, the more successful (i.e., dominant) individuals are watched, groomed, and greeted with deference. In playgroups of young children, the more successful individuals are similarly distinct: they are watched more by their peers, imitated, and tend to be sought as playmates. Until age four or five, this status is achieved by aggressively assertive behavior, which is commonly viewed by adults as selfish. Effective socialization tames aggressive behavior as children start to attend to the needs and wishes of their peers.

Hawley has pointed out that this transition, which is accomplished more effectively by some children than others, is usefully understood in evolutionary terms. Social dominance requires skill in dealing with others, but it serves the same basic purpose as the antagonistic tactics of toddlers. Those children who discover how to cooperate and manipulate (as well as deceive) get the attention, admiration, and friendship of their peers. In short, they garner social resources of the very kind that make for success in the adult world. Interestingly, but not surprisingly, by the third grade those children whose tactics do not change drop in the social hierarchy and are shunned by their peers.

---

bees, and it was implicit in the earlier parts of this chapter.

When we use the word "purpose" to characterize our own behavior, however, it generally means that we have some immediate goal *in mind*. Thus the purpose of going to the bank might be to withdraw enough cash to buy groceries. In common language, purpose therefore implies conscious awareness of what we are planning or doing. But possessing conscious awareness is a statement about how *our* minds work, and in the larger world of organisms it is not a necessary condition for achieving reproductive success. For example, the purpose of color in a flower is to attract animal pollinators, and of course plants do not have mental states of any kind. When we observe complex, goal-directed behavior in a bird or mammal, however, particularly a pet with which we share living space, we can easily attribute wants and desires, and it is a short step from there to imagine (without any evidence) that the animal has the intelligence for consciously purposeful behavior.

Here, then, is the source of the confusion. When we see purpose in the behavior of an animal we seem to be saying "that's a sensible thing to be doing." When a squirrel is chased by a dog, it climbs a tree. We consider that sensible because that is how we would save ourselves if we were a squirrel. This understanding of purpose includes an intuitive recognition of how evolution has crafted squirrels. Consequently, when we think of animal behavior as sensible or purposeful, it is because it is very likely to be adaptive. Whether adaptive behavior

involves foresight, planning, and other higher cognitive abilities, however, is a separate question. The neural control of behavior varies greatly from jellyfish to humans, which prompts the question What is intelligence? Or to frame the issue more broadly, What do we know about animal cognition, and how do we know it?

THE PROBLEM OF KNOWING THE MIND OF ANOTHER

Philosophers tell us that we can never know if our sensations and feelings, our perceptions and intuitions are shared by any other person. When we perceive the color red, our mind draws on its unique history of experiences and associations, to which no one else has the same access. Our mind is thus a private and personal part of our being. Similarly, we are told, we have no equivalent access to the mind of any other person.

In order to function in the presence of other people, however, this is not a very useful way to think about the world. Human brains are all built on the same basic plan, and they all process information by the same general means. People around the world share a common suite of emotions, which are both evoked and expressed in frequently predictable ways. This is because all human brains share a common evolutionary heritage.

In our social relationships we tend to project our innermost feelings onto others. If events make us angry or sad, jealous or joyful, we tacitly assume that other people will have similar responses under similar conditions. If we did not, we would be incapable of empathy or sympathy. Furthermore, we know that how we feel about other people depends on how they behave toward us, and we recognize (usually) that our own behavior is likely to influence how others will treat us. We look for signs in other people that will tell us what they are thinking, because this provides us with indications of how they are likely to behave toward us in the future. We are sensitive to misleading signals, particularly if we perceive them as deliberate deceit, and we are distressed if our own intentions have been misread by others. We are thus practiced, even skilled, in making inferences about the minds of fellow humans.

The proposition that the minds of other humans are essentially very like our own is thus not only supported by knowledge of brain architecture and function and evolutionary principles, it is reinforced by everyday experience. This reinforcement is so potent that the behavior of nonhuman animals frequently arouses our empathy, and we find it easy to attribute our own familiar inclinations and emotions to them. Although we may feel certain when our dog is happy or fearful, on reflection we realize that whether the tail is wagging or curled between the hind legs indicates an emotional state (Fig. 11.11), and we really know nothing about what Spot is "thinking."

Just as it can be hard to read all the signals from humans in unfamiliar cultures, where custom and context frequently shade meaning, knowing the mind of another species imposes even greater difficulties. We do not have the same evolutionary history as other species, and their brains, like the rest of their anatomy, differ in many details from our own. It is therefore impossible to infer the mental processes of another species with anything like the confidence that we have in dealing with other humans. When an animal obtains sensory information through channels with which we are unfamiliar or lives in a hostile environment, our insight fails us completely. We can have no inkling what it is like to be a bat catching insects by echo-location or a porpoise cruising in the bow wave of a ship.

What mental activity, if any, other animals experience is therefore a difficult question to answer, and it has traditionally been shunned by science on the grounds that there is no way to study it. If the situation were totally impossible, we would have no insights into the evolution of human mental abilities, but as we will discuss shortly, comparative study of other primates is providing fascinating and useful new information.

WHAT IS INTELLIGENCE?

The common understanding of intelligence involves individual comprehension, reasoning, and judgment. Simple as this sounds, psychologists continue to wrestle with the concept of intelligence. Much effort has gone into measuring intelligence by tests designed to distinguish among individuals and to distribute the results on a numerical scale. This places emphasis on variation among individuals rather than on differences in the sensory input that is occurring. As a consequence, until relatively recently study of the cognitive processes that underpin intelligence have not received as much experimental attention as various measures of intelligence. Furthermore there has been little agreement on the attributes that comprise intelligence, which, incidentally, makes intelligence tests a blunt tool with which to formulate social policy. As long ago as 1923 an eminent psychologist observed that "intelligence is what the tests test."

IQ tests are a good example of a narrow conception of intelligence. They are designed to predict success in school, and have been consistently modified, based on white, middle-class school children, to serve this purpose. It is naïve to suppose that the complex and varied intellectual faculties of any individual can be properly captured by a single number. Solving real-world problems invariably draws on memory of earlier experiences, comparisons with the current situation, and inferences about the future. Consequently, both context and knowledge (as well as beliefs) *must* be important to finding intelligent solutions.

The psychologist Howard Gardner has suggested that the common measures of intelligence are arbi-

Attitude of alarm

Threatening at a distance

All-out attack

Strong conflict between aggression and fear

Weak conflict between aggression and fear

Defensive

Inferiority

Timid social approach

Approach to prospective mate

Figure 11.11 Not just mammals communicate information about the internal state of the nervous system. Here is an example that is probably less familiar than the profiles of dog faces in Figure 7.8.

trarily restrictive, and that intelligence should be thought of as having a number of dimensions such as linguistic intelligence (understanding and generating language); logical intelligence (solving problems, including mathematics); spatial intelligence (a variety of tasks, from getting around in your environment to playing chess); musical intelligence; social intelligence (understanding and relating to other people); and intrapersonal intelligence (understanding oneself). This broad view of intelligence draws some support from the observation that local damage to the brain can have its primary effect on one or another of these abilities: e.g., left hemisphere, language; right hemisphere, spatial abilities; parts of the frontal lobe, interpersonal relations. Another psychologist, Robert Sternberg, has formulated a view of intelligence that draws attention to several cognitive processes that likely underlie the multiple intelligences just enumerated: we acquire knowledge and learn how to do things; we perform specific tasks; and we need to decide what to do and monitor and evaluate our actions, all in context-dependent ways.

It seems that psychologists are still discussing what constitutes human intelligence and how it can best be studied. If human intelligence is truly a multidimensional attribute of the mind, then we are likely to get only modest evolutionary insight by framing questions about animal behavior in terms of relative degrees of intelligence. Behavior that is purposeful, as described above, may look intelligent, but that comparison confounds adaptive effectiveness with presumed processes of cognition. Moreover, behavior must be understood in the ecological context in which it evolved. In thinking about the evolution of mental capacities, we therefore need a wider lens than intelligence through which to view animal behavior.

ANIMAL COGNITION: THE PHYSICAL ENVIRONMENT

To what sensory cues do animals attend? What use do they make of that information? What inferences do they make about their environment? The answers to these questions provide a richer understanding of animal cognition than judgments of intelligence.

Like "purpose," the words we use to frame the answers can be easily misunderstood. We say that a foraging honeybee knows where it collected nectar the day before because it returns to the same patch of flowers (Chapter 7). We do not suppose, however, that the bee or any other animal possesses human mental processes. For example, saying that an animal knows about a source of food is a shorthand way of saying that information from previous experience has been stored in memory and can be recalled to influence future behavior. Similarly, to say that an alarm call signifies danger

is an objective statement about the external conditions that cause an animal to call, such as the presence of a particular kind of predator. The observation tells us something about the sensory information that causes an animal to signal other members of its species. What the listeners know, however, must be inferred from observations of their behavior. Once again, to know is to have access to information.

Many animals move around in their environments. Some birds migrate from one continent to another, navigating by celestial cues to which they are developmentally programmed to respond. Many animals forage by repeatedly visiting favored locations in their territories. That they remember their surroundings is illustrated by their capacity to travel by the most efficient, straight-line route (e.g., Fig. 11.8). In this regard, primates seem to have no exceptional cognitive abilities, but humans are able to extend their navigational skills by using place names, maps, and instruments. Being able to store information about previous finds and the conditions that prevailed at that time have been important in the evolution of avian and mammalian brains.

If monkeys and apes watch an experimenter place a familiar object under an opaque cover, these primates will lift the cover to reveal the object. Like human infants older than about eight months, these animals therefore know that objects still exist when they cannot be seen. Hunting lions are probably able to project the movement of their prey even when it is temporarily out of sight, and it is likely that many animals have this capacity when it is ecologically important.

What do animals know about how the movement of physical objects can affect other objects? Some animals use objects as mechanical aids; for example, one species of Galapagos finch uses twigs or thorns to extract insects from crevices in bark. Sea otters float on their back, hold a stone on their belly, and use it as an anvil to break mollusk shells. Raccoons hold food in water to wash off debris. Hands capable of grasping are a general characteristic of primates, and consequently most species show more versatility in manipulating objects than do other animals. But like other animals, most primates show little foresight in solving novel problems that require understanding physical relationships such as fitting two sections of an interlocking pole together so that it is long enough to reach a desirable object like a piece of fruit. Capuchin monkeys are able to use objects in this fashion, but they seem to find solutions by random trial and error. There is a single observation of a captive chimpanzee, however, who seemed to solve this problem quickly.

Chimpanzees in the wild are nevertheless unique among nonhuman animals in using simple tools (Fig. 11.12). What makes these observations different from

FIGURE 11.12 Chimpanzee fishing for termites.

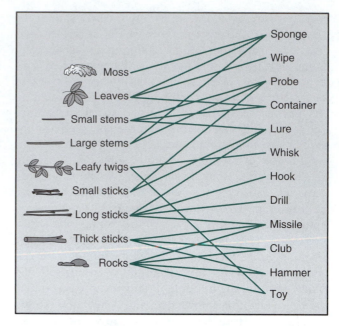

FIGURE 11.13 Chimpanzees use a variety of objects as simple tools.

ANIMAL COGNITION: THE SOCIAL ENVIRONMENT

Being a social primate in a group of several score individuals places a special demand on the brain. For example, rhesus macaques and many other primates live in groups that contain numerous adult males and females and their offspring. There are dominance relations within and between the sexes at all ages. These relationships determine access to mates and food, and they often shift over time (Chapter 13). Offspring of dominant females tend to "inherit" (in a social rather than genetic sense) the status of their mothers. Monkeys behave as though they are aware of their place in these hierarchies. For example, when a juvenile who is not with its mother cries in distress, other adult females recognize whose offspring it is and immediately look in the direction of the mother. Or a monkey may hesitate to challenge a smaller individual for a piece of food if it recognizes the other as a member of a more dominant matriline (individuals related through its mother) whose members will come rapidly to its support if it shrieks. Similarly, if *A* attacks *B*, *B*'s relatives may turn on relatives of *A* as well as *A*. Furthermore, monkeys and apes characteristically reconcile with each other after fights by grooming not only their opponents but the opponent's close kin. These primates thus recognize each other as individuals in a network of kin and dominance relationships and know how other individuals are likely to behave toward them.

The fact that many primates are aware of complex social networks has important implications for the ca-

tool use by other animals is the variety of objects chimpanzees use and deliberately modify for specific tasks (Fig. 11.13). Twigs are stripped of leaves and used as probes to fish termites or ants from their nests; larger sticks are used as levers to pry or, along with stones, are thrown as weapons; leaves are used as sponges to obtain drinking water from the hollows of trees or to wipe sticky foreign matter from their bodies or as sandals in climbing trees with thorny trunks. Some chimpanzees break hard nuts by placing them on a rock or in a depression in a piece of wood and striking them with another stick or stone. This appears to be a difficult skill to master, and it is usually learned by observation and practice. An adult female has been observed helping her youngster, however, by placing the anvil in a more effective position. As a rare, perhaps unique example of behavior that looks like teaching, this is potentially interesting. As described below, tool use by chimpanzees is socially learned.

Chimpanzees make medicinal use of some plants. When ill with an upset stomach they eat leaves that the local people also have found useful for such ailments. The leaves are not tasty, and chimpanzees do not eat them except when they are sick.

pacity of their brains. For one monkey (or human) to have a knowledge of thirty other individuals is much less demanding than to know how each of those thirty fits into the web of associations with everyone else in the group. The demands of knowing about others only in terms of one's personal interactions with them increases in direct proportion to the size of the group, but the number of possible pair-wise relationships that are possible among all members of the group increases much more steeply as the group becomes larger (Fig. 11.14). In fact, with a group size of fifty, more than 1200 different pair-wise interactions are possible, and with 1500 individuals this number has grown to over a million. Observations suggest that effective interactions between all members of a social group, humans included, are no longer possible when the size of the group approaches a hundred. It is not accidental that historically humans name individuals in ways that convey information about patterns of kinship.

The demands of storing and processing such extensive information about social relations was likely a major factor in the evolutionary enlargement of the cortex of primates. This inference is supported by data on the relative size of the neocortex. Among species of primates, the Pogarithm of the ratio of neocortex to the rest of the brain increases approximately linearly with the logarithm of mean group size (Fig. 11.15): thus a five-fold increase in neocortex ratio is associated with more than a fifty-fold increase in average group size.

Michael Tomasello and Josep Call have suggested that the social environment can be viewed as an extension of the physical environment, carrying the addi-

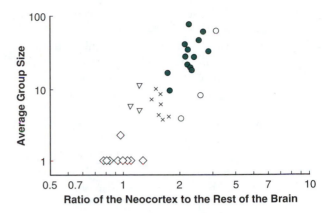

FIGURE 11.15 The size of the cortex relative to the rest of the brain is bigger in species that live in large groups. *Open circles*: great apes. *Filled circles*: polygynous anthropoids. *Crosses*: monogamous anthropoids. *Triangles*: diurnal prosimians. *Diamonds*: nocturnal prosimians.

tional complication that, unlike plants and nonliving objects, other animals behave spontaneously. A piece of fruit will wait for a few days to be picked, and a foraging monkey need only know where and when to find it. Another monkey, however, is likely to respond to the forager's presence in ways that may thwart its immediate goals. One evolutionary outcome of group living is therefore behavioral propensities that reduce unpredictability. Knowing its place in a dominance hierarchy enables an animal to take advantage of others when it can and avoid injury when it can't. Interactions between individuals thus become selective and fine-tuned by social knowledge.

THE ISSUE OF INTENTION

Most animal behavior is goal-directed. Recall the alarm calls of ground squirrels described in Chapter 6. They have the effect of alerting other members of the colony to an approaching predator. From the nature of the responses, we can infer that those who respond do so because their knowledge of the world has been importantly altered. In this example, the alarm call conveys the information that a specific kind of danger is nearby and causes the responder to search for a predator and run for cover. The behavior of the caller is purposeful in the sense of being adaptive. The caller has a goal, so the behavior is intentional. But intention also suggests something about cognition, and that idea requires further discussion.

Consider a couple of examples of primate behavior: a monkey solicits the help of a relative in an altercation it is having with another individual or a male chimpanzee knocks his knuckles on a branch, inviting a female to engage in sex. As Tomasello and Call point out, such behaviors imply that animals not only behave

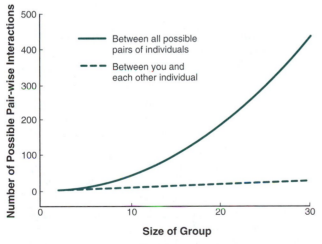

FIGURE 11.14 The number of possible interactions between all pairs of individuals increases with group size much more rapidly than the number of possible pair-wise interactions any particular individual can have. For one to keep track of all relationships in the group therefore involves much more information than simply remembering one's own social interactions.

with intention (they intend a particular outcome), but they can also recognize other individuals as capable of acting with intention. The monkey and the chimpanzee are each trying to influence the goal-directed behavior, and in this sense, the intention of another individual.

What sort of cognitive processes are involved? Do the monkey and the chimpanzee intend simply to alter the behavior of the other individuals, recognizing them as animate agents that can be influenced? Or do nonhuman primates recognize, as people do, that other individuals have beliefs that can be changed? From the standpoint of effective evolutionary adaptation, the distinction is not essential. But as part of the effort to trace the origins of human cognitive processes it is an important question. What does the evidence suggest?

THE ALARM CALLS OF VERVET MONKEYS

Dorothy Cheney and Robert Seyfarth's studies of the alarm calls of vervet monkeys illustrate how behavioral experiments can sometimes reveal the limits to what an animal knows. These studies are also important because they were done with free-living monkeys in their native habitat. Vervet monkeys (Fig. 11.16) are common in East Africa. They have several different vocalizations for specific common dangers: a bark for leopards, a sort of cough for eagles, and a characteristic chatter for pythons. Each of these calls elicits a different and appropriate behavior from the listeners. For leopards the monkeys climb onto small branches where the big cats cannot pursue them; for eagles they take refuge in dense underbrush; and for pythons, from which the monkeys can readily escape, they stand on their hind legs and peer about to locate the source of

FIGURE 11.16 Vervet monkey (*Ceropithecus aethiops*).

the danger. These calls and the different behavioral responses they evoke are learned during the first several months of life.

What do these observations say about the monkeys' understanding and their intentions? First, the calls are not involuntary responses to states of fear, because calling depends on circumstances. Calls are made when other monkeys are present to hear, and a solitary monkey chased by a leopard may remain silent.

Calls change the information the listeners have about the world because they refer to distinct classes of danger from predation and because the responses of the recipient monkeys are also conditional. For example, an eagle call may either send the hearer scampering into the brush or stimulate it to look up in an effort to locate the danger, depending on whether it is relatively safe in its present location. The calls therefore do not automatically trigger specific behaviors; rather they attract the hearer's attention to specific classes of predation. A monkey then decides what to do, depending on its assessment of the danger and how exposure can be minimized with the least effort.

Cheney and Seyfarth investigated what information the monkeys extract from calls by playing tape recordings of individual vervets. In one series of experiments they played one monkey's leopard alarm call while the group was feeding. Initially it caused the group to climb into trees. When it was repeated, with no sign of a leopard about, the monkeys habituated and ceased to respond, just as the shepherds in Aesop's fable of the boy who cried "wolf." If another monkey's leopard call was then played, however, the troop responded as though there really were a leopard in the immediate vicinity. Likewise, if the first monkey's eagle call were then played, the group responded as if an eagle had been sighted overhead.

From this behavior we can conclude that a listener processes sensory information about predators in different ways. Furthermore, the monkeys are able to distinguish one caller from another. They not only recognize who is calling, they are able to adjust their behavior to both the caller and the specific source of danger to which the call refers. Such observations provide less insight, though, about what the *caller intends*. Does it call to *change the behavior* of other monkeys, or does it call to alter what the other monkeys *believe*?

In fact, there is no evidence that when a monkey utters an alarm call he intends that other monkeys understand his state of mind. Cheney and Seyfarth describe an example of vervet behavior that illustrates this point in dramatic fashion. A subordinate monkey watched the approach of a strange male. Males often move from one troop to another, but their arrival can depress further the social status of the least dominant resident males. On one occasion such a resident male gave an alarm call, and Cheney and Seyfarth describe how this call sent the approaching newcomer scampering up a tree. The resident male, however, continued to move about on the ground calling "leopard."

Now if you were a newcomer treed by such a ruse, it would not take you long to figure out that the fellow who is shouting alarms must be deceiving you about the presence of danger, for otherwise he would promptly climb a tree himself. Likewise, if you were the resident male and had any hope of fooling the newcomer, you would know that strutting around in the open yelling "danger" would be unlikely to deceive for very long. In short, to make the ruse convincing, you would have to persuade the newcomer that you really want him to conclude *that you believe* there is a leopard on the prowl. The newly arrived vervet, however, was more simply fooled. He attributed no significance to the caller remaining on the ground. Similarly, the caller had no understanding that by not getting into a tree he should be giving the game away.

A number of other studies, many in laboratory environments, support the same conclusion. There is no evidence that monkeys have a "theory of mind" in which beliefs are attributed to other individuals.

OBSERVATIONS OF APES

Because the great apes—chimpanzees, bonobos, gorillas, and orangutans—are our closest living relatives, during the last forty years considerable effort has been made to study them in the wild (Chapter 13). Here, however, we will consider a few observations that bear on their mental abilities, and in particular the question of whether, unlike monkeys, they infer what other apes believe about the world. This is an issue that remains controversial, for reasons that will become clear presently.

The primatologist Frans de Waal reports an instance of a chimp at the Yerkes Primate Center being provided with some particularly tasty food just before other animals were to join him. The chimp who had received this treat immediately hid it before the others could see him with it. While the other chimps were present, he took surreptitious peeks at his cache, but only returned to retrieve it when he would not be pestered to share it. This behavior not only indicates a recognition of how other animals in the group were likely to treat him if they saw the food, it also shows foresight and planning to deal with the situation. It is also an example of deception, and some have suggested the deception requires a knowledge of what other individuals are likely to believe.

Another example comes from de Waal's study of the colony of chimpanzees at the Arnhem Zoo in the Netherlands. Male chimpanzees contest with each other for dominance, and the dominant male may require an alliance with another male to preserve his position. These associations are fluid, however, and as new males mature and older males age, the social rela-

tionships change. On one occasion, where an up-and-coming male was hooting and displaying, the face of the dominant male assumed a characteristic expression called the "fear grin," which is observed when chimps are anxious or fearful. It seems to be a response that is under control of the autonomic nervous system, thus involuntary, like the sweaty palms of anxious or fearful people. In this instance the chimp tried to cover his face with his hand. The dominant male's position would become more vulnerable if a rival detected any sign of weakness, and his response was both defensive and deceptive. His behavior therefore suggests a cognitive process that weighs stored information about the likely effect his facial expression would have on others.

Chimpanzees are also capable of observing another chimp trying to reach food and inferring how they can help, such as by passing a stick or holding a large branch to serve as a ladder. Monkeys can be taught to help one another, but the relationships are not reciprocal without additional training.

Do these observations indicate that chimpanzees and other apes act to influence the beliefs of other individuals? Let's consider an account that is based on an actual laboratory experiment with an orangutan and the alternative ways the results can be interpreted. The orangutan sits behind a transparent partition. By looking through the partition, it is able to see an experimenter as she enters the room and places an edible treat in one of two opaque containers. It is possible in such an experiment to train the orangutan to point to the container that has the food. The experimenter then leaves the room and a colleague enters. When the orangutan correctly indicates by pointing, this second handler then removes the treat and gives it to the orangutan. In principle, the orangutan's mastery of this task is no different from many examples of one animal manipulating another to achieve some desired goal, although in this case the manipulation involves a member of another species.

Now consider a more difficult situation. The second person must use a tool, for example a small rake, in order to push the food to the edge of the container where it can be reached. The orangutan watches this process for a number of times. Then a third person enters the room between trials, and the orangutan watches while he hides the rake under a piece of cloth in the corner of the room and then leaves. Shortly thereafter the second experimenter returns, the orangutan points to the box with the food, the person looks about for the rake, but does not see it and seemingly is unable to extract the food. After a few such trials, an orangutan that was tested in this fashion pointed spontaneously to where the rake was hidden.

One interpretation of this kind of experiment is that the orangutan has a mental process similar to what a human would experience in such a situation. Put yourself in the position of the orangutan. You would likely reason that the person who got you the food with the rake the first time does not understand where the rake is now located and you need to change his understanding. In other words, you would want to alter what the other person believes.

Did the orangutan want to alter what its handler believed? Perhaps, but there is an alternative interpretation that does not require that the orangutan have any knowledge of the person's mind. The orangutan may have wished to direct the handler's behavior so that it would be given the food. It recognized, though, that removing the food from the box required the use of the rake as a tool. By pointing to the hidden tool it was directing the human's actions, albeit with a sophisticated understanding of the physical relations of objects and causal events, but without any comprehension of another creature's thoughts.

Virtually every observation or experiment that can be interpreted as showing an effort of one animal to alter the mind of another can also be interpreted as an effort to change the behavior of the other without assuming that it has a "theory of mind." Think about the two examples of deceit described earlier. In each case the chimpanzee behaved as though he knew how others would interpret what they saw, taking the food away in the first example, or detecting weakness and becoming more aggressive in the second. Alternatively, however, each of those chimpanzees might simply have known from experience how adult males would be likely to behave toward him when he has food or is feeling fearful. Temporarily hiding food so that others will not see you with it is nevertheless a sophisticated manipulation of the physical world undertaken to prevent a behavior of other animals. The chimpanzee in the second example was not just fearful; at some cognitive level it recognized that having a facial expression that it could not control was an undesirable signal to present to another male, and it behaved so as to prevent the situation from getting worse.

Psychologists and primatologists are divided on whether apes act to control the understanding of other apes. Some who have the more conservative view argue that, along with the capacity for language, the ability to make mental images of what another individual is thinking is a unique feature of *Homo sapiens*. The problem is that absence of conclusive evidence is not the same as evidence against, so the most cautious interpretation is that we do not yet understand the bounds of primate cognition.

AWARENESS OF SELF

One of the features of our own mind is that we have a keen sense of self, and knowing about ourselves provides the basis for inferring the beliefs of others (Chapter 9).

Experiments with mirrors have suggested to some that the great apes have a concept of self that no other primate except humans possesses. If a parrot, an intelligent bird by other measures, views itself in a mirror, it reacts to its image as if it were seeing another bird. Dogs rarely take any interest in their own reflection, and if they take notice at all, they quickly lose interest after a sniff provides no additional information. Monkeys show interest for awhile, as though they were seeing another monkey, and they may try to peer around the mirror to see what is behind. With a little experience, however, chimpanzees, bonobos, and orangutans behave as though they see themselves. They make faces in the mirror, decorate themselves with leaves, and try to examine parts of their anatomy like their bottoms or the insides of their mouths, that they are ordinarily unable to see. In a now classical observation, a chimp familiar with mirrors had a spot of water-soluble red paint placed on its forehead while it was sleeping. On awaking it saw its reflection in a mirror and immediately touched its forehead and examined its fingers. This simple experiment seems to say that the chimpanzee understood that the image in the mirror represented itself.

Chimpanzees identify other chimpanzees by their faces, and Lisa Parr and Frans de Waal have shown that chimpanzees can frequently relate sons and mothers in photographs of unfamiliar individuals from distant locations. Faces therefore convey very specific information. A conservative interpretation of the mirror experiment is that the animals are simply examining parts of their bodies that they cannot otherwise see, but it is hard to escape the conclusion that they know it is their *own* body they are seeing.

APES AND LANGUAGE

Are apes capable of language? This question has been studied by several research groups, using chimpanzees, bonobos, orangutans, or gorillas. Some investigators have taught the animals to sign, as deaf people communicate. Others have taught them to associate objects with arbitrary visual symbols. Most psychologists largely lost interest in this work when it became clear that apes are so limited in syntax that they are incapable of human language. Several research groups have persisted in their studies, however, revealing that apes possess a remarkable cognitive capacity for abstract representation that would never have been detected by observing the animals in nature. Moreover, the apes are able to communicate by signing simple two-word phrases with their human trainers or with other apes who have been similarly taught, sometimes displaying frustration when novice animals do not respond.

Particularly striking results were obtained with Kanzi, a bonobo studied by Sue Savage-Rumbaugh.

When it was realized that Kanzi was learning to sign simply by watching the relatively unsuccessful efforts to teach her mother, the training regimen was altered. Kanzi simply listened to the caretakers talking about what was going on and the activities that Kanzi enjoyed. As these people spoke, they pointed to symbols. Kanzi (and several other apes) have been successfully taught by this passive means. Kanzi could understand 150 spoken English words when he was six, and on hearing a simple command he was able to use word order to distinguish the subject of the sentence from the object of the verb. Stunning as these observations are, they are comparable to the accomplishments of a two-year-old child. It is nevertheless interesting that this limited ability of apes to learn to communicate symbolically appears to be greatest when they are young, and like human children they learn simply by observation.

These language studies illustrate another important point. During the last several decades there have been a number of studies of the cognitive capacities of apes that have been reared in captivity in close daily association with humans. These animals frequently show abilities that are not obviously displayed by apes in the wild, suggesting the presence of cognitive capacity that becomes apparent only in this very special cultural environment. Earlier in this chapter we pointed out that the social environment of primates has likely been an important factor in the evolution of the neocortex. The comparative study of apes and monkeys, however, shows that this cannot be the entire story. Except for mothers and their young offspring, orangutans are solitary, yet when kept with humans and studied they seem just as clever as chimpanzees. As chimpanzees are the only apes known to use tools extensively in the wild, this is further evidence for a latent ability in the brains of apes that is not necessarily apparent in their normal day-to-day activities in their natural environment. The same might also be said, however, for parrots, dolphins, and most humans. Brains that evolved to draw on experience and adjust behavior on a contingent basis appropriate for the individual have emergent capacities that are not, in themselves, adaptations. Evolution can of course exploit those latent capacities further. Human language is such an example, for sound production has required physical changes in the larynx as well as additional changes in the brain.

DO ANIMALS HAVE CULTURE?

Culture—the inventing, copying, and passing on of ideas, technologies, and beliefs—is the hallmark of the human species (Chapter 14). But to what extent are we unique? Is it possible to find the precursors of our capacity for culture in other species?

Some argue that language or teaching are required for culture, thus defining culture as exclusively a capacity of humans. A biologically more useful definition of culture, however, is the existence of distinct learned behaviors in different populations of the same species. As many animals are able to learn, and the development of species-specific skills such as hunting can involve the participation of parents, perhaps we should not be surprised to find learned behaviors that differ between populations of the same species. The songs of birds, for example, can differ from one geographic region to another, reflecting learned variants on a common theme. More interesting, however, are examples of novel behaviors that have appeared in one or a few individuals and then spread to others in the social group. A now famous example is potato washing in a troop of Japanese macaque monkeys on the island of Koshima. Sweet potatoes were placed along the beach to supplement their diet. About a year later a two-year-old female named Imo (Japanese for "potato") began to wash the sand off her potatoes by holding them under water with one hand and brushing them with the other. In time this behavior spread to other young members of the troop. About two years later Imo was using another new cleaning technique, this time for wheat that had been scattered on the beach. Instead of picking up the wheat grains one by one, she seized a handful of sand and wheat, tossed it on the water, and skimmed off the floating wheat grains after the sand had sunk.

What cognitive processes were involved in this discovery? All monkeys brush dirt and debris from their food, and the beneficial effects of dropping a potato into the water may have been discovered by accident. Similarly, it is not clear that other monkeys imitated her actions with a clear intention of producing clean food. The custom spread sufficiently slowly that it is quite possible that in social interactions with her they discovered the benefits for themselves, and repetition of the behavior was thereby reinforced. Nevertheless, useful behaviors were discovered and spread within and between generations. We can see here the early origins of a capacity for culture.

Not surprisingly, the best evidence for the rudiments of culture and cultural evolution among animals is found in chimpanzees. Recently a group of scientists who study chimpanzees at the seven major field sites in Africa compiled their observations on behavioral variations. Of the sixty-five behaviors that they cataloged, seven occur at all sites and another sixteen have been observed so infrequently that they did not occur regularly at any site. The remaining forty-two, however, occur at some sites and not others, either in all members of the same age and sex or at least repeatedly in a subset of individuals. Three of these 42 cases could be accounted for by special ecological circumstances. For example, chimpanzees only make their night nests on the ground where they are not vulnerable to leopards and lions.

There are thus thirty-nine known examples of arbitrary, socially transmitted behaviors that are present in some chimpanzee groups and not others. Moreover, each of the seven chimpanzee populations has its own unique combination of tools (Fig. 11.13) and the techniques for using them, as well as variants in how the animals communicate, cooperate in hunting, respond to the behavior of other species in the area, and use medicinal plants against infectious agents.

What are some examples? The primate behaviorist Christophe Boesch has described evidence for social learning in catching ants. This "ant dipping" is carried out in different ways in two populations of chimpanzees. At Tai in West Africa the chimps place a stick near the entrance to a nest of army ants, wait until the ants crawl on it, and then pick them off with their lips. In Gombe, however, they use a longer stick (ca. 0.8 m), strip off the bark, and shove the stick into the entrance of the ant nest. They catch the ants by grasping the stick at one end, quickly running their closed hand along its length, then popping the accumulated ants into their mouth. Although there is no ecological factor that would prevent one group from using the other's technique, the second way is four times as efficient as the first. If these techniques were regularly learned anew by individual discovery, we would expect the efficient method to have been found by both groups. Instead, it appears that imitative learning can constrain ant-dipping to a social norm that is not maximally effective.

Arbitrary variation among chimpanzee populations in the attention-getting behavior of males suggests that it, too, is acquired by social learning. Dominant males make their mating intentions known to estrous females by loud displays such as shaking saplings or dragging branches. Less dominant males must use less conspicuous attention-getting displays so as not to provoke an attack from a higher-ranking male. In one population the less dominant males pick leaves and tear them in order to signal their interest. In another population they knock their knuckles on the trunks of small saplings. There are obviously many ways of making low-level noise to attract attention of others in the vicinity, and the exclusive use of one method by members of a population very likely reflects a local custom established by social learning. This certainly is the way in which humans learn and transmit many of their cultural traits.

In hominid evolution, when the rewards of new discoveries were great, we would expect selection to have favored those individuals that were better able to understand why their actions had the consequences they did. A mind that can form mental images of cause and effect would be adept at this process and would open the door to the explosive growth of culture, to which we will return in Chapter 14.

GENETICS AND HUMAN BEHAVIOR

The notion that "genes are destiny" is a prevalent lay view deeply imbedded in our culture but so simplified as to be a caricature of nature. One reason for this widespread view is the way students are introduced to genetics in school. In order to illustrate how Mendel's rules follow from the independent segregation of alleles associated with the movement of chromosomes during meiosis, introductions to genetics emphasize clear examples of inheritance based on single genes. In these examples, phenotypes are determined by a few discrete combinations of alleles.

The idea of genetic determination is commonly reinforced in newspaper accounts of inherited disorders with severe consequences such as muscular dystrophies. These sorts of examples convey an overly simplified view of genetics, because many aspects of phenotype depend in complicated ways on the actions of numerous different genes. Consequently a mutation in one gene can be associated with a variety of phenotypic outcomes, depending on what alleles are present at other loci and how the expressions of these genes are affected by the environment that the organism experiences (Chapter 4).

Another misconception is that developmental outcomes can be apportioned between genes and environment—between nature and nurture—and in the case of human behavior, with a strong predilection for the latter. As we have argued above, brains and behavior have evolutionary histories, and genes affect behavior through their influence on the formation of the nervous system during development.

Genetics is a powerful tool in understanding biological phenomena, so let us explore what difficulties it encounters and what insights it provides in understanding human behavior. Genetics offers two different approaches: one derived from the study of populations and the other from knowledge of particular gene products and the pattern of inheritance of specific alleles.

IDENTICAL TWINS

Is it possible to find evidence for genetic influences in the most general aspects of human behavior, in features so inclusive that individual differences can reasonably be viewed as natural variation in the genetic instructions? One interesting example comes from studies of personality differences in monozygotic and dizygotic twins. Monozygotic or identical twins develop from a single fertilized egg and separation into two genetically identical embryos early in development. Dizygotic twins, on the other hand, develop from two independently fertilized eggs and are genetically no more alike than siblings that result from separate pregnancies.

Personality has no simple definition, and it is compounded of several traits. In order to measure personality differences among individuals, questionnaires have been devised that characterize individuals simultaneously on several different scales such as agreeableness, conscientiousness, extraversion, neuroticism, and openness.

Two studies of personality differences between mono- and dizygotic twins of both sexes have been carried out in recent years. Most of these twins were reared together with their genetic parents, but some 1000 were reared apart by adoptive parents. The principal finding is that the personalities of monozygotic twins are more likely to be similar to each other than are the corresponding traits of dizygotic twins, even if the twins are reared apart. A useful way to analyze a large amount of data with a number of different variables is to employ statistical techniques that resolve the *sources of variation in the population* into genetic and environmental components (Chapter 4). In these studies on twins, more than 40% of the *variation* in personality is due to genetic influences (heritability = 0.4) and the remainder to a combination of environmental influences and unidentified factors, in about equal proportions. These studies of twins thus lead to the conclusion that aspects of an individual's personality are strongly influenced by their complement of alleles.

A related and very interesting aspect of this work is that shared environments account for only about 7% of the total variation in personality differences. A traditional source of criticism of twin studies (advocated by those convinced of the absolute supremacy of the environment) is that identical twins behave more alike because they are treated more similarly than nonidentical siblings. The modest influence of shared environments

not only undercuts this hypothetical objection, it points up the uncertainty about which environmental influences are important and why. A number of the psychologists who have contemplated this problem believe that in important ways children as well as adults make their own social micro-environments in the activities they pursue and in the friends they seek. Viewed this way, an individual's genotype comes to influence their social environment (through behavioral choices), which in turn influences the development of the behavioral phenotype. Once again we confront the inseparability of nature and nurture.

BEHAVIOR GENETICS

Genetic experiments are necessary to establish whether a particular phenotypic character is inherited. Matings between people, however, cannot be arranged like crosses between fruit flies, so indirect genetic analyses must be used. For example, one might examine the members of a family in several generations to see whether the pattern of occurrence of a trait is consistent with Mendelian rules. This can be fruitful if there is only a single genetic locus involved (Fig. 11.17), but it will likely fail if the trait is based on several genes. In the past decade it has been possible to extend this kind of analysis by looking at all members of a family for regions of DNA where a specific difference in base sequence correlates with the presence of the phenotypic character.

There are three important aspects of this kind of work to consider further: *(i) conceptual*: what biological processes do the results of the experiments address; *(ii) technical*: how certain can one be that a genetic interpretation is likely to be valid, or the only interpretation; and *(iii) practical*: to what social uses should genetic information be put?

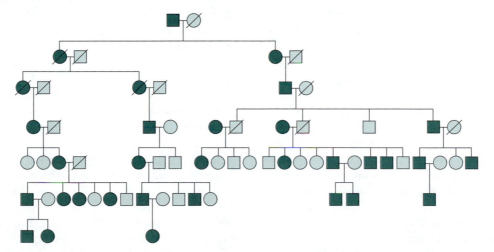

FIGURE 11.17 Pattern of inheritance of the autosomal dominant allele that causes Huntington's disease. Circles are females, squares are males. Dark symbols are afflicted individuals; lighter symbols are individuals with no symptoms. This family is from Venezuela.

(i) A Conceptual Pitfall in Behavior Genetics: Genes "for" Behaviors. Consider the kinds of phenotypes that have been studied in genetic analyses of human behavior, for here lies a problem. Most have attracted attention for medical reasons such as mental retardation, schizophrenia, Tourette's syndrome (involving loss of control of movement and speech), or senile dementias. Or alternatively, the individuals may exhibit behavior that is unacceptable for social reasons: aggression, criminal propensities, or other rather general features that may well resist consensus definition and do not represent a definable medical problem. Assuming for the moment there is a finding of genetic involvement (but see point *ii* below!), the common presentation in the lay press is that a gene for alcoholism, or a particular kind of mental illness, or for a certain sort of criminal behavior has been identified.

Such language is very misleading. What may have been discovered is a genetic mutation that leads to an alteration in some physiological or neurological process. What has *not* been discovered is a gene that codes for a behavior. Brains cause behavior, not genes, and functioning brains are the result of the process of development in which many genes acting sequentially both affect and are affected by events, many external to the genome, external to any particular neuron, and even external to the organism (Chapter 10). Furthermore, brains change during the lifetime of an individual, most notably by a process we call learning.

A simple analogy may help you to remember this very important distinction. Computers are based on the designer's plans for how the various components of the computer can be assembled to make a machine on which you can write documents, solve problems, and communicate with other computers. In a sense, the designs are like the genome. For the design to become functional there must be a process of assembly in which the parts are put together (analogous to development), and the installation of software (crudely analogous to learning). Now imagine you are sitting in front of your new computer trying to finish a paper for class, and the mouse responds erratically when you push it about and depress its buttons. Have you discovered a mutant "mouse gene," a feature of the design that codes for the mouse? Most assuredly not. The problem could be a bad electrical connection, faulty software (e.g., the wrong mouse driver), a failure in one the computer's chips, or any of a large number of problems. Your troubles are to be found in the "physiology" or the "biochemistry" or the "neurology" of the computer, perhaps even an error in assembly. Clearly, wherever the fault is ultimately located, there will be some feature of the design that relates to it, for the design requires that the right instructions be present in the mouse driver, and it requires that all electrical connections in the circuits conduct, and that all the transistors

and memory locations work properly. Moreover, a failure in the hardware can be manifest in a variety of ways at the level of the computer's "phenotype," but the specifics of the malfunction may not lead you directly to the source of the problem, even if you are an experienced computer "physician." (Incidentally, computer "physicians" have a relatively easy job: they generally heal their patients by the simple technique of "organ transplants.")

Back, now, to humans. The condition known as phenylketonuria is a well-understood example of a phenotype that is caused by an inherited defect in metabolism. Behaviorally, individuals with this disorder have severe mental retardation. The genetics is simple and involves a recessive allele on one of the autosomal chromosomes. The gene codes for an enzyme (phenylalanine hydroxylase). Phenylalanine is one of the twenty amino acids found in proteins. It is therefore a normal constituent of the diet and is used in assembling your own proteins, but it is also used elsewhere in the biochemistry of the body. If there is an excess of phenylalanine, it is modified by the enzyme phenylalanine hydroxylase. In victims of this genetic disease, the phenylalanine hydroxylase is defective, phenylalanine concentrations increase in the blood, and for reasons that are less well understood, this interferes with the proper development of the nervous system. Interestingly, if high levels of phenylalanine in the blood are detected early enough, the behavioral deficit can be prevented simply by limiting the intake of phenylalanine in the diet.

Should this gene be considered a gene for intelligence? This label provides no insight, for the mutation is one of countless ways in which the development of a normal brain can be hindered. Moreover, the phenotype can be defined in terms of different levels of organization, from molecular (enzymatic) to behavioral. This gene is better viewed as one of many whose protein product is required (in this case seemingly indirectly) for a completely functional brain to come into existence. Note further that although the mutant form of this gene causes an enzymatic defect, the phenotypic consequences for behavior can be averted by modifying the *environment*.

In principle, mutations in genes coding for ion channels or enzymes involved in the metabolism of neurotransmitters are likely to have behavioral consequences. Chapter 9 provided several examples that have clinical importance. Genetic contributions to the commonly recognized psychiatric disorders such as schizophrenia are not simple, however. Multiple genetic loci are involved, and there are strong environmental interactions, none of which is adequately understood.

To summarize, there are innumerable ways in which genetic mutations can alter the development of nervous systems, and thus behavior. The mutations that have been identified so far either influence some basic function of the working brain or disrupt the normal course of its development. No single gene or gene product so far known is the "cause" of a complex behavior in the sense that it specifies complex neuronal architecture. To speak of genes for particular behaviors is therefore conceptually misleading. It is possible for a single genetic mutation to perturb behavior by interfering with the development of a specific region of the brain or the metabolism of a particular class of neural transmitters. But in such cases, the behavioral alteration may be simply the most obvious among many phenotypic manifestations of the mutation.

(ii) Technical Pitfalls in Behavior Genetics.

In addition to the conceptual trap of thinking that there are genes "for" behaviors, there are several technical pitfalls in the analysis of genetic influences on behavior. Consider first how one defines the behavioral trait, the presumed phenotype that is suspected of being heritable. Although many medical problems with a genetic basis are readily identified, others are gray. Forms of mental illness can be particularly elusive because their symptoms can vary from almost normal to chronically dysfunctional, and even when severe, it is not always clear that a given set of symptoms represents a single disease state. If we turn to behavior that is recognized because it is antisocial, excessive aggression, for example, the problem of distinguishing from the normal is even more problematic. Genetic studies are problematic if the phenotype cannot be identified unambiguously.

There are several reasons why human behaviors can be so hard to characterize in ways that make the search for genetic underpinnings useful or appropriate. Consider aggression again. We all know that individuals vary greatly in their aggressiveness, that aggression takes many forms in addition to physical violence, and that aggression is more likely to occur among individuals who are denied opportunities or are mistreated. The line between the normal and the abnormal is therefore likely to be arbitrary, to depend on the social and cultural context, and to resist a consistent definition. Additionally, a concept like aggression involves both feelings and actions, motivations and behaviors, and therefore, like personality, is an emergent feature of the brain.

The search for genes involved in psychiatric disorders is further complicated by the frequent participation of many genes, each making a small contribution to the outcome and each influenced by genes at other loci as well as by environmental factors such as diet or toxins like lead. Identifying any single gene in studies of family lineages therefore becomes difficult, and interpretations of the data on an individual family can be led seriously astray by only one or two missed diagnoses of phenotype.

(iii) Moral and Policy Issues.

Suppose a mutant gene is identified in an infant, one whose phenotypic effect may lead to serious illness later in life. Perhaps, for reasons unknown, the phenotype is not always expressed and some individuals with the allele are never troubled with the disorder. Perhaps this outcome depends on what other genes are present. Alternatively, maybe the absence of phenotypic expression requires an as-yet-unknown environmental factor, and if the factor were to be identified, the symptoms of the disease could be prevented in every individual who bears defective copies of the gene.

What should be done with this information? What are the psychological effects on an individual if he or she must live with the knowledge that they are doomed to an early and unpleasant fate? Or that only *maybe* they going to contract the disorder? And how do we know how any particular individual will handle that information? What is to prevent insurance companies from denying them coverage or employers from failing to hire them? On the other hand, suppose a preventive therapy is discovered in the near future. What, then, are the consequences of not having made information about their susceptibility available to carriers of the gene? To what extent might a specific set of genes (if any!) make an individual prone to criminal behavior? Would knowledge of that information help to prevent the antisocial behavior? Or would it simply stigmatize him so that the behavior becomes more likely? Would such knowledge lead to an expanded class of legal defenses in which individuals are claimed not to be responsible for their behavior? Or perhaps to arguments that punishments should be *increased* on the premise that individuals more prone to antisocial behavior require more severe deterrence.

These are among the kinds of questions that are currently being raised about research on the genetics of human behavior. They are important issues, and they illustrate how our societal institutions and our beliefs can be challenged by scientific knowledge. There are no simple answers to these questions. Indeed, because they involve conflicts of human interests at so many different levels and in so many different contexts, perhaps the best that can be achieved is informed judgment, constantly honed by an ongoing process of discussion. The knowledge and the opportunity that science can provide, however, will certainly change; that is the nature of science. So as these issues enter the public dialogue, and increasingly they will, each of us must bring to them understanding of their inherent complexity and how no single sphere of human understanding can address them in isolation.

SYNOPSIS

Charles Darwin extended the concept of homology to behavior and argued that

> The difference in mind between man and the higher animals, great as it is, certainly is one of degree and not of kind. We have seen that the senses and intuitions, the various emotions and faculties, such as love, memory, attention, curiosity, imitation, reason, etc., of which man boasts, may be found in an incipient, or even sometimes in a well-developed condition in lower animals. Charles Darwin, *The Descent of Man*, 1880, D. Appleton & Co., p. 126.

During the first part of the twentieth century the study of animal behavior traveled along separate paths: evolutionary theory lacked the conceptual understanding necessary for meaningful analysis of animal behavior, and psychology was preoccupied with discovering the rules that govern learning. In recent years, biology and psychology have developed a common understanding that behavior cannot be partitioned into learned and inherited components. As nervous systems develop, the actions of genes and the actions of environmental influences are so intertwined that it is nonsense to think of either as independent of the other. This understanding is an extension of general principles that operate during development, with implications for the word "instinct." For example, various adaptive (thus "purposeful") behaviors may or may not require external programming (learning) for their expression. Likewise, some experiences with the external world (learning) are a normal part of development. Furthermore, natural selection has influenced the kinds of learning that occur in different species. Examples of all of these phenomena are present in humans.

Understanding the cognitive abilities of other species is difficult and largely depends on observing how individuals behave under natural conditions in the presence of particular stimuli. Despite these problems, there is evidence that apes are more intelligent than monkeys, both in dealing with novel problems and in what they seem to know about the intentions if not the minds of other individuals.

The application of genetics to behavior is still in its infancy. Behavior is an expression of the activity of brains, and brains in turn are products of development. Many genes must act at appropriate places and in regulated sequences for brains to develop. The causal relationships between genes and behavior are therefore indirect, complex, and frequently indeterminate, even though small differences in behavior can have large effects on reproductive success. Single mutations that disrupt aspects of neural function or development have phenotypic expression in behavior, but it is a misconception to conclude that there are specific genes "for" behaviors.

Knowledge from molecular genetics will continue to enlarge our understanding of evolution, development, and the nature of behavior and of disease. Some of this knowledge will raise practical and moral issues about how best to use new information. Wise decisions will require that those affected understand the possible consequences of alternative courses of action.

QUESTIONS FOR THOUGHT AND DISCUSSION

1. What is the evidence that behaviors, like morphological structures, can be adaptations?

2. What is the evidence that animals are prepared, unprepared, or even counter-prepared to learn specific behaviors?

3. What do the experiments on navigational and food source learning in bees, alarm call and song acquisition in birds, and language acquisition in humans suggest about the distinction between "nature and nurture" or "instinct and learning"? Can these distinctions be made for any behaviors? Explain.

    How do these examples of behavioral development relate to principles of morphological development introduced in Chapter 10?

4. Explain in your own words the possible meanings of "know," "purpose," and "intention" as applied to the behavior of nonhuman animals.

SUGGESTIONS FOR FURTHER READING

Alcock, J. (1998). *Animal Behavior*. Sunderland, MA: Sinauer Associates. A popular introduction to animal behavior and a general reference for the first part of Chapter 11.

Byrne, R. (1995). *The Thinking Ape: Evolutionary Origins of Intelligence*. New York, NY: Oxford University Press. An accessible review of primate cognition. Readers interested in the subject should compare the conclusions with those of Tomasello and Call.

Hauser, M. D. (1996). *The Evolution of Communication*. Cambridge, MA: M.I.T. Press. A searching review of animal communication from the multiple perspectives of behavior, neurosciences, linguistics, psychology, and evolution.

Tomasello, M., and Call, J. (1997). *Primate Cognition*. New York, NY: Oxford University Press. A careful overview of primate cognition, summarizing both laboratory experiments on captive animals and field observations under natural conditions. Some of the interpretations are more conservative than in Byrne.

PART FOUR

Our Place in Nature

After the rain at Ndutu, Tanzania. Landscapes stir the emotions, and in many cultures they are a common theme in painting and photography. Biologists have begun to explore these connections (e.g. Orians and Heerwagen, 1992). Landscapes—in life as in art—also provide information about habitat, including possible sources of cover, and storms can signify both danger and the coming of precious water. Most of the evolution of humans took place on the savannas of East Africa, and the characteristic shapes of its acacia trees tend to evoke positive feelings even in individuals who have never visited the region.

12

The Physical Record of Human Origins

DECIPHERING HISTORY

Discovering our origins is a complex puzzle that illustrates several important features of science as a process for understanding nature. First, science is ongoing, always unfinished, and many questions about our evolution remain unanswered. Nevertheless, numerous discoveries have been made since the publication of *On*

Photo: Skulls of two of the several species of *Australopiticus*, hominids that lived in Africa from more than 5 to less than 2 million years ago. These fossils illustrate two lineages of australopithecines: lightly built "gracile" forms (above) and heavier "robust" forms (below). These species lived at the same time, but from their dentition the robust species seems to have been more dependent on coarse plant food. The amount of interaction between contemporaneous lineages of hominids is one of the interesting unresolved questions in physical anthropology.

the Origin of Species in 1859 and *The Descent of Man and Selection in Relation to Sex* in 1871. In Darwin's time only a few human fossils were known and recognized as relevant to human origins, but since then thousands have been found.

Because the question of human origins is a historical problem, the record can never be complete. As is true for all organisms, it is very unlikely that the remains of any individual will be covered and preserved in sediments. It is even more unlikely that they will then survive thousands or millions of years of geological alteration and finally be discovered by trained paleontologists. The fossil record is therefore frustratingly fragmentary.

As evidence slowly accumulates, it is constantly reinterpreted in the light of new knowledge, and because the record is always incomplete, alternative interpretations are common. In this respect, the study of human origins is like all historical problems, whether they are in politics, art, literature, or any other aspect of culture. Nevertheless, as information accumulates, parts of the puzzle start to connect, and the outlines of the historical picture become clearer. Certainty is never absolute, but as knowledge grows, understanding deepens.

The second point about the scientific process illustrated by the search for our origins is that science is a human enterprise, and interpretations can be influenced by the cultural milieu of the time. This has been particularly true of human origins, because beliefs and understandings about humanity are based on many different sources besides scientific investigation. Many nineteenth century anatomists found it difficult to reconcile their religious or political beliefs with new geological and biological evidence of human evolution, and this is still true for many people. More recently, anthropologists have been understandably eager to infer how early hominids lived from findings of fossil teeth, bones, and simple tools. While these inferences are frequently clever and plausible, they can still be influenced by cultural beliefs. For example, how the damaged skulls of ancient hominids are interpreted may depend on whether one thinks that conflict is an old rather than recent feature of human behavior.

Finding human fossils is very difficult and time-consuming, and many early fossil hunters honored their efforts by assigning their finds to a new genus or species. The synthesis of natural selection and genetics that emerged in the 1930s and 1940s (Chapter 5), however, changed paleoanthropology profoundly. The importance of natural variation was made apparent, and with the discovery of increasing numbers of fossil specimens, the proliferation of taxonomic names has been tamed as experts have been able to clarify and agree on many of the relationships. Even with this greater knowledge, however, the classification of fossils involves a conceptual issue that can be easily overlooked:

different criteria are often used in defining living and extinct species. One important conception of a species—as a population of potentially interbreeding organisms—emerged during the "modern synthesis." It is a useful concept for analyzing microevolutionary processes in living animals, but paleontologists cannot know whether the animals represented by fossils—sampled from a few locations and sometimes separated by hundreds of thousands of years—were capable of interbreeding. Fossil species must be defined by anatomical criteria, and since the evidence is often fragmentary, it is open to multiple interpretations. For example, because many lineages become extinct, there can be uncertainty whether a particular fossil is ancestral to any more recent form. Furthermore, with clearly distinct yet similar and seemingly related fossils that come from different geological times, interpretation can be difficult. Their differences may have arisen as gradual alterations in a single lineage, or changes may have occurred in small isolated populations that then expanded and displaced older forms.

A deeper conceptual point is that the categories used to classify animals and plants are subjective constructs. The standard hierarchical scheme of classification—Phylum, Class, Order, Family, Genus, and Species—is based on perceived degrees of relatedness, but the criteria used to define these categories in anatomically very different organisms must inevitably be different. Just as different features are used to classify ships and airplanes, different criteria must be used to define taxonomic families of trees and mammals. Although classifications reflect evolutionary relationships, they all require particular interpretations of the data. As we proceed, we will enlarge on this point with specific examples.

Discoveries about our evolutionary history also illustrate how science has become an interdisciplinary activity and how different approaches can yield totally unanticipated findings. To identify and interpret fossils requires considerable knowledge of anatomy. Finding sites likely to yield new fossils also requires an understanding of geology. Once found, fossils must be dated by correlating the features of the rocks in which they are located over wide areas, sometimes even separate continents. Alternatively, physical techniques are employed, such as measuring the proportions of element isotopes that have decayed constantly with time (Box 2.2). The interpretation of human fossils also requires knowledge of the artifacts of human activities, such as tools, habitation sites, and the remains of animal and plant foods. In recent years, molecular biology has added an entirely new approach to tracing and dating biological evolution, and comparative linguistic analyses have added to our knowledge of human migrations. Our understanding of human evolution will continue to improve by bringing together information and techniques from these varied sources.

"MISSING LINKS" AND HUMAN ANCESTORS

The notion of a "missing link" in human evolution is a misleading simplification. The term came into use at the end of the last century with the growing realization that chimpanzees, gorillas, and orangutans are our closest living relatives. It is associated, however, with a pair of misconceptions. The first is that we are descended from one of these living apes. Humans are of course not descended from any of the living primates; like humans, all of them are at the tips of branches of the evolutionary tree, and all have evolutionary histories as old as our own.

The second misconception is that evolution consists of progress from "less well-adapted" ancestral species into "better-adapted" descendants. In this view, our chimpanzee-like ancestor entered one end of a corridor about 6 million years ago, had all of its features slowly perfected toward a final outcome, and then emerged at the other end as a modern human. This is not an accurate description of the evolution of any group of organisms. The fossil and archaeological record of hominids now shows several different branches of hominids, each with a unique combination of adaptations that enabled it to occupy a particular niche. Moreover, the survival of our lineage to the present was never inevitable.

HUMANS AND APES: GENETICALLY CLOSE BUT PHENOTYPICALLY DISTINCT

From the first efforts to understand the origin of humans in an evolutionary framework it has been clear that those animals that bear the greatest similarity to humans are the primates, monkeys and apes in particular. Like people, primates have their eyes set in the front of the face with largely overlapping visual fields and binocular vision. They have hands with opposable thumbs capable of grasping, and they have an arm that rotates in the shoulder socket allowing them to reach overhead, climb, and swing by their arms. Among the primates, the apes bear the closest resemblance to humans. They are large, without a tail, and their molar teeth have a pattern of ridges shared only with humans and some related fossils. Much of this is relatively obvious to a casual observer and led to the early expectation of an evolutionary relationship between humans and apes.

This relationship was borne out by other lines of evidence. Among the primates, apes and humans appear most recently in the fossil record. Moreover, the fossil series that appears to lead to modern humans (see the following section) reflects a continuous change from apelike features—departure from a primarily ar-

boreal life, a new ability to walk erect, a change in diet, and most important, a steady enlargement of the brain.

By comparing the degree of similarity of the DNA or proteins of humans and apes it is possible to construct a tree of evolutionary relationships (Fig. 12.1). For example, the extent to which single strands of DNA from different species will pair together through complementary bases to form a hybrid double helix is a measure of shared genes (Box 3.2), and the more similar the genetic material of two species, the closer are their evolutionary relationships (Chapter 5). The remarkable finding revealed by this analysis of primates is that we share about 98.4% of our DNA with the two species of chimpanzee. Moreover, by this measure humans are more closely related to chimpanzees and bonobos than either humans or chimps are to gorillas. In fact, the genetic difference between humans and chimpanzees is less than the difference between the two species of gibbons, those long-armed apes of southeast Asia that swing from tree to tree in such spectacular fashion.

What are we to make of this fascinating finding? Most importantly, it confirms by independent means the inference that the great apes are our closest living relatives. But more than that, it has changed our understanding of the timing of the evolutionary divergence. We used to think that the human lineage diverged a long time before the separation of chimpanzees and gorillas, but the DNA evidence indicates that it came later, about 5 to 6 million years ago and well after the separation of the orangutan lineage. By this measure we are truly among the apes.

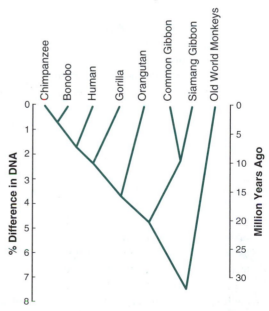

FIGURE 12.1 An evolutionary tree of "higher" primates based on DNA hybridization (Box 3.2). By this genetic criterion, we and chimpanzees are more closely related than either species is to gorillas.

Another reason that this new molecular evidence is surprising is the great phenotypic difference between apes and us. In terms of our anatomy, we walk erect, are relatively hairless, and are weaker, with smaller jaws and canine teeth, and we have much larger brains relative to the size of the body. More dramatically, we have language and a vastly more elaborate behavioral repertoire in the form of extensive cultures. These great differences in phenotype are clearly based on a very small fraction of our DNA. Thus one cannot infer the amount of genetic change from morphological divergence, and vice versa.

Based on the pronounced phenotypic differences between apes and ourselves, we have traditionally been placed in our own family, the Hominidae, whereas the apes are put in another family, the Pongidae. This example illustrates a characteristic of schemes of classification. Although all reflect in some measure evolutionary relationships, some emphasize a mixture of genealogy and obvious phenotypic differences, whereas another, *cladistics*, bases its conclusions on strict attention to genealogy. The latter employs a variety of criteria to decide which characteristics should be used in determining the locations of branch points in the evolutionary tree, of which the DNA evidence in Figure 12.1 is one example. The traditional placement of humans in a separate family, however, emphasizes phenotypic divergence at the expense of strict attention to genealogy. Taxonomic classifications are thus human creations and can have different implications for understanding evolutionary history.

SOME SNAPSHOTS OF HUMAN EVOLUTION

All of the paleontological evidence points to Africa as the place of origin of humans, starting about 6 million years ago with our last common ancestor with chimpanzees. Although many details of our evolutionary history remain unclear, the general pattern of hominid evolution shown in Figure 12.2 is a reasonably secure interpretation that is based upon a great deal of anatomical, paleontological, and archaeological evidence. All of this evidence has been gathered since Darwin, and most of it in the last thirty to forty years.

The tree in Figure 12.2 is just the hominid branch of the much larger tree of primates. The primates that gave rise to both humans and the apes of today are called dryopithecines, forest-dwelling creatures whose fossils are known from Africa, Europe, and Asia. The numerous fossils are at present divided into about three genera and eight species. One of the Asian genera is believed to have given rise to orangutans, and in Africa the group is considered to be the forebears of both apes and humans. A gap of about 3 million years separates

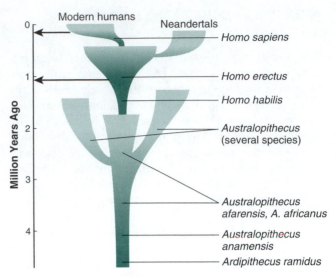

FIGURE 12.2 A summary interpretation of the evolution of *Homo sapiens* from *Australopithecus*. The arrows on the scale indicate the approximate times at which *Homo erectus* and *Homo sapiens* appear to have made major migrations out of Africa. See the text for further details.

the last known fossil dryopithecines from the first fossil that is considered human, *Australopithecus*. The paleontologist's work is never done!

LUCY AND HER FRIENDS WALK ERECT: THE AUSTRALOPITHECINES

Between 1973 and 1975, expeditions to Africa yielded several landmark fossils that are important for their age and for the story they tell. In the Afar Triangle of Ethiopia near the Red Sea, the paleoanthropologist Donald Johanson found a knee joint of a creature that clearly walked erect on two feet. Dating of volcanic rock above the fossils by radioisotope methods (Box 2.2) showed that the animal lived more than 3.2 million years ago (m.y.a.). Subsequently his expedition found nearly half of the skeleton of an individual of the same type, a female about 3 feet tall (to which he gave the name Lucy, Fig. 12.3). They also found the remains of more than a dozen individuals who apparently perished together in a flash flood—they are referred to popularly as "the first family."

In 1978, Mary Leakey, a paleontologist well known for her and her husband's finds of ancient hominids, made an even more remarkable discovery. At Laetoli near Olduvai Gorge in Tanzania her group found footprints that were left in wet volcanic ash 3.5 million years ago. The footprints are clearly hominid, lacking the opposable great toe of apes, and they were made by two individuals that were likely walking together. At the time of their discovery, these finds were the earliest known evidence of bipedal walking in hominids.

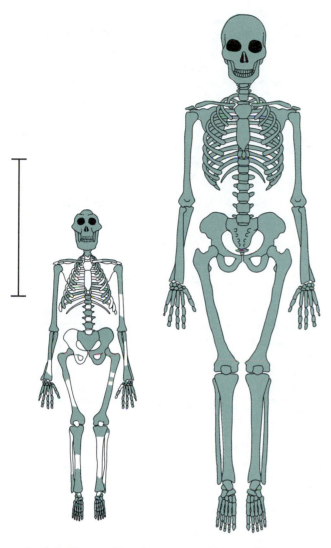

Australopithecus afarensis   Homo sapiens

FIGURE 12.3 Line drawing of Lucy (*left*) and an average modern female skeleton (*right*). The shaded portions of Lucy are bones of *Australopithecus afarensis* that have been found; the remainder is reconstruction. Specimens of *A. afarensis* vary in size, as shown by the scale on the left.

These fossils, and many younger ones of the same sort known from several sites in Africa, have been assigned to the genus *Australopithecus*. The name means "southern ape" and was given in 1925 by the anatomist Raymond Dart, who first found their fossils in South African limestone cave deposits. Australopithecines lived in Africa for about 2.7 million years, so it is not surprising that their remains provide evidence of evolutionary change and diversification. There are two broad categories of australopithecines: the smaller and more lightly built "gracile" forms and the larger, more "robust" forms. Among the gracile forms two species are generally recognized: *A. africanus*, of which the original South African find is an example, and *A. afarensis*, of which Lucy is the prime specimen. One of

the two gracile forms likely gave rise to our genus (*Homo*) between 2.8 and 2.2 million years ago. The robust forms (*A. robustus*, *A. boisei*) continue in the fossil record to more recent times, but they eventually became extinct without leaving any descendants.

Australopithecines are a mosaic of ape features, human features, and specializations of their own. Like later hominids, they were bipedal, but they did retain apelike adaptations for climbing in trees: relatively shorter legs, greater mobility of the wrists and ankles, and long curved fingers and toes. Also like later hominids, they had reduced canines and enlarged enamel-coated molars. But unlike both apes and later hominids their jaws and molars were relatively much larger—which suggests that hard, gritty plant food was a significant part of their diet. Males were 50% larger than females, perhaps more in some species. In this respect australopithecines are more suggestive of gorillas and orangutans (Chapter 13) or hamadryas baboons (Fig. 6.9) than chimpanzees. Finally, australopithecines were entirely apelike in one way: the relative size of their brains.

The adaptations of australopithecines for bipedalism and the plant and animal fossils found associated with them have been interpreted as supporting the "savanna hypothesis" for human evolution. According to this reading of the evidence, the divergence of hominids started when a forest-dwelling, long-armed (brachiating) ape that ate primarily soft plant food ventured bipedally onto open savannas to gather hard plant foods and hunt animals, using tools for both activities. However, it is very difficult to know from the sediments in which fossils are found how much time early hominids spent in various habitats. River, stream, or lake sediments are the most likely place for fossils to form, but many animals visit water only to drink, and they spend more time in dense rain forests, open woodlands, savannas, or combinations of these habitats. On the other hand, no fossils of the great apes are known from the same sedimentary rocks that yield early australopithecine fossils. It is therefore reasonable to infer that apes and early hominids were not sharing the same habitat. Nevertheless, no paleoecological evidence rules out the possibility that bipedalism first evolved in hominids within the forest. Bipedal locomotion would free the shoulders, arms, and hands to become more efficiently specialized for carrying food and using tools in hunting, defense against predators, and intergroup aggression, regardless of the habitat. Bipedalism, however, preceded the enlargement of the brain that characterizes later hominid evolution.

UNDERSTANDING THE ORIGINS OF AUSTRALOPITHECINES

During the past several years dramatic new fossil finds in Africa are bringing our knowledge of hominid evolution closer to the estimated time, 5 to 6 million years ago, when hominids diverged from a chimpanzee-like ancestor. One of the finds is of fossil leg, lower face, and jaw bones from 4 million-year-old river sediments in northern Kenya by Meave Leakey, another member of this famous family of paleoanthropologists. These fossils represent a new species, *Australopithecus anamensis*. The remains are fragmentary, but they tell much about this creature. The jaws and skull features are very ape-like, but unlike the teeth of apes, the enamel of the molars is relatively thick, as is characteristic of the teeth of all subsequent hominids. The nearly complete tibia (the larger of the two lower leg bones) is equally telling. In its length and articulations with the ankle and knee it is unlike that of apes and closely resembles that of fully bipedal hominids of a million years later. Furthermore, the end of the upper arm bone (the humerus) is similar to that of humans in an important feature. It lacks the deep hollow found in apes that makes the elbow joint more stable for knuckle walking by allowing the lower arm bone (the ulna) to lock in place. In brief, *A. anamensis* looked like a chimpanzee but walked erect, and its diet included significant amounts of hard and/or gritty plant food. Plant and animal fossils associated with those of *A. anamensis* suggest an environment of dense thickets and narrow stretches of forest along rivers, surrounded by bushes and open savanna.

In 1992 and 1994, bones of the skull, arm, hand, leg, and foot of a still older hominid were found by the paleoanthropologist Tim White and his coworkers in 4.4 million-year-old sediments in Ethiopia. This hominid is so different from later australopithecines and so close to our common ancestor with chimpanzees that it has been given a new genus and species name, *Ardipithecus ramidus*. In its thin dental enamel and strongly built arm bones it is apelike, but it resembles later hominids in having reduced canines. The foramen magnum, the hole in the skull through which the spinal cord connects to the brain, is at the base of the skull instead of toward the back, as in apes and other mammals. This position strongly suggests habitual upright walking. If this very early hominid was largely bipedal, it means that walking erect was an early adaptation associated with the divergence of hominids from chimpanzees, and that the change in diet, toward small and hard or gritty plant food, came later. Most of the plant and animal fossils associated with those of *Ardipithecus* are woodland species, suggesting that this early hominid spent more time in forests than did later australopithecines.

THE FIRST MAKERS OF STONE TOOLS: *HOMO HABILIS*

Under the watchful eyes of the famous paleoanthropologists Louis and Mary Leakey, the sedimentary rocks at Olduvai Gorge in East Africa yielded many important clues to early human evolution. Between 1960 and 1964, the Leakeys found fossils of a 1.6 to 1.9 million-year-old hominid with a cranium 50% larger than that of australopithecines and with smaller, more humanlike teeth. Louis Leakey named it *Homo habilis*, or "handy man," because of its large brain size and because he thought it had made the crude stone tools found in the same sediments. Specimens with similar traits are now known from other sites in East and South Africa, most notably a 2.5 million-year-old skull discovered in 1972 at Lake Turkana (East Africa) by a member of a team led by Louis Leakey's son Richard. Stone tools similar to those found at Olduvai have also been found elsewhere, the oldest from 2.5 million-year-old deposits in Ethiopia.

The anatomy of *Homo habilis* suggests that it evolved from a gracile australopithecine. *H. habilis* had smaller teeth than australopithecines, suggesting that this species relied less on coarse plant food. The most important feature of evolutionary change in the lineage from *Australopithecus* through *Homo*, however, was a progressive increase in brain size, both in absolute terms and relative to body size (Figs. 12.4 and 12.5).

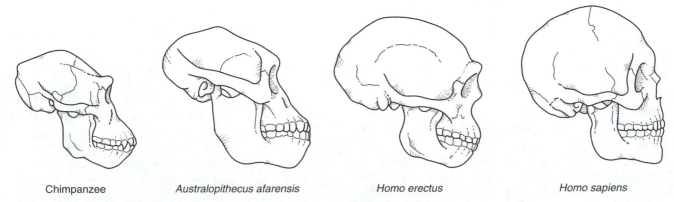

Chimpanzee   *Australopithecus afarensis*   Homo erectus   Homo sapiens

FIGURE 12.4 Skulls of *Pan paniscus* (chimpanzee), *Austraolopithecus afarensis*, *Homo erectus*, and *H. sapians*. Not to scale.

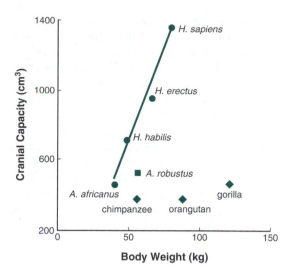

Figure 12.5 Increase in brain size as a function of body weight in the evolution of *Homo sapiens* from the likely ancestor *Australopithecus (sp)*. The fossil remains known as *Australopithecus robustus* are believed to represent a collateral evolutionary branch that did not give rise to present-day humans. The cranial capacities of three of the great apes living today are shown for comparison.

Where the Olduwan tools and bones have been found in association, some of the bones bear cut marks, indicating that *Homo habilis* used stone choppers, scrapers, and hammers to remove meat and marrow. These archaeological sites have been interpreted as living places, but the evidence is not decisive. The anthropologist Richard Potts argues that they were butchering sites. If these stone tools were time-consuming to make and inconvenient to carry while hunting, they may have been cached at a few locations where game was carried for butchering.

It is actually difficult to tell from the available evidence whether the meat *Homo habilis* obtained was from hunting or scavenging. Hunting large and elusive mammals on the savanna was surely difficult for a small biped, and scavenging from the kills of large predators would have been dangerous. Scavenging, however, is a successful tactic employed on the present-day savanna by jackals and hyenas. Small groups of early hominids armed with rocks or pointed sticks could have been successful competitors at scavenging. Anthropologists have long hypothesized that the open savanna offered new opportunities for obtaining food, which led to natural selection for the supposed increased intellectual demands of cooperative hunting. But there are problems with this interpretation. First, although *Homo habilis* and other early hominids likely hunted small animals, there is no compelling evidence that they regularly hunted wildebeest or the larger antelopes. Second, male chimpanzees hunt cooperatively (Chapter

13) and share meat (but not fruit) so hunting, by itself, is not likely to have been the main cause for an increase in brain size.

Olduvai Gorge is interesting for an additional reason. It is the site of fossils of *Australopithecus boisei*, the largest of the robust australopithecines. The average body weights of the males and females of this species are estimated to be about 110 and 70 pounds respectively. *A. boisei* coexisted in the same area as *H. habilis* for a long time before becoming extinct. From its heavy jaws and teeth it is clear that *A. boisei* ate much coarser plant food than *H. habilis*, so the two species may not have competed much for this resource. The two species may have competed for other resources, however. Competition between ancestral and collateral species could well have been responsible for the disappearance of all but one hominid lineage, as bigger brained and culturally more versatile descendants evolved.

FIRST STEPS OUT OF AFRICA: HOMO ERECTUS

Homo habilis seems never to have left Africa, but this hominid did give rise to another one with a still larger brain that did spread throughout much of Europe and Asia. *Homo erectus* is its name, not because it was the first to walk erect, but because of the rules of priority in assigning scientific names. The oldest *H. erectus* fossils from Africa and central and south Asia are 1.8 million years old, and the youngest ones from these areas and Europe are about 200,000 years old.

Homo erectus differed from its ancestor *H. habilis* and was more like modern humans in having a larger brain (50% larger than *H. habilis*), a smaller less-protruding face, and greater stature (up to 6 feet). Compared with australopithecines there was a smaller difference in the body sizes of males and females, which began to approximate the average sexual dimorphism characteristic of later humans.

The molar teeth of *Homo erectus* were smaller than those of earlier hominids, suggesting less dependence on course or uncooked plant food. *Homo erectus* differed, however, from both earlier and later hominids in features that are thought to be specializations of the mouth and teeth for biting and tearing. The face was broad and flat, and the front, top, and back of the skull were reinforced with thickened bony ridges to which jaw and neck muscles attached.

Homo erectus continued to use simple stone choppers of the same sort as *H. habilis*, but except for the east Asian population, they also made biface tools by knocking flakes off two sides of a stone until it was flat and had sharp edges. Hand axes were the most common tools of this sort. Examples have been found at many sites in association with collections of animal bones, some from species as large as elephants and rhinoceroses, and at some sites, *Australopithicus robustus*. Cut marks on the bones and microscopic wear patterns on the axes indicate that the axes were used in butchering. This conclusion is supported by successful efforts of anthropologists to use the axes themselves.

Many such sites are also associated with the remains of fire. One such African site is 1.6 million years old. Physical analyses of the underlying soil and burned bones indicate that these were campfires, which generated more heat than grass or stump fires. *Homo erectus* may have been the first hominid to cook food on a regular basis. The possible evolutionary implications of this cultural practice will be considered further in Chapter 13.

Both tools and anatomy indicate that *Homo erectus* processed meat. This hominid also expanded over a much wider geographic area than its predecessors, and it must have been a successful hunter, especially in colder environments in which plant food was not available in the winter. Two recently explored sites on the island of Flores in the Malay Archipelago (present Indonesia) suggest that *Homo erectus* had greater technological abilities than previously supposed. These sites have been dated to between 880,000 and 800,000 years old. At that time and since, there has been about twelve miles of open water to the west, hindering migration from Asia. *Homo erectus* must therefore have constructed rafts or boats to make this crossing.

A 350,000-year-old shore site near the mouth of a river in France was occupied repeatedly and seasonally by hunters or fishers who constructed groups of about twenty shelters whose stone bases were subsequently covered and preserved by drifting sand. No fossil evidence speaks directly to the first appearance of human language, but it is possible that sophisticated forms of vocal communication were in use by this time.

THE ORIGIN OF HOMO SAPIENS

Modern humans, *Homo sapiens*, are believed to have evolved from *Homo erectus*. Physical anthropologists have shown that anatomical changes from *H. erectus* to modern humans include an increase in brain size, heightening of the forehead, lightening of the skull, loss of the eyebrow ridges, reduction in the size of the face and teeth, protrusion of the chin, an increase in height, and the development of a longer, more slender trunk and limbs. The first or "archaic" forms of *H. sapiens* began to appear in Africa about 400,000 years ago and somewhat later in Eurasia. Their brains were 20% larger than those *H. erectus* but were still 20% smaller than those of modern humans. Although the stone tools of archaic *H. sapiens* were generally similar to those of *H. erectus*, more sites are known where tools, cut bones of large animals, and remains of campfire hearths are found together. These early people are believed to have hunted big game.

One well-documented group of archaic humans known as the Neandertals, lived in Europe, the Middle East, and Asia between 150,000 and 30,000 years ago, during the time that the northern Eurasian continent was emerging from the last ice age. The Neandertals had several features that distinguish them from modern humans: they were short (females averaged 5 feet 3 inches, 110 pounds and males 5 feet 6 inches, 140 pounds) and had short, heavily muscled arms and legs. Their skulls retained and exaggerated many of the features of *H. erectus*, and the nose and teeth were large and protuberant. Their brains, however, were about 7% larger than those of present-day humans. Their compact body form was likely in part an adaptation for minimizing heat loss in a cold climate, but the reason for the "heavy-duty construction" of their skeleton and skull is not known. Perhaps their way of life included wielding heavy objects to kill large prey at close quarters, carrying heavy loads, and extensive chewing of soft but tough animal hides. They appear to have buried their dead along with other objects, which suggests a mental awareness that is clearly human.

What were the evolutionary relationships between *H. erectus*, archaic *H. sapiens*, the Neandertals, and modern humans? By themselves, the fossils do not answer this question, and there has been disagreement on precisely where and when modern *H. sapiens* evolved from *H. erectus*. One view—the "multi-regional hypothesis"—is that gradual changes in all or most of the Eurasian populations of *H. erectus* led through archaic *H. sapiens* to modern humans. An alternative view, which gains support from genetic data (described below), is that modern humans are derived from an African population of archaic *H. sapiens*, which in turn was derived from *H. erectus*. This population increased in size, and starting sometime between 200,000 and 100,000 years ago pushed out of Africa, spread geographically, and displaced previous populations. The oldest fossils of anatomically modern humans, between 120,000 and 100,000 years old, are known from Africa and the Middle East. The oldest known fossils of modern humans in Europe are those of Cro-Magnons, who spread from eastern Europe westward between approximately 45,000 and 30,000 years ago.

How do the Neandertals fit into this picture? In some interpretations, the Neandertals are considered to be *H. sapiens* and ancestral to modern humans. Some anthropologists, however, have placed them in a separate species, *H. neanderthalensis*. The confusion arises in part because the Neandertals of western Europe appear anatomically more distinct from modern humans than fossils from further east. The important issue, though, is not the scientific name they are given but whether the Neandertals were assimilated or were displaced following the arrival of another population of *Homo*. Based on anatomical evidence, the possibility of interbreeding remains controversial. As we will see shortly, however, the genetic evidence suggests they represent a distinct lineage. Nevertheless, as anatomically modern humans spread into Europe from the Middle East, the Neandertals disappeared with an east to west progression. The Cro-Magnons and their immediate ancestors thus may have been responsible for this extinction, either indirectly through competition, directly by conflict, or some combination of both.

The expansion of modern humans brought many cultural innovations. These improvements included elaborate clothing, bags, baskets, ropes, complex shelters, and more varied and finely crafted tools made from a greater variety of materials (bone, ivory, antlers, stone). These new hominids also transported raw materials for tools over long distances, fished, hunted game to the size of wooly mammoths, performed ritual burials, and made frequent use of caves as shelters.

The caves provide the most spectacular evidence of Cro-Magnon technological and artistic abilities. At Lascaux in France and Altamira in Spain, the walls and ceilings of deep caves are decorated with finely executed, colorful paintings of game animals, many of which are now extinct. These sites are so far from the cave entrances that artificial light (from fire) and ladders or scaffolding were needed to make the paintings. These striking artistic creations may have served a ceremonial or ritual function, perhaps reflecting a desire to assert control over the animals that were essential sources of food and skins. In every known respect there is ample reason to see Cro-Magnons as intellectually and emotionally very much the same as ourselves.

GENETICS, MIGRATIONS, AND THE APPEARANCE OF LINGUISTIC DIVERSITY

Until recently, the study of human evolution was based entirely on the historical record of fossils and archeological findings. Understanding history by examining the available archives is infinitely better than speculating on the basis of preconceptions, but historical records, as we have already noted, are never complete. Furthermore, to paraphrase one of the advocates of molecular genetic techniques, a fossil may or may not have left descendants, but all of our molecules had ancestors. Consequently, the recent use of molecular genetics to study human origins opens an exciting new window on the subject. In science, new and different perspectives frequently initiate the questioning of assumptions and the reinterpretation of older data, but a richer and more accurate understanding generally emerges. Study of the last several hundred thousand years of evolution of *Homo sapiens* is currently in such a dynamic state.

ANCESTORS, DESCENDANTS, AND "MITOCHONDRIAL EVE"

First some basic reasoning about genetic ancestry. You have two parents, four grandparents, eight great-grandparents, and sixteen great-great-grandparents (Fig. 12.6A). Each of those sixteen ancestors has contributed to your genes with a probability of $(1/2)^n$, where $n = 4$, the number of generations that separate you. The chance that a gene came from any particular one of your great-great-grandparents is thus 1/16. This mixing of genes results from meiosis and sexual reproduction.

There are other possible patterns of inheritance, however. Remember that all of your mitochondria came from your mother in the cytoplasm of her egg (Fig. 10.1), and that mitochondria (Chapter 3) contain some DNA because they arose through a symbiotic association with microorganisms early in the evolution of cells. Your mother's mitochondria came from *her* mother, and so forth; consequently all of your mitochondria descended from just *one* of your eight great-great-grandmothers through an uninterrupted line of maternal inheritance. Unlike nuclear DNA, mitochondrial DNA of parents does not mix, and this feature of its inheritance makes it a valuable tool for tracing maternal lineages.

In another respect, however, the spreading fan of ancestors in Figure 12.6A is potentially misleading. Suppose (conservatively) there are three human generations per century. That means that there have been about sixty human generations in the last 2000 years. If the number of your ancestors doubled with each generation, as you trace backwards you should have had 2^{60} ancestors at the time of Christ. But 2^{60} is a *very* large number. In fact, it is nearly a billion times larger than the entire human population today! Something is obviously amiss with this reasoning.

As the genealogy extends back in time, it does not spread continuously like the forking spokes in Figure 12.6A. Because of marriages between near and distant cousins, as we go further back in time, the fan looks more like a web of relationships. Put another way, any two people in the web will be able to claim at least one ancestor in common in the not-too-distant past. When people research their genealogy, they are fascinated to discover how they are related to a myriad of other individuals.

Figure 12.6B suggests another way to look at ancestors and descendants that at first may seem paradoxically at odds with *A*. The two disks represent populations at two slices in time. The later (lower) disk is larger, because human population has been increasing. Everyone alive at the later time must have had ancestors at the earlier time. Less obvious but equally true, not everyone alive at the earlier time left descendants who lived at the later time. Some lineages just died out. So even though the population expanded with

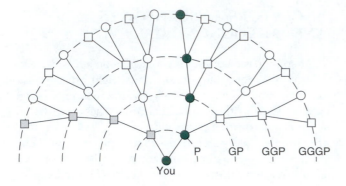

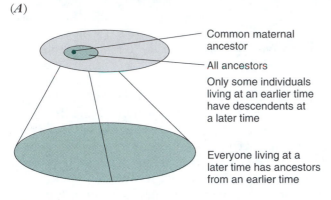

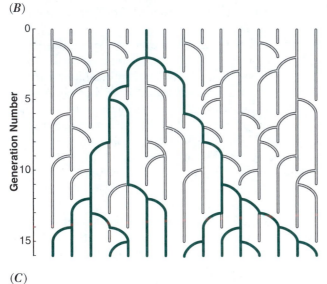

FIGURE 12.6 **(A) Nuclear genes are inherited from a spreading fan of immediate ancestors (males represented by squares, females by circles).**

Mitochondrial genes are inherited through a single, maternal line of descent (colored circles). Y chromosomes, like surnames, are inherited through paternal descent (gray squares). (B) Everyone has ancestors, but not everyone leaves descendants. (C) Maternal lines of descent disappear until only one survives. This can happen very quickly in a small population.

time, chance dictated that the ancestors of the later population were just a subset of everyone living at the earlier time. If we took an even earlier slice of time, the pool of ancestors would be even smaller.

Now consider the inheritance of mitochondrial DNA. Figure 12.6C shows how in a small population (which for ease of illustration stays the same size in successive generations), chance terminates one maternal line after another until only one remains. At this point, everyone in the population can trace their mitochondrial DNA to a single common maternal ancestor. It may seem surprising that one maternal line should grow at the expense of all the others, but remember that until about 1800 human populations were increasing very slowly. This means that for most of our evolution, people were reproducing very close to the replacement rate, so on average each woman would produce one surviving daughter. If the average number of daughters who survive to reproduce is one for each mother, there will be many women in each generation who leave no daughters and whose mitochondrial lineage therefore ends. As a result, we all have a common maternal ancestor who gave us our mitochondrial genes.

The *most recent common maternal ancestor* is sometimes referred to as "mitochondrial Eve." Despite the biblical reference, mitochondrial Eve was one woman in a population of thousands. In Figure 12.6B she is present as one of the many people in the subpopulation who were fated to become the ancestors of the later population. The mixing of nuclear genes that takes place in each generation preserves them for longer spans of time than mitochondrial genes. That is why "mitochondrial Eve" not only had many contemporaries, some of these individuals produced descendants via their nuclear genes. However, if all of those descendants trace their lineage back in time *entirely through women*, the path leads eventually to a single common maternal ancestor.

Surnames are analogous to mitochondrial genes, because (with the exception of hyphenated names that recognize two families) they do not mingle at marriage. Like Y chromosomes, surnames descend through the male line. Surnames frequently drop out of branches of a family when there are no sons, and in China, where local populations have been stationary, occupying the same locations for several thousand years, the number of surnames has decreased. Moreover, in genealogical study, identifying your most recent common ancestor with other individuals through shared paternal descent requires a more limited search than tracing all paths of genetic relatedness to other individuals.

With this background, we can now consider how mitochondrial DNA has contributed to our understanding of human evolution. The late evolutionary biochemist Alan Wilson and his colleagues examined a segment of mitochondrial DNA from 243 individuals from around the world, which yielded 182 slightly different sequences, each reflecting the independent accumulation of neutral mutations. The construction of evolutionary trees from molecular data was discussed in Chapter 5. With 182 terminal branches, there are many possible evolutionary trees that can be drawn, but with the help of computers the search can be narrowed to the trees that have the fewest number of mutational steps. Their evolutionary tree is shown in Figure 12.7.

The deepest branches are the oldest. Note that the very earliest fork leads to Africans on one limb, and everyone else, including some other Africans on the other. This means that the most recent common maternal ancestor of all humans, who is located at the root of the tree, very likely lived in Africa. The possibility of fitting the data with other trees made this interpretation provisional, but more recent work from other laboratories on other regions of mitochondrial DNA supports the conclusion that "mitochondrial Eve" was African.

How long ago did this individual live? The answer requires calibrating the rate at which mutations accumulate in mitochondrial DNA. From the percentage difference between chimpanzee and human mitochondrial DNA and dated fossils that provide an estimate of the time when these two lineages split, our most recent common maternal ancestor lived less than 200,000 years ago (120,000–300,000 years ago by various estimates). Recent study of genes on the Y chromosome lead to a similar conclusion. This is a very interesting number, because the oldest fossil remains that appear to be clearly ancestral to contemporary humans are somewhat older, and they were also found in Africa.

We can conclude from this evidence that there has been more than one exodus of hominids from Africa. These data suggest that the fossils of *Homo erectus* from China and Indonesia, some of which are about a million years old, as well as a recent finding from Europe that appears to be even older, were not ancestors of modern humans. Modern humans arose in Africa much later, sometime in the last several hundred thousand years, and subsequently spread to the four corners of the earth. As we shall see, evidence from nuclear genes supports this conclusion.

MITOCHONDRIAL DNA AND THE MYSTERY OF THE NEANDERTALS

In the summer of 1997, a group of molecular biologists working jointly in Germany and at Pennsylvania State University reported a remarkable discovery. They recovered and amplified a region of mitochondrial DNA from an arm bone of the original Neandertal skeleton that had been found in 1856 in the Neander Valley (Tal) in Germany and had been in a museum since its discovery. The sequence of DNA was 379 base pairs in length and could be compared with the corresponding DNA of humans living today. On average, contemporary humans differ from one another at eight of these 379 positions, but they differ from the Neandertal DNA at an average of 25.6 largely different positions. We saw earlier that modern human mitochondrial DNA sequences have been diverging for something like 150,000 years. By the same reasoning, the Neandertal sequences must have diverged from a common maternal ancestor with us about 550,000 to 690,000 years ago, likely also in Africa. The Neandertals became extinct about 30,000 years ago, but they were present in Eurasia for at least 70,000 years before that, and they thus coexisted with modern humans in at least part of their range for about 70,000 years. Whether the two groups were ecologically or temporally separated during any of that period is not clear, but the genetic

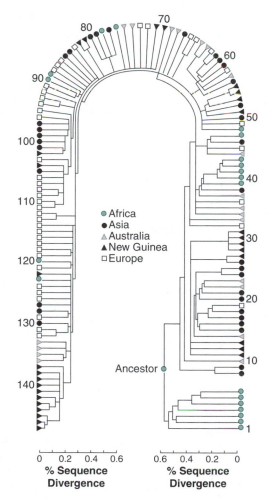

FIGURE 12.7 Tree of mitochondrial DNA of living humans. There are two main branches: one (lower right) consisting of a group of Africans, and the second, larger branch consisting of other Africans and everyone else. This branching pattern indicates that modern humans had a common maternal ancestor in Africa.

data are consistent with the interpretation that Neandertals were a separate species. This is an area of study that clearly requires more data.

This work on DNA obtained from very old skeletal remains is a landmark example of the convergence of molecular biology and anthropology. What is the future of this technique? The science fiction film *Jurassic Park* was based on the idea that nuclear DNA in the cells of dinosaur blood might be present in the stomachs of fossil mosquitoes trapped in amber (fossilized pine resin) and could be used to grow new dinosaurs. Quite aside from the currently insurmountable problem of growing an organism from its DNA without an appropriate parental egg to provide the necessary cellular environment (Chapter 10), there is the additional difficulty that outside of a living organism, DNA, like all organic compounds, degrades (oxidizes) with time. For reconstructing the evolutionary past from DNA in fossil material, mitochondrial DNA has an advantage over nuclear DNA simply because there are many more copies in each cell. (There is one nucleus but many mitochondria.) Even so, unless the remains of the organism have been frozen since death, DNA that is older than 50,000 years will likely always be so degraded that identifiable sequences cannot be recovered.

NUCLEAR GENES ALSO POINT TO AFRICAN ORIGINS OF CONTEMPORARY HOMO SAPIENS

For over thirty years, geneticist Luca Cavalli-Sforza, has collected data on the frequencies of different alleles of nuclear genes in populations of people from around the world. Figure 12.8 is a tree of genetic relatedness of people from nine geographical regions, based on genes at 110 different loci. Many of the alleles of the 110 genes are thought to be little influenced by natural selection, and geographical differences in allele frequencies therefore likely represent chance events in founder populations (e.g., genetic drift, Chapter 4). The gene responsible for the A, B, and O blood groups that are important in blood transfusions provides an example. There are three alleles at this locus, each coding for a variant of a protein in the membranes of red blood cells. Some other genes, however, show marked geographical variation because of strong natural selection. For example, the absence of a particular protein on the surface of red blood cells provides protection from malaria by making it difficult for the parasite *Plasmodium* to enter the cell. The corresponding allele of this gene occurs in highest frequency in sub-Saharan Africa and is almost absent in the rest of the world.

The tree of genetic divergence (Fig. 12.8) is readily interpreted as reflecting a historical pattern of human migration from an origin in Africa. Although there are some remaining ambiguities in the genetic data regarding the inhabitants of southeast and northeast Asia, the pattern of branching is otherwise consistent with archaeological evidence. Evidence of modern humans (100,000 years ago) is known from the Middle East and Africa, and this is probably about the time of their arrival from Africa. At this time they would also have contained a number of the mitochondrial lineages that continued forward to the present and that can be traced backward to a woman who lived tens of thousands of years earlier.

Archaeological evidence shows that Australia and New Guinea were populated 35,000 to 60,000 years ago. The Aboriginal peoples now living in these areas differ

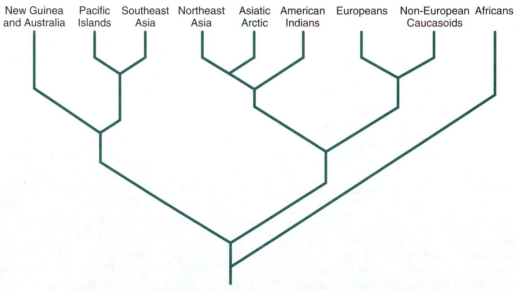

FIGURE 12.8 The close correspondence between the tree of genetic relatedness and geographical proximity of contemporary groups of people.

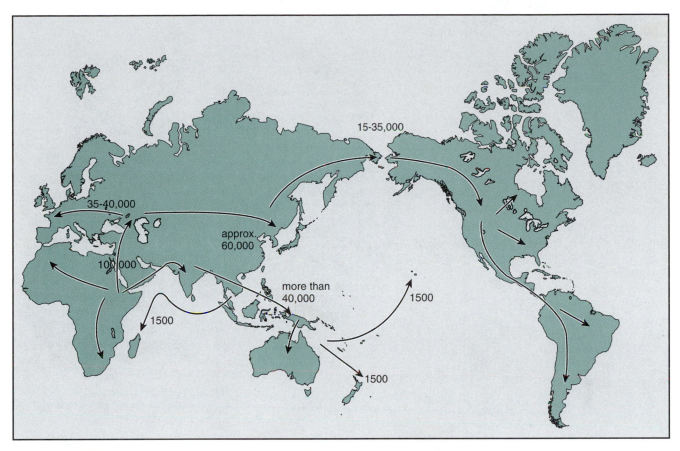

FIGURE 12.9 Migrations of modern humans out of Africa, beginning about 100,000 years ago.

genetically from their neighbors in southeast Asia by about 60% of the genetic difference between contemporary Africans and non-Africans. Europe was populated from the east 35,000 to 40,000 years ago, and the genetic distance between contemporary Europeans and Asians is about half the reference distance between Africans and non-Africans. Finally, the Americas were populated from northeast Asia 15,000 to 35,000 years ago (the date is still quite uncertain), and the corresponding genetic distance between Amerinds and the Mongoloid peoples of northeast Asia is about 30% of the African:non-African distance. The archeological estimates of the times of first habitation by modern humans in the Middle East, Australia, Europe, and North America thus correspond to sequential branching of the genetic tree. These migrations are summarized by the map in Figure 12.9.

LINGUISTIC EVIDENCE SUPPORTS THE GENETIC AND ARCHEOLOGICAL DATA

The relationships among the human languages of the world form a specialized subject investigated by linguists, who examine the similarities between words in different languages. Joseph Greenberg has suggested that certain words such as numbers and names of conspicuous body parts like fingers or ears are learned early in childhood, do not change much with time, and can provide the basis for systematically examining the relationships between languages. On a small scale you are familiar with this subject. You know that the Romance languages, e.g., French, Italian, Spanish, all have similarities because they are derived from Latin and are more like each other than any is to Dutch or German. As described in a book by Merrit Ruhlen, languages thus fall into groups based on their degree of relatedness. Morris Swadesh and Robert Lees have built on these ideas and constructed "evolutionary" trees for language that are similar to phylogenetic trees in biology, even though the underlying causes of change in the two systems are very different.

A fascinating discovery is that the coarser features of linguistic diversity (Fig. 12.10) are similar to the tree of genetic diversity (Fig. 12.8). This implies that the migrations of humans between 100,000 and 30,000 to 15,000 years ago not only led to enough geographical separation to allow detectable shifts in gene frequencies, they also coincided with the cultural divergence of languages.

FIGURE 12.10 A tree of major language groups is similar to the tree of genetic divergence in Figure 12.8.

It is important to understand that differences in genes are not responsible for the differences in languages; there is no causal relationship between genes and the language a person speaks. There are several thousand languages in the world, and many more that have ceased to exist. Like the many religions of the world, different languages are cultural variants, and every individual is capable of speaking any language fluently. To speak without an accent, however, usually requires learning during that sensitive period of childhood when languages are acquired naturally, or at least before puberty (Chapter 11).

GENETICS AND THE MIGRATIONS OF HUMANS SINCE THE DISCOVERY OF AGRICULTURE

The earliest farming started about 11,000 years ago in a region of the Middle East known as the Fertile Crescent. Similarly, agriculture began independently at several other places in the world. The biologist Jared Diamond has compellingly traced the consequences. In each of these areas, different foods were adopted because each of these parts of the world had a somewhat

different climate and was occupied by different plants and animals. Because of the winter rains and long hot summers, the Fertile Crescent was particularly rich in species of annual plants that put much of their energy into seed production and die at the end of each growing season, rather than making inedible woody structures that enable them to live for many years. Their production of large numbers of nutritious seeds made these abundant plants particularly suitable for human food. Furthermore, this region also contained many large herbivorous (or omnivorous) animals that could be domesticated for milk, meat, hides, or for plowing and transportation. Their domesticated descendants include cattle, sheep, goats, pigs, and horses. The Fertile Crescent was not only the earliest center of cultivation, for these ecological reasons its impact on subsequent history has been particularly important.

The development of agriculture and the domestication of animals allows population density to grow, and this in turn leads to the creation of increasingly complex social structures with internal divisions of labor. The Fertile Crescent was therefore the center of migration of people, whose westward movement has been extensively analyzed. Over a period of about 4000 years, this Neolithic population expanded until all of Europe was occupied. The movement was slow—about half a mile per year—and the resident hunter-gatherers were either displaced or absorbed. By the end of this migration, the population density of Europe had increased at least ten fold.

Although archeologists can track the spread of culture by the artifacts they find, it is hard to distinguish the roles of trade, warfare, and migration in the diffusion of culture. The geneticist Cavalli-Sforza has used a sophisticated statistical technique (known as principal components analysis) to analyze the distribution of genes in the present-day population of Europe. Allele frequencies for a large number of genes are plotted on a map of Europe, and the analysis produces "genetic landscapes" that look like coarse contour maps. The statistically most important of these landscapes looks remarkably like a map based on archaeological data showing the spread of agriculture from the middle East (Fig. 12.11).

Among the different centers of agriculture, the Fertile Crescent was uniquely favorable. The Americas, for example, presented a different ecology. The earliest cultivar was squash, but the indigenous cereal crop is corn. The wild ancestor of corn is remarkably different from the form that ultimately was developed through

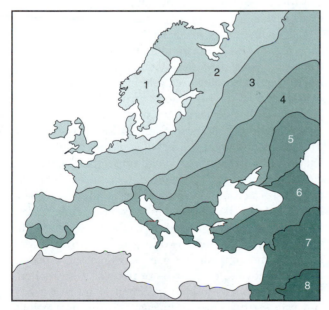

FIGURE 12.11 A statistical analysis of the genetic diversity of Europeans shows that the principal component of variation can be accounted for by an early migration of people from the region of the Fertile Crescent, corresponding to the spread of agriculture that started approximately 10,000 years ago. Think of the shaded patterns as though they were contours on a topographic map. Instead of representing different elevations, they convey degrees of similarity of the present-day inhabitants at ninety-five different genetic loci.

centuries of cultivation and selection. Moreover, in the thousand years after humans first reached the Americas, most of the large mammals in the Western Hemisphere became extinct. This may have been due to extensive hunting of species that had no evolutionary experience with human predators. Whatever the reason, there were practically no large herbivores left to domesticate. The llama, derived from the guanaco, was used by the Incas in the Andes, but the Aztecs and other groups of Indians further north had no domesticated animals larger than dogs and turkeys. The familiar horses of the Plains Indians were descendants of animals brought to the Americas in the sixteenth century by the Spanish. Ecological happenstance has therefore had a large influence on the start of agricultural economies, their subsequent rate of change, and the impact of one human population on another.

RACE IS MORE A SOCIAL THAN A BIOLOGICAL CONCEPT

People like to put things in categories, and out of this predilection has come one of the most pernicious of human concepts, that of race. Race can be employed to identify groups of people different from the speaker's, often with the intention of drawing unfavorable comparison's, if only by inference. Thus Winston Churchill referred proudly to the "British race," implying an intrinsic superiority that actually had more to do with cultural chauvinism than biological differences. More often, race refers to some combination of skin color and presumed homeland of origin, as in white (European), black (African), or yellow (East Asian).

A closer anthropological look, as well as the genetic analyses described in the previous section, tells quite another story. The people of the world vary in depth of melanin pigmentation in the skin, in the color and texture of hair, and in other superficial features such as the shape of the nose and lips. However, these characteristics do not all vary together or in a simple way with geography. For example, there is enormous variation in the physical characteristics of the native peoples of Africa, and dark skin color is found elsewhere in the world, in the native peoples of Australia and Polynesia.

Everyday conceptions of race are based on what we notice, and skin color and shape of nose are two examples. These features are adaptations of the body surface to extremes of light and temperature, and we notice them because they are so superficial. In the words of an eminent geneticist, they are only "skin deep." Melanin in the skin protects the body from the harmful, short wavelength radiation of sunlight, and northern Europeans are relatively pale. The reason appears to be associated with nutrition. People who are heavily dependent on cereal crops for their diet also require sunlight so that they can make vitamin D. In northern latitudes, where less sunlight is present, lighter skins have been selected during the last 10,000 years. From India west to Europe and North Africa, extent of skin pigmentation is correlated with latitude. Eskimos are descended from Mongoloid peoples. Despite their Arctic environment, however, they have not become light-skinned because there is ample vitamin D in their diet of meat.

The emphasis on genetic diversity in the preceding sections may leave an incorrect impression. Of the genetic variation that has been identified in *Homo sapiens*, more than 85% fails to correlate with traditional definitions of race. Viewed another way, if two people from anywhere in the world are chosen at random, they will differ on average by 1 DNA nucleotide in 1000. One nucleotide in 1000 is only 0.1% of our DNA, but it corresponds to about a million nucleotides in our genome. This is the variation on which conclusions about human migration were based. Furthermore, if we compare the average difference between parents and children (or between brother and sister), we find that it is about half of the difference between any two people selected randomly from the same area. Considered this way, humans are genetically quite similar to one another.

Not only do the world's people not fall neatly into a few racial categories, there are no substantive genetic differences in abilities between people from different parts of the world. People certainly differ from one another in intrinsic abilities in athletics, music, creativity, and a variety of other ways. Nevertheless, none of these characteristics has been shown to vary with geographical origin in ways that can be attributed to genetic differences. Whatever genetic variation lies behind individual differences in these abilities, it seems randomly sprinkled around the world.

Race is therefore not a useful biological concept as it applies to humans, but it continues to be widely used in society for historical-cultural reasons. In American society, it is tightly coupled to the former practices of slavery and segregation. The notion that "one drop of blood"—i.e., any African ancestry—defines a person as black is a convention with no basis in genetics, and it has contributed to a polarized society.

Seeing others who are different as inferior and therefore not worthy of respect frequently provides a "rationale" for economic or military domination of one group by another. This has been common when two cultures with very different technologies come in conflict, and those with the greater military or economic power seek to justify their self-serving actions. The biologist Jared Diamond has pointed out that differential advantage in these conflicts has nothing to do with the intellectual capacities of the two groups. The reasons why technology has developed at different rates in different parts of the world can be traced to those ecological circumstances in prehistory to which we referred earlier.

"Racism"—the unfair or hostile treatment of others who are perceived as different—is found in all human cultures and has been with us through history. In the following chapters, we will consider further why the propensity for such human behavior has evolved. Out of such deeper understanding we may hope to create cultures in which such attitudes hold no sway.

SYNOPSIS

Understanding the evolution of humans from a common ancestor that we shared with chimpanzees and bonobos has become an interdisciplinary field, involving paleontology, anthropology, archeology, molecular genetics, and even linguistics.

During the past several decades, numerous fossils have been discovered in east Africa, which is believed to be the geographical origin of *Homo sapiens*. Our appearance today is the result of the same evolutionary processes that have shaped other organisms, and several other lineages of hominids appear in the fossil record, only to have gone extinct. More than 1.5 million years ago, *Homo erectus* left Africa and spread east into China and Indonesia. Modern humans, however, arose much later, again in Africa, and starting about 100,000 years ago this lineage began to migrate over the rest of the earth, reaching the Americas at least 15,000 years ago.

Agriculture was invented 10,000 to 12,000 years ago in the Middle East, independently and possibly somewhat later in Africa, China, and the Americas. Agriculture permitted a ten fold increase in population density and led to expansions of population from the original locations where it was practiced.

Much of this understanding is based on paleoanthropology and archeology, but molecular biology has led to some important conclusions, some of which could not have been reached by other means. The historical diffusion of populations can be inferred by analyzing allele frequencies in people living today. Similarly, study of mitochondrial DNA supports the view that modern humans originated in Africa, and it confirms the approximate date at which these people began to move to other parts of the world. Mitochondrial DNA from a museum specimen suggests that Neandertals were a separate lineage that became extinct. Whether they interbred with modern humans requires further evidence.

Genetic analyses fail to support the idea that contemporary human races are distinct categories of people with characteristic intellectual, artistic, or other abilities. Genetically, *Homo sapiens* is a relatively homogeneous species. The differences that catch our eye are superficial; the most prominent is skin color. Lightly pigmented skin is a relatively late adaptation to climate.

Our human proclivity for categorizing each other, however, has to be understood. We make the categories socially important by incorrectly supposing them to have deep biological significance, and this contributes greatly to social conflict. This behavior is so widespread that it requires further evolutionary analysis, to which we will return in the final chapter.

QUESTIONS FOR THOUGHT AND DISCUSSION

1. Although they are often called "missing links," fossil animals in the lineages between ancestors and descendants are almost never precisely intermediate in their traits, and often they possess structures that neither the ancestor nor the descendant have. In what ways does the fossil history of humans illustrate this generalization?

2. The fossil remains of some australopithecines show pronounced sexual dimorphism, with the males 50% larger than females. Drawing on general biological knowledge of sexual selection, mating systems, and social structures of other animals, what reasonable inferences can you make about the social organization of these species?

3. What hypotheses can you formulate to account for the extinction of Neandertals? How might these ideas be tested?

4. What is the difference between the inheritance of mitochondrial and nuclear DNA, and what advantages and limitations do comparisons of mitochondrial DNA offer for establishing genetic relatedness?

SUGGESTIONS FOR FURTHER READING

Boyd, R., and Silk, J. B. (1997). *How Humans Evolved*. New York, NY: W.W. Norton & Co. A wide-ranging, well-illustrated, and detailed account of human evolution.

Cavalli-Sforza, L., and Cavalli-Sforza, F. (1995). *The Great Human Diasporas: A History of Diversity and Evolution*. Reading, MA: Addison Wesley. An eminent population geneticist provides genetic and linguistic evidence for the major migrations of humans during the last 100,000 years. Written for a general audience.

Diamond, J. (1997). *Guns, Germs, and Steel: The Fates of Human Societies*. New York, NY: W.W. Norton & Co. An extremely engaging and readable account of why some cultures developed agriculture, domestic animals, and technology faster than others. Diamond makes convincing connections between geography, ecology, and human history.

Lewin, R. (1988). *In the Age of Mankind: A Smithsonian Book of Human Evolution*. Washington, DC: Smithsonian Institution. A lavishly illustrated, balanced, and readable account that is still timely.

13

Exploring Our Behavioral Heritage

RECONSTRUCTING BEHAVIOR

In the last chapter we described the many sources of evidence—fossil, molecular, archaeological, and cultural—that are filling in the story of how hominids evolved from a chimpanzee-like ancestor. The story began in Africa about 6 million years ago, but the distribution of modern humans throughout the rest of the world took place during the last 100,000 years. We now turn to a set of more difficult historical questions. What sorts of social organization and mating systems did early hominids have, and what was their social behavior like?

In response to these sorts of questions, biologists and anthropologists have suggested many hypotheses for how human behaviors evolved, each based on possible similarities between early hominids and present-day animals or human groups. Some of these models are based

Photo: A chimpanzee medicating itself. When parasitic worms infect these nearest relatives of humans, they ingest food that they normally do not eat. This chimpanzee is about to roll up a leaf that is covered with tiny sharp hooks. Many of these leaves are swallowed whole early in the morning, before other foods are eaten, and they pass through the intestines largely intact. They are expelled with many worms stuck to and trapped inside them. Several of the other plants chimpanzees ingest when suffering from gastrointestinal infections contain antimicrobial and nematodicidal compounds, and these plants are also used by local humans to effectively treat the same kinds of infections.

on a single behavioral adaptation postulated to have been the primary cause of the divergence of hominids from apes, such as hunting, gathering plant food, or scavenging the kills of large carnivores. Other models propose similarities between the behavioral ecology of hominids and a single living primate species, such as baboons or chimpanzees, or of a contemporary human culture, such as hunter-gatherers. Based on one or some combination of these models, speculations have been made concerning general human behavioral dispositions. At one extreme is the "killer ape" hypothesis, argued some years ago by Robert Ardrey. In this view our ancestors, like lions or hyenas, were savanna-dwelling, highly territorial carnivores that used their hunting adaptations against competing hominids. At the other extreme is the "peaceful egalitarian" hypothesis, modeled on contemporary hunter-gathers such as the Ju/'hoansi (!Kung Bushmen) of Africa's Kalahari Desert. Still other ideas about early human behavior, based upon comparative cultural and psychological studies, suggest that our ancestors passed through stages in which mating was entirely promiscuous, or in which women dominated men, or in which property, work, and social influence were equally shared among members of the group. Some of these ideas are based more on wishful thinking than a close attention to the available evidence.

There is a multitude of hypotheses because reconstructing the social evolution of animals, especially extinct species, is much more difficult than reconstructing their physical evolution. One obvious problem is that fossils seldom reveal much about an animal's behavior, and other information must be used to infer how extinct animals lived. A basic approach is to compare the ecology, social behavior, and mating systems of phylogenetically close species in order to discover what factors have influenced social evolution. This information can then be combined with evidence from fossil animals and the ancient environments in which they lived.

A second difficulty is that social organization often evolves more rapidly than physical traits. Thus two species may be morphologically very similar and eat very similar foods, but their social structures and behavior may be quite different because of a slight difference in the distribution of their food or because they are subject to different kinds of predators.

Natural selection alters behavior by shifting the ways in which sensory information is evaluated. One sensory modality may increase in importance, such as hearing in bats. But the process can also be much subtler, involving only interpretation. For example, the presence of another individual can evoke very different responses, depending (in part) on whether the species is solitary or social. Changes in cognitive mechanisms for interpreting sensory information readily respond to natural selection, likely because they require only small changes during development of the nervous system

(Chapter 11). When we speak of behavior evolving, this is the kind of change to which we are referring.

Social behavior is plastic for another reason—one that plays out in the life of each individual. Mammalian brains—particularly primate brains—are able to generate behavior that is contingent on immediate conditions. Tracking and responding to changes in the physical and biological surroundings occur on time scales from fractions of a second through hours, days, and years. The strike of a snake, the rising and setting of the sun, the change of seasons, and the impact of a prolonged drought not only occur in different time frames, they have vastly different implications and require wholly different responses. The flexibility that is expressed in the life of each individual is an example of the "range of reaction" (Chapter 4)—the degree to which phenotype can change with different environmental conditions. When phenotype includes the capacity for foresight, planning, and conscious choice, it must also have accurate and detailed memories.

Modifications in behavior occurring by natural selection over many generations also enable a lineage of organisms to track and respond to environmental changes. In what follows we will consider *evolutionary* causes of social structure, building on ideas introduced earlier (Chapter 6). One aim is to see how behavior of animals living today, particularly primates, can give us clues about the lives of extinct hominids. In comparing living and extinct species, we will apply some familiar reasoning. For example, evolved similarities between two species can have different explanations (Chapter 2). Two species may have adapted to similar ecological niches and therefore have evolved similar behaviors independently (*convergent evolution*). Alternatively, behavioral traits may have been retained during descent from a common ancestor.

FACTORS THAT INFLUENCE SOCIAL EVOLUTION

To develop and test hypotheses about which factors are most important in social evolution and how they interact, a great deal of information must be gathered on the ecology, mating system, and social behavior of many animals. It takes many years of careful observation—often under trying and dangerous conditions—to gather such data. Major features of the mating system are related to sex differences in parental investment and sexual selection (Chapter 6), but much fieldwork suggests that two ecological factors are additionally important in affecting social evolution. These are (*i*) the quality and distribution of resources—food, water, safe places to breed and rest—and (*ii*) the risks and costs of attacks by predators and competitors.

The influence of resource distribution on social evolution is evident in the strategy often used in an Easter egg hunt or similar games, in which children must find eggs and candy that have been hidden evenly over a large area. Sometimes the children follow each other and squabble over their findings. More often they quickly perceive that they can pocket more of the goodies by distancing themselves from each other instead of foraging in a group. The distribution of resources required by most species of birds—small insects, seeds, and safe places to nest—is similar to that of Easter eggs and candy. And similarly, the most efficient way to exploit resources that are distributed in this way is for a mated pair of birds to collaborate in defending a territory that is large enough to sustain them and their growing offspring. This is particularly true for birds that hatch in a naked, helpless condition requiring much parental care and is likely the reason why more than 90% of bird species breed as isolated pairs and forage in their surrounding territory.

A different distribution of resources can lead to the evolution of very different social behaviors. Suppose, for example, that an animal's food is not evenly distributed throughout its environment but instead is "patchy," concentrated at fixed locations in amounts that will support many individuals feeding at the same time. Such a resource might be defendable, if not by an individual, perhaps by a group. Communities of chimpanzees behave this way. Resources can also be seasonal, depending on rainfall or elevation, and can require animals to move to different parts of their range during the course of a year. The herds of antelope, wildebeest, and zebra on the Serengeti in east Africa migrate in this manner.

How the distribution of resources, the benefits and costs of competition for access to resources, and sexual selection can all interact to influence social evolution is illustrated by lions, the only member of the cat family that is not predominantly solitary. Lions were once distributed throughout the Middle East and south Asia as well as Africa, but they are now found only in the savannas of Africa. Like 95% of mammals, lions are polygynous, reflecting the greater parental investment of females (Chapter 6). They live in social groups or "prides" that consist of two to seven males (usually two), two to seventeen females (average seven) and their dependent young (Fig 13.1). Males are expelled from their

(A)

(B)

FIGURE **13.1** (A) Female lions resting by a water hole.

(B) Male and two female lions, showing sexual dimorphism.

natal pride, and they cooperate with one or more other males, often brothers or cousins, to expel the resident males and kill the cubs of another pride. The new males then mate with the females and thus sire their own offspring. Most female lions remain in their natal pride, so the stable core of the group is full- or half-sisters and cousins. It was once thought that lion sociality evolved because of the advantages of cooperative hunting for their large and elusive prey, such as antelope, wart hogs, wildebeest, zebras, and buffalo. But long-term studies of Serengeti lions by the behavioral ecologists Craig Packer and Anne Pusey and their students show that lions hunting in groups do not eat better than those hunting alone. This is because lions do not, in fact, cooperate in hunting unless the prey is difficult to capture, such as zebra and buffalo. But they do help eat the prey, whether they helped capture it or not! Additionally, female lions do most of the hunting and most of the defense of their hunting territory against the incursions of neighboring prides. Male lions, with their greater size and protective manes, are not as fast and maneuverable as females. They are much better at the close-quarter and frequently lethal fighting they must do to take over a pride, to defend against usurpations by other males, to steal prey captured by females, and to thwart attempts by females to prevent them from killing their cubs sired by previous males. In brief, lion sociality is not based on the advantages of cooperative hunting but upon the advantages of cooperative and lethal competition against other lions.

Lion sociality also shows how factors "internal" to a species can profoundly affect its social evolution. Lions are the top predators in their ecosystem, so their evolution is not affected by the external threat of predation. But their adaptations for hunting and killing also equip them to inflict considerable costs on their main competitors, other lions. Thus, this is an animal in which intraspecific competition combined with lethal weaponry has had a freer-than-usual rein to affect social evolution. As we will discuss later, many evolutionary anthropologists think that similar intergroup aggression may have played a significant role in hominid social evolution. Such behavior is found in ourselves and in our nearest relatives, chimpanzees. By about 2 million years ago our hominid ancestors were using primitive tools, but it is not known at what point they began to use tools as weapons against each other.

Many other features of an animal's natural history may interact in complex ways with the distribution of resources and predation in influencing its social evolution. These include the size, stability, and composition (age, sex) of the social group, the pattern of dispersal of offspring, the degree of genetic relatedness among members of the group, and the relative and total investment by each parent in offspring. These additional traits are themselves the products of previous evolutionary processes, caused by the same sorts of factors. In other words, social evolution is like other kinds of historical change in the natural world: it has multiple and complex causes, and the directions it takes in the future depend upon where it has been in the past.

This discussion of the possible influences of resource distribution, predation, and competition illustrates the kinds of evidence and reasoning required to account for social evolution and to reconstruct the social behavior of extinct animals. We now turn to a consideration of how these ideas apply to our nearest relatives, the great apes. What are the differences and similarities between their social behaviors and mating systems, why have they evolved, and what do they suggest about continuities between the social behaviors of their ancestors and ours?

SOCIAL EVOLUTION AND BEHAVIOR IN OUR NEAREST RELATIVES, THE GREAT APES

Our nearest living relatives, the great apes, exemplify the contrast between the conservatism of morphological traits and the relative ease with which social behavior can change. There are four great apes—chimpanzees, bonobos, gorillas, and orangutans (as well as the more distantly related gibbons)—and although they are similar in morphology and ecology, all are quite different from each other in social behavior and mating system. It is not yet known how differences in their ecology have led to such large differences in their social behavior, but detailed field studies over the past twenty years have provided good working hypotheses. This work has also shown some interesting similarities between the social behavior of great apes and us. We will describe some of these findings for each of our near relatives, seeing how they might help us understand how ancestral hominids lived as well as the roots of some of our own behaviors.

ORANGUTANS

Orangutans were distributed throughout east and southeast Asia during the Pleistocene epoch, but they are now found only in the rainforests of Borneo and Sumatra (Fig. 13.2). They spend most of their time moving slowly through the high crowns of trees, using both their arms and legs. Males average 183 pounds, females 82 pounds, and this "quadru-manual clambering" is how these large mammals distribute their considerable weight among relatively small branches. Their preferred food is fruit and seeds, but when that is not available, they eat leaves and bark too. They also eat ants and termites.

FIGURE 13.2 Male orangutan.

Orangutans are the least social of all the day-active primates. Except for females with one or two dependent offspring, occasional small groups at common feeding sites, and male-female mating consortships of five to eight days, they spend all of their time alone. The feeding ranges of females are 0.4 to 3 square miles and overlap each other. Males range over larger areas that include the ranges of females. The large ranges, slow movements, and eight-year birth interval of females makes mate-guarding by males a poor reproductive strategy. Males mate with more than one female and females mate with more than one male.

A promiscuous mating system is not necessarily free of male-male competition, and mating need not be random. A female orangutan's successive young may be fathered by different males, but those males were likely the winners of fights with other males. A female may have some choice as to who the father will be by accepting or avoiding a male's advances at the time of her maximum fertility. Alternatively, a male may coerce an estrous female into mating with him.

There is evidence for both female choice and male coercion in orangutan mating. In size, appearance, and behavior, the large mature males seem to be a typical product of sexual selection in mammals. They are more than twice the size of females, range over much larger areas than females, have exaggerated secondary sexual traits, announce their presence from a distance by pushing over dead trees and calling loudly, and are very combative toward each other. Violent fights often result when two males meet, especially if a fertile female is nearby, and most adult males show scars and disfigurements from these encounters. Females are attracted to the large, loud, and aggressive males, and from the relaxed way in which they engage in sex with these males during consortship, they seem to be willing partners.

Coercive mating is part of a second kind of male mating strategy that has evolved, likely in response to the pressures of mating competition. Some males stop growing and do not develop the vocal apparatus or other secondary features of the large males. They nevertheless mature sexually and then gain matings with females by force. Their smaller size enables them both to avoid combat with the larger, slower males and to overtake females, who almost always attempt to escape. Once caught, the female resists by writhing and squealing or struggling violently and trying to bite the male during copulation. According to the independent observations of several researchers, between one-third and one-half of orangutan copulations are of this violent sort.

Forced copulation by subadult orangutans is the physical equivalent of rape in humans. The evidence strongly suggests that in these primates it is an evolved mating strategy that enables them either to impregnate a female directly or to demonstrate the futility of resistance at another time when she is fertile. This hypothesis awaits further tests, such as determining the proportion of offspring that results from forced copulations and whether the number of these smaller males is determined by the frequency of mating competition from larger males.

GORILLAS

Gorillas are the largest living primate and are found in equatorial Africa as three subspecies. The mountain gorilla (Fig. 13.3), living at the base of the Virunga Volcanoes in Rwanda, is the best known, due in part to the pioneering observations of Dian Fossey and the ongoing research at the center she founded there. The food of mountain gorillas is primarily leaves of herbs, vines, shrubs, and bamboo, but they also eat the stems, pith, roots, blossoms, and fruits of these and other plants. Their preferred habitat is where these foods are abundant: in the thick shrub and ground-level vegetation found in humid, open-canopy forests located along rivers, forest edges, areas of secondary forest growth, the sides of mountains, and bamboo stands.

FIGURE 13.3 Male silverback gorilla.

Gorillas live in stable bisexual groups consisting of an older "silverback" male (so called because of the broad streak of gray hair running down the middle of his back), zero to two younger, black-backed males (usually sons, brothers, or cousins), three to nine mature females, and a similar number of their immature offspring. Both males and females usually leave their natal group. The groups are small, and this avoids inbreeding. Emigrating females either attach themselves to another established group or they join a lone silverback. Young emigrating males either live alone or assemble in bachelor groups for several years until they become fully mature. They then attempt to acquire females by attacking established groups or inducing lone females to join them.

The feeding ranges of gorilla groups overlap slightly, but the groups usually keep apart. There is no fighting when they do come in contact, although the silverback males will often display hostility by beating their chests, calling, and charging. There is fighting, however, when lone males attempt to come close to a group. Silverback males are twice the size of average females and can reach a height of six feet and weigh over 400 pounds. Although there is virtually no overt aggression within a group, the silverback male is clearly dominant. He determines the movements of the group, monopolizes mating with females, suppresses minor disputes between females, and guards them and their small young.

Conceptions of gorilla behavior have changed with time. Fifty years ago they were thought to be fierce and dangerous, largely because their large size was dramatized in circus publicity. When ethologists succeeded in studying them in their native habitat in Africa, quite another picture emerged. The quiet and relaxed pace of gorilla life, the rarity and low level of aggression within the group, and the willingness of this large and powerful animal to allow human observers into close contact, even to groom and be groomed, led to an image of peaceful vegetarian giants. Neither of these pictures is quite right.

Male gorillas may be gentle and protective toward their own young, but they are killers of the infants of others. In fact, one of every seven gorillas born ends a victim of male infanticide, and about 40% of infant mortality in the Virunga Volcano gorilla groups was attributed by Dian Fossey to this cause. These killings occur when a mother and her infant are unprotected by a silverback, either because he has died and all the infants of the group are vulnerable, or because the female and her infant are momentarily out of his protective range. Lone raiding males charge, wrest the infant from its mother, and kill it. The response of the mother to this act seems unexpected and even repugnant: she leaves the group and joins the male that killed her infant. Some people may find this upsetting, but remember that natural selection, not human expectations, designed gorilla behavior. To cast the outcome in evolutionary (not cognitive) terms: a successful infanticide seems to demonstrate to the female that the raiding male will be an adequate and perhaps even better protector of her future offspring than the present silverback, and that he will father sons who will also be successful at this tactic for acquiring mates and reducing competition.

Why do gorillas live in social groups in the first place? The evidence suggests that a combination of three factors is involved. First, females join males in order to gain protection, mainly from the aggression of other males, but also from predation by leopards and humans who hunt them in all parts of their range. Females never travel alone, and when they switch groups, it is when two silverbacks are in close proximity. The importance of predation is also suggested by the formation of bachelor male groups, even though this must increase competition for females. A second and related factor is that a combination of size, experience, and skills at aggression enable some males to monopolize the mating of several females, and we would expect females to be attracted to such males instead of lone individuals. By mating with a male who has demonstrated such qualities, a female is likely to leave more descendants. If the qualities of their polygynous fathers are to any extent heritable, which they very likely are, her sons are also more likely to have above-average reproductive success. This self-reinforcing process of sexual selection, involving male-male competition and female choice, lies behind the considerable sexual dimorphism in gorillas.

The third factor that likely influences group living in gorillas is the distribution and nature of their food. Food is abundant, the different items vary little in quality, and they are found in patches that are large enough to support feeding groups. This combination of attributes lessens competition for food and has thus made it possible for the mating system of gorillas to evolve. If food were sparsely and evenly distributed, as it is for orangutans, females would have to forage alone so as to avoid feeding competition, and a monopolistic male strategy of guarding females from other males would not work. Gorillas clearly show how internal social pressures (an aggressive male mating strategy) combined with permissive ecological conditions (abundant food that varies little in quality) can interact to produce a particular social system.

CHIMPANZEES

Our closest relatives among the great apes, the chimpanzees, include two species. *Pan troglydytes*, the common chimpanzee, ranges across equatorial Africa from the east coast almost to the west coast, whereas *Pan paniscus*, the bonobo (or pygmy chimpanzee), is restricted to an area south of the Congo (Zaire) River. Common chimpanzees primarily inhabit rain forests, but they are also found in forest-savanna mosaics and open woodlands. In a few parts of Africa they even venture from open woodlands onto grassland savannas. The common chimpanzee is the best known of all the great apes, as it has been studied continuously for ten to forty years by five different research groups at different sites. The most familiar of these was established by Jane Goodall at the Gombe Stream National Park in Tanzania in 1960.

The most stable large-scale social unit of the chimpanzee is a community of twenty to a hundred individuals that occupy a recognizably defined area for many years. These areas are two to fourteen square miles in densely wooded regions to as much as two hundred square miles in sparsely wooded habitats. Related males remain in their natal community and maintain its long-term continuity. Females, on the other hand, usually emigrate. Female emigration is also found in gorillas and, historically, in most human cultures, but in almost all other primates it is the males who leave. The integrity of communities is maintained by constant patrolling of the borders and hostile displays by groups of males.

Chimpanzees eat mainly plant food, and although there is regional and seasonal variation, fruit is the major component of the diet, followed by leaves, seeds, piths, flowers, bark, and nuts. They also eat a variety of animals, including insects and small mammals. The small mammals—monkeys and the young of wild pigs, baboons, and antelope—are hunted by elaborate cooperation among two to six chimpanzees, almost always adult males. They surround the prey and position themselves so as to cut off escape. Once caught and killed, the prey is dismembered, and pieces are shared, the dominant male usually retaining a larger share. Pieces of the carcass are frequently given to other chimpanzees that beg, and males often give pieces to females after copulating with them. In some areas, chimpanzees hunt monkeys so frequently and effectively that they kill 15 to 30% of their prey population in a year.

The members of a community are never all together at any one time. They move about and feed in small groups of two to six individuals, but in a social pattern of fusion and fission, groups make contact and associate for a while, then part company, sometimes forming new groups. This pattern of associations is probably determined by the distribution of their food. Food is found in patches that are usually neither large nor dense enough to support feeding by large numbers of chimpanzees at one time. Nevertheless, one area might have more high-quality patches of food than another, and females do establish distinct but overlapping core feeding areas and attempt to displace other females who intrude. When a particularly rich source of food is discovered, such as a large tree in fruit, larger aggregates of animals form temporarily.

The groups in which chimpanzees travel within the range of their community may include any combination of ages and sexes, but there are consistent patterns. The most stable groups are individual females with their dependent young, often accompanied by their older adolescent young. Adult males spend twice as much time together as females do, and adult females spend longer continuous periods alone than do males. There are no consistent, long-term associations between individual adult males and females.

As we discussed in Chapter 11, chimpanzees make a variety of tools. Branches are stripped of leaves and carried to termite mounds where they are used to fish the insects out of their holes (Fig. 11.12). Leaves are chewed and used as sponges to retrieve water from narrow pockets in tree trunks after a rain. Leaves are used to wipe feces and blood from their hair or as bowls for scooping and drinking water. Branches and stones are used as hammers and anvils in cracking open hard fruits and nuts (Fig. 13.4). Leafy branches are grasped in the feet and hands and used as sandals and gloves in climbing trees with thorny branches. Sticks are used to dig up roots, open nests, and pry open food containers. Stones are thrown to hit or intimidate opponents or intruders, and sticks may be used as clubs to hit or scare others. Given the complexity and variety of tools made and used by chimpanzees, it is not surprising that there are cultural variations from one region to another— different tools may be used for the same purpose or a different technique may be used with the same tool.

Mating among chimpanzees is promiscuous. The female menstrual cycle is thirty-six days, and during the sixteen days before ovulation her genitalia swell and become pink. During this time she is eager to copulate with males, and they with her. Males are able to detect within a day or two, probably by smell, when the female is likely to ovulate, and this increases their interest in mating (Fig. 13.5). The most dominant or alpha male (or a coalition of dominant males) uses aggressive threats to monopolize mating with a fertile female. Males also use threats, sometimes successfully, to coerce females into an exclusive mating consortship, away from other members of the community, for a few days to weeks. Females may also show a preference for particular males. A fertile female, however, usually solicits copulations with many males, a tactic that serves to obscure paternity. A male who remembers copulating with her would therefore not be inclined to harm her infant, for he might be its father. In fact, males are equally tolerant and protective of all young born within their community, but they will kill the young likely to have been fathered by males outside their group.

FIGURE 13.4 (A) Chimpanzee cracking nuts.

(B) A chimpanzee has used a small rock as a wedge to stabilize its anvil.

(C) Chimpanzee fishing for termites.

(*B*)

(*A*)

(*C*)

Like male orangutans, male chimpanzees show an evolutionary history of sexual selection. They are larger than the female, and they constantly strive to dominate other individuals, using loud displays and physical threats (Fig. 13.6). This behavior starts in late adolescence with attempts to intimidate females. As soon as males become big enough, they attack females, hitting them, throwing them to the ground, and stomping on their backs. The males have large canine teeth that they use with lethal effectiveness in killing prey, but they do not bite the females with them. In time, a growing male is able to dominate all the females, and he continues this behavior throughout adult life, sometimes without apparent provocation. (In the extensive study of the

FIGURE 13.5 As the time of ovulation approaches, the genital region of female chimpanzees enlarges and becomes pink. While in this state they are followed by males. The female is the second from the right.

captive chimpanzee colony at Arnhem in the Netherlands, however, once a male had reached alpha rank his behavior toward the females changed. He intervened on their behalf when younger males were bullying them, and the ethologists who studied this colony concluded that female support was important to the male in maintaining alpha status.)

The older a male becomes, the more he is able to dominate other males. If he does become the alpha male, which is not likely to occur until he is twenty to twenty-five years old, he retains this status for three to ten years. As in other mammals, striving for dominance by male chimpanzees is driven by direct and indirect reproductive rewards. The alpha male is able to take meat from others, displace others from rich feeding sites, and gain more matings with fertile females.

There is another aspect to male chimpanzee relationships that may seem incompatible with all of this conflict and competition. Adult males are much nicer to each other than they are to females or than females are to other unrelated females. They spend twice as much time together in groups, and they groom each

FIGURE 13.6 The dominant male on the left is displaying; his hair is erect, making him look larger than he is, and his threats have induced the other chimpanzee to behave submissively.

other (Fig. 13.7), exchange friendly greetings (calls, kissing, embracing) and share meat much more frequently. But this "male-bonding" behavior has a function. Males need each other for success in competition against other males, and it is crucial in two contexts. First, one of the routes to high status, besides being the most aggressive individual, is forming and manipulating alliances with other males. These alliances consist of two to three individuals who groom each other, share meat, and support each other in fights, either with other males or while intimidating females. The shifting exchanges of support and the social maneuvering that characterize these relationships have astonished many observers by their complexity and subtlety. The ethologist Frans de Waal has described the dynamics and consequences—sometimes lethal—in his aptly titled book *Chimpanzee Politics*. Figure 13.8 shows an interesting interaction between a juvenile and an adult male.

A second context in which cooperation is crucially important to male competition is in the relations between communities of chimpanzees. These relations are hostile—even lethally so. Groups of five to six male chimpanzees not only patrol the borders of their community territory, they also carry out expeditions into adjacent communities for the purpose of killing neighbors. From the start of one of these organized raids, the males behave in a characteristically different way. They group tightly together, travel quickly into the adjacent community, and ignore even choice food. They move silently, listening intently for the sounds of one or two isolated neighbors. If a single male is found, they surround it, pin it down, beat, stomp and bite it, and twist its limbs unmercifully. Most of the victims of witnessed attacks were either found dead or disappeared shortly thereafter because of crippling and infected wounds. Older females and young are also killed, but younger females without young are generally not harmed; however, they usually return with the raiders to their home territory and remain with them.

The odds in these attacks are overwhelmingly in favor of the raiders, and they return largely unscathed. If attack poses risk of injury because the raiding party has encountered two or more members of the neighboring community, however, the foray quickly dissolves into a hasty homeward retreat. Humans frequently use this same strategy of maximizing cost to enemies at minimal cost to invaders when one group attacks another.

Lethal raiding by chimpanzees is frequent enough to have reproductive consequences. The behavior has been seen in several widely separated populations, and in the two well-documented cases, all of the males in two small communities were eliminated by the larger adjacent community over a ten-year period. Lethal raiding eliminates genetic competitors, opens new feeding territories, and provides reproductive females. It is not hard to imagine how such consequences of intergroup aggression constitute a powerful selective force for this behavior.

What role have female reproductive strategies and female-female interactions played in chimpanzee social evolution? Until about thirty years ago, primate behavioral studies were heavily influenced by the cultural myth that females "by nature" are (and therefore should be) the passive pawns of male maneuverings. Then it was found that among monkeys such as baboons and rhesus macaques, female behaviors make a

FIGURE 13.7 Two adult males in close company grooming each other.

FIGURE 13.8 The adult male chimpanzee is in the process of losing his alpha status and is behaving as if in distress. The young male appears to be offering reassurance.

great difference to their reproductive success. In these primates it is males who emigrate and females who remain in their natal area, compete with each other for dominance, and make alliances with sisters and female cousins to advance their status and that of their matrilines. There is considerable variation in reproductive success among these female monkeys, and most studies have shown that the higher their rank in the female hierarchy, the more surviving offspring they have. Higher-ranking females gain better access to high-quality food and suffer less stress from aggression that is directed at them.

From low numbers of adult female relatives and the infrequency of female contacts in chimpanzee communities, it initially appeared that female dominance had few reproductive consequences. Recently, however, the results of twenty-five years of observations of female dominance interactions and reproductive success at the Gombe station have shown how subtle and infrequent behaviors can have significant effects on reproductive success. It takes much time to acquire such information about an animal whose long schedule of maturation, reproduction, and death is very similar to our own. The ethologists Anne Pusey, Jennifer Williams, and Jane Goodall analyzed dominance relations among female chimpanzees by noting the occurrence and direction of the "pant-grunts" by which both males and females express submission to more dominant individuals. Several measures of reproductive success correlated with the dominance status of females. A greater proportion of offspring survived to seven years of age; daughters first gave birth at a

younger age; and more surviving offspring were produced over a twenty-two-year period. By all of these measures, the more dominant females were 50 to 100% more successful than subordinate individuals. It remains to be seen whether such large effects will be found in other populations.

A female chimpanzee's dominance over other females apparently affects reproduction in several ways. The high mortality of infants of low-ranking females was partly attributable to the infanticidal behavior of one high-ranking female and her daughters who snatched, killed, and ate the infants of several lower-ranking females. This is not an isolated pathological behavior because other high-ranking females have been seen attempting to grab the newborns of low-ranking females, and other females have been observed eating another female's infant.

The younger age at which daughters of high-ranking females reach sexual maturity is doubtless related to their faster growth. This in turn is likely due to the presence of more high-quality food in the core feeding areas of dominant females. Better nutrition likely also accounts for the better survival of high-ranking females and their offspring.

We have already discussed the evidence from DNA comparisons that humans and chimpanzees diverged 5 to 6 million years ago. Furthermore, there is reason to believe that chimpanzees have changed relatively little since that time. Chimpanzees are adapted to a range of environments that have existed in Africa during the last 10 million years—rain forests, forest-savanna mosaics, and open woodlands. Of the living primates, chimpanzees are therefore likely the most similar to our common ancestor. Early hominids are believed to have moved into the drier woodlands and open savanna, a habitat now occupied by another primate, the baboon. These large monkeys were therefore an object of early study among biological anthropologists, but because they are so distantly related to hominids, interest in them has slackened. Chimpanzees, however, have the behavioral flexibility to range into this habitat, where they are faced with the same challenges and opportunities as early hominids. Some chimpanzee populations do cross open areas to reach fruit trees and other edible plants, termite mounds, and small mammals. No long-term studies of these populations have been carried out, but there are reports of root digging, termite fishing, chasing small antelope, and spending the night in fig trees in the open savanna.

BONOBOS

The impression given by early studies of gorillas and chimpanzees was that they are happy vegetarians who lead quiet and peaceful lives, devoid of selfishness, jealousy, and hostility toward one another. But these impressions were revised when the results of long-term observations were reported. Orangutan males seriously injure each other in fights and forcibly copulate with females. Gorilla males kill each other's infants. Chimpanzee males batter females and carry out murderous raids into adjacent communities, kill infants likely fathered outside their community, and continuously use aggression and social maneuvering to gain dominance over rivals. Female chimpanzees dominate each other to the degree that they derive reproductive advantage, and they sometimes try to kill the newborns of subordinate females in their own community. There is little doubt that natural selection has been at work in shaping primate behavior and that these conflicts improve the relative reproductive success of the winners.

But natural selection has also molded a second kind of chimpanzee, the bonobo, that is very different in its behavior from any of the other great apes. Bonobos live in the humid swampy rain forests in an area of central equatorial Africa bordered on the west, north, and east by the Congo (Zaire) River. They are approximately the same size as chimpanzees, but they have a more slender frame and limbs, longer legs, a shorter clavicle, and smaller molars (Fig. 13.9).

The preferred food of bonobos, like that of chimpanzees, consists of the ripe fruit of trees and vines. Unlike chimpanzees, however, when fruit is not available they rely on the shoots, pith, and stems of herbaceous plants. Such vegetation is abundant in the rain forests in which they live. Bonobos also consume more different kinds of invertebrates than chimpanzees. They will eat meat—they will grab hidden infants of ungulates—but they have not been observed hunting cooperatively.

Bonobos live in communities similar in size to those of chimpanzees, but the size of their feeding parties is many times larger, consisting of thirty to fifty individuals. These are mixed groups, and contacts between individuals of both sexes and all ages occur frequently.

The social interactions of bonobos differ from those of chimpanzees in several notable ways. Grooming among adults is most frequent between males and females, and it is least frequent between males. Food sharing is frequent among females and rare among males. Although there are linear dominance hierarchies among males and females, and males do engage in the same sorts of noisy displays as chimpanzees, the level of dominance aggression is much lower. The effect of dominance on reproductive success is unknown for either sex. Males do not form alliances to gain dominance over others, and they do not dominate females of equal rank. In fact, females form alliances with each other, and they are not only able to thwart male attempts at dominance, they are also able to displace males from desirable food. Most of the encounters between bonobo communities so far observed have seemed tense, and fighting and wounding do occur. During one meeting, however, the two groups did mingle, and there were matings between members of the different communities. To date, no lethal raids across community borders have been observed.

Bonobos are especially different from chimpanzees in their sexual behavior. They copulate more promiscuously, more frequently, for longer periods of time, and starting at a much earlier age. Unlike chimpanzees, bonobos frequently copulate face-to-face, and their matings are accompanied by mutual gazing and soft vocalizations (Fig. 13.10). They also use signals—arm and hand gestures, body postures, facial expressions, and quiet calls—to communicate the position they desire.

Bonobos do not differ consistently from chimpanzees in inter-birth interval, age of weaning of young, or total number of offspring produced by a female over her lifetime. As in chimpanzees, frequent sex is likely part of a female strategy to obscure paternity and lessen male aggression that may harm them or their young. But female bonobos have carried these tactics even further. Sex is not only more frequent and elaborate, but as in humans, ovulation is concealed. Unlike male chimpanzees, male bonobos seem unable to tell when ovulation occurs, and attempting to monopolize mating with a female during the entire time when she may be fertile is not feasible, especially when the female actively mates with many males.

FIGURE 13.9 Female and male bonobos.

FIGURE 13.10 Bonobos copulate in a variety of positions.

Much of the sexual activity of bonobos has no direct reproductive function. As in chimpanzees, juvenile females copulate before they are fertile, and older females engage in sexual activity with each other. They embrace ventro-ventrally, appose their clitorises, and make rapid lateral movements of the pelvis (Fig. 13.11). This homosexual behavior, clinically described by ethologists as "genito-genital rubbing" but called "hoka hoka" by the local African Mongandu people, occurs among all the female members of a community. It is more frequent between adolescent females who have just joined a community and older resident females with whom they are establishing a long-term friendship. It is also more frequent during periods of high excitement, such as when a rich food source is discovered or two groups meet. The behavior of bonobos during this activity leaves little doubt that it is enjoyable, and two anatomical changes have apparently evolved to enhance the pleasure. The bonobo clitoris is both larger and more ventrally located than in other apes.

All of this suggests that "hoka hoka" functions to establish and maintain bonds between unrelated females. These bonds are clearly important, because female alliances are effective in thwarting male aggression and even enabling females to dominate males when choice food is available. Whether this female-female sexual behavior helps younger individuals to advance in their dominance hierarchy and whether female dominance status affects reproductive success (as it clearly does in chimpanzees) remain unknown.

When and why did bonobos and chimpanzees become so different? Comparison of bonobo and chimpanzee DNA shows that they had a common ancestor 1.5 to 3 million years ago. Other comparisons of the two species, with each other and with gorillas, suggest that bonobos are more specialized than chimpanzees and therefore less like their common ancestor. The bones and skull of bonobos are lighter and there is less sexual dimorphism. The teeth of bonobos are reduced in size, and their cheek teeth have high shearing edges for chewing the soft buds and stems of herbs that comprise an important part of their diet. They are confined to an area of rain forests south of the Congo River where such vegetation is stable and abundant (Fig. 13.12). Furthermore, they do not have to compete for it with gorillas, as they would on the other side of the river. Taken together,

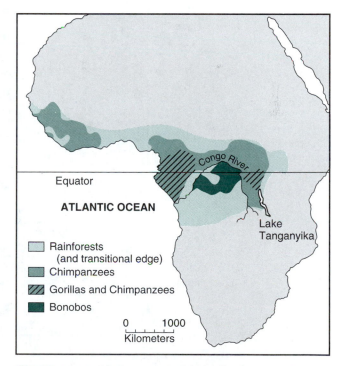

FIGURE 13.12 Chimpanzees and gorillas have overlapping ranges north of the Congo river, whereas bonobos are found south of the river. Ecological aspects of their distributions are thought to have influenced their social structures in important ways. See the text.

FIGURE 13.11 Female bonobos mutually rubbing their genitals.

this evidence suggests that the bonobo lineage split from the chimp line and adapted to an unexploited environment.

HUMAN SOCIAL VARIATION: UNITY WITHIN DIVERSITY

When human societies are compared with respect to the production of food and goods, how kin are classified and treated, where a married couple lives after marriage, who inherits what, religious beliefs, and the degree of political complexity and socioeconomic stratification, there seems to be considerable variation. A somewhat different perspective emerges, however, with criteria that are useful in comparing different species of primates. For example, look at Figure 13.13. Individuals of each sex can be found exclusively with non-kin, with no other member of the same sex, or with kin (and perhaps non-kin as well). When mating occurs, the

consortships can be temporary, or males and females can pair in exclusive associations.

For example, adult orangutans of both sexes are solitary, associating only briefly for mating. Either the male or the female may mate with other individuals at other times, so their associations are not exclusive. Or consider gorillas. An adult male does not ordinarily tolerate the presence of another silverback, and the females in his group are usually unrelated to each other. A number of the females in a group of chimpanzees are also unrelated to each other, because, like gorillas, female chimpanzees move between groups. The males, however, are characteristically in the presence of male kin. Chimpanzee mating is promiscuous.

The vast majority of human societies consist of many conjugal (nuclear) families in which family members share the same dwelling and the different families associate in the same community. Families are united by bonds of marriage, kinship, friendship, or cooperation and reciprocal exchange. Even if the genetic kin

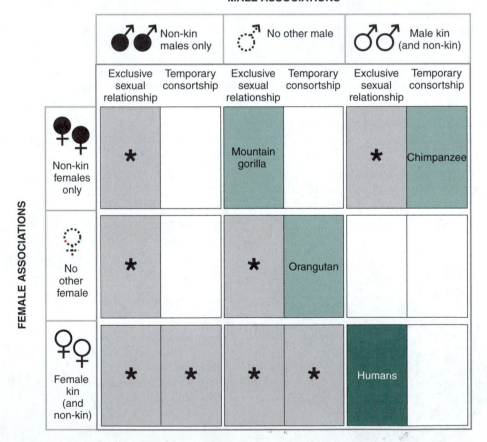

FIGURE 13.13 A matrix of the principal social relationships of primates. Within-sex associations for a typical group of individuals are shown along the edges of the matrix, and the major reproductive (between sex) associations are represented by the larger cells in the center. The colored cells correspond to apes and humans; the gray cells with stars represent associations characteristic of other primates. Humans occupy one small corner of this social space.

of married individuals do not live in the same community, consanguinal ties are often maintained. By the criteria used to compare other animals, most long-term sexual relationships in humans, especially those that produce children, are exclusive, and socially prescribed to be so.

A striking feature of Figure 13.13 is that humans seem to occupy only a small part of the "social space" that is filled by other species of primates. This point is reinforced by Figure 13.14, which is a more detailed look at the social relationships of just gorillas and humans and indicates some of the variation that is present in both species. As described above, the most common social group of gorillas consists of one silverback male

and a number of unrelated females. Less frequently there are groups with one silverback and several related females. Sometimes there are polygynous groups consisting of two or more males who are related and females who are either related to each other or not. Each of these social structures is represented by a separate cell in Figure 13.14.

Still other social arrangements of gorillas are part of their life cycle. After leaving their natal group, males may travel alone or form bachelor groups with either related or unrelated males. If a male is able to breed successfully, the first step is to acquire a single female. These relationships are shown in the bottom row of cells in Figure 13.14.

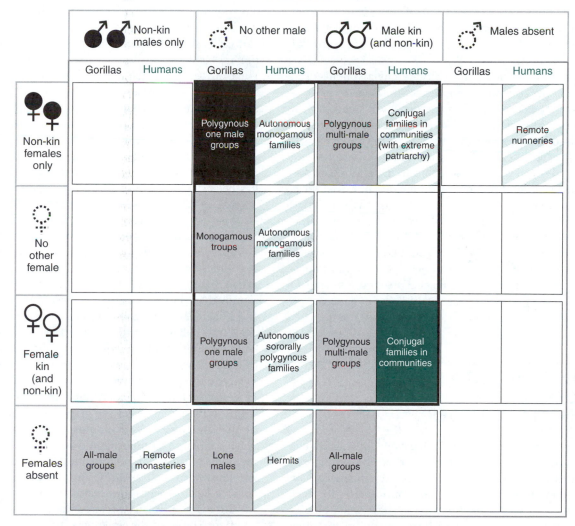

MALE ASSOCIATIONS

	Non-kin males only		No other male		Male kin (and non-kin)		Males absent	
FEMALE ASSOCIATIONS	Gorillas	Humans	Gorillas	Humans	Gorillas	Humans	Gorillas	Humans
Non-kin females only			Polygynous one male groups	Autonomous monogamous families	Polygynous multi-male groups	Conjugal families in communities (with extreme patriarchy)		Remote nunneries
No other female			Monogamous troups	Autonomous monogamous families				
Female kin (and non-kin)			Polygynous one male groups	Autonomous sororally polygynous families	Polygynous multi-male groups	Conjugal families in communities		
Females absent	All-male groups	Remote monasteries	Lone males	Hermits	All-male groups			

FIGURE 13.14 Long-term groupings of humans (*green*) compared with those of gorillas (*gray*). All reproductive associations are located within the heavy black border. The two cells with darker shadings identify the typical social structures of each species. All of the other associations of gorillas (*light gray*) occur regularly, but as described in the text, the alternative social structures of humans (*light green stripes*) are numerically and evolutionarily unimportant. As in Figure 13.13, the major within-sex associations that occur in groups of individuals are shown in different rows and columns.

What corresponding variations on human social structure do we find? In one, a woman's ties with her relatives are broken, either because of the great distances between her natal community and that of her husband, or because the society requires that such ties be severed (represented in Fig. 13.14 by the cell labeled "extreme patriarchy"). All the other examples are ethnographic oddities. In a small minority of societies, families live alone for most of the year because the resources they require (forage for domestic animals, game, farmland, plant food) are sparse and patchily distributed. Even in these cases, however, ties with relatives are maintained, and during some of the year families live together in a community.

Long-term, single-sex groupings of humans are not a part of the life cycle, as they are in male gorillas. Religious orders are very seldom sufficiently isolated and economically independent of the societies from which they were derived to be truly autonomous. Hermits are uncommon and rarely spend the majority of their lives alone.

Humans occupy a diversity of environments, and they vary greatly in their modes of subsistence and production. Nevertheless, humans occupy a small part of the range of social states found in nature, and there is no ethnographic or historical evidence that human societies have ever existed with mating systems and modes of social organization akin to many of those displayed by other primates. A feature of humans, however, is that intellectual capacity has enabled us to exploit a variety of ecological opportunities in a relatively short period of history. Political and economic networks have increased in size and complexity as agriculture and technology have developed and communities have grown. But with respect to emotions and to reproduction, the impact of different habitats on human social systems has been remarkably small.

EVOLUTIONARY RECONSTRUCTIONS AND HYPOTHESES

If there were no evolutionary heritage shaping human behavior, one might wonder why there have not been societies in which women organize raiding parties to kill women in neighboring communities and take back their men as husbands. Or why we have been able to imagine but never create enduring societies in which there is no sexual jealousy, no prescriptions against incest and rape, no distinction between genetic relatives and nonrelatives, no differences between individuals in their social influence on others, and no universal sharing of food procured, goods produced, and wealth accumulated. Groups have tried, but even with strict

behavioral regimens reinforced by religious teaching, success has been elusive. One nineteenth century sect, the Shakers, sought to implement their goals by forbidding sex. Without a steady influx of like-minded recruits, such a society was certain to be temporary.

Our brief survey of the hominid fossil record and of social evolution among the great apes is a cautionary tale that should guide our efforts in reconstructing hominid ways. In spite of the general similarities in both diet and morphology that are shared by the great apes, their mating systems, social organizations, and social behaviors are all different in important ways. On close examination it appears that these different patterns of behavior have evolved, at least in part, in response to differences in the variety and distribution of foods and how these ecological factors interacted with earlier suites of morphological and behavioral characters. Application of this reasoning to humans, however, is made difficult by an overlay of relatively recent ecological and cultural diversity that conceals most of our evolutionary history.

Can we nevertheless identify some common features of ape and human social systems that might give us insight into the evolution of hominid behavior? The general approach to evolutionary reconstructions from comparative data is to look for similar traits in two or more species that have been retained from a common ancestor. If there are good reasons for thinking a trait has been phylogenetically conserved, then it was likely present in all of the species along the line of descent. Because we are more closely related to chimpanzees and bonobos than to the other great apes, and because chimpanzees have changed least since divergence from our common ancestor, traits that we share with chimpanzees are more likely to have been present in all hominids than traits we share only with other apes.

What are the similarities in the social organizations and behavior of humans and apes? First, human males in almost all societies form alliances with each other. Alliances can be based on kinship, but often they involve unrelated men and are based on the expectation of behavioral reciprocity. Among the nonhuman primates, only chimpanzees are known to form coalitions of unrelated males. As in chimpanzees, human male alliances function in hunting and in conflicts with other members of the species. Only humans and chimpanzees are known to raid neighboring communities to kill competitors and gain reproductive females.

In many primates, related females cooperate in competition with the members of other matrilines. Only humans and bonobos, however, commonly form coalitions of unrelated females. In both chimpanzees and bonobos, sex itself lies behind single-sex coalitions. For the male chimpanzees, it is access to females. For female bonobos, it is defense against male aggression, using female-female sexual activity for social bonding.

With the evolutionary expansion of the hominid brain and increased capacity for communication, planning, and sharing of goals, single-sex alliances could be based on a wider range of behavioral reciprocity than the promise or exchange of sex. There is nevertheless a striking similarity between chimpanzees and humans in the reasons for male-led attacks of one group on another.

This male-led intergroup aggression has been built on a more basic property of primate social groups, for all tend to be wary of outsiders. In most primates, there is significant movement of individual males between groups, while the females remain in their natal communities. Depending on the breeding system, however, the arrival of a new male can lead to serious conflict. In chimpanzees and bonobos, on the other hand, males remain in their natal communities and it is mostly the females who emigrate. Similarly, in more than two-thirds of human societies, it is primarily the females who move to neighboring communities, and in fewer than one-fifth do females remain while males emigrate. In the cases in which males do emigrate, the distances are small and ties with related males in the home area are maintained. Except for the transfer of females, the communities of chimpanzees are closed and hostile to outsiders. Both historical and contemporary evidence for intergroup conflict of humans suggests that this was likely an important feature of hominid social evolution as well (Chapter 14).

Another common characteristic of most primates is the systematic sexual coercion of females by males. This behavior was not invented by primates, for aggressive behavior by males is frequently associated with sexual selection. Among the great apes this coercion includes forced copulation in orangutans, infanticide among gorillas, and battering in chimpanzees. All of these tactics, as well as more subtle ones, are sometimes used by human males.

Inferences about continuities between ape and hominid mating systems are on less firm ground. One problem is the distinction between mating and paternity. Mating may be promiscuous with respect to the number of individuals with whom an individual copulates, but if one male is able to mate more often with females near the peak of their fertility, the mating system might be more polygynous than promiscuous with respect to paternity. Thus, by the criterion of number of copulation partners, chimpanzees and bonobos are certainly promiscuous, but it is known that dominant male chimpanzees gain a disproportionate share of matings with fertile females. If genetic tests of paternity confirm that dominant males are more often fathers than are subordinate males, the mating system would be polygynous with respect to paternity, even though it may be promiscuous with respect to copulations.

Difficulties are also encountered in trying to classify and compare human mating patterns. Most marriages in most human societies are monogamous, but most societies allow polygyny (Chapter 14). Polygyny in humans requires resources, so it is usually practiced by wealthy and powerful men. Although no human societies are as promiscuous as chimpanzees or bonobos, humans in many societies have sexual relations before marriage as well as after marriage with partners other than their recognized spouse. Furthermore, a man and woman may be married monogamously, have children, and then divorce. If they remarry, the reproductive consequences of this "serial monogamy" may be equivalent to polygyny and polyandry.

Despite these variations in mating practices, the modal pattern of lifelong sexual relationships between most humans in most societies—as well as the modal pattern of paternity—is much closer to monogamy than it is to either polygyny or polyandry. Considering the largely polygynous mating system of gorillas and the promiscuous-polygynous system of chimpanzees, it is likely that mating in the earliest hominids was less monogamous than it is now. In this view, monogamy likely increased with the evolutionary changes that included the substantial increase in male parental investment, which in humans, as you recall, is unusually large for a mammal.

Some of the secondary sexual characters of humans and the great apes are correlated with their social organization and mating systems in ways that reinforce the conclusions we have been discussing. In gorillas and orangutans, competition between males for females is intense, males are much larger than females, and the mating systems are largely polygynous. Human males are somewhat larger than females (8% taller and 20% heavier, on average), whereas male chimpanzees and bonobos are about a third larger than females. Humans are mainly monogamous with polygynous tendencies, whereas chimpanzees and bonobos seem to be promiscuous with polygynous tendencies.

The relative sizes of the testes of humans and the great apes also reflect different evolutionary histories of mating competition (Fig. 13.15). Surprisingly, the testes of a 450-pound male gorilla are smaller than a man's. Furthermore, the testes of male chimpanzees are twice the size of those of humans, even though the average body weight of a male chimpanzee is only 100 pounds.

The size of the testes determines the rate of sperm production. Although a male gorilla may have a harem of several females, the females are not ready to mate until three to four years after giving birth, and even then they mate only a few days a month. A male silverback gorilla therefore mates infrequently. Moreover, as long as a silverback maintains control of his harem, he experiences minimal competition in fertilizing the females in his group.

FIGURE 13.15 Species differences in sexual dimorphism (*open rings*) and accessory sexual structures (*darker symbols*). Area of each ring is proportional to average body weight. Male gorillas and orangutans are about twice as heavy as the females, but the difference is smaller for humans and chimpanzees. As explained in the text, testes size of men is intermediate between that of chimpanzees and gorillas. What hypotheses can you suggest to account for the species differences in average breast size and penis size, both greatest in humans?

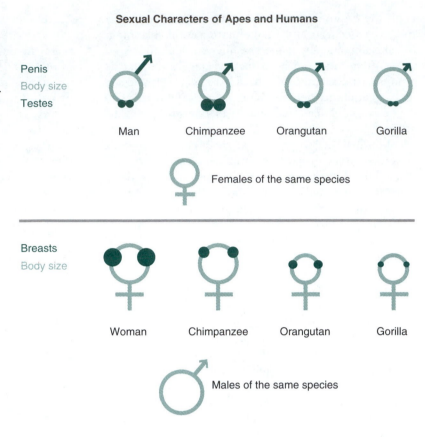

Sexual Characters of Apes and Humans

Penis
Body size
Testes

Man   Chimpanzee   Orangutan   Gorilla

Females of the same species

Breasts
Body size

Woman   Chimpanzee   Orangutan   Gorilla

Males of the same species

The sex lives of chimpanzees are very different from those of gorillas. At any time in a chimpanzee community there is very likely one or more receptive females mating promiscuously. Males therefore not only compete for priority in mating, they must also mate frequently and compete in producing sperm. The more sperm a male can make, the more likely he is to become a father.

The intermediate size of human testes suggests intermediate sperm requirements. Arguing just from testes size and sexual dimorphism, it is clear that competition among human males has been evolutionarily important. From this perspective, exclusive monogamy therefore appears to be more of a cultural ideal than an accurate description of human mating. But if, as the evidence suggests, the human lineage has evolved in the direction of monogamy, what influences have been at work? Humans differ from the other primates in the large amount of male parental investment associated with reproduction, and this correlates with the human offspring's long period of dependence on its parents. Sex appears to play a dual role in humans, just as it does in bonobos. It exists not only for the reproductive act, but because it is pleasurable, it can also be used to foster relationships. Female bonobos use it this way in coalitions with each other, and humans use sex to strengthen the bonds between couples. The concealment of ovulation in human females plays a comple-

mentary role. A male who does not know when his mate is fertile is more likely to be present more of the time, if only to assure his own paternity.

Recently, Richard Wrangham, James H. Jones, Greg Laden, David Pilbeam, and Nancy Lou Conklin-Britain, anthropologists with diverse backgrounds in behavioral ecology, paleontology, functional morphology, and nutrition, have pooled their knowledge and proposed an interesting if speculative hypothesis that may account for several seemingly unrelated features of human evolution. Recall from Chapter 12 that *Homo erectus* differed from its predecessors in (*i*) an increase in body size, (*ii*) reduction in the size difference between males and females to the ratio found in modern humans, (*iii*) reduced size of jaws and molars, indicating a change in diet, (*iv*) occupation of a wider range of more open habitats, and (*v*) the use of fire.

The central idea that possibly ties these changes together is the controlled use of fire for cooking food, specifically the underground storage organs of plants. Tubers and corms provide a concentrated source of nutrition that is available in open habitats in central Africa but is relatively uncommon in forest environments. Cooking makes a larger proportion of plant food digestible, and it can destroy some plant toxins. Cooking would therefore decrease the need for larger jaws, teeth, and intestines for processing and digestion as well as make a greater variety of plant food available. This improvement in nutrition would in turn allow for larger bodies and a greater ability to populate new locations.

Wrangham and his colleagues further point out that in virtually all cultures women gather food and are the primary cooks. They hypothesize that in australopithecines and early *Homo*, both males and females foraged for plant food, but that in *Homo erectus* women became the primary foragers and cooks of such food. With fires to tend, however, *H. erectus's* food was likely to be prepared at fixed locations where tubers, nuts, and other supplies would be brought for cooking and sharing. If males were guaranteed a reliable fallback source of food that they could get by dominating (scrounging from) women, they would be free to hunt meat, a more dangerous and less certain strategy, but one that provides a higher-payoff source of nutrition.

Coercive confiscation of food by men would favor those women who could make long-term alliances with males. Women could encourage these alliances by extending the period of their sexual receptivity beyond the time of estrus, thus increasing their sexual attractiveness and forming long-term sexual bonds. Males, in return for a reliable source of food and increased likelihood of paternity, would guard against food raids by other males as well as supply meat. This suggestion is consistent with other ideas to account for the evolutionary movement toward monogamy in humans. The occurrence of such behavioral reciprocity over extended times provides greater male parental investment in offspring.

In primates, the amount of sexual dimorphism does not simply reflect male-male competition. It tends to be greater the larger the primate, and in this respect, the size difference between men and women is smaller than it might otherwise be. However, among primates there is a consistent inverse relationship between sexual size dimorphism and the proportion of a female's estrus cycle during which she is sexually receptive: the greater the female's receptivity, the smaller is the size difference between males and females. Wrangham and his colleagues found that modern humans fit this pattern, and they conclude that the reduction in sexual size dimorphism in *H. erectus* coevolved with an increase in female sexual receptivity.

In this hypothesis, the innovation of cooking by *Homo erectus* had profound and possibly rapid evolutionary consequences. It led toward monogamy in multi-male, multi-female groups, changes in body size, dentition, and sexual dimorphism, a large alteration in the reproductive cycle of females, and the capacity of the species to extend its range. If the hypothesis is even partly correct, it illustrates a very large impact of a cultural innovation on genetic evolution.

These issues remain topics for speculation, discussion, and further investigation, and we will see more of them in Chapter 14.

SYNOPSIS

Understanding the behavioral evolution of hominids is more difficult than deciphering the physical record of their evolution. Compared with bones, behavior leaves few fossils.

The social structures of animals are influenced by sexual selection, parental investment, and the mating system. Sexual selection has affected humans and the great apes in different ways, but various forms of coercive male behavior are present in all species. On close examination, however, closely related species frequently differ markedly in their social structures because of differences in the availability of resources, the effects of predation, or intraspecific conflict. Chimpanzees, bonobos, gorillas, and orangutans illustrate this variation.

Of the great apes, chimpanzees appear to have evolved the least from the common ancestor we share with them and bonobos. Consequently, there is particular interest in studying their behavior in order to see what features of human behavior may have been present in earlier hominids. The use of tools, the formation of alliances among unrelated males, and cooperative behavior, including attacks on neighboring communities, are all practiced by both humans and chimpanzees.

Bonobos, although more specialized than chimpanzees, have a promiscuous mating system in which sexual selection and the coercive behavior of males has been reduced by the formation of female alliances. Sexual behavior, including homosexual activity and copulations between adults and juveniles, is used extensively by all members of bonobo communities to further relationships between individuals. Bonobo females, like human females, have concealed ovulation.

Human behavior, like that of any species, is unique. No human society could hold together for long if it were based upon the social lives of bees, elephant seals, porpoises, lions, gorillas, or chimpanzees. It may be amusing to suggest that we might learn from bonobos to make love, not war, but it is most unlikely that humans could adopt their promiscuous behaviors. (It is no accident that many of the sexual practices of bonobos, particularly between adults and young, are not lawful.) It is nevertheless possible to glimpse some of the origins of our human nature through the detailed study of our closest primate relatives.

QUESTIONS FOR THOUGHT AND DISCUSSION

1. In this book we have discussed the influence on social evolution of the genetics of sex determination, sexual selection, and ecological factors. How do these factors interact? Consider why social groups

(communities) of chimpanzees form. Which individuals cooperate socially, what resource do they seem to be defending, and from what? In the light of your reasoning, propose some hypotheses about why the earliest human communities may have formed.

2. Recent finds of early hominid fossils suggest that bipedalism evolved in the human lineage before the evolution of heavier molar teeth for chewing coarse and hard plant food. In what ways might bipedalism have been of net advantage to a chimpanzee-like ancestor of the hominids?

3. Humans have frequently tried to design communities with different, sometimes novel, social conventions. If these experiments are at variance with evolved human nature, they do not last. Can you think of examples and identify what was overlooked in their design?

   Alternatively, what would comprise your ideal social system? What differences in the evolutionary history of humans might have made your ideal system easier to implement? Assuming you could rewrite evolutionary history, is your ideal even a theoretical possibility?

SUGGESTIONS FOR FURTHER READING

deWaal, F. (1982). *Chimpanzee Politics: Power and Sex among Apes*. New York, NY: Harper and Row. The first of deWaal's several books on primate behavior describes the shifting alliances and dominance hierarchies of male chimpanzees in a captive colony living in a zoological park in the Netherlands.

Kinzey, W. G., ed. (1987). *The Evolution of Human Behavior: Primate Models*. Albany, NY: SUNY Press. Important discussions of the bearing of primate field studies on human behavioral evolution, especially those by Wrangham and by Tooby and DeVore.

Packer, C., and Pusey, A. E. (1997). "Divided We Fall: Cooperation among Lions", *Scientific American*, (May): 52–59. A summary account of an extraordinary and detailed long-term study of lion predation and sociality.

Rodseth, L.; Wrangham, R. W.; Harrigan, A. M.; and Smuts, B. B. (1991). "The Human Community as a Primate Society", *Current Anthropology*, *32*: 221–254. A classic paper in biological anthropology that compares the mating systems, social behaviors, and group compositions of humans and other primates.

Smith, E. A., and Winterhalder, B., eds. (1992). *Evolutionary Ecology and Human Behavior*. New York, NY: Aldine de Gruyter. A collection of twelve original articles that summarizes current theory and research in primate, early hominid, and human evolutionary ecology.

Smuts, B. B.; Cheney, D. L.; Seyfarth, R. M.; Wrangham, R. W.; and Struhsaker, T. T., eds. (1986). *Primate Societies*. Chicago, IL. University of Chicago Press. Individual chapters on diverse primate species, their behavior, ecology, social organization, mating systems, communication, and cognition. A good introduction to the study of nonhuman primates.

Wrangham, R., and Peterson, D. (1996). *Demonic Males*. New York, NY: Houghton Mifflin. An engaging account of continuities in the social behavior of great apes and humans, with emphasis on lethal raids by male chimpanzees on neighboring groups.

14

Viewing Human Cultures in an Evolutionary Context

Photo: Decapitation scene on a black-on-white pottery bowl from the Mimbres region of southwest New Mexico, 1,000 to 1,130 A.D. The man is wearing a horned serpent-head costume and an arrow holder. In one hand he is carrying a head and in the other a fending stick used to deflect darts thrown by opponents with axalatls. Both written history and archeological evidence indicate that violent conflict between groups has a very ancient role in human history.

If animals could speak, we might hear some of them arguing that their unique abilities set them above as well as apart from the rest of nature—even to the point, perhaps, of asserting that they had been specially created. Bats might boast of the sonar they use to locate food and navigate. Sperm whales would proudly describe how they maintain neutral buoyancy in different ocean layers by changing the proportions of the oil in their heads between the liquid and solid phase, thus enabling them to lie still in ambush for prey. Elephants would boast of their multipurpose trunks, which they use as a drinking straw, arm, hand, sensory probe, water hose, snorkel, and weapon. We tend to believe that our uniquely developed capacity for culture gives us a special and elevated status in nature.

There are many reasons for thinking that however unique and complex they may be, bats' sonar, whales' mechanisms for adjusting buoyancy, and elephants' trunks are the products of gradual evolutionary processes driven by natural selection. And there are many reasons for thinking that both the human capacity for culture as well as many of the particular forms it takes in different environments are also, directly or indirectly, products of natural selection. In this chapter we will explore the relations between human culture and other aspects of biology, and we will argue that a deeper understanding of the nature and history of culture can be achieved by viewing its recurrent patterns as the products of a brain molded by the same selection processes that have driven the social evolution of other animals.

WHAT IS CULTURE?

Human cultures may be thought of as human inventions—knowledge and associated behaviors—that are transmitted socially. Specific languages, beliefs, technologies, and arts contribute to culture. So do systems of exchange, law, morality, religion, kinship, marriage, inheritance, socioeconomic stratification, and political organization. These features of human societies vary widely in their details, and one of cultural anthropology's tasks is to document, compare, and interpret these differences. Although cultural anthropologists disagree among themselves about the significance of cultural variation, the most important features of culture are usefully viewed as behavioral rules that regulate its members' interactions with each other and with their environment.

Some features of culture, though arbitrary, are automatically passed across generations. For example, one language serves as well as another, and specific languages are transmitted by a learning program that develops during early childhood (Chapter 11). Other cultural details may reflect local ecological conditions or simply chance historical events. Still other details, such as the division of labor and stratification of wealth and power, have intensified with the development of agriculture and the formation of larger groups with extensive divisions of labor. And some cultural practices are adopted simply for pleasure. At their bases, though, cultures are the products of human minds, and these minds, in turn, have been shaped by evolutionary processes.

CULTURAL AND BIOLOGICAL EVOLUTION COMPARED

In earlier chapters we showed that biological evolution includes the origin of new gene alleles by mutation, their transmission during reproduction, and their changes in frequency over time caused by natural selection. Cultural traits, however, are creations of the phenotype, the brain, that can be passed to other individuals during a single lifetime. Because there are no known mechanisms by which environmentally induced changes in an organism's phenotype (such as a suntan or learning how to tango) can modify its genes in a directed way, cultural evolution is largely independent of natural selection and changes in gene frequencies. This distinction is critically important, but it obscures another truth. Although we have enormous, indeed unprecedented, flexibility to change our behavior as the circumstances require, that does not mean that our brain is free of evolutionary "design" that can bias cultural choices. These relationships are summarized in Figure 14.1.

In cultural evolution there are no units comparable to genes. Genes are distinct and stable physical entities that replicate with great accuracy. Cultural traits, in contrast, are often diffusely defined behaviors that can be rapidly modified and combined in a variety of ways. Because they depend upon the activities of nervous systems for their origin, retention, modification, and transmission, cultural traits are emergent properties of highly organized biological systems. Cultural traits can be, and are, *selected*—selected, however, by the frequently conscious choices of people, a process quite different from natural selection of genes in biological evolution.

Another way in which cultural evolution differs from biological evolution is in the origin of new information. Mutations of genes are rare and random events, but new cultural traits arise frequently and can be modified in consciously directed ways. Individuals can invent, alter, or adopt cultural traits either to fulfill a useful purpose or simply for fun. Sometimes a cultural trait of no apparent use can sweep through society like a contagious disease—for example, the recent practice of wearing baseball caps backwards instead of forward, as they were functionally designed to be worn.

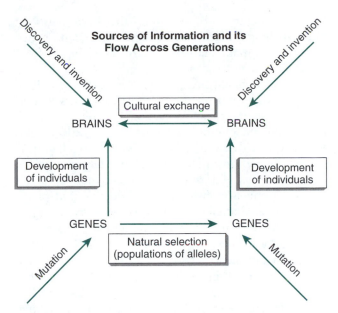

Sources of Information and its Flow Across Generations

FIGURE 14.1 The relationships between the various kinds of information that are transmitted across generations in cultural and biological evolution. Genes act through a hierarchy of levels, and the emergent properties of the brain arise through the processes of development. Brains, in turn, provide somatic mechanisms for transmitting and changing behavior across generations, frequently with consequences for reproduction.

Cultural change is based on learning and memory, processes that allow the transmission of traits both within and across generations at rates that greatly exceed the changes in gene frequency that result from natural selection. You are experiencing this feature right now: the knowledge described in this book and the language required to communicate it do not have to be discovered anew in each generation. Once technological invention provided mechanisms for accumulating and transmitting information—printing, followed by telegraphs, telephones, radios, television, computers, satellites, and the World Wide Web—the potential rate of cultural evolution was enormously increased.

These features of cultural traits—goal-directed change, very high rates of replication, and the cumulative inheritance of acquired information—combine to make cultural evolution much more rapid than genetic evolution and therefore capable of proceeding independently. For example, there is no evidence that the diversity of human cultures and languages that has appeared during the last ten thousand years has been accompanied by significant genetic changes in the neural mechanisms underlying human mental faculties. If provided with the proper learning environment, humans from anywhere in the world can acquire each other's language and culture with equal facility.

CULTURE AND THE SOCIAL SCIENCES

Culture is commonly viewed as a socially transmitted system of symbols or signs that express, integrate, and give meaning to the experiences of a human social group in coping with their physical and social environments. Although these symbols originate as expressions of human experience, in the view of many social scientists the differences among cultures are largely arbitrary.

However, many cultural practices have functional implications that make them seem adaptive in a biological sense. For example, expectations concerning the behavior of others are absolutely necessary for individuals to benefit from living in a social group, and their counterparts are evident in the social behaviors of other primates (Chapter 13). That cultures characteristically have codified behavioral rules is an extension of these expectations, made elaborate through the human capacity for language. The ultimate reasons for moral and legal codes will be considered later, but at this juncture we consider a more basic issue involving the relationships between brain, mind, and culture.

A recent introductory textbook on cultural and social anthropology presents the following as a commonly held and " . . . fairly consistent interpretation of what humans are like . . . ."

> . . . the human personality and the human mind are plastic, malleable, and almost wholly molded by the life experiences of the individual. Because this experience takes place within a cultural context, the resultant personality is largely explainable by social factors. Humans, then, are tabula rasa—blank slates—upon which the imprint of their social environments is indelibly printed. [There are] certain innate drives, such as sex and hunger, but [they are] diffuse in form and few in number. Granted that some human institutions serve these biological needs, no other propensities of the psyche seem to much influence culture. Rather, the determinism goes the other way. Culture is not the expression of the mind, but mind is the product of culture and social activity. You are what you do.
>
> Robert Murphy (*Cultural and Social Anthropology, An Overture*)

This extreme "environmentally deterministic" view of the brain and its development asserts that culture and cultural change are not influenced by "propensities of the psyche"—the cognitive and psychological mechanisms built into the human mind by natural selection. The view that the mind is largely the product of culture leads to the further supposition that each culture can only be analyzed and understood in its own terms, as a set of arbitrarily related and changing symbols that affect, but are little affected by, the material factors of subsistence, production, and reproduction. This effectively dissociates the study of culture from the natural sciences and commits it to documenting the diversity of cultural symbol systems, how they influence one another, and the many arbitrary ways in which they can be interpreted.

Cultures can influence each other through contact or be descended from a common ancestral culture, and both of these processes have played a role in cultural change. But we will argue that insights from evolutionary and developmental biology deepen our understanding of many functionally important cultural practices. We have already discussed several lines of evidence that support this view. Growing knowledge of how the brain develops and functions and how it evolved indicate that the human mind did not spring fully formed as an autonomous organ ready to be programmed in any arbitrary fashion (Chapters 10 and 11). Comparison of the social behaviors of the great apes and humans reveals patterns that require evolutionary analysis to understand fully (Chapter 13). But before expanding on how evolutionary social theory enlightens many human cultural practices, it is instructive to consider why such a wide separation between the "cultural" and "biological" study of culture became so prevalent.

THE CULTURAL GAP IN THE STUDY OF CULTURE

We know that our unique *capacity* for culture originated through natural selection—on this point, there is general agreement. This vague postulate, however, is only the first step in applying evolutionary theory to account for general features of human cultures and their histories. In fact, attempts to incorporate Darwinian theory into the social sciences have had a long history of failure.

The reasons for this failure are many and complex. The most important one, we think, is that until the last generation, biologists had not begun to develop a comprehensive and predictive theory of animal social evolution that could be extended to the unique cultural evolution of humans.

A related reason is that science, like all realms of human thought, can be strongly influenced by the social environment. In our culture, philosophy and religion have had strong connections to each other and have in turn influenced the much newer science of human behavior. The statement that ". . . culture is not the expression of the mind, but mind is the product of culture and social activity . . . " provides an example. The idea of the *tabula rasa* is found in the writings of the seventeenth century philosopher John Locke, who argued that our ideas are a product of our sensory experience, and as individuals we start as blank slates. As you have seen in Chapters 10 and 11, this early view of behavioral development is biologically incomplete. Not only has the human brain evolved to analyze particular kinds of information in particular ways, it develops

through programmed interactions that have their own internal design, albeit tuned, adjusted, and propelled onward by specific and specifically timed inputs from the environment.

But why should such an old idea continue to have such importance in one branch of the social sciences? We could point to an educational system in which meaningful integration of knowledge is little taught, but that would tell only part of the story. Unfortunately, the path from Darwin's *On the Origin of Species* and *The Descent of Man* to our present thinking about human origins is strewn with misguided and ideologically motivated attempts to use evolutionary theory in understanding human cultures and human history.

The best known of these mistakes is Social Darwinism, a nineteenth century theory of social change that saw inequities in wealth and power within and between societies as the result of differential survival of those who are the most skilled, creative, and adaptable. An English social scientist, Herbert Spencer, began to elaborate these ideas before Darwin published the *On the Origin of Species*, but he was quick to exploit Darwin's concept of natural selection as scientific validation for his thesis. Uncritical belief in this connection has hindered both biology and social science.

First, Spencer believed in an inevitable, goal-directed progress in the evolution of human societies. As biological evolution is not directed toward a goal, this recognition alone should have falsified claims of a causal connection. But around the turn of the century, as the science of genetics began to develop, notions of genetic differences in behavior and intelligence were also incorporated into this social doctrine to explain why some people enjoyed more economic success than others. Inequities in wealth and opportunity were then seen as the "natural"—and therefore justifiable—outcome of a "natural" selective process.

One problem with this argument is the false doctrine that what is natural must be good and/or right. Another difficulty is that because of circumstances unrelated to their abilities, many people start the race for "success" at different points along the course: some get to start well down the track, but many are not even in the stadium when the gun sounds. Nonetheless, Social Darwinism was widely adopted as a social philosophy and used to justify low wages, inhumane working conditions, and the inheritance of vast wealth, especially during the first decades of the twentieth century.

Social Darwinism is not the only example of how theories of heredity and evolution have been distorted to justify and implement political ideologies. Eugenics, the belief that only some people should be permitted to reproduce, drew its inspiration from the same source. Francis Galton (a cousin of Charles Darwin who coined the term "eugenics") argued that intermarriage of men and women in the upper social classes would produce gifted individuals. During the first part of the twentieth century, eugenics arguments were used to oppose immigration of eastern Europeans into the United States and to legalize sterilization of individuals who were deemed mentally ill, retarded, or epileptic. During the Nazi Holocaust, millions of people were exploited, imprisoned, and exterminated in the name of fallacious ideas about the genetic inferiority of several ethnic groups. Along with Social Darwinism, these dismal episodes have made many people wary of any proposed connection between biology and human culture and ready to cling to a modified but recognizable version of the Judeo-Christian doctrine that humans, even if not specially created, are nonetheless separate from and "above" all other animals. In this view, culture is therefore not only what makes us unique, it has liberated us from a supposed "biological determinism" of our animal past.

True as this history is, it can nevertheless hardly be claimed that science either causes or justifies such brutal behavior when the perpetrators don't even have the science right. It is very likely that Hitler would have created death camps if Darwin, Galton, and Spencer had never lived, for similar brutal behavior has been present throughout human history. The institution of slavery, the justifications for colonialism, the rationale for religious persecution, and the concept of racism are all reinforced by beliefs that some other group is inferior and unworthy of respect. Knowledge of evolutionary biology provides no justification for this kind of behavior, but, it does offer help in understanding *why* people act these ways as well as the flawed premises on which they base their behavior.

Another reason for the gulf separating the study of culture from other aspects of biology is that the unique nature of human cultures contributes to the sense that "biological" factors have had little impact on cultural evolution. Each culture is an amalgamation of historically accumulated and socially learned behavioral instructions that is greater than the parts in use at any one time, and it extends, mostly unchanged, from long before to long after an individual's lifetime. Extensive division of labor in many societies adds to the diversity of this information, so that no one individual can assimilate all of it, much less imagine that they might influence the whole of it, or understand why its unfamiliar components change.

A related consideration that seems to discount the role of biological factors in cultural evolution is that most of the present diversity in human cultures was likely created during the last 5,000 to 10,000 years, and there is no evidence that it was significantly affected by changes in gene frequencies. If so much cultural evolution has gone on in the absence of biological evolution, then, it is reasoned, why is it necessary to consider "biological" factors in explaining cultural change at all?

Notice how an important and confusing proposition has crept unnoticed into this question. The word "biological" is not synonymous with "genetic"—the brain, after all, works by biological principles—but in everyday discussions of human behavior, "genetic" and "biological" are frequently used as if they were interchangeable. Let us therefore rephrase the query in a more appropriate, certainly less ambiguous way. How has biological evolution influenced cultural choices? Can the relationship be turned around? Are there instances where cultural choices have altered gene frequencies?

PROCESSES OF CULTURAL CHANGE

In broad terms it is easy to make the case that biological evolution influences many cultural choices. Brains provide an extra-genetic mechanism for the replication of cultural traits that affect survival and reproduction, so such traits are likely to change in frequency independently of genes. But if brains are biased toward adopting cultural traits with positive effects, these traits can increase in frequency as fast as learning and transmission allow. Early vaccination of children against infectious diseases is an obvious modern example. Consider another with somewhat different implications. When cultural choices contribute to success in competitive striving among humans, the selective retention and transmission of new behaviors can also be adaptive in the evolutionary sense of increasing reproductive success. So accepting your group's belief that the members of a neighboring band are no better than the animals you both hunt can give a psychological advantage in conflict, complementing the technological advantage of using your bows and arrows against their spears. Or to pick a less obvious and more complex example with contemporary social and economic salience, in the development of new pharmaceuticals in our market-driven economy, devastating illnesses that occur only in poor, underdeveloped, tropical countries are largely neglected.

The conscious motivation for adopting any particular cultural trait need not be related to whether it is reproductively advantageous. Just as antibiotic resistance in bacteria can increase in frequency despite bacteria's ignorance of biochemistry, so in humans a beneficial cultural trait may increase in frequency because those who practice it, though possibly unaware of its causal effect, have more offspring (or garner more resources) than those who don't. When the offspring learn the cultural trait from their parents and practice it in their turn, culture can reinforce natural selection.

A partly hypothetical example of how such a link could be established is provided by the metabolism of vitamin D in humans. In Chapter 12 we saw how a combination of diet, available sunlight, and the need for vitamin D have led to an evolutionary loss of pigmentation in the skin of certain groups of people living in northern latitudes. Because vitamin D is required for the absorption of calcium from the intestine and for the deposition of inorganic salts during the growth and maintenance of bones, a deficiency during childhood leads to soft bones and bowed and twisted spines and legs. This condition, known as rickets, was a common childhood disease in seventeenth century London, where an inadequate diet interacted with smoke pollution and long winters, shielding infants and children from sufficient exposure to sunlight. The adult equivalent of rickets, osteomalacia, is found in some Muslim women who are required at all times to remain clothed to the point that only their eyes are showing.

It is therefore reasonable to postulate that early human populations in the northernmost latitudes of Europe and Asia were sometimes deficient in vitamin D. Suppose that some couples decided to place their infants and children outdoors in the sunlight for part of every day, even during the winter. They might do so for any of a number of reasons, real or imagined. Perhaps fresh air seems to be healthier than constantly breathing the smoke from cooking fires, or the sun is believed a benevolent deity that favors those that it sees every day. Their reasons for this custom could be quite independent of its beneficial effect in reducing the incidence of rickets, but those who practice it would experience significantly greater reproductive success.

This example shows how the differential replication of a cultural trait can be linked to a metabolic function. It also shows how this process can take place in basically the same way that natural selection causes differential replication of genes. But this description probably understates the likely role of cultural transmission *within* generations. Humans pay very close attention to what others are doing and saying, and if some arbitrary cultural practice subtly but noticeably benefits skeletal development where rickets is common, its use will spread, whether or not the causal relations are understood.

As another hypothetical example, it seems likely that early in hominid evolution small differences among individuals in the ability to exploit novel situations or to manipulate the behavior of others would have had significant positive effects on reproductive success. The relatively rapid evolution of the human brain—which is still poorly understood—thus very likely involved a reciprocal reinforcement between continually increasing behavioral capacities and natural selection. The accelerating rate of cultural change during the last several thousand years, however, has further obscured this likely relationship by occurring so much faster than most evolutionary change.

Some cultural practices are nevertheless known to have altered the frequencies of particular genes. The enzyme required for digesting lactose, the common sugar of milk, is present in all human infants, but its gene is expressed in adults only in those populations with a long history of eating dairy products. As a consequence, many adults from cultures in the Far East and those societies in Africa where milk and cheese are not regularly consumed fail to produce the enzyme required for their digestion. Similarly, Pima Indians of the American southwest desert have adapted to a diet of coarse plant foods by increasing their body's ability to extract and sequester limited supplies of sugar. Their metabolism, however, is ill suited for a diet of twentieth century, sugar-rich foods, and they now have an uncommonly high incidence of adult diabetes.

THE ADAPTIVE ROOTS OF CULTURAL PRACTICES

Cultures exist and cultures change because most of the time most humans adopt and modify ideas, beliefs, and ways of doing things when they perceive them to be in their best interests. Our perceptions of interests range over the practical and pleasurable in all possible combinations—trying to find a job, deciding how to spend free time, choosing new fashions in clothes—but in general, the goals that motivate our behaviors involve the following:

- surviving—obtaining food, shelter, clothing, medical care; providing defense against intruders; accumulating reserves for hard times

- access to and control of resources and, when conditions permit, accumulating possessions; controlling land or other assets, frequently involving cooperation of individuals in a larger social world of competition

- finding a mate or mates—involving criteria for a choice of mate, intersexual and intrasexual competition, and economic exchange

- gaining social acceptance and influence; forming marital, social, and political alliances; supporting and adopting common goals and attitudes

- leaving genetic descendants, arranging marriages, and supporting the physical and social well-being of children; influencing the behavior of offspring; guarding against cuckoldry or other potentially disruptive sexual behavior; allocating assets after death

- minimizing the disruptive effects of competitive striving within social groups—codifying and sanctifying moral and legal systems; conforming to religious rules and secular laws

Most explanations of human history and behavior allude to such goals. For example, in our culture, competition between individuals within groups is said to be a matter of "keeping up with the Joneses," "getting ahead," "winning," "making it," and "becoming successful." Groups of individuals—clans, tribes, states, nations—are said to engage in exploration, colonization, and conquest in order to expand their trade, control transport routes, capture the resources and labor of others, or prevent others from exploiting them.

These kinds of incentives lie immediately behind the behaviors of both individuals and groups, but they do not explain *why* humans are so motivated in the first place. Such motivations are *proximate* causes of behavior, but to account for their virtually ubiquitous presence, we need to explore the complementary role of *evolutionary* causation.

However we formulate these motivations for behavior, they are metaphors for efforts of the brain to pursue goals that in the past have had a positive impact on reproductive success. It is not that there are genes for specific behaviors; we trust you were convinced in earlier chapters that is an inappropriate and quite useless idea. A more suitable formulation is that during its evolution the human brain has been crafted by natural selection to assess information and motivate behavior in ways that tend to enhance the reproductive success of self and close kin. But because the cortex has considerable flexibility in how it manages the details, there is enormous variation among cultures. Our task, therefore, is to not be overwhelmed by the variety of cultural differences, but to explore how consistencies in the most important cultural practices can be interpreted in an evolutionary context. Seen this way, culture—in all its diversity—becomes an expression of the human behavioral phenotype.

This goal may seem too broad. After all, many of the things that people do seem to have very little connection with reproductive success, and some activities—rock climbing, parasailing, and entering celibate religious orders—can be "counter-reproductive." Do these exceptions preclude an evolutionary analysis of recurrent patterns in human behavior?

They don't, and there are two related reasons why. First, during most of our evolutionary history our ancestors were living in physical and social environments very different from those we experience today (Chapter 12). Second, motivations for alternative behaviors are frequently in conflict: whether we act for ourselves or others, or how we allocate our time and our economic resources. For alternatives that require a conscious decision, there are likely to be psychological benefits (reinforcements) associated with either choice. In our hunting and gathering ancestors, curiosity and the urge to experience novel situations likely had a positive ef-

fect on exploiting new environments and competing for resources, even though at present these motivations may induce some individuals to jump off high bridges while tied to the end of a huge rubber band.

Similarly, appetites and tastes for sweets, fats, and salt doubtless drew our ancestors to these necessary items, especially in times of scarcity. But in the environments in which we now forage—supermarkets and fast-food restaurants—these tastes and appetites subtract from reproductive success by leading us to overeat and thus suffer increased susceptibility to heart disease and stroke.

As a third example, contraception now enables people to invest more in fewer children and expend resources on other pleasurable activities. None of this, however, invalidates an evolutionary perspective. It simply illustrates that the various motivations with which we are endowed can influence reproductive success in ways that depend on the environment in which we find ourselves. Pursuing some goals can actually reduce reproductive success, but we do many of these things because in our present environment they are fun.

If cultural evolution has nevertheless been heavily influenced by people living in social groups pursuing their genetic interests, we should be able to find evidence in the patterns of behavior that have been studied by ethnographers, psychologists, sociologists, and historians as well as in laws and religious codes. We should be able to see how major cultural practices influence reproductive success, and how the interests of individuals are adjusted for group living. To interpret this information, we should be able to draw on principles that have been informed by studying kin selection, reciprocal altruism, parent-offspring conflict, sexual selection, and differential parental investment in other species. And we should find evidence that our minds have been molded by many of the same interests that are manifest in the behaviors of other social animals. We now consider that evidence, starting with a subject that has attracted wide interest.

INCEST TABOOS AND CULTURAL NORMS

The subject of incest illustrates the complex relationship between biological and cultural evolution, but we need to ground the discussion on three clear observations. First, virtually all societies have strictures against sexual relations between very close relatives, although marriage of first cousins is frequently encouraged. Exceptions have been documented, but they usually involve important individuals such as royal siblings and the retention of wealth and power.

Second, knowledge of genetics makes clear why inbreeding leads to weak, sickly, or even lethal phenotypes. Consider a brother and sister who both carry a recessive allele for a serious disorder that they have inherited from a parent. Neither has symptoms of the disease because the other normal allele at that locus is dominant (Chapter 4). If these siblings mate incestuously, ¼ of their children will have *two* copies of the mutant gene (i.e., be homozygous) and will therefore suffer from the disease. On average, everyone carries a couple of lethal alleles (as well as other deleterious but nonlethal recessive genes) in the heterozygous state, so the probability of undesirable phenotypic consequences from close inbreeding can be greater than ¼.

Third, when we look at other species we find that evolution has generated behavioral mechanisms to suppress inbreeding. One sex, typically males, disperses from its natal group, but in some primates (e.g., chimpanzees) it is primarily the females that emigrate. Female primates will not mate with their sons, and they resist copulation with their brothers. In other words, strong behavioral biases against mating with individuals recognized as kin minimize incest.

Do humans have an instinctive aversion to incest? Freud thought not, partly because of the universal presence of incest taboos. Out of this reasoning he manufactured the Oedipus complex, the unsatisfied sexual interest of young sons for their mothers in competition with their fathers. It is difficult to understand how this bizarre idea of prepubertal sexual desire became so popular, as the only evidence came from the subjective analysis of dreams. But in Freud's view, incest taboos were necessary to keep family structures from unraveling.

In 1891, however, a Finnish anthropologist Edward Westermarck had proposed that people who grow up together from infancy subsequently show no sexual interest in each other. Although he was suggesting a mechanism against brother/sister rather than son/mother incest, the idea of such a learning bias seemed to challenge Freud's theory, and Westermarck's hypothesis was rejected for more than seventy years. Then two independent studies confirmed the idea that for humans the first two and a half years of life are a critical period of development in which subsequent sexual interest in close playmates is inhibited. As it is often put, boys and girls who "share the same potty" as little children do not become romantically interested in each other after adolescence. The specific details of the process are not understood, but it may well be a general epigenetic mechanism at work in all primates.

The anthropologist Arthur Wolf studied over 14,000 Taiwanese marriages in which the woman had been adopted at a very young age by the family of the intended groom and raised with the intimacy of a sister. (This was formerly a custom in parts of China to assure the availability of a wife when the sex ratio had been skewed.) The results of this study support Wester-

marck's concept, for in cases where the girls were younger than two and a half years when adopted, they later tended to resist the prearranged marriages, as wives they engaged in adultery at three times the rate of women in "normal" marriages, they had fewer than half as many children, and their divorce rate was three times the average.

The second study, also a "natural experiment," was done by the anthropologist Joseph Shepher and colleagues on 2,767 marriages of Israelis who had grown up in kibutzim, where children are reared communally in what amounts to large extended families. Few of these marriages involved individuals who had shared early childhood experiences in the same location, again suggesting that early childhood associations in a critical period influence later preferences of a sexual partner.

If there is an evolved developmental "instinct" to suppress incest, why do incest taboos exist in virtually every culture? The answer to this question clarifies an important aspect of culture. For most people the idea of an incestuous relationship is repelling. This is an emotional enabling mechanism that develops in the brain to motivate alternative behavior. When incest occurs, it is therefore seen by others as deviant. Behavior that is merely unpredictable or mysterious, let alone deviant, can be unsettling, and it prompts societies to codify behavior. Individuals who then do not conform to the group norms are punished. Culture thus often reinforces the emergent features that natural selection has built into the human brain. Furthermore, natural aversions can be culturally manipulated toward personal goals. In many societies, older polygynous men guard their junior wives from the sexual interest of their sons by telling these sons that the younger wives are also the son's "mother."

A possible complementary explanation for incest taboos is that humans have frequently noted that children of marriages between close relatives are often abnormal, physically or behaviorally. Here the evidence is more ambiguous. In a study of sixty different cultures, the anthropologist William Durham found that a third were clearly aware of this possible outcome. But in their mythologies, incestuous relationships, although viewed as out of the ordinary, more often produce heroic figures.

REPRODUCTIVE COMPETITION AND THE MATING SYSTEM

In the previous chapter we discussed one universal feature of the human mating system: males and females form conjugal families to mate and rear offspring. This marriage relationship is a cooperative alliance that is socially sanctioned by religion, law, and verbal agreements, and in principle almost always entails exclusive sexual relationships. Practice may depart from principle, however, frequently with an asymmetry that mirrors polygyny. Marital alliances also entail investment in offspring by both the mother and the father (or, in circumstances where paternity is often uncertain, the mother's brother instead of the husband), and it defines and legitimizes the social status of offspring. Like concealed ovulation in females, these cultural arrangements help to minimize sexual competition between males, protect females from male mating aggression, and maximize certainty of paternity, all desirable for the stability of social groups that contain both men and women. But within these boundaries, there are important associated differences among cultures in how many mates and offspring the two sexes have, how marriages are arranged, regulated, and terminated, and the relationships between married men and women.

MONOGAMY AND POLYGYNY

As was discussed in Chapter 13, on the basis of sexual size dimorphism (about 1.08), we might expect humans to have somewhat polygynous tendencies accompanied by male-male competition. Additionally, as humans are unusual mammals in the long period of dependence of the young and the large parental investment ordinarily made by males, it should not be surprising that some of this competition is expressed over material resources.

The diagram in Figure 14.2 shows that polygyny, in which a man is married to more than one woman at the same time, is allowed in 83% of nearly 900 human societies. This graph does not tell the whole story however, because the majority of human marriages are monogamous, even where polygyny is legally possible.

Who are the polygynous men, and what are the features and consequences of human polygyny? The anthropologist William Irons was the first to demonstrate a direct quantitative relationship between "cul-

tural" and "biological" success. Rich Yomut Turkmen, a group of nomadic herders in Iran, have more wives and surviving children than poor men. Since the publication of that study in 1979, essentially the same correlation has been found in every preindustrial society that has been similarly studied. The men who are better hunters or who have more land, herds, wealth, male relatives, or higher political rank obtain more mates and produce more children than men who have fewer of these assets.

The anthropologist Laura Betzig has documented the incidence of the more extreme forms of polygyny in a variety of human cultures in various parts of the world, using historical records as well as contemporary data. She, too, finds that the vast majority of polygynous men are both wealthy in land, crops, herds, or money, and powerful in terms of social, political, or religious influence. Wealth and power provide the means either to attract women as wives or concubines, or (depending on the culture) to hold women in harems or servitude. In every one of the first recorded civilizations—Aztec Mexico, Inca Peru, ancient Mesopotamia, Egypt, India, and China—powerful men mated with many women and passed their power on to a son by one legitimate wife.

In these extremely polygynous situations, enhanced reproductive success of powerful and wealthy men is achieved at the expense of other men. Historically, harems were guarded so that the despot had exclusive access to the women, an arrangement that was made maximally effective by castrating the (male) guards. In Dahomey in west Africa prior to French conquest at the end of the nineteenth century, the king not only had hundreds of wives, concubines, and slaves, he kept a small army of women soldiers who were expected to remain virgins while in his military service. In such a court, there were inevitably women who never visited the royal bed but who were denied any chance of marriage to another man. This is a doubly effective strategy in reproductive competition: one man fathers many children and prevents others from fathering any.

Excluding large numbers of men from reproduction is a source of potential unrest, and it is little wonder that the most polygynous societies have also been the most hierarchical, authoritarian, and despotic. Powerful men at the top—often related—have at their disposal great wealth and the repressive political, religious, legal, and military systems needed to control those individuals with much less. Order is kept by random and arbitrary acts of terror and by severe punishments for adultery. Betzig's list of societies in which despotism served extreme polygyny (defined as ten to a hundred simultaneous conjugal relationships) includes the ancient Babylonians, Egyptians, and Hebrews; Imperial Rome; the Bemba, Fon, Mbundu, Azande, and Ashante of Africa; the Khmer of southeast Asia; the

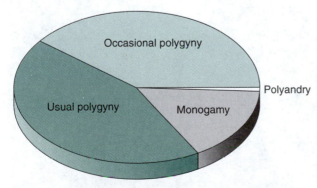

FIGURE 14.2 Marriage arrangements allowed in 849 different cultures. Although polygyny is possible in most cultures, monogamous unions are more common.

Natchez and Aztecs of North America, and the Inca of South America. These examples of polygyny are dramatic and show how extreme it can become when conditions allow, but they do not define the incidence of human polygyny worldwide.

GLIMPSES OF PRE-AGRICULTURAL SOCIETIES

Affluence and power go hand in hand in societies in which people can strive to accumulate material resources such as land and crops, livestock, access to game or water, precious metals or other forms of wealth, and in more recent times, manufactured goods and their means of production. But in many preagricultural societies that depend upon foraging (hunting, gathering) or on planting temporary gardens for food (horticulture), there is little or no division of labor between men and few or no possibilities for accumulating wealth. Because there are no distinct classes, and status is not defined in terms of material wealth, societies have been characterized as "egalitarian," and some of them—particularly hunter-gatherers—have been portrayed as free of violent competition. However, in the preagricultural societies that have survived to recent times, men can and do compete for wives and for material resources that directly affect reproduction, and they cooperate with other men, usually relatives, in order to achieve these ends.

The Ju/'hoansi (!Kung San) of the Kalahari region of Africa are a well-studied foraging society. Some men are polygynous, either because they have two wives at the same time or because they have successive monoga-

mous marriages. These polygynous men are respected and considered leaders because they are good hunters, effective arbitrators and spokesmen for the group, or both. Long-term studies have shown greater variation in reproductive success among men than among women, a hallmark of polygyny. Furthermore, more men than women die before reproducing. Despite the early reputation of the Ju/'hoansi as "the gentle people," the anthropologist Richard Lee was told of twenty-two murders over thirty-five years in one population of 1,500, a rate that is four times that found in the United States in recent years. The large majority of the killers and victims were men, and the motives were mostly sexual jealousy, adultery, and revenge for previous killings. As we will discuss later, this is largely an outcome of sexual competition among men.

The Yanomamö Indians of the Venezuelan Amazon region are the best-studied example of reproductive striving and polygyny in a preagricultural society, thanks to the remarkable work of the anthropologist Napoleon Chagnon. In the region he studied intensively, Chagnon found that fights within villages and raids (warfare; Fig. 14.3) between villages account for approximately 30% of deaths among adult males, a rate that is similar to those reported by anthropologists among other peoples living in comparable environments around the world. Most of the fights within villages and the cycles of revenge killings in intervillage raids start over matters directly related to reproduction: sexual infidelity or suspicion of it, seduction, sexual jealousy, forcible appropriation of women, failure to deliver a girl promised for marriage, and (rarely) rape.

FIGURE 14.3 Yanomamö war party preparing to leave their village.

The social status and marriage possibilities of male and female Yanomamö are defined by their membership in patrilineal descent groups, and the reproductive success of a man depends to a large extent on the numbers and support of his male kinsmen. The leader of the largest patrilineal descent group in a village is invariably the village headman, and these headmen are usually polygynous, having up to six wives simultaneously and up to a dozen over their lifetime. One man had forty-three children by eleven wives. Furthermore, violence among the Yanomamö pays reproductively. Chagnon found that "unokais," men who have killed or participated in killing enemies, had significantly more wives and offspring over all ages than men who had not.

Many investigators of primitive warfare initially rejected Chagnon's hypothesis that violence and warfare among Yanomamö was about competition for women. They found it difficult to imagine that warfare in primitive societies could occur unless there were a shortage of material resources. If there is enough game to hunt, plant food to gather, or land to cultivate or graze, everyone should have enough to eat and reproduce, so what could possibly be gained by fighting? One group of investigators argued that Yanomamö population densities were limited by the amount of animal protein that the rainforest can provide, and that warfare functioned to defend each village's hunting territory. They argued that warfare and even their practice of infanticide of newborn girls (see below) serve to regulate population size.

Chagnon and his colleagues, however, found no medical evidence for protein deficiencies among the Yanomamö. In fact, they found that the daily consumption of animal protein by these and other Amazonian Indians is greater than that in many developed countries in Europe and Asia. Of course, it could be argued that the sufficiency of protein in the Yanomamö diet shows the effectiveness of warfare and infanticide in preventing overhunting. But the argument that men kill each other and women kill some of their newborn daughters in order to preserve monkeys and peccaries for others to hunt is, to say the least, implausible. Such altruism does not accord with the evidence that humans are much more likely to exploit rather than conserve natural resources. When Chagnon suggested the hypothesis to Yanomamö men that they were fighting about meat, they replied, "Even though we do like meat, we like women a whole lot more!"

Chagnon's studies of violence, warfare, and polygyny among the Yanomamö illustrate an important point about human reproductive striving, especially in primitive societies. Whether or not the resources required for sustenance are abundant, a human male's reproduction is limited by the number of women with whom he can mate, and sexual selection has clearly operated in human evolution. Men who can use their physical assets and their social intelligence to gain a greater share of material resources and social power for themselves and their relatives will outreproduce the males that are less successful in this competition. And as the words of the Yanomamö men testify, their behavior is supported by deep-seated emotional rewards.

The soils of tropical rainforests are quite poor, and plots of land can become depleted after a few cycles of crops. Nevertheless, as Chagnon has documented, although the Yanomamö move about, their garden plots frequently remain in use for decades and are abandoned either because of overgrowth by other vegetation or because the village is under constant threat of attack from a more powerful neighboring village. It may be that, as in lions (Chapter 13), lethal competition keeps their population density within the bounds that the environment can support, but there is no evidence that this is the function of the killing. If Yanomamö population density is limited by warfare, *it is an effect, not a function*, of cooperative and competitive behaviors that enhance the reproductive success of individuals.

In sum, humans compete for and gain disproportionate shares of "the means and ends of production" in post-agricultural societies, but this is an extension of competition for the means and ends of reproduction already present in preagricultural societies. The relationship between these kinds of competition is frequently masked by the rich array of proximate motivations and psychological rewards with which the human brain is equipped, but these features of the mind with their overlay of emotions would not have evolved if they did not support the furtherance of reproductive success.

WIVES IN POLYGYNOUS UNIONS

Discussions of how sexual selection influences male and female reproductive strategies tend to put a disproportionate emphasis on male behavior. Women may be more constrained in the most patriarchal and polygynous societies, but they are seldom passive pawns in the machinations of men. It is, after all, they who prefer wealthier men, even if this requires being a second or third wife. The wives of a wealthy polygynous man—especially the first or favorite wife—stand to gain a Darwinian advantage over women monogamously married to poorer men. If her son is at or near the front of the line for paternal beneficence and inheritance, he too will become polygynous and be a more effective agent for the propagation of his mother's genes than the children she might have had by marrying a poorer man.

The anthropologist Monique Borgerhoff Mulder found that women in a patrilineal, agricultural, and pastoralist society, the Kipsigis of Kenya, preferred to marry the man who offered the largest amount of land per wife, even if he already had one or more wives. But they also preferred marriages in which there were few co-wives who had already produced children. Women who most successfully fulfilled these criteria in their marriages produced more children than those who were less successful.

Polygynous marriages are frequently marked by competition between wives for the husband's attentions and, ultimately, for how he apportions his property and support among his children. Such competition requires the same intelligence and social skills as the competitive scheming of men. Indeed, greater social abilities may be required of women, as they usually must continue to use words when men turn to weapons.

As a measure of the intensity of co-wife competition, however, women in these polygynous marriages have often killed or conspired to kill their co-wives, their co-wive's children, and even their husbands for perceived unfairness in how he has distributed his affection and wealth. The seriousness of such conflict is also evident in the architecture of many of the dwellings of polygynous rulers around the world: separate as well as guarded quarters and households for each wife or concubine (Fig. 14.4). Among the Yanomamö, women are much less often injured or killed in conflicts than men, but they are verbally and socially as engaged as men. In *Ax Fight*, a film of a dispute among the Yanomamö made by Napoleon Chagnon and Timothy Asch, the women are active instigators and participants in a violent confrontation that nearly results in a homicide.

POLYANDRY

Polyandry, in which more than one male mates with a female, is rare in mammals, and it is found in species with high paternal investment in offspring. It is also rare in humans, present in a few percent of the nearly 900 societies on which the survey in Figure 14.2 is based. The Thongpa, land-owning farmer-pastoralists of the arid, cold Himalayan foothills of Tibet, illustrate the circumstances in which people sometimes opt for polyandry. The land in their environment is poor and rocky, the growing season is short, and the available land is occupied. Situations arise in which dividing an inheritance of family land between brothers leaves so little to each that neither will be able to support a family. If a younger brother is unable to find work elsewhere, two brothers sometimes agree to share a wife in a polyandrous marriage. These arrangements (which are made by men) are not considered ideal and are avoided where possible. There is frequently conflict between the brothers over sexual access to the wife, and the older brother usually dominates the relationship.

These features of human polyandry are readily understood in an evolutionary context. Both its rarity and the common occurrence of friction between the brothers are consistent with the norms of monogamy and polygyny, in which males seek exclusive sexual access to their mates. The ecological and economic circum-

FIGURE 14.4 Aerial view of the harem section of the Topkapi Palace in Istanbul. The sultan's mother, wives, children, and concubines along with their guards and servants lived within the secluded, separate, and inaccessible rooms and courtyards of the harem. Islamic law allows a man four wives, and although it is said that most of the sultans preferred to stay loyal to a few women, the number in the harem was a symbol of royal wealth and power. Moreover, those not chosen by the sultan as mates were frequently later married to his officials and political allies, which increased his power by solidifying connections and incurring debts.

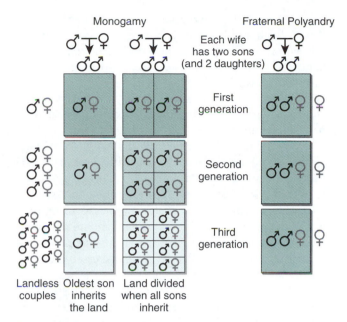

Figure 14.5 An economic/ecological reason for fraternal polyandry. Suppose land is inherited through the male line and wives have more than one son. (For purposes of illustration, each family in this diagram has two sons and two daughters. The female symbols represent wives or potential wives, not sisters from the same family.) In monogamous marriages in which the eldest son inherits the land, other sons and their wives must seek a livelihood elsewhere. If the land is divided equally among all sons, however, in only a few generations the parcels become too small to support a family. Fraternal polyandry occurs in some situations where there is no new land to be settled and no other alternatives for work available to younger sons. Two brothers sharing one wife prevents the land from being divided and maintains an approximately constant population density on the land. There are problems with this social system, however, including fraternal conflict over the wife. As the diagram suggests, there are also likely to be unmarried women who do not have an opportunity to marry.

stances in which human polyandry occurs are so pressing that the reproductive success of one, if not both brothers, appears to be jeopardized. It is therefore adopted as an expedient, linking the number of people the land can support to the inheritance of that land, even though the disadvantages are clear to all (Fig. 14.5). It is also likely no accident that the only known instances of human polyandry involve brothers and not unrelated men, because each will have a degree of relatedness of 0.25 with his brother's children.

ARRANGING AND DISSOLVING MARRIAGES

Many of the ways in which marriages are arranged involve the use of material resources to control and uti-

lize the reproductive capacities of women. In 66% of the world's societies for which ethnographic information is available, a man acquires a wife in direct exchange for bride service (he works for her kin), for bride-price (he or his kin give money, land, goods, or livestock to her family), or one of his female kin is offered in exchange (Fig. 14.6). Among many of the 24% of societies in which no such direct exchanges occur, a debt is nonetheless incurred: if a man acquires a wife from outside his social group, his kin or social group are expected to reciprocate in kind later.

There is good evidence from many societies that bride-price is affected by reproductive considerations: youth, beauty, health, plumpness, and strength are cited as important in the amount of bride-price negotiated by the parental families at the time of marriage. In her study of reproductive success, wealth, and marriage practices among the Kipsigis of Kenya (discussed earlier), Monique Borgerhoff Mulder found that early maturing girls (younger at menarche) produce more children than later maturing ones, and they command a higher bride-price. Girls classified as "plump" were more expensive than those classified as "skinny," a preference that makes sense in the light of growing evidence for an association between a woman's nutritional status and reproductive performance, especially in environments in which there is periodic scarcity. Additionally, the Kipsigis say that plump women remain healthier and better able to work during the dry season.

Bride-price increased with distance between bride's and groom's residence, and this was found to be related to the loss of the bride's help to her mother. Finally,

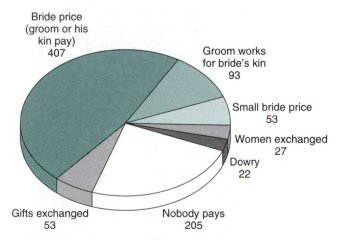

FIGURE 14.6 Economic exchanges that take place at marriage, based on a survey of 860 societies. In 67% of the cases the husband or his family must pay bride-price, offer a woman in exchange for the wife, or the husband must work for his wife's family in exchange for his wife. In 30% of the cases there is no cost or there is a mutual exchange of gifts. Dowry (payment by the bride's family) occurs in only 3% of the cases.

pregnant women and women with children were less expensive, the former likely because the groom cannot be sure of paternity and the latter because these additional children are considered a strain on family resources and a source of increased conflict when inheritance must be divided between genetic and adopted sons.

The anthropologist John Hartung found that bride-price is paid in 77% of polygynous societies but in 38% of monogamous ones. He argued that this association between bride-price and polygyny reflects a parental reproductive strategy. Families compete for resources (property, livestock, wealth) and then use these resources to obtain a first wife for the son. Purchase of the wife compensates her family for the risk that if she is followed by additional co-wives, inheritance to her children will be diverted in part to others. The ability of the groom's family to pay bride-price is also an assuring display that there is sufficient wealth to support more than one wife and their children. In monogamous unions, bride-price and the display of wealth are less important because there will be no unrelated children with whom the bride's children must share inherited resources.

In many of the societies in which there is exchange or purchase of women as wives, these transactions also enable families to forge advantageous economic, political, and military alliances within and between their communities. In such societies women's additional work outside the household (in agriculture, animal husbandry, producing goods) may be crucially important to family income and sustenance, so it might be argued that these exchanges are largely economic compensation for future labor. We could as well describe "marriage" in a monogamous pair of birds as an "economic" union: the male establishes territory and builds nests, and both the male and the female gather food for the young. But the question would remain "economic for what"? The answer for birds is clearly "to produce offspring," and ultimately it is the same for humans. Furthermore, in marriage systems employing economic exchange, it is men, not women who pay for a spouse. There are no human societies in which women think and speak of men as property—for example, a mother "giving" her son or daughter in marriage—or in which men are exchanged between families for labor, resources, or political alliances.

Dowry, found in about 4% of human societies, is another kind of exchange that is associated with marriage. It occurs in Africa, Europe, India, and other parts of the world, but is less common than formerly. A wife or her family gives property or wealth to their daughter and her new husband, ranging from a trousseau through furniture, jewelry, money, livestock, servants, land, and, for prospective queens, entire provinces with their inhabitants. The proximate purposes of dowry include helping to establish the new household economi-

cally, guaranteeing favorable treatment of the wife by the husband and his family, and providing support for the wife if the husband should die. Although the husband can use the property or wealth during the marriage, he is usually obligated to return it if the couple later divorces. In some societies the dowry served as a reciprocal exchange to bride-price, cementing relations between families.

Why would the bride's family wish to transfer wealth to the new husband, along with their daughter? Part of the answer is that dowry is a form of insurance. It is in the interests of the bride's family to assist the husband in establishing himself as well as assuring their daughter has resources if the husband should die or the marriage should dissolve. There may also be more subtle factors at work. In Chapter 6 we discussed examples of how polygynous animals manipulate the sex ratio if reproductive prospects are more favorable for one sex than the other. This phenomenon, known as the Trivers-Willard effect, appears to be present in human societies. The anthropologist Mildred Dickemann has examined the practice of dowry in socioeconomically stratified societies in India and Europe prior to the end of the nineteenth century. These societies had wealthy men at the top and subsistence poverty among the dispossessed at the bottom. There was little chance for males to move up through the barriers of inherited wealth and status. Under these circumstances, a family of moderate means might have more grandchildren by buying their daughter a place with a higher-status male whose children would benefit from his socioeconomic position.

A prediction of this model of reproductive competition has recently been dramatically confirmed by molecular genetic techniques: the DNA sequences of Y chromosomes within Indian castes are significantly more uniform than are the DNA sequences of X chromosomes. In other words, Y chromosomes in males have not moved across caste lines as much as X chromosomes in females have.

The anthropologists Steven Gaulin and James Boster have also presented evidence that the practice of dowry generally functions to improve the reproductive prospects of daughters and their descendants by allying them in marriage with wealthier males. Dowry is rare: it is found in only 38 (3.6%) of 1,066 societies studied. However, 34 of these 38 societies are found in the 340 (32% of the total) that are socioeconomically stratified and 29 of the 38 are monogamous. Their findings suggest then that dowry is a tactic used in competition among women (with the help of their families) to monogamously marry a higher-status man so that their wealth will be concentrated in her children and not be distributed to the children of other wives, as in polygyny. As John Hartung notes, the reproductive payoff of dowry is realized when a daughter's son is born into a wealthy, long-lived lineage, in which males typically augment their reproductive success by serial monogamy and extramarital affairs. Thus, whereas brideprice in economically stratified polygynous societies enable families to place sons in a reproductively advantageous position, in economically stratified monogamous societies, dowry enables families to do the same thing for grandsons.

Laura Betzig has shown that another way to explore whether reproductive competition is a primary cause of the cultural evolution of marriage practices is to look at the reasons for divorce. People may marry for one set of reasons, but what conditions must be met in order for a marriage to last? In a cross-cultural study, she found that adultery was the most common cause of divorce, significantly more common than any other grounds except sterility. Actual or suspected infidelity by the wife was two to twenty-five times more likely to lead to divorce than infidelity by the husband. (This double standard is discussed further below.) Cruelty or maltreatment—instigated by the husband 90% of the time—ranked third, and it was very often prompted by the wife's adultery or a perceived risk of it (see the further discussion of male sexual jealousy below). In other words, divorce is initiated mainly by men, and a marriage in which a man is threatened with investing in offspring that are not his own can have a powerful impact on his perspective.

Furthermore, in most of the societies in which there is bride-price or exchange of women as wives, a man whose marriage does not produce children or whose wife was unfaithful is entitled to a refund of his payment or to a replacement of his wife, often by a younger sister. These practices and motivations associated with divorce reinforce the Darwinian view that human marriage is a reproductive union designed to convert material resources into offspring, and that cultural practice has frequently emphasized the interests of men relative to those of women.

CHOOSING MATES

Human ethnic groups differ in average height, color of skin, color and texture of hair, and in the shape and size of eyes, nose, and lips. And they differ even more widely in culturally acquired features such as hairstyle, dress, ornamentation, and deliberate physical alterations of the face or body. Confronted with such variety in human physical appearances (and sometimes repelled by the unfamiliar), students of culture have usually concluded that there are no universal standards of beauty and that the criteria used in choosing a mate are entirely learned and vary arbitrarily among cultures.

Superficial physical traits such as shape of nose or texture of hair are unrelated to reproductive potential, but variations may be considered attractive where they occur frequently, at least partly because they are familiar. Dress and ornamentation can identify someone as a member of a particular group, culture, or class, and although this information may say little about a person's physical capacity to have children, it can supply important clues about their economic or social position, which can affect reproductive potential. Before accepting that the criteria for selecting a mate are entirely arbitrary, however, we should ask from an evolutionary perspective whether there are in fact features that indicate reproductive promise and that can be shown to influence the choice of mates in most or all cultures. If natural selection has shaped our psychology to foster reproductive success, we should be able to identify such features.

The psychologist David Buss has surveyed the criteria used in choosing mates by 10,000 men and women in thirty-seven cultures on six continents and five islands around the world (Fig. 14.7). Of the three most important features men seek in a mate, youth and physical attractiveness are strongly associated with reproductive potential, and the third, sexual fidelity, is also important, as in all species where males make a substantial parental investment. The younger a woman is, the longer she can bear children, and in all thirty-seven cultures that Buss studied, men wished to marry women younger than themselves. Sometimes the age difference is a generation or more, but the reverse—an older woman marrying a younger man—is usually considered anomalous.

Several aspects of physical attractiveness indicate high reproductive potential because they are correlated

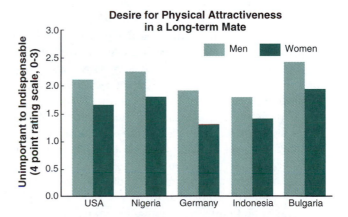

FIGURE 14.7 Both men and women value physical attractiveness in a potential spouse, but statistically these qualities are more important to men than to women. These five examples are from a study of 37 different cultures in which 19 different variables were examined. The sample sizes in this figure varied from 143 to 1,491, and in each case the difference between men and women was statistically significant (p < 0.001—i.e., the probability was less than 1 in 1000 that the observed differences were due to chance).

with normal development, good metabolism, and resistance to disease. These include smooth and clear skin, full and lustrous hair, good posture, an even and flowing gait (a corollary of well-formed joints), as well as body, facial features, and teeth that are well formed, normally proportioned, and symmetrical. Cross-cultural research on the attractiveness of women's body shapes have previously focused on fatness versus thinness, and in environments in which food is often or periodically scarce, men may show a preference for plump women. Such conditions were likely frequent over much of our hunting-gathering past, where plumper women would have had more reserves to sustain offspring during lean times. As described above, among the Kipsigis plumpness increased the bride-price.

Recent research shows that the distribution of fat is an important indicator of a woman's health and ability to invest in offspring. After puberty, a woman's estrogens stimulate fat deposition in the upper thighs, hips, and breasts, which produces the "gynoid" (hourglass) body shape of mature women. In contrast, testosterone in men stimulates muscle development in the upper body and fat deposition (if any) in the abdomen, which leads to a more masculine or "android" shape. The ratio of waist to hip circumference is an index of these differences in body shape that is easily measured. In most human populations the waist:hip ratio of women is lower and overlaps little with that of men. Fat reserves in women's hips and thighs are preferentially used during late pregnancy and lactation, and their presence very likely reflects the energy requirements of reproduction during the early hunting-gathering period of subsistence, where women needed energy to sustain fetal development and lactation during food shortages. Interestingly and importantly, women with lower waist:hip ratios exhibit earlier pubertal levels of sex hormones and higher fertility. Women with higher waist:hip ratios have significantly increased risks for diabetes, hypertension, cancer of reproductive tissues, gallbladder disease, heart disease, and stroke.

The psychologist Devendra Singh studied the criteria used by U.S. Caucasian, African-American, and Indonesian men and women of different ages in evaluating body shapes by asking them to rank the physical attractiveness and infer the age, health, and personality attributes of twelve women who were viewed as line drawings. The figures represented combinations of four waist:hip ratios (0.7, 0.8, 0.9, 1.0) and three categories of body weight (average, underweight, overweight). Men and women in all three study groups and of all ages ranked average-weight figures with low waist:hip ratios as more attractive, healthy, young-looking, desirable, and capable of having children than underweight and overweight figures or figures with higher waist:hip ratios. Underweight figures were assigned the lowest reproductive potential. Within each weight category, the

evaluations declined with increasing waist:hip ratio. From these findings, Singh proposed that human males share an unconscious "template" or prototype of the attractive female body that is used as a "first filter" in evaluating prospective mates. If this is true it can be overidden by environmental conditions because recent studies by the anthropologists Adam Wetsman and Frank Marlowe (and others) on non-industrialized societies have shown that a woman's weight is important and the waist:hip ratio is not.

The idea that the sensory system filters information in specific ways was encountered in Chapter 9. Sensory signals are often used socially to reinforce a message about the body. For example, a variety of cultural practices in dance or dress emphasize a woman's hips and buttocks or bust, frequently with overt sexual intent. The epaulets and shoulder boards of military officers signify rank and status not only by their specific insignia, but also by artificially emphasizing the width of the shoulders.

In his cross-cultural survey of long-term mate preferences, Buss found that in spite of cultural variation, sexual fidelity tops the male's list of desirable features in a mate, even in Western countries where birth control methods are available. We expect to find concern about certainty of paternity in monogamous species where male parental investment in offspring is high (Chapter 6), and human males are no exception. A cuckold is a husband whose wife has had an affair with another man. The word refers to the behavior of the European cuckoo, a parasite that lays its eggs in the nest of another species. Like the victim of the cuckoo, the cuckold who is unaware of his mate's dalliance risks investing in genetically unrelated offspring. Not surprisingly, and as we saw in the discussion of divorce, human males are prone to sexual jealousy, with a strong interest in knowing that their mate has not been impregnated by another man. One common manifestation is the male's wish that his mate be a virgin at the time of marriage. Indeed, in *most* human societies there is usually much more male concern about female virginity as a criterion for marriageability than the reverse, and in *all* human societies the same is true concerning sexual fidelity after marriage. Sexual jealousy colors male-female relations in several ways, as we saw in the discussion of polygyny and will meet again in following sections.

What should a woman look for in a husband to assure her reproductive success? For humans, like all other mammals, reproduction is physiologically much more demanding for females than for males. Given her large parental investment, the most important requirement for a female's reproductive success is material resources, which historically has usually meant reliable support by a male. In most post-agricultural societies, therefore, a woman is frequently interested in her future mate's willingness and ability to provide support

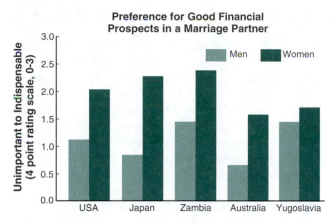

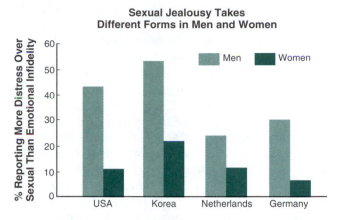

FIGURE 14.8 In contrast to physical attractiveness (Fig. 14.7), women rate the economic prospects of a potential mate as more important than do men. All of the differences between men and women are statistically significant (p < 0.02 for Yugoslavia; p < 0.001 for the other cultures).

FIGURE 14.10 Men and women have statistically different feelings about their partner's infidelity. Men (as illustrated here) are more disturbed about the physical act of sex, whereas women report greater distress about emotional infidelity. These differences occur cross-culturally.

for her and her offspring. In his cross-cultural survey of criteria for choosing mates, Buss found that women from all continents, political systems, mating systems, and ethnic groups place the most value (twice as much as do men on average) on good financial prospects, as indicated either by present wealth, or the intelligence, ambition, and social status needed to acquire it (Fig. 14.8). Women also attached great importance to signs of willingness to invest in them: kindness, generosity, and sincerity of interest. Although women thought physical attractiveness desirable, they found it much less important than do men, and less important than financial prospects. Also, women prefer to marry men older than themselves, doubtless because the status and wealth men gain by competition and inheritance increases with age (Fig. 14.9).

Psychological studies indicate that female sexual jealousy tends to take a somewhat different form than it does in men. Distress over infidelity is less focussed on the act of sex itself and more on the betrayal of commitment. One source of concern is that the male's resources will be diverted from her to his children by another woman (Fig. 14.10).

Nothing in this description of sexual differences in mate choice should be read as diminishing the importance of individual personalities in establishing stable, long-term bonds between men and women. This kind of compatibility is a particularly important criterion in societies where women choose freely, and it provides the basis for deep affection.

CONTROLLING MATES

The psychological motivations and impulses that reflect a male's concern with paternity have an understandable evolutionary origin, but in humans they have become entwined with male-male cooperation in acquiring and controlling resources. The invention of agriculture and the consequent growth and specialization of social structures accelerated this latter process, and in many societies women have to depend entirely upon men for their basic material needs. This condition is seldom, if ever, met in any other animal.

We have seen from the ways that humans arrange and dissolve marriages that men use resources to gain wives and invest in offspring, and the more resources they control, the more they can control the cultural processes associated with marriage. How humans guard and control mates in cultures around the world also illustrates this interaction between reproductive competition and the control of resources. Indeed,

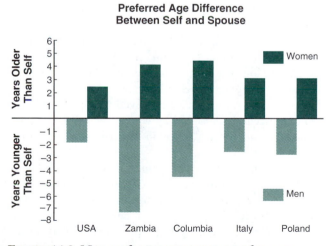

FIGURE 14.9 Men prefer to marry women a few years younger than themselves, whereas women prefer men a few years older.

the ubiquity and cultural variety of these practices are astonishing.

Probably the most conspicuous evidence is the double standard found in moral attitudes and legal codes concerning sexual activity. Men are everywhere more concerned about restricting women's sexuality than the reverse, and both men and women behave in the same asymmetrical way toward their daughters and their sons. Although it is true that men are less discriminating than women in their sexual activity—men, after all, are the consumers in the trade known as prostitution—the control of resources by men complicates an evolutionary analysis. Those who control resources have the power to make and enforce societal rules, and the rules therefore reflect their interests. Once such monopolies develop historically, it becomes in the reproductive interests of females to adopt the same standards.

In most societies, sex before marriage is considered more normal and permissible for boys than for girls. In all but the most recently codified laws concerning adultery, more restrictions have been put on a woman's than a man's sexual freedom, and the consequences of disobeying (punishment, ostracizing) are more severe for a woman than for a man. This double standard in sexual morality and its expression in written codes is an extension of another behavior that goes much further back: the threat and use of violence against females (Chapter 13).

The level and frequency of male violence against women varies from one society to another, but wherever records have been kept and careful observations made, male sexual jealousy and proprietary behavior are the most important factors. When such emotions verge on the pathological, anything that allows a woman contact with others—in our society, having a car, a job, friends, a telephone—can become a threat. We will consider spousal homicide explicitly in a later section, but we note here that it represents the more publicized tip of the iceberg of male sexual proprietariness.

In a class discussion of this topic one of our students expressed doubt that such male aggression against women is evolutionarily rooted in mate guarding. The women she had helped over several years at a local women's shelter told her that the last thing on their minds was having an affair with another man. All they wanted was some contact with the outside world. Another woman student in the class raised her hand to reply: "Well, I guess the violence works!" In thinking about this exchange it is important to keep two points in mind. First, as we will document in the section on spousal homicide, male sexual jealousy can be so powerful an emotion that in some individuals it is expressed in the absence of any objective cause and can destroy the bond that has kept two people together. Second, women may harbor equally strong emotions about sexual infi-

delity, but legal systems have traditionally not been active in protecting their interests.

"Female circumcision" is an especially brutal kind of violence used to control women's sexuality. Part or all of the clitoris is removed (clitoridectomy), or the clitoris and part or all of the labia minora (the inner lips of the vagina) are removed (excision). In some cultures, after clitoridectomy and excision the outer lips of the vagina (the labia majora) are stitched shut (infibulation), leaving only a small opening to pass urine. The stitches are removed after marriage to allow intercourse. These procedures have been carried out with crude cutting implements and without anesthesia on some 110 million currently living women (as girls ages four to twelve) in twenty-eight countries, mostly across the center of Africa. Besides being extremely painful, these operations leave scar tissue and obstructed passages that can lead to reproductive and urinary tract infections, excess bleeding, pain during intercourse, and complications during childbirth.

The term "female circumcision" is in quotation marks because it is a misnomer. "Circumcision" means "cutting around" and in males it refers to removal of the foreskin of the penis for reasons of hygiene, religion, or both. The structure in males that corresponds (developmentally) to the clitoris is the penis. Both structures contain nerve endings required for orgasm. Clitoridectomy and excision are thus genital mutilations that reduce or abolish feelings of pleasure during sex, and "male circumcision" has no such effects. Many incentives are constructed around female circumcision: in some cultures people believe that if the woman's genitalia are not removed they will continue to grow to a grotesque encumbrance, and in others it is thought that the vaginal glands should be removed because they produce unclean secretions. The procedure is also a mystical rite of passage to adulthood and is the basis of the woman's reputation for chastity, cleanliness, and purity—all required for a good marriage. In fact, the colloquial word in Arabic for clitoridectomy, *tahara*, means "to purify," and girls spared the surgery by parents opposed to it will automatically acquire a reputation for being promiscuous. The procedure is done by women, not men, which illustrates how a custom of long standing that mutilates evolved features of the female body can be adopted by women because in a particular social context it is in the interests of their inclusive fitness.

Many other ways of coercing and constraining women's freedom to mate and choose mates have been devised, especially in socioeconomically stratified and highly patriarchal societies in which women have little power. The anthropologist Mildred Dickemann notes that in such societies, of which there are examples in the Mediterranean basin, a family's honor is closely identified with the chastity, sexual fidelity, and modesty of its women. Male relatives defend and enforce these qualities, even killing their female kin (as well as the women's lovers) who have not conformed. Beginning at or before puberty, women are concealed by body-covering clothing and veils, and they live, eat, and socialize separately from men, both relatives and strangers. These practices and their meanings are status-graded: the higher a family is in the socioeconomic ladder, the greater is the seclusion, veiling, and reputation for honor of its women. Dickemann argues that this gradation in the value of female chastity with social status is further evidence for its mate-guarding function. Women are expected to marry at or above the level into which they were born, and the more wealth a male commands for investment in offspring, the more concerned he should be about his certainty of paternity.

We should expect mate guarding by women to be different from that by men. The main threat to a woman of her mate's infidelity is that the diversion of his sexual interests to another woman will be followed by the diversion of his resources to her and their children. Using physical aggression to prevent this outcome is ordinarily not an option, both because women are not as strong as men and because they are legally and socially more constrained than men in using violence. However, keeping a husband's sexual attentions from wandering is an alternative, and there are two distinctive features of human female sexuality and reproductive physiology that seem to have evolved for this purpose. First, unlike almost all other mammals, which will copulate only around the time of ovulation, women can be receptive throughout their 28-day reproductive cycle. Second, and again unlike almost all other mammals where the time of ovulation is advertised by sexual swellings or olfactory signals, a woman's ovulation is concealed (Chapter 13). As in bonobos, copulation therefore serves more than one purpose. As a pleasurable act, its frequent repetition is encouraged, which helps to keep her mate close, both physically and emotionally. From the male's perspective, copulations must be frequent enough to ensure production of his own offspring, and staying close reduces the possibility that his mate will take an interest in other men. Concealed ovulation is also likely related to the unusually long period of dependence of human offspring. Seen in this light, concealed ovulation appears to be an evolutionary tactic of women to foster bonding with a male and secure his paternal investment.

Women have additional means of guarding and controlling mates. In most cultures married women are more concerned than men to retain the appearance of youthfulness and sexual desirability, and this is manifested in a variety of ways, including the support of entire industries in many Western countries offering cosmetics, cosmetic surgery, clothing, and weight-watching programs. Such attention to appearances may

also attract other potential mates and signal class and status, but retaining a husband's interest is usually a strong motive. Women also marshal social opprobrium and legal sanctions against their husband's infidelity, although in most cultures, as we have discussed, men are more empowered to use these controls against their wife's infidelity.

PERSPECTIVES ON THE ORIGINS OF PATRIARCHY

There are many ways in which women's access to material resources and economic or political power are restricted in cultures around the world. From a basic understanding of sexual selection and parental investment, males should be expected to compete with each other for access to the reproductive potential of females, but by the same reasoning, males and females have somewhat different reproductive interests, which, while strongly overlapping, are not identical (Chapters 6 and 13). One feature that seems to set humans apart from other primates is the capacity of males, as a group, to control females, as a group, made possible by the unprecedented human capacity to manipulate and control resources. Patriarchy appears, then, to be a very old human institution that has developed from and built upon a mating system that is unusual for a mammal in its approach to monogamy and its large amount of male parental investment.

Patriarchy is a social structure that primarily serves the reproductive interests of men, although it has frequently been justified by appeals to the "natural order," as manifest in religious traditions and supported by custom and law. Even as the societies of Europe and North America became more democratic, the political and economic rights of women have lagged behind those of men. Women were not allowed to vote, are frequently not paid the same wages as men, have been excluded or restricted from political, military, religious, bureaucratic, and economic hierarchies and from social and professional organizations, and have been denied access to education. Furthermore, women have been socialized (along with men) to believe, through a variety of cultural myths, that these inequities are justified because they stem from gender differences created by God (a very powerful male image in several of the world's major religions) or by biological evolution (historically argued by male scientists).

Some feminists have argued that historically "scientific" views of female nature are but one among many kinds of socially constructed beliefs that, along with cultural practices, were devised by men to gain dominion over women. They locate the beginnings of these efforts at domination in the development of agriculture and find their fullest expression in capitalism. They either deny any influence of evolutionary heritage in the appearance of patriarchy or fail to see its relevance. In the absence of any compelling explanation for why men should strive for such dominion, however, these arguments are inadequate. *In fact, views of human nature that exclude all evolutionary perspective are the ideas most likely to be arbitrary social constructs.*

Interestingly, however, many leading feminists correctly perceive that men's efforts to control women's sexuality and reproduction by controlling their access to resources are central to patriarchy. In this important aspect, therefore, feminist explanations are actually convergent with those developed from the concepts of parental investment and sexual selection, and they illustrate how an understanding of cultural evolution can be enriched by exchange between disciplines.

The anthropologist Sarah Hrdy has suggested that current views of evolutionary social theory can nevertheless be cast in ways that still seem to support traditional views of human sexual relations. The emphasis on male-male competition and female choice in the evolution of mating systems can reinforce and extend the idea that women are by nature (and therefore *should be*) less competitive, less easily aroused sexually, and less interested in sex. As female mammals, women are obligated physiologically to invest much more in offspring than males, and this is likely the reason why most mammalian mating systems are polygynous and sexual selection acts mainly on males. But as Hrdy documents, this does not mean that female primates do not compete with each other for resources and mates and that they are not managing their own reproduction and manipulating males. In fact, as long-term observations of primates accumulate, such behavioral strategies are being found among the females of all group-living primates. Females are competing and cooperating with each other for access to food, forming alliances with each other and with males to thwart male aggression, mating promiscuously during much of estrous but selectively when most likely to conceive, mating with males from other groups, exchanging matings for food or protection, manipulatively mating when they are pregnant or not ovulating, and having orgasms. It is clear that female primate sexuality, especially our own and that of our nearest relatives, chimpanzees and bonobos (Chapter 13), includes a complex repertoire of assertive, manipulative, and deceptive behaviors that serve other functions besides conception.

Several evolutionary anthropologists with an interest in feminist perspectives, Barbara Smuts, Sarah Hrdy, and Patricia Gowaty, argue that male attempts to control female sexuality by controlling resources started well before the development of agriculture and capitalism. The precursors to such behaviors are present in chimpanzees, where related males form alliances with each other to acquire and defend resources (females, food-producing territory) from other groups

of chimpanzees. However, chimpanzee males cannot individually or collectively monopolize ("own") access to food *within* their communities because of its nature and distribution. Aggression is their primary means of controlling female reproduction, but bonobo females are able to circumvent this bullying (Chapter 13).

The hunting-gathering mode of subsistence, thought to have typified much of our evolutionary history, may have presented more opportunities for men to control access to one resource, meat, but among the surviving hunter-gatherers the importance of women's contributions to subsistence prevents male monopolies, and these societies are relatively egalitarian. The invention of animal husbandry and agriculture certainly presented new possibilities because livestock and land can be controlled and made heritable along male lines. The later development of capitalist industrial economies added real estate, factories and their laborers, manufactured goods, and their convertible equivalent, currency, to the controllable resources. Although the degree to which men control women by controlling resources varies considerably among societies, and there is nothing inevitable about patriarchy, it is nonetheless under these conditions that male strategies for controlling and guarding mates have reached their fullest cultural expression.

The psychologist David Buss argues that in considering the evolution of patriarchy, many feminists and evolutionary biologists have overstated the importance of male reproductive interests and their implementation through aggressive coalitions. It may be that men unite in some ways to control women and exclude them from positions of power, but most human competition is between members of the same sex. Men compete mainly against other men, individually and in groups, to gain a disproportionate share of resources. Traditionally, women also compete with each other, using physical attraction, economic incentives, and social manipulation to gain access to high-status men. In other words, male and female reproductive strategies have been engaged in a long and complex coevolutionary process. From this perspective, patriarchy may be seen as a cultural manifestation of sexual selection, in which male-male competition has also led to male control of most of the resources. Reproductively, the outcome is a standoff. Women, the ultimate source of the competition, are nevertheless able to gain access to resources controlled by males without having to kill each other.

The risk in pushing comparative evolutionary arguments too far is that each species is physically and behaviorally unique and has its own evolutionary history. We are neither bonobos nor chimpanzees, and there are features of our sex lives that have to be understood as uniquely human. As we discussed above, our mating system is much closer to monogamy than the polygyny of bonobos, chimpanzees, and gorillas, and in women ovulation is concealed.

Another confusing factor is that we are currently in a period of accelerated social change. Technology has made it possible for women in many parts of the world to have unprecedented control over their reproduction, and political changes are granting women greater access to employment in areas not previously available to them. For complex reasons, however, in some societies the nuclear family that has existed through most of human history is no longer a stable or necessarily prevalent social institution. Some of these changes are clearly good in that they extend opportunity for fulfillment to many more individuals. Others, however, are creating new social challenges that will require further adjustment and accommodation.

WHY DO HUMANS KILL EACH OTHER?

It used to be said that humans are the only species that regularly kills its own kind. This most assuredly is not correct. In previous chapters we have seen that in polygynous mating systems there is extensive male-male conflict, although it is usually not lethal. We encountered examples of infanticide in several species that clearly serve the killer's reproductive success. And we saw that male chimpanzees conduct lethal raiding parties into the territory of neighboring groups. Do these phenomena have parallels in human behavior?

YOUNG MALES

In discussing the occurrence of different marriage practices we saw evidence of male-male competition and conflict. This pattern, however, is not confined to a selected subset of "primitive" societies. The psychologists Martin Daly and Margo Wilson have provided a window on sex differences in a compilation of nearly 11,000 same-sex homicides—excluding deaths from warfare—from thirty-five different societies on five continents. Their data range from 1,514 cases in the thirteenth and fourteenth centuries to modern industrial societies, but the uniformity is striking. Men killing men accounted for 95.2 ($\pm 1.9\%$ of the cases; women killing women only 4.8%, a ratio of 19:1. If cases of adults killing children are removed, limiting the data to altercations between adults, the ratio rises to 34:1.

The numbers probably do not surprise, for they are consistent with the impressions we get from the daily news of crime. What is important, however, is that they are consistent across cultures and history. Seeking causes for this sex difference in our "violent society," in different legal systems or modes of punishment, or in any other feature of a particular society is therefore not very useful.

Most of this violence is perpetrated by young men. Figure 14.11 shows homicides committed by men of different ages for one year in an American city and for a decade in all of Canada. The two populations differ dramatically in the overall rate of homicides, but the age profiles of the killers are very similar. About a quarter of these homicides were incidental to another crime such as robbery, and the killer did not know the victim. In roughly another quarter the two individuals were related. In about half the cases, however, killer and victim knew each other, although they were not related. A number of sociological studies of violence and homicide, including a 1969 Presidential Commission report of thirteen volumes, agree that the principal cause of this last category of male-male mayhem is altercations over what frequently appear to be trivial matters. A spilled beer, a small debt, a perceived insult to a sister, an argument over who is the better man; any one of these or countless other matters can escalate into lethal violence. Why this is so was clearly put by Daly and Wilson:

> A seemingly minor affront is not merely a "stimulus" to action, isolated in time and space. It must be understood within a larger social context of reputations, face, relative social status, and enduring relationships. Men are known by their fellows as "the sort who can be pushed around" or "the sort who won't take any [abuse]," as people whose word means action and people who are full of hot air, as guys whose girlfriends you can chat up with impunity or guys you don't want to mess with.
>
> In most social milieus, a man's reputation depends in part upon the maintenance of a credible threat of violence. Conflicts of interest are endemic to society, and one's interests are likely to be violated by competitors unless those competitors are deterred. Effective deterrence is a matter of convincing our rivals that any attempt to advance their interests at our expense will lead to such severe penalties that the competitive gambit will end up a net loss which should never have been undertaken.
>
> The utility of a credible threat of violence has been mitigated and obscured in modern mass society because the state has assumed a monopoly on the legitimate use of force. But wherever that monopoly has been relaxed—whether in an entire society or in a neglected underclass—then the utility of that credible threat becomes apparent.

As suggested in the last sentence, there are many factors that affect the nature of male-male competition and the likelihood that their social conflicts will become lethal. Alcohol clearly lubricates brawls, and low socioeconomic status that leaves young men with little sense of a future seems to play an aggravating role. The latter in fact lies behind the homicidal rampages known as *amok* that occur occasionally in southeast Asia. In our society, illicit commerce in drugs and the ready availability of firearms also contribute, but they are

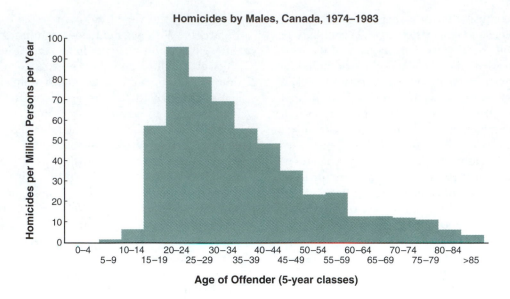

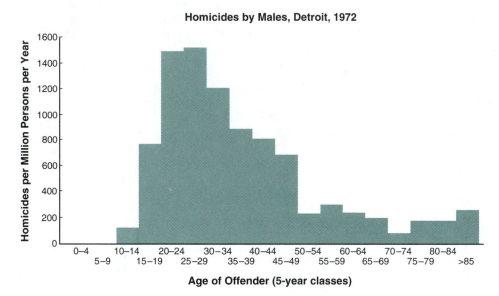

FIGURE 14.11 Most murders are committed by young men.

proximate aggravating factors in the underlying struggle for status and respect of peers.

From time to time there have been assertions that other cultures exist with fundamentally different sex roles or no male violence. The popular writings of the anthropologist Margaret Mead claimed as much for several societies in New Guinea, but like her earlier work in Samoa, her research has not stood the test of time. None of these societies seems to be free of polygyny, male sexual jealousy, male-male competition, and lethal strife.

The concepts of sexual selection and parental investment have broad application and considerable value in understanding animals, including humans. Those who would argue that members of a particular human society are living in ways that seem to contradict general biological principles therefore have an enormous burden of proof. Either the people of such a culture have a unique cognitive and emotional endowment—a prospect with vanishingly small likelihood!—or there is a special array of ecological and historical factors at work that demand extensive study. The problem in the past is that there has been little understanding and much misunderstanding of the biological principles. Margaret Mead's work was done with the conviction that any social structure whatsoever is likely to be found if one looks for it. Furthermore, the ease with which she seemed to find what she was looking for—

three appropriately contrasting societies within a few score miles of each other—illustrates how interpretation of data can be strongly influenced by the cultural biases of the investigator. With the benefit of hindsight, it also shows the importance of knowledge that bridges disciplines.

SPOUSAL ALTERCATIONS

As we saw above, in roughly a quarter of homicides the offender and victim are relatives. In their analysis of Detroit police records, however, Daly and Wilson found that only 22% of these cases involved genetic relatives, whereas in 78% of cases the individuals were relatives by marriage. In fact, two-thirds of all the cases involved spouses, either by legal or common-law marriage. Living together provides opportunities for dispute, but there is a disparity in homicide rate between spouses and genetic relatives living in the same household, so factors other than proximity are at work. Moreover, similar data are found in other societies.

In evolutionary terms, marriage is an alliance for producing offspring, and we have seen examples of cultural conventions that emphasize this relationship. The expectation that bride-price will be forfeited if a union proves infertile is one example. We have also seen that males have a genetic interest in assuring their paternity of offspring, and that in guarding and sequestering women, human males show they are not an exception. Only very recently and primarily in the Western democracies have the laws on adultery approached symmetry. Historically, the legal importance of an adulterous affair depended on the married status of the woman, not the man. In fact, the aggrieved party was considered to be the woman's husband. Depending on the culture, he might be owed compensation from the family of his wife's lover, or the law might look with an understanding eye if he killed the other man. When a woman's adultery is the provocation, English common law views as manslaughter a homicide that in other circumstances would be tried as murder. These may be the rules of patriarchal societies, but they also seem to acknowledge a particular power of male sexual jealousy. Moreover, psychological studies suggest that men and women have somewhat different feelings about adulterous behavior of their partner. As mentioned previously, whereas men may be obsessed with thoughts of the sexual act itself, women are more likely to be concerned about commitment.

The single most important factor in precipitating spousal homicide is sexual jealousy. In the vast majority of cases, husbands kill wives and not *vice versa* (Fig. 14.12), and husbands also kill their male rivals in about equal numbers. The power of the underlying emotions is conveyed when a man earnestly yet paradoxically proclaims his love for the woman he has just slain. Although women sometimes kill their husbands or their female rivals because of adulterous affairs, charging their husband with infidelity occasionally leads to their own murder. Furthermore, women sometimes kill their husbands as a result of chronic abuse. These homicides occur everywhere; they are not associated with any particular part of the world or any culture.

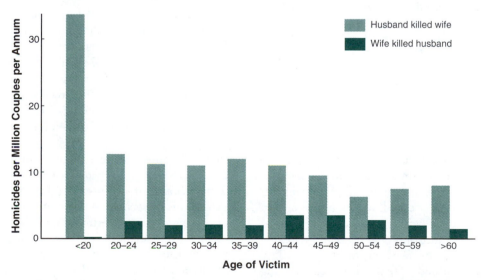

FIGURE 14.12 Spousal homicides within legal marriages. Wives, especially teen-age wives, are at much greater risk from husbands than vice versa. Data for Canada, 1974-1983.

FRATRICIDE

In an extensive comparison of nearly 2,500 modern cases from several different cultures with data from thirteenth century England, Daly and Wilson showed that the average degree of relatedness (r) between co-conspirators in murder was two to fifty times that between the killers and their victims. This is another measure of how the danger of lethal conflict decreases, and the willingness to collaborate against enemies increases with genetic relatedness.

There are nevertheless instances of rivalry between brothers that end in murder, for which the biblical story of Cain and Abel provides a familiar allegory. The conditions for fratricide occur in societies in which inherited resources are not divisible without greatly diluting their future value—the family farm, business, or positions of power. Thus in agricultural societies characteristically one son, usually the first, inherits the land, and the others are expected to seek their fortunes elsewhere, in colonies, military campaigns, or a religious order. Historically, bloodletting between full and half brothers has been common in the noble families of feudal societies, such as medieval Europe, where alliances within large opposing factions were based upon reciprocity rather than kinship, and the payoffs for the victors and their leader in productive lands and strategic and taxable assets were very large.

Conversely, fratricide is rare or absent in preagricultural societies that subsist by hunting, fishing, and gathering or by a combination of foraging and horticulture. In these societies, most of the resources upon which subsistence depends are not fixed, durable, and costly to replace, as are agricultural lands, buildings, and livestock, and they are neither inherited nor fought over within the group. Under such conditions, an individual's status, support, marriage possibilities, and safety from enemies depends largely on the number of male kinsmen he has. Under most circumstances, fratricide would seriously disrupt the cohesiveness of such groups of male kinsmen.

INFANTICIDE

The death of a child strikes us as the worst horror that can beset parents, and we find it difficult to comprehend how any parent could actually kill their own offspring. Such happenings, we feel, must be rare and so abnormal as to be pathological.

Both history and anthropology, however, expose a different reality. Greek and Roman societies practiced infanticide by the simple expedient of abandoning sickly or deformed babies in wild areas to die of exposure. In ancient Roman society a father had the legal right to dispose of an infant, which was one means of restricting his parental investment in offspring that he suspected he had not fathered. Although fourth century Roman law and church teaching forbade infanticide, in Europe the practice of leaving infants to perish of exposure continued. By the eighteenth century the frequent sight of abandoned babies in public places stimulated governments to create foundling hospitals where unwanted infants could be left for care. The load was so heavy and the physical condition of the babies so desperate, however, that an estimated two-thirds of the infants died anyway. In 1833, 164,000 babies were left at foundling hospitals in France; in the latter part of the nineteenth century, a hospital in Moscow was receiving 17,000 infants each year.

Comparative data from other societies tell a similar story. Significant rates of infanticide have been documented in Chinese, Hindu, and Arab cultures as well as in many societies studied by anthropologists. Present-day estimates of infanticide are as high as 40% of all births in one New Guinea tribe, and numbers in the range of 10 to 25% have been reported elsewhere. In the United States, several children die from abuse each day, which amounts to a couple of thousand every year. The finding of a newborn in an urban dumpster is an echo of Victorian London. Some recent data on the killing of children in a Western culture are shown in Figure 14.13.

Viewed in this broader perspective, infanticide is a phenomenon that requires further understanding. It is not restricted to a few cultures, and although the incidence varies widely, the circumstances provide important clues for why this is a recurrent human behavior and how knowledge of evolutionary processes can help us to understand it. Under what conditions do mothers kill their infants, and do those circumstances make sense in an evolutionary context?

Infanticides by unwed adolescent mothers are reported in the news media in ways that suggest depravity: a young girl conceals her pregnancy, gives birth, abandons her child, and then tries to resume her normal life. But these descriptions also hold clues to a deeper understanding. The predicament of these young women is usually desperate: they are alone, without family or other support, and either they do not want to marry the father of their child or he has abandoned them. The other feature of these mothers is that they have most of their reproductive years ahead of them. Throughout our evolutionary history—probably even now in many societies—the lifetime reproductive success of a teenage single mother and her descendants are likely to be enhanced, not diminished, by ceasing investment in an infant that is without male support and is perceived as not wanted by her family. Indeed, youth and inappropriate social circumstances are among the most important proximate reasons why human mothers kill their newborn in other cultures.

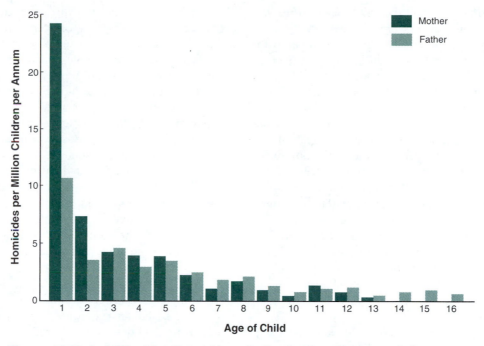

FIGURE 14.13 **A child's risk of homicide at the hands of mother versus father as a function of the age of the child. Risk is higher for infants than for older children, and mothers are about twice as likely as fathers to take an infant's life. Data are for a 9-year period in Canada, 1974–1983.**

Another factor that can lead to infanticide is the physical condition of the newborn. In many societies, infants that appear physically or mentally defective are often attributed to the influence of evil spirits or associated with evil omens and are abandoned or destroyed. In our own society, many parents love and care for handicapped children, but not all. Sadly, studies have shown that handicapped children are two to ten times more likely to be severely abused than are normal children raised in the family household.

One might argue that knowing the evolutionary reasons why infanticide occurs is not necessary in order to construct social policies in which the abuse and killing of helpless children is minimized. For example, compelling reasons for reducing poverty and improving education can be cast entirely in the realm of proximate causes of behavior. This is true, but viewing infanticide as due solely to a failure of moral teaching minimizes the scope of the problem. If human reproductive behavior were not so complex, there would be no need for social and religious codes that attempt to regulate it.

There are other circumstances of maternal infanticide that make sense in an evolutionary context. In some environments, humans subsist upon food that is dispersed, seasonal, hard to procure, and hard for infants to digest. For example, among the Ju/'hoansi (!Kung San) of the African Kalahari desert, the men

hunt for elusive animals, and the women gather coarse plant food. Under these conditions of sparse food and hard work, a mother can sustain only one infant at a time until it is weaned at age three or four. Ovulation was usually suppressed in Ju/'hoansi mothers by lactation (nursing) and low body fat. Infant mortality data show that the birth of twins or too short intervals between births seriously endangered the survival of both infants. Experience and observation taught these mothers that in this difficult situation the better choice was to leave the smaller twin or the younger infant to die.

STEPPARENTS

In folktales from Ireland to India, stepmothers and stepfathers have a bad reputation. The stories of Hansel and Gretel or Cinderella are familiar examples. Stepparents have no genetic relationship to their spouse's children, and the normal basis for parental love is absent. Recent profiles of such families confirm that there are frequent difficulties in establishing deeply loving relationships, particularly between stepfathers and the wife's previous children. (When the children are young, they are ordinarily with their genetic mother, so the stepparent is likely to be the man.) For stepfathers (or a resident male sexual partner) the situation requires considerable parental investment in another man's offspring, and the child may (in the eyes of the man) compete for the woman's attention.

A Yanomomö man who marries a woman with children from a prior marriage can insist that the children be killed. This practice may seem more like lion or langur than human behavior, but it reflects a deep-seated male concern about paternity that has identifiable counterparts in our society. In North America, young children in a household with one stepparent are seven to forty times more likely to be victims of abuse than are children living with both genetic parents. The chance that an infant will suffer death at the hands of a stepparent is about a hundred times greater than when it is living with both genetic parents (Fig. 14.14).

The risk to children from members of the household falls dramatically after the first year of life. This is understandable, given the particular circumstances that lead new mothers to kill their babies. The risk from stepfathers also drops, presumably as relationships in the family group stabilize.

The emotional motivations of adoptive parents and stepparents can be quite different. Adoptive couples are anxious to have children on which to express the parental love that humans quite naturally feel. Their commitment is mutual. To the extent that a new stepparent has that same desire, his family relationships may be normal and richly rewarding. But the nature of stepparenthood has a different basis: two adults are attracted to each other, but their feelings about existing children may be very different.

The value of offspring to the lifetime reproductive success of a parent increases with the cumulative parental investment the young have received. Losses cut early are less severe. If drought strikes, it is better to abandon a nest of eggs, or even to forego reproduction for a season, than attempt to rear the clutch in the face of insurmountable odds. This is the broader context in which human mothers are capable of making decisions at the time of their offspring's birth that would be even less probable with older children. Whether those decisions are socially and legally acceptable, however, depends on the culture in which they are made. In our society, infanticide is not only considered a crime, it is a violation of religious teaching. These strictures, however, do not prevent some mothers from assessing their prospects in different terms. As we move back in time toward the moment of conception, the cumulative parental investment by the mother is smaller, and society is far from consensus concerning the rights of the zygote or embryo relative to the rights of the mother. Remember that "rights" in this context are socially derived, whether they are based on religious teaching or secular law. This is inevitably so, because religions, like laws, vary from one culture to another.

DIFFERENTIAL INVESTMENT IN SONS OR DAUGHTERS

Parents sometimes wish for offspring of a particular sex: for example, to extend the family name, for labor, or simply for balance in the family. The anthropologist Mildred Dickemann has presented evidence that selective care of offspring may be used by human parents to suppress investment in the sex with lower reproductive prospects. She reasoned that in highly stratified societies, such as medieval Europe, imperial China, and feudal north India, high-status sons (with many wives or concubines) were likely to leave more descendants than daughters. Furthermore, the cost of dowry required to marry a daughter to a still higher-status male was frequently prohibitive. Indeed, she found in these societies that female infanticide—as well as non-reproductive roles for daughters—was socioeconomically graded: higher-status daughters were more frequently neglected or placed in religious orders, whereas such practices were not found in lower-status families. In fact, lower-status families in these societies could benefit from bride-price brought by their daughters.

A similar pattern of status-graded, sex-preferential investment was found by the anthropologist Eckart Voland in his study of detailed genealogical records of landowners and peasants kept over two centuries in the northern coastal region of Germany. Among the landowners the mortality of female infants was significantly higher than that of males, and the reverse pattern was found for those who owned no land. This

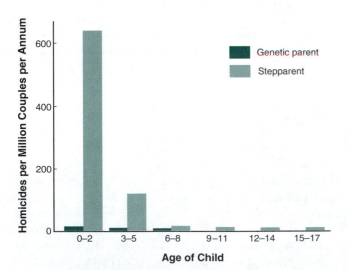

FIGURE 14.14 The risk of death at the hands of a stepparent is much greater than from a genetic parent and declines rapidly with the age of a child. As young children ordinarily stay with their mothers, the stepparents in these data are virtually all male.

manipulation of the sex ratio seems to have been affected indirectly—by feeding and caring more for one sex than the other—rather than by direct infanticide. Voland also found in the same records an important corollary to the survivorship data that supports the hypothesis that sex ratio was being manipulated to maximize descendants: among the landed families more sons married than daughters, and among the landless families, the reverse was true.

Another reversal of the usual pattern of male-biased investment was found by the anthropologist Lee Cronk among the Mukogodo, one of several tribes of Maasi-speaking pastoralists in central Kenya. Cronk found a significant correlation between wealth in livestock (cattle, sheep, goats) and reproductive success among these tribes. The Mukogodo have the smallest herds, and they are looked down upon by others because they had previously spoken another language and subsisted by hunting, gathering, and beekeeping. The consequence of this lower wealth and status is that the bride-price for a Mukogodo women is relatively low, so non-Mukogodo men can purchase them, and the bride's family gains. But a Mukogodo man cannot easily afford to buy a non-Mukogodo wife: he is poorer, and he must pay a higher bride-price for a non-Mukogodo woman than for a woman from his own tribe. Thus daughters bring their families wealth, and their marital and reproductive prospects are better than those of sons. Just as theory predicts, Cronk found that the sex ratio among children less than four years of age was biased significantly toward girls (ninety-eight girls to sixty-six boys), even though the majority of their mothers, the primary caretakers, stated a slight preference for sons. There is no evidence for infanticide by Mukogodo mothers or fathers, and corporal punishment of children is rare. However, there are significant differences in the care given to boys and girls. Daughters are more frequently brought to health clinics and dispensaries when ill, and they are more often nursed, held, and kept close than are sons. Measures of growth performance also indicated that girls are on average better nourished than boys. Among the non-Mukogodo people in the same area Cronk found evidence of the opposite bias: sons are favored over daughters.

CONFLICT BETWEEN GROUPS: WARFARE

Virtually every nation in the world maintains military forces for defense against other states. Centralized authorities regulate economic relations and adjudicate internal disputes, thus maintaining the group unity upon which military effectiveness depends. State religions and historical accounts celebrate past conflicts and call for sacrifice and obedience in the pursuit of group interests. Furthermore, written records of state-level societies, from Central and South America to Europe, the Middle East, and the Far East document the pervasive role of warfare in the formation and expansion of states and in improving their technologies of weaponry, transportation, communication, and construction. How did this begin?

Understanding the history and significance of warfare has been influenced by two contrasting concepts of human nature expressed in Western thought. One, found in the writings of the English philosopher Thomas Hobbes (1588–1679), is that humans are by nature entirely selfish, and that social order requires the presence of authority. Civilization thus provides the means for people to live together in relative harmony, and in Hobbes' words, life in an earlier state of nature was "solitary, poor, nasty, brutish, and short."

The French philosopher Jean-Jacques Rousseau (1712–1778) had a very different view of human nature. He thought that technologically primitive people lived in a peaceful, uncomplicated state of nature in which they were ruled by gentle if self-serving inclinations. Only the trappings of civilization, such as ownership of property, rules of marriage, and civil and religious authority, brought strife and conflict. Thus was born the notion of the "noble savage."

During the nineteenth century, when the competition among the major European powers for colonial expansion was at its height, a Hobbesian view of technologically primitive peoples provided a convenient rationale for pursuing European economic and political interests. Bringing civilization to the uncivilized was seen not only as morally justified, in an excess of communal self-deception it was elevated to a duty as the "white man's burden."

At the present time, some anthropologists maintain that strife among primitive societies is an outcome of their interaction with more technologically proficient groups. This neo-Rousseauian view denies the occurrence of warfare prior to the formation of large and complex social units capable of raising armies.

Neither of these philosophical views of human nature and human prehistory accurately reflects reality. Both underestimate the richness and diversity of social relationships that anthropologists have documented in every culture, irrespective of their technologies. Technology enables larger societies to form, but as we saw in Chapter 12, evidence from genetics and evolution shows that human nature and intelligence do not vary geographically and are very unlikely to have changed significantly during the last 10,000 years. Humans everywhere possess the same basic desires, psychological reward systems, capacity for reasoning, and the ability to acquire complex language. Furthermore, individual social relationships are richly nuanced everywhere one looks. Whatever genetic variation in these traits exists is very likely the same among Australian Aborigines as in random samples of Italians, Swedes, or Chinese. As described in Chapter 12, there is no evidence for the alternative, and the reasons why some cultures have developed technology faster than others are not to be found in the genes.

Rousseau, however, was further off the mark than Hobbes. The archaeologist Lawrence Keeley has assembled extensive evidence that warfare among humans did not originate recently. The oldest reliable archaeological evidence comes from a variety of Paleolithic sites in Europe and North Africa dating from approximately 30,000 years ago and consists of skeletons with embedded projectile points, skulls broken by stone axes, and mass graves containing men, women, and children. Following the institution of agriculture about 12,000 years ago, larger settlements started to appear, and additional evidence of violent death is also found in the Middle East and North America. Excavations in Europe of ditch and palisade fortifications from this period have yielded thousands of arrowheads around their perimeters, with particular concentrations at the gates. This distribution is not consistent with hunting; it indicates that the structures were being defended from attack. In one case the skeleton of a man was found in the bottom of the ditch under burned rubble. He had been carrying an infant and had been shot in the back with an arrow.

Although fortifications and mass burials of people of all ages that died violently provide persuasive evidence of warfare, some anthropologists interpret ditch and palisade structures as mere symbols, not functional fortifications. Some have also argued that primitive warfare is neither comparable to, nor the precursor of, state-level warfare because intergroup aggression in the primitive societies that still exist usually involves small numbers of individuals and appears to end after a lot of display and a few casualties. As Keeley has pointed out, however, the ethnographic evidence shows that ambushes, raids, and occasional massacres occur repeatedly, and their cumulative effects are greater than the formalized warfare between large "civilized" states in which major clashes tend to occur only a few times a century. Figure 14.15 shows that deaths from warfare, expressed as a percentage of the population per year, are actually higher in technologically primitive societies than in large literate cultures with complexly stratified social structures. The mortality from warfare in primitive societies may actually be larger than suggested by these data, because the numbers for modern warfare include subsequent deaths from wounds as well as collateral civilian casualties. Such information is less readily available from ethnographic studies of primitive warfare.

Another difference is that modern warfare has provision for surrender whereas in primitive war only fertile women are likely to be taken prisoner. Primitive warfare is no-holds-barred conflict, tactically like the guerilla warfare in which modern armies can prevail

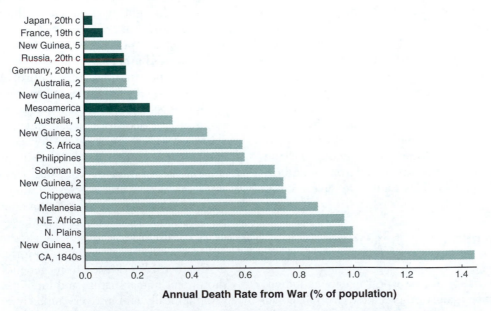

Annual Death Rate from War (% of population)

FIGURE 14.15 Estimated annual mortality from warfare in twenty cultures. The darker bars represent societies with organized military forces. The others engage in "primitive" warfare. Most of the unfamiliar names of the cultural groups have been replaced by geographic designations.

against a determined enemy only with massive logistical superiority. Examples of recent failure include the French and Americans in Vietnam and the Russians in Afghanistan.

Rousseau's view of conflict in primitive societies is wrong by another measure. As Keeley has documented, only a small number of societies (5 to 10%) have no history of war. These tend to be groups that live in marginal environments with few close neighbors, although these conditions by themselves do not assure a peaceful life. Moreover, the evidence does not support the more recent claim that primitive societies only took up war as a result of contact with more technologically advanced cultures. Periodic conflict between human groups has been going on everywhere in the world and for as long (at least!) as the archaeological trail can be followed.

THE CAUSES OF WAR

What causes war? It is useful to frame this question in an evolutionary context (Box 14.1). When all of the proximate causes such as the judgment or emotional state of political leaders are stripped away, what remains are economic issues that have an influence on reproductive success. For example, in primitive societies a conflict over resources such as a source of water, access to a fishing stream, or exclusive use of a deposit of salt or metal used in trade can all lead to warfare. Alternatively, the wheels of war may start to turn when one group kidnaps another's women or steals another's possessions such as livestock.

Commerce does not eliminate frictions between groups. When two groups engage in trade or exchange women in marriage, disputes can arise over whether a trade was fair, proper bride-price was paid, or a woman has been mistreated. Any one of these or similar issues can evoke anger, jealousy, wounded pride, a sense of unfairness, or a desire for retribution. An attack on one man may bring retaliation from his kin. Escalated emotions then mix with calculations of risk and benefit, and an ambush or other relatively low-cost tactic may be planned. In many cultures ritualistic mutilation (commonly beheading) of an adversary who has been killed has a spiritual function such as drawing on the strength that the victim possessed in life or disabling him in an afterlife.

The discovery and adoption of agriculture allowed the growth of denser and larger human populations as well as creating fixed and valuble resources in the form of farmland, stored crops, and livestock. This in turn produced new incentives for intergroup conflict, and surplus production made possible the construction of fortified settlements as well as support and provisioning of military forces. A number of the archeological findings described above correlate with the expansion of population from the center of agricultural invention that was located in the Middle East (Chapter 12).

Seen this way, the evolutionary (ultimate) causes of war between modern nations are also conflicts over relative access to resources. For example, the origins of World War II in Europe are frequently traced to the devastating effects of the Treaty of Versailles on the German economy during the 1920s. These conditions allowed a political leader to rise to power by exploiting emotions of pride (intensified to a conviction of racial superiority), group solidarity (embellished with xenophobia), and a desire for revenge (drawing on a nationalistic interpretation of history). Additional proximate factors were the physical and emotional exhaustion of Britain and France (reflected in their political leadership), as well as independent political and economic events occurring in the Soviet Union, Japan, and the United States.

Societies that are particularly prone to warring with their neighbors are not genetically predisposed to violence. The Scandinavians who terrorized northern Europe a thousand years ago are today paragons of peace. What propels people to warfare are ecological and economic circumstances and the interplay of human passions, in some cases sustained by a cultural tradition of enmity toward neighbors. Times of hardship can heighten competition, and demographic expansion of one group can disturb boundaries and relationships with others.

History suggests that cycles of warfare may be difficult to break without changes in economic or ecological relations. Just as disputes between individuals are often hard to resolve without the intervention of a third party (friends, a legal system), outside powers (for example, Imperial Rome or colonial Britain) have sometimes imposed peace on others, although frequently becoming the object of enmity themselves.

Warfare is thus a collective behavior that enables human groups to appropriate the economic or reproductive resources of other groups and to defend themselves from exploitation. It is neither inevitable nor (usually) rationally desirable, but it has occurred so often that it is clearly a common component of the human behavioral repertoire. In the last fifty years the technology of atomic weapons and their delivery by guided missiles has made warfare potentially catastrophic on a scale that was unimaginable before 1945. Although humankind has created effective methods for resolving disputes nonviolently within societies, the processes for dealing with disputes between societies are still largely in the hands of small numbers of political leaders, and the results are frequently driven as much by passions and prejudices as by cool reason. But as we saw in Chapter 9, that paradox of the human mind is part of our evolutionary heritage.

Box 14.1
Proximate and Evolutionary Causes of War

Many anthropologists reject all-encompassing explanations of primitive warfare on the grounds that it is complex, multicausal, and differently motivated in each society. The following passage from a book on warfare in tribal societies exemplifies this thinking:

> . . . different analysts could look at one case of war and conclude that it is a conflict over political status, women, natural resources, or trade goods; an expression of witchcraft beliefs, cognitive orientations, pent-up frustration, rules of conflict, or belligerent personalities; a quest for prestige, revenge, security, power, trophies or wealth; a consequence of residence patterns, level of political evolution, men's organizations, sovereignty, or inadequate conflict resolution mechanisms; and that it is generated by individual decisions, the functioning of societal subsystems, or cultural selection. Possibly, all of these conclusions could be correct. Each could accurately identify one aspect of the multiple interactions involved in that particular case of war. This complexity must be recognized and dealt with. Not having done so up to the present is one reason that anthropological analyses of war tend to be particularistic and eclectic.
>
> Brian Ferguson
> (*The Anthropology of War*)

This passage can be cast in an evolutionary framework of explanation by separating the phrases that refer to evolutionary causes, proximate causes, and enabling conditions. "Political status," "prestige," and "power" all refer to the proximate motivations that lead males, acting alone or in coalitions, to try to dominate others and thereby control access to resources. Given our earlier discussion, the reference to "women" needs no further explanation. "Natural resources," "trade goods," "trophies," and "wealth" all refer to defendable resources that can be used to obtain wives and invest in offspring and relatives. "Witchcraft beliefs," "cognitive orientations," "pent-up frustration," "belligerent personalities," "revenge," and "inadequate conflict resolution" refer to proximate psychological mechanisms or their cultural manifestations that regulate attitudes toward enemies, tolerance of competition, and the maintenance of a credible threat toward enemies. "Security" and "sovereignty" refer ultimately to maintaining access to resources by retaining possession of land, trade routes, or strategic sites. "Residence patterns," "level of political organization," and "men's organizations" refer to the numbers and proximity of allies, their willingness to unite in battle, and whether the group's resources and military organization can support a war.

Unpacking the catalog of causes this way makes clearer the usefulness of evolutionary thinking. To illustrate this point further, consider an imaginary colony of highly intelligent bees in which a scholar bee wonders what causes her peers to forage for nectar and pollen:

> . . . different analysts could look at one case of foraging and conclude that it is caused by an attraction to gaudy colors, hunger, a sweet proboscis, an aesthetic taste for colored objects with radial or bilateral symmetry, thirst, hive claustrophobia, protein shortage, a fondness for floral odors, love of sisters, a belligerent queen, inadequate greed-restraining mechanisms, belief in the medicine bee's saying that nectar contains substances that prolong life, or in the queen's proclaimed revelation that making honey is an act of religious devotion.

The evolutionary explanation of foraging for nectar and pollen in bees is, of course, that it is an adaptive process for providing food, thereby contributing importantly to reproductive success. All of the imaginary musings of the "scholar bee" are not alternatives to this explanation of evolutionary cause. They are possible proximate mechanisms that could motivate, trigger, or facilitate foraging behavior in this fanciful world of superintelligent bees.

The rephrasing of the causes of war suggests that warfare—primitive or technologically sophisticated—is ultimately about resources. War is a form of behavior in which males (historically males) form coalitions to carry out aggression against other groups of males. In primitive cultures, killing an enemy may be a rite of passage. Similarly, in nation states that are under serious external threat, military service for young men can be both a moral and a social necessity. Regardless of the cognitive rationale, however, the evolutionary payoff for (surviving) members of successful groups is differential access to resources and reproductive success. Military leaders in primitive societies as well as in early state societies were among the most polygynous men, and their rank determined the number and quality of captured women they received as wives or concubines. As Sarah Hrdy notes, the term "trophy wife" is historically derived from this practice, and it signified ownership of a trophy of war.

We can see a shadow of the evolutionary origins of warfare in the intergroup raids of chimpanzees (Chapter 13). Because humans have higher intelligence, language, and culture, and because they utilize a much greater variety of resources in a wider range of environments than chimpanzees, there are more proximate causes of conflict, and they interact in more complex ways.

SOCIAL INTELLIGENCE AND MORAL SYSTEMS

BEYOND THE APES: SOCIAL NETWORKS AND RECIPROCAL EXCHANGES IN HUMANS

People are intensely social, and every stable group of humans contains networks of relationships. Some are based on kinship or marriage, some on personal friendships, some on shared interests like religion or recreational activities, others on formal arrangements associated with school, employment, or government. These webs of human associations are one of our distinguishing features as a species, but we can see their origins in monkeys and apes. These primates recognize others as individuals, and from their behavior they appear to know how others fit into the social structure of kin-lines and dominance hierarchies. As we saw in Chapter 9, there is a part of the frontal cortex that is particularly involved in social relationships. And as we discussed in Chapter 11, fostering social intelligence in groups of up to a hundred individuals appears to have been one of the early factors leading to the enlargement of the primate cortex.

Compared with other primates, humans have remarkable social abilities. We not only remember the faces and voices of many dozens of different individuals, we develop complex emotional attitudes toward each of them. Some of our information about other people is based on direct experience, and if the encounter was especially pleasant or unpleasant, our memories linger.

Our capacity for language greatly enlarges our social knowledge. It enables us to learn about the intentions, attitudes, and beliefs of others and to communicate such information about ourselves. It also enables us to manipulate others by falsifying or conveying such information selectively. Furthermore, it makes possible the spread of information about others through the common custom of gossip. Political speeches, advertising, sermons, and classroom lectures can all be used for the same purpose, just with a larger audience. And words really are cheap. At a cost of little energy they can bring great benefit or inflict much damage.

In Chapter 6 we met the concept of reciprocal altruism—low-cost behavioral exchanges that work to the long-term benefit of both individuals. The evolution of this kind of behavior requires that individuals interact repeatedly, and in humans these exchanges cement those parts of our intricate social networks that are not based on kinship.

The capacity for self-awareness, glimmerings of which seem to be found in the great apes, enriches the power and flexibility of reciprocal exchanges. When we mentally construct and test the possible consequences of our behavioral choices, we find ourselves at center stage. Our brains make us each a playwright in a continuously running theater of the self. Through observation and conversation we try to ascertain the motives, intentions, and strategies of other actors, and we use this information to anticipate how our words and deeds will influence their behavior, sometimes indirectly through others. We see a possible outcome—perhaps something beneficial or simply pleasurable to us, our kin, or friends, or something costly to competitors or enemies. We design a plot to move us toward that goal, scripts are imagined, screened in our mind's eye, modified, and screened again until we think the desired outcome is likely. Sometimes this may take seconds, other times weeks, and often we are only dimly aware or even oblivious of what we are doing.

Interestingly, when the interactions occur between individuals with very different backgrounds and disparate positions in the social structure, there is enormous scope for misunderstanding. A corporate industry in personnel management has been established to deal with this problem, and an important function of the legal profession is to formulate agreements so as to minimize misunderstanding.

As we have seen, the evolution of hominids likely involved much cooperation within small groups in hunting, gathering, and tool making, punctuated by both exchange and intense competition with neighboring groups. Many of the details are not known, but this broad view of our history is consistent with features of our psychology. The emotions of love, friendship, compassion, trust, and gratitude speak directly of the importance of reciprocity and cooperation between individuals, for they enable us to maintain, regulate, and benefit from reciprocal exchanges. Ambition, jealousy, and envy motivate striving to increase the relative proportion of benefits accruing to self and kin, either directly by increasing one's own share or indirectly by decreasing the shares of others. Feelings of guilt and obligation help to maintain fair and balanced exchanges.

These emotions are all subject to refinement during childhood development, but they are not simply cultural creations. Humans seek love and friendship and become disturbed if they do not find them. The solitary play of young children, using toys as props, is characteristically very social in its imaginings. Nor should we be surprised by the interplay of selfish motivations with feelings of generosity toward others. Because social groups consist of individuals who are each genetically unique, one person's interests are never identical with another's, and groups are therefore bound to have internal conflicts.

A behavioral propensity for reciprocal altruistic exchanges is subject to cheating. If benefits can be derived without having to pay, selfish behavior will often occur. By the same argument, however, we should also expect that people possess psychological mechanisms to defend against cheating. Thus we are sensitive to any evidence of lying, and a general suspicion of individuals whom we do not know sharpens our skill at detecting cheaters. Not only do we look for signs of dishonesty, in our theater of the imagination, we explore other people's possible motivations. Moralistic indignation and expressions of unfairness and injustice help to discourage prospective cheaters, and if accompanied by threats of aggression, sometimes intimidate them as well.

When cheaters are exposed, they may become distrusted and disliked by other members of the group. If they are excluded from future reciprocity, their position in the social system is seriously compromised. Feelings of shame, remorse, and guilt suppress cheating and motivate those caught either to desist or improve their skills at avoiding future detection. Evidence of contrition and acts of reparation may allow their reinstatement to a position of trust, thus avoiding their exclusion from future exchanges.

An ability to exaggerate or mimic such emotions as shame or indignation can deceive others into believing that the actor needs or deserves more than is necessary or appropriate. Insincerity accompanied by feelings of guilt, however, is not easy for many people, for they may be exposed by uncontrollable activity of their autonomic nervous system. The trembling hand, the sweaty palm, the shifty gaze, or the blushing face can give the game away. As Robert Trivers has pointed out, deception of others is easier if one has first deceived one's self. If the speaker really believes that his actions and motives are just and sincere, his efforts become more convincing. Self-deception is frequently unconscious. It is most readily achieved when no effort is made to understand the interests of the other party.

THE PSYCHOLOGY OF GROUP MEMBERSHIP

For much of human evolution, people existed in groups of fewer than a hundred related individuals and with tense or hostile relations between groups. Judging by the behavior of the few remaining hunter-gatherer peoples, friendly contact between neighboring groups was likely facilitated when individuals could be identified who had relatives in both groups. One of the outcomes of this evolutionary history is that people tend to be wary or suspicious of strangers, particularly when they seem to have no legitimate purpose in being close. Individuals also tend to be uneasy when they find themselves in the presence of others who look different, or dress in unfamiliar fashion, or who speak a language that they cannot understand.

A second and related feature of this evolutionary history is that people also tend to identify with their group in ways that are independent of close degrees of relatedness. Ethnic identity based on historical or presumed distant genealogical ties, patriotism and nationalism based on geographical and political associations, and religious affiliation are all familiar examples in present-day nation-states.

A less-admirable side of this behavior is that people can readily assume that their own group is superior to others in any of a variety of ways such as intelligence, morals, manners, ethnicity, ancestry, religion, patriotism, education, appearance, or accent. All cultures have terms of approval for in-group members and denigration for members of other groups. The out-group can be from another continent, another part of the country, or another neighborhood. Moral and legal systems frequently prescribe double standards of justice, one for treatment of in-group members and another for out-group members. It is thus a short mental step from group pride to xenophobia, racism, ethnocentrism, classism, or sexism.

The psychology of group identity can also be a potent motivator. When groups are in conflict, the notion that one's own group has the higher claim to moral authority or legal right is an effective spur to action. In war it is therefore customary for both sides to claim that the deities are on their side. Put another way, the pursuit of material advantage at the expense of others is easier when one feels earnest righteousness, so unfounded belief in in-group superiority is an important manifestation of self-deception.

It is easy to see how natural selection fostered cognitive processes that made group identity so important. Those individuals who belonged to groups that succeeded because of cooperation and balanced reciprocity in times of intergroup conflict or ecological stress would be more likely to survive and reproduce. Understanding the evolutionary origins of the darker side of this human behavior and recognizing that it is neither adaptive nor morally justified in our shrinking world is a necessary first step to minimizing warfare and other forms of ethnic conflict.

THE GENESIS OF MORAL AND LEGAL SYSTEMS

Every discussion of human nature and human societies, whether it is scientific, philosophical, or religious, has struggled with the paradox that humans strive in self-interested fashion to achieve personal goals and yet live in social groups where individual aspirations can only be met through cooperation and sacrifice toward the common good. Although the interests of each genetically unique individual do not coincide with those of any other, natural selection has fostered psychological adaptations that enable group living. As Richard Alexander has discussed, the same fundamental conflict is also responsible for the creation of a variety of cultural rules. Because internal conflicts detract from achieving collective goals that ultimately benefit the reproduction and well-being of group members, humans have invented moral systems, behavioral norms, the concepts of right and wrong, and all of their legal and religious manifestations.

These systems are necessary because the conflicts are inevitable and there are no simple rules for resolving them. First, disagreements are so varied that no simple formula could produce an appropriate resolution in every circumstance. Second, the interests of the two parties frequently differ in a qualitative way, so "splitting the difference" is seldom a realistic solution. Third, overlap of interests is less likely in large groups where there are few ties of kinship. And fourth, conflicts invariably involve individuals with unequal power—physically strong and weak, men and women, leaders and followers, corporations and employees— and different cultures view such asymmetries in different ways. Even efforts to define individual rights— freedom of speech, freedom to bear arms, reproductive freedom, to pick examples familiar from our culture— inevitably have implications for other members of the group and continue to be sources of both discourse and discord.

Some rules for regulating behavior in human social groups are embedded in religious teaching while others derive from secular authority. Sometimes behavioral strictures are stated as moral imperatives (such as never telling a lie), a form intended to identify boundaries between permissible and punishable behaviors but unsuited for resolving individual conflicts. As societies have become larger, more complex, and economically stratified, the rules are frequently made by an oligarchy whose interests conflict with those of the ruled. Large democracies are no exception: consider lobbyists working to see that legislation gives advantage to the usually wealthy "special interests" that they represent. Inevitably, behavioral codes mean different things to different people at different times, and even the codes themselves may be used as weapons in gaining advantage. For example, in this country the use of wealth to influence the electoral process is currently protected by the individual right of freedom of speech.

Some general features of moral systems are shared by many cultures, because the basic features of human nature occur everywhere. For example, the importance of kin, and by extension, of group, the scope of mating behavior, the obligations of parental investment, competition for status, and the need for social contracts (even if informal) all drive the cultural invention of religious and secular rules. The details of the rules differ among societies, and they frequently reflect the ecological circumstances or the power structure in which the culture developed. Laura Betzig has provided a vivid example of the latter. The stratified relation between power and reproductive opportunities was quantified in the Inca empire of Peru: minor chiefs were entitled to seven wives, those with authority over a hundred people had eight, governors who ruled 1,000 people had fifteen, and a major king had 700.

Moral systems frequently permit or promote behavior toward other social groups that is ordinarily unacceptable within the group. The Hebrew Commandments—"love thy neighbor" along with the other "shalts" and "shalt nots"—were originally prescriptions encouraging the in-group unity that enabled the Israelites to conquer neighboring groups. Among derivative Christian religions these commandments and associated beliefs have been interpreted at various times to suggest a universal brotherhood of the human species—that "thy neighbor" refers to all humans. However desirable, such a social arrangement is remarkably different from anything that either natural selection or cultural invention has produced thus far.

Note that although individuals are ordinarily proscribed from killing one another ("taking the law into their own hands"), the state not only organizes intergroup warfare, it frequently reserves the right to take the lives of citizens if their transgressions are deemed excessive.

On page 331 we discussed the common occurrence of patriarchy and its relation to sexual selection. With this background it is clear why systems of laws and justice have historically been created, revised, interpreted, and enforced by men. Moreover, as power accrues with maturity, it is usually older men who have these roles. Across all human cultures, laws and customs regulating marriage, adultery, owning and inheriting resources, labor, education, voting, dress, the practice of religion, and access to power or wealth have historically reflected men's efforts to control economic resources, limit the competitive efforts of other men, and control women, ultimately all as means of achieving greater relative reproductive success. The admonition about coveting thy neighbor's wife is an example, but, like the biblical stricture against killing, it was also specifically addressed to males of the same religious group. It promotes internal harmony by reducing adultery, thus damping the fires of sexual jealousy.

The relation between cultural evolution and the slower and older evolution of human nature by natural selection is complex and imprecise, but it provides a general explanation for why the practical implementations of moral reasoning—systems of laws and justice— have never been considered complete and fair by everyone in any social group. It also helps us understand why conflicts of interest within and between groups persist, and why most humans are more concerned about what resources they have *relative to others* than they are about what *absolute amounts* they have. Natural selection designed our minds, and it is a process driven by *relative* reproductive success.

In our ability to reason abstractly and emotionally about the costs and benefits of our behavior to self, kin, and other members of our social group, we seem to be as distant from our nearest primate relatives as we are in our ability to create and use language. We are the only moral animal, but this circumstance imposes a unique obligation. We alone have the capacity to consciously shape the future of our own society as well as its impact on the physical well-being of our planet.

RELIGIONS

One of the striking features of human societies is the virtually universal presence of religions. Anthropologists have not hesitated to discuss the significance of religious belief, but biologists have usually avoided the subject. If we are to understand the human mind as a product of evolution, however, we cannot ignore this cardinal feature of humanity.

Like other features of culture, religions are diverse, and through most of history they have been as transient as the cultures that created them. During the last two thousand years, however, several of these belief systems have expanded to become the "major" religions of the world, coincident with the economic and political success of their parent cultures. The question of most general interest is not how or why religions differ from one another, but why have they arisen so frequently and why are they so important to people?

There is a coherent answer to this question that is entirely consistent with our understanding of evolution and of the human mind. Consider the functions that religions serve. First, they anchor the moral authority required to enforce behavioral rules. People everywhere desire and value social stability, and expectations for social behavior based on religious belief provide continuity that descends through generations relatively immune to changes in secular leadership. Second, religions help to satisfy a facet of human nature that we first met in the opening vignette in Chapter 1: curiosity about the world. Although we are the first species with the cognitive ability to consider our own origins, we did not evolve to understand ourselves, and the exploratory method that we call science is a relatively recent invention. But our ability to observe the rising and setting of the sun, the slow movement of the stars in the night sky, the passing seasons, and the ephemeral nature of life itself have filled us with both wonder and fear. We seek significance, and when we cannot see clear answers, we invent convincing stories. Every cul-

ture has accounts of origins replete with agents of creation as well as beliefs about the meaning of death. We are such social creatures that more often than not we populate these stories with spirits and deities, sometimes in the forms of other animals, but characteristically with very human emotions. The curiosity of the Haida's trickster raven; the lust of Apollo for the nymph Daphne; and the jealousy of the Hebrew God of the Old Testament are examples of increasing familiarity. Sacrifices and supplications characterize human relations with the deities, as though these gods were personages of high status with the authority to grant favors. This kind of appeal to unseen and unknown causal agents is the work of a human mind that has evolved to function in a social environment and finds explanations for many commonplace events in the actions of other humans. There is thus a logic to explaining the unknown through belief in deities who possess the familiar wants and emotions of people.

Religions thus supply some very important needs for both individuals and societies. They provide rules and incentives to regulate the socially disruptive, egoistic strivings of individuals and mobilize them toward common goals that require cooperation and reciprocity. Religious rituals also supply a sense of continuity with the past as well as sanctifying the milestones of life: birth, puberty, marriage, and death. They nurture billions of people with love and hope and provide a spiritual release that is both real and comforting.

Faith can be so powerful as to sustain beliefs that are manifestly false. When not in a religious context, this important phenomenon goes by other names, such as self-deception. It appears to be an evolved capacity of the mind that contributes to well-being in more than one way and is therefore in need of further understanding.

There is another face to religion that cannot be praised. Very often in history religion has been a token by which one group identifies itself at the expense of another. The Old Testament contains explicit commands to the Israelites to eliminate surrounding groups, identified by their different religious practices. Through the Christian and Muslim worlds, the word "heathen" and "infidel" have been virtually calls to arms. The destruction of the Aztec and Inca civilizations was justified by religious arguments. It is not necessary for religions to explicitly advocate violence in order for them to be powerful symbols of ethnic identity that perpetuate conflict between groups. Consider, for example, the recent wars in Bosnia and Kosovo and the "troubles" in Northern Ireland.

We gain much useful understanding by exploring ourselves, our history, and our aspirations through the prism of evolutionary diversity. As for what will happen in the future, we need to use those extraordinary if imperfect brains that natural selection has fashioned if we are to deal effectively with the profound changes that are now taking place. Population is growing and generating serious environmental consequences, but basic human aspirations are intact, old animosities continue and the rapidly evolving technology of conflict—from guns to bombs and missiles—risks putting warfare beyond our slowly evolved abilities to regulate it. The ethical challenges are clearer when we recognize that the human mind is a product of the evolutionary process and that we are *required* to understand our evolved behavioral propensities if our descendants are to live in health and harmony with each other and the rest of nature.

SYNOPSIS

In *Homo sapiens*, organic evolution has produced a species that creates extensive behavioral traditions and through language and teaching is able to transmit the associated ideas to succeeding generations with unprecedented efficiency. Human culture is therefore capable of changing at a rate that is many times faster than biological evolution. When considered along with the unique richness of human thought and the enormous diversity of human activity, culture has traditionally seemed disconnected from biology. This isolation has been further encouraged by the false claims of Social Darwinists and eugenicists in the first half of the twentieth century.

Culture, however, is the work of human brains that have been produced by natural selection, so it is inevitable that our evolutionary heritage should be reflected in cultural practices. We are equipped with motivations that have proved useful during our evolutionary history, and we assess sensory information in prescribed and identifiable ways. Furthermore, we bear the stamp of a moderate amount of sexual selection. These features are evident in many patterns of social behavior, including marriage practices, mate guarding, inheritance, male violence, infanticide, responses to and control of the sex ratio, patriarchy, altruism, self-deception, nepotism, group identity, and intergroup conflict in the form of warfare.

Humans form extensive associations based on friendship and informal expectations of reciprocity, greatly extending cooperative behavior beyond networks of close kin. There is an inherent tension, however, between the requirements of group living and selfish inclinations that derive from the genetic uniqueness of each individual. Consequently, as groups have become larger, moral systems and codes of conduct have become formalized through religious teaching and secular authority. The economic and reproductive interests of most members of the group—especially those in economic, political, and military control—are best served when the group is free of internal discord. "Do unto others as you would have others do unto you" is a formula for suppressing self-indulgence in the cause of group harmony and success.

Archaeological evidence shows that warfare between groups is an old and recurrent feature of our species. Competition and conflict between groups is likely much older than the archaeological record, however, for it is has left its mark on the human psyche. People are suspicious of strangers and have strong urges to identify with their own ethnic or religious group. Moreover, they frequently assert the superiority of their group as a rationale for securing resources at the expense of another group. Herein lie the origins of racism and similar forms of discrimination.

Scientific theories improve understanding of the world by synthesizing information and by discovering relationships between observations. They are coherent and inclusive descriptions that allow us to see order and meaning amid diversity or even what appears to be chaos. Evolutionary theory is one of the most successful, both for the power and scope of its conception. It requires us to see ourselves as part of nature, not only in terms of our anatomy and biochemistry, but also in the unique features of our human nature that have previously seemed remote from scientific explanation. It shows us how the evolution of mind has influenced the formation of human cultures. It also tells us we will have to use wisdom in determining our own future.

QUESTIONS FOR THOUGHT AND DISCUSSION

1. How would you integrate genetic data (Chapter 12) with information on the development of behavioral phenotype (Chapter 11) and evolved psychological propensities (Chapter 14) in an informed discussion of race and racism? What implications are there for educational and social policy?

2. The comedian George Burns said that "there will always be a battle of the sexes because men and women want different things: men want women and women want men." Discuss this comment in light of what you know about evolutionary social theory.

3. Some anthropologists have proposed that the function of warfare in humans is to limit population density and prevent overexploitation of the environment. What are the difficulties with this hypothesis?

4. Recall the question at the end of Chapter 7 about a hypothetical, highly intelligent species with a mode of sex determination similar to bees or termites. Extend your answer to that question by suggesting what moral and legal systems might arise.

5. The theory of evolution is a "paradigm shift"—a major change in how nature is understood. Why is the scope of this theory as broad as that characterization implies?

SUGGESTIONS FOR FURTHER READING

Alexander, R. (1979). *Darwinism and Human Affairs*. Seattle, WA: University of Washington Press. A wide-ranging and important discussion of the relevance of Darwinian theory to human behavior and history, especially the final chapter, "Evolution, Law and Justice".

Alexander, R. (1987). *The Biology of Moral Systems*. New York, NY: Aldine de Gruyter. An in-depth discussion of why moral systems and their underlying emotions and mental abilities evolved and the relevance of an evolutionary perspective to contemporary societies.

Arnhart, L. (1998). *Darwinian Natural Right: The Biological Ethics of Human Nature*. Albany, NY: SUNY Press. Of particular interest to students of political science and philosophy.

Betzig, L., ed. (1997). *Human Nature: A Critical Reader*. New York, NY: Oxford University Press. A collection of eighteen classic studies, with updated evaluations by the authors, of how Darwinian social theory explains patterns of human behavior in traditional and modern societies and over recorded history.

Betzig, L.; Borgerhoff Mulder, M.; and Turke, P., eds. (1988). *Human Reproductive Behavior: A Darwinian Perspective*. Cambridge, U.K.: Cambridge University Press. A collection of twenty-one original papers by biologists and anthropologists on how evolutionary social theory applies to human mating and parenting behavior.

Brown, D. E. (1991). *Human Universals*. New York, NY: McGraw Hill. What the study of human behaviors, societies, and cultures tells us about what all humans have in common.

Buss, D., and Malamud, N. M., eds. (1996). *Sex, Power, and Conflict: Evolutionary and Feminist Perspectives*. New York, NY: Oxford University Press. Twelve essays by natural and social scientists on the convergence of evolutionary and feminist perspectives in understanding male-female relationships in humans.

Buss, D. M. (1999). *Evolutionary Psychology: The New Science of Mind*. Needham Heights, MA: Allyn & Bacon. Brings some of the techniques of social psychology to an evolutionary analysis of peoples' beliefs and desires.

Chagnon, N. A. (1997). *Yanomamö*, 5th edition. San Diego, CA: Harcourt, Brace & Co. A remarkable and uniquely detailed case study of a technologically primitive people, told with warmth and understanding by an anthropologist who has lived among them and learned their language.

Daly, M., and Wilson, M. (1983). *Sex, Evolution, and Behavior*, 2nd ed. Belmont, CA: Wadsworth Publishing Co. Provides the background in biology required to understand evolutionary social theory and its application to human social and cultural evolution. Replete with the basics and is still timely.

Daly, M., and Wilson, M. (1988). *Homicide*. New York, NY: Aldine de Gruyter. An evolutionary analysis of civil and historical records of why people commit murder.

Hrdy, S. (1999). *Mother Nature: A History of Mothers, Infants, and Natural Selection*. New York, NY: Pantheon Books, Random House. A clear, engaging, and extensively documented discussion of maternal behavior in human females, from its evolutionary roots to the present.

Keeley, L. H. (1996). *War before Civilization: The Myth of the Peaceful Savage*. New York, NY: Oxford University Press. An archeologist has compiled extensive data on warfare from prehistory to the present.

Smuts, B. (1994). The evolutionary origins of patriarchy. *Human Nature*, 6, 1–32. Extreme and institutionalized male dominance over women is traced to the combination of our evolutionary history of intrasexual male competition and the development, beginning some 15,000 years ago, of controllable and heritable resources.

Bibliography

There are many texts on cell and molecular biology, genetics, developmental biology, and neurobiology to which interested students can turn for more detailed treatment of these subjects, but the literature connecting the biological and social sciences in an evolutionary framework is more diffuse. Consequently, the following sources emphasize material in the later chapters of the book and include a mixture of sources we have found useful as well as a selection of recent books.

Chapter 2

Dennett, D. C. 1985. *Darwin's Dangerous Idea: Evolution and the Meaning of Life*. New York: Simon & Schuster.

Moorehead, A. 1969. *Darwin and the Beagle*. New York: Harper & Row.

Ruse, M. 1998. *Taking Darwin Seriously*. Amherst, N.Y.: Prometheus Books.

Thomson, K. S. 1995. *The Beagle: The Story of Darwin's Ship*. New York: W.W. Norton.

Chapter 4

Boag, P. T., and Grant, P. R. Intense natural selection in a population of Darwin's finches (Geospizinae) in the Galapagos. *Science*, 214 (1981): 82–85.

Cavalli-Sforza, L., and Bodmer, W. F. 1971. *The Genetics of Human Populations*. San Francisco: W.H. Freeman.

Grant, B. R., and Grant, P. R. Evolution of Darwin's finches caused by a rare climatic event. *Proceedings of the Royal Society of London, Series B*, 251 (1993): 111–117.

Grant, P.R. Natural selection and Darwin's finches. *Scientific American*, 265 (October 1991): 82–87.

Karn, M. N., and Penrose, L. S. Birth weight and gestation time in relation to age, parity and infant survival. *Annals of Eugenics*, 15 (1951): 206–233.

Lerner, I. M., and Libby, W. J. 1976. *Heredity, Evolution, and Society*, 2nd ed. San Francisco: W.H. Freeman.

Ridley, M. 1993. *The Red Queen: Sex and the Evolution of Human Nature*. New York: Penguin Books.

Simpson, G. G., and Beck, W. 1965. *Life: An Introduction to Biology*. New York: Harcourt, Brace and World.

Smith, J. M. 1998. *Evolutionary Genetics*. New York: Oxford University Press.

Weiner, J. 1995. *The Beak of the Finch*. New York: Vintage Books, Random House.

Chapter 5

Dawkins, R. 1982. Replicators and Vehicles, King's College Sociobiology Group (eds.), *Current Problems in Sociobiology*, 45–64. Cambridge: Cambridge University Press.

Dawkins, R. 1989. *The Extended Phenotype*. New York: Oxford University Press.

Freeman, S., and Herron, J. C. 1998. *Evolutionary Analysis*. Upper Saddle River, N.J.: Prentice Hall.

Futuyma, D. 1998. *Evolutionary Biology*, 3rd ed. Sunderland, Mass: Sinauer Associates.

Hrdy, S. B. 1977. *The Langurs of Abu: Female and Male Strategies of Reproduction*. Cambridge, Mass.: Harvard University Press.

Li, W. H. 1997. *Molecular Evolution*. Sunderland, Mass: Sinauer Associates.

Chapter 6

Austad, S. N., and Sunquist, M. E. Sex ratio manipulation in the common opossum. *Nature*, 324 (1986): 58–60.

Axelrod, R. 1984. *The Evolution of Cooperation*. New York: Basic Books.

Bateman, A. J. Intra-sexual selection in *Drosophila*. *Heredity*, 2 (1948): 349–368.

Brown, J. L. 1987. *Helping and Communal Breeding in Birds: Ecology and Evolution*. Princeton, N.J.: Princeton University Press.

Clutton-Brock, T. H., Albon, S. D., and Guinness, F. E. Maternal dominance, breeding success and birth sex ratio in red deer. *Nature*, 308 (1984): 358–360.

Clutton-Brock, T. H., Guinness, F. E., and Albon, S. D. 1982. *Red Deer: Behavior and Ecology of Two Sexes*. Chicago, Ill.: University of Chicago Press.

Cosmides, L., and Tooby, J. 1992 Cognitive adaptations for social exchange. In Barkow, J. H., Cosmides, L., and Tooby, J. (eds.), *The Adapted Mind: Evolutionary Psychology and the Generation of Culture*, 163–228. New York: Oxford University Press.

Cronin, H. 1991. *The Ant and the Peacock*. Cambridge, U.K.: Cambridge University Press.

Darwin, C. 1871. *The Descent of Man and Selection in Relation to Sex*. London: Murray.

Fisher, R. A. 1958. *The Genetical Theory of Natural Selection*, 2nd ed. New York: Dover Press.

Gigerenzer, K., and Hug, K. Domain-specific reasoning: social contracts, cheating and perspective change. *Cognition*, 43 (1992): 127–171.

Hamilton, W. D. The genetical evolution of social behavior I, II. *Journal of Theoretical Biology* (1964): 1–52.

Hamilton, W. D., and Zuk, M. Heritable true fitness and bright birds: A role for parasites. *Science*, 218 (1982): 384–387.

Hrdy, S. B. 1977. *The Langurs of Abu: Female and Male Strategies of Reproduction*. Cambridge, Mass.: Harvard University Press.

Sherman, P.W. Nepotism and the evolution of alarm calls. *Science*, 197 (1977): 1246–1253.

Sherman, P. W. Alarm calls of Belding's ground squirrels to aerial predators: Nepotism or self preservation? *Behavioral Ecology and Sociobiology*, 17 (1985): 313–323.

Silk, J. Local resource competition and facultative adjustment of sex ratios in relation to competitive ability. *American Naturalist*, 121 (1983): 56–66.

Smith, J. M. 1978. *The Evolution of Sex*. Cambridge, U.K.: Cambridge University Press.

Thornhill, R., and Gangestad, S. W. Human fluctuating asymmetry and sexual behavior. *Psychological Science*, 5 (1994): 297–302.

Thornhill, R., and Moller, A. P. Developmental stability, disease and medicine. *Biological Reviews*, 72 (1997): 497–528.

Trivers, R. L., and Willard, D. E. Natural selection of parental ability to vary the sex ratio of offspring. *Science*, 179 (1973): 90–92.

Wason, P. Regression in reasoning. *British Journal of Psychology*, 60 (1969): 471–480.

Wilkinson, G. S. Information transfer at evening bat colonies. *Animal Behavior*, 44 (1984): 501–518.

Wilkinson, G. S. Reciprocal food sharing in the vampire bat. *Nature*, 308 (1984): 181–184.

Wilson, E. O. 1975. *Sociobiology: The New Synthesis*. Cambridge, Mass: Belknap Press of Harvard University Press.

Woolfenden, G. E., and Fitzpatrick, J. W. 1984. *The Florida Scrub Jay: Demography of a Cooperative-Breeding Bird*. Princeton, N.J.: Princeton University Press.

Zahavi, A., Zahavi, A., and Balaban, A. 1997. *The Handicap Principle: A Missing Piece of Darwin's Puzzle*. Oxford, U.K.: Oxford University Press.

Chapter 7

Hölldobler, B., and Wilson, E. O. 1990. *The Ants*. Cambridge, Mass: Belknap Press of Harvard University Press.

Seeley, T. D. 1985. *Honeybee Ecology: A Study of Adaptation in Social Life*. Princeton, N.J.: Princeton University Press.

von Frisch, K. 1967. *The Dance Language and Orientation of Bees*. Translated by L. E. Chadwick. Cambridge, Mass: Belknap Press of the Harvard University Press.

Chapter 8

Billing, J., and Sherman, P. W. Antimicrobial functions of spices: Why some like it hot. *Quarterly Review of Biology*, 73 (1998): 3–49.

Ewald, P. W. Evolutionary biology and the treatment of signs and symptoms of diseases. *Journal of Theoretical Biology*, 86 (1980): 169–176.

Hahn, B. H., Shaw, G. M., De Cock, K. M., and Sharp, P. M. AIDS as zoonosis: Scientific and public health implications. *Science*, 287 (2000): 607–614.

McGuire, M., and Troisi, A. 1998. *Darwinian Psychiatry*. New York: Oxford University Press.

Ridley, M. 1993. *The Red Queen: Sex and the Evolution of Human Nature*. New York: Penguin Books.

Stevens, A., and Price, J. 1996. *Evolutionary Psychiatry: A New Beginning*. New York: Routledge.

Chapter 9

Bounds, D. M. 1999. *The Biology of Mind: Origins and Structures of Mind, Brain, and Consciousness*. Bethesda, Md.: Fitzgerald Science Press.

Dowling, J. E. 1998. *Creating Mind: How the Brain Works*. New York: W.W. Norton.

Gazzaniga, M. S. 1992. *Nature's Mind: The Biological Roots of Thinking, Emotions, Sexuality, Language, and Intelligence*. New York: Basic Books/Harper Collins.

Gazzaniga, M. S. 1998. *The Mind's Past*. Berkeley, Calif.: University of California Press.

Kolb, B., and Whislaw, I. Q. 1995. *Fundamentals of Human Neuropsychology*. San Francisco: W.H. Freeman.

Purvis, D., Augustine, G.J., and Fitzpatrick, D. 1997. *Neuroscience*. Sunderland, Mass: Sinauer Associates.

Chapter 10

Bock, G. R., and Carde, G., eds. 2000. *Evolutionary Developmental Biology of the Cerebral Cortex*. Chichester, N.Y.: Wiley.

Bonner, J. T. 1995. *Life Cycles*. Princeton, N.J.: Princeton University Press.

Dawson, G., and Fischer, K. W., eds. 1994. *Human Behavior and the Developing Brain*. New York: Guildford Press.

Gehring, W. J., and Ruddle, F. 1998. *Master Control Genes in Development and Evolution: The Homeobox Story.* New Haven, Conn.: Yale University Press.

Halder, G., Callaerts, P., and Gehring, W. J. Induction of ectopic eyes by targeted expression of the eyeless gene in *Drosophila. Science,* 267 (1995): 1788–1792.

Johnson, M. H., ed. 1993. *Brain, Development, and Cognition: A Reader.* Malden, Mass.: Blackwell.

Wolpert, L., Beddington, R., Brockes, J., Jessel, T., Lawrence, P., and Meyerowitz, E. 1998. *Principles of Development.* New York: Oxford University Press.

Chapter 11

Barkow, J. H., Cosmides, L., and Tooby, J., eds. 1992. *The Adapted Mind: Evolutionary Psychology and the Generation of Culture.* New York: Oxford University Press.

Boesch, C. Teaching among wild chimpanzees. *Animal Behavior,* 41 (1991): 530–532.

Bouchard, T. J., Jr., Lykken, D. T., McGue, M., Segal, N. L., and Tellegen, A. Sources of human psychological differences. The Minnesota study of twins reared apart. *Science,* 250 (1990): 223–228.

Cheney, D. L., and Seyfarth, R. M. Vervet monkey alarm calls: Manipulation through shared information? *Behavior,* 93 (1985): 150–166.

Cheney, D. L., and Seyfarth, R. M. 1990. *How Monkeys See the World: Inside the Mind of Another Species.* Chicago, Ill.: University of Chicago Press.

Cummings, M. R. 1997. *Human Heredity: Principles and Issues,* 4th ed. New York: Wadsworth.

de Waal, F. 1986 Deception in the natural communication of chimpanzees. In Mitchell, R. W., and Thompson, N. S. (eds.), *Deception: Perspectives on Human and Nonhuman Deceit,* 221–244. Albany, N.Y.: SUNY Press.

de Waal, F. 1989. *Peacemaking among Primates.* Cambridge, Mass: Harvard University Press.

Dunbar, R. I. M. Neocortex size as a constraint on group size in primates. *Journal of Human Evolution,* 22 (1992): 469–493.

Eibl-Eibesfeldt, I. 1989. *Human Ethology.* Hawthorne, N.Y.: Aldine de Gruyter.

Garcia, J., McGowan, B. K., Ervin, R. R., and Koelling, R. A. Cues: Their relative effectiveness as a function of the reinforcer. *Science,* 160 (1968): 794–795.

Gould, J., and Marler, P. Learning by instinct. *Scientific American,* 256 (January 1987): 74–85.

Harlow, H. F., and Harlow, M. K. Social deprivation in monkeys. *Scientific American,* 207 (November 1962): 136–146.

Mange, E. J., and Mange, A. 1999. *Basic Human Genetics.* Sunderland, Mass: Sinauer Associates.

McGrew, W. C. 1992. *Chimpanzee Material Culture: Implications for Human Evolution.* New York: Cambridge University Press.

Parr, L. A., and de Waal, F. B. M. Visual kin recognition in chimpanzees. *Nature,* 399 (1999): 647–648.

Pinker, S. 1994. *The Language Instinct.* New York: William Morrow & Co.

Plomin, R., DeFries, J. C., and McClearn, G. E. 1990. *Behavior Genetics: A Primer.* New York: W.H. Freeman.

Seligman, M. E. P. On the generality of the laws of learning. *Psychological Review,* 77 (1970): 406–418.

Stacey, P. B., and Bock, C. E. Social plasticity in the acorn woodpecker. *Science,* 202 (1978): 1298–1300.

Whiten, A., Goodall, J., McGrew, W. C., Nishida, T., Reynolds, V., Sugiyama, Y., Tutin, C. E. G., Wrangham, R., and Boesch, C. Culture in chimpanzees. *Nature, 399* (1999): 682–685.

Wrangham, R. W., McGrew, W. C., deWaal, F. B., and Heltne, P.G., eds. 1994. *Chimpanzee Cultures.* Chicago, Ill.: Chicago Academy of Sciences.

Chapter 12

Brown, G. D., Jr. 1995. *Human Evolution.* Dubuque, Ia.: W. C. Brown Publishers.

Caccone, A., and Powell, J. R. DNA divergence among hominoids. *Evolution,* 43 (1989): 925–942.

Cavalli-Sforza, L., Menozzi, P., and Piazza, A. 1994. *History and Geography of Human Genes.* Princeton, N.J.: Princeton University Press.

Darwin, C. 1871. *The Descent of Man and Selection in Relation to Sex.* London: Murray.

Diamond, J. 1992. *The Third Chimpanzee.* New York: Harper Collins Publishers.

Greenberg, J. 1987. *Language in the Americas.* Stanford, Calif.: Stanford University Press.

Johanson, D. 1996. *From Lucy to Language.* New York: Simon and Schuster.

Johanson, D. C., and Edey, M. A. 1981. *Lucy: The Beginnings of Humankind.* New York: Simon and Schuster.

Jones, S., Martin, R., and Pilbeam, D., eds. 1992. *The Cambridge Encyclopedia of Human Evolution.* Cambridge: Cambridge University Press.

Krings, M., Stone, A., Schmitz, R. W., Krainitzki, H., Stoneking, M., and Paabo, S. Neanderthal DNA and the origin of modern humans. *Cell,* 90 (1997): 19–30.

Leakey, M., and Walker, A. Early hominid fossils from Africa. *Scientific American,* 276 (June 1997): 74–79.

Leakey, R. E. 1994. *The Origin of Humankind.* New York: Basic Books.

Lewin, R. 1998. *Principles of Human Evolution.* Malden, Mass: Blackwell Science.

Ruhlen, M. 1994. *The Origin of Languages.* New York: Wiley.

Sibley, C. G., and Ahlquist, J. E. DNA-DNA hybridization evidence of hominoid phylogeny: Results from an expanded data set. *Journal of Molecular Evolution,* 26 (1987): 99–121.

Sibley, C., Comstock, J. A., and Ahlquist, J. E. DNA hybridization evidence of hominoid phylogeny: A reanalysis of the data. *Journal of Molecular Evolution,* 30 (1990): 202–236.

Stoneking, M., Bhatia, K., and Wilson, A. C. Rate of sequence divergence estimated from restriction maps of mitochondrial DNAs from Papua New Guinea. *Cold Spring Harbor Symposia on Quantitative Biology , LI Molecular Biology of Homo sapiens* (1986): 433–439.

Swadesh, M. 1971. *The Origin and Diversification of Language.* Chicago, Ill.: Aldine Atherton.

Trinkaus, E., and Shipman, P. 1992. *The Neandertals: Changing the Image of Mankind.* New York: Alfred A. Knopf.

White, T. D., Suwa, G., and Asfaw, B. *Australopithecus ramidus*, a new species of early hominid from Aramis, Ethiopia. *Nature,* 371 (1994): 306–312.

Wilson, A., and Cann, R. L. The recent African genesis of human genes. *Scientific American,* 266 (April 1992): 68–73.

Woldegabriel, G., White, T. D., Suwa, G., Renne, P., De-Heinzelin, J., Hart, W. K., and Heiken, G. Ecological and temporal placement of early Pliocene hominids at Aramis, Ethiopia. *Nature,* 371 (1994): 330–333.

Chapter 13

de Waal, F. 1989. *Peacemaking among Primates.* Cambridge, Mass: Harvard University Press.

de Waal, F. 1997. *Bonobo: The Forgotten Ape.* Berkeley, Calif.: University of California Press.

Diamond, J. 1992. *The Third Chimpanzee.* New York: Harper Collins Publishers.

Foley, R., ed. 1984. *Hominid Evolution and Community Ecology.* London, U.K.: Academic Press.

Fossey, D. 1983. *Gorillas in the Mist.* Boston, Mass: Houghton Mifflin.

Goodall, J. 1986. *The Chimpanzees of Gombe: Patterns of Behavior.* Cambridge, Mass: Belknap Press of Harvard University Press.

Parr, L. A., and de Waal, F. B. M. Visual kin recognition in chimpanzees. *Nature,* 399 (1999): 647–648.

Pusey, A. E., Williams, J., and Goodall, J. The influence of dominance rank on the reproductive success of female chimpanzees. *Science,* 277 (1997): 828–831.

Rodseth, L., Wrangham, R. W., Harrigan, A. M., and Smuts, B.B. The human community as a primate society. *Current Anthropology,* 32 (1991): 221–254.

van Schail, C. P., and Hrdy, S. B. Intensity of local resource competition shapes the relationship between maternal rank and sex ratio at birth in cercopithecine primates. *American Naturalist, 138,* (1991): 1555–1562.

Whiten, A., Goodall, J., McGrew, W. C., Nishida, T., Reynolds, V., Sugiyama, Y., Tutin, C. E. G., Wrangham, R., and Boesch, C. Culture in chimpanzees. *Nature,* 399 (1999): 682–685.

Wrangham, R. W., de Waal, F. B. M., and McGrew W. C. 1994. The challenge of behavioral diversity. In Wrangham, R. C., McGrew, W. C., de Waal, F. B. M., and Heltne, P. G. (eds), *Chimpanzee Cultures.* Cambridge, Mass: Harvard University Press.

Chapter 14

Abusharaf, R. M. Unmasking tradition: A Sudanese anthropologist confronts female "circumcision" and its terrible tenacity. *The Sciences,* 23 (March/April 1998): 23–27.

Ardrey, R. 1961. *African Genesis: A Personal Investigation into the Animal Origins and Nature of Man.* New York: Dell.

Barkow, J. H., Cosmides, L., and Tooby, J., eds. 1992. *The Adapted Mind: Evolutionary Psychology and the Generation of Culture.* New York: Oxford University Press.

Betzig, L. 1986. *Despotism and Differential Reproduction: A Darwinian View of History.* New York: Aldine.

Betzig, L. Causes of conjugal dissolution: A cross-cultural study. *Current Anthropology,* 30 (1989): 654–676.

Borgerhoff Mulder, M. Kipsigis women's preference for wealthy men: Evidence for female choice in mammals? *Behavioral Ecology and Sociobiology,* 27 (1990): 255–264.

Buss, D. Sex differences in human mate preferences: Evolutionary hypotheses tested in 37 cultures. *Behavioral and Brain Sciences,* 12 (1989): 1–49.

Chagnon, N., and Irons, W., eds. 1979. *Evolutionary Biology and Human Social Behavior: An Anthropological Perspective.* North Scituate, Mass: Duxbury Press.

Chagnon, N. A. Life histories, blood revenge, and warfare in a tribal population. *Science,* 239 (1988): 985–991.

Cronk, L. Intention vs. behavior in parental sex preferences among the Mukogodo of Kenya. *Journal of Biosocial Science,* 23 (1991): 229–240.

Cronk, L. Preferential investment in daughters over sons. *Human Nature,* 2 (1991): 387–417.

Dickemann, M. 1979. Female infanticide, reproductive strategies, and social stratification: A preliminary model. In Chagnon, N. A., and Irons, W. (eds.), *Evolutionary Biology and Human Social Behavior: An Anthropological Perspective,* 321–367. North Scituate, Mass: Duxbury Press.

Dickemann, M. 1981. Paternal confidence and dowry competition: A biocultural analysis of purdah. In Alexander, R., and Tinkle, D. (eds.), *Natural Selection and Social Behavior,* 417–438. New York: Hiron Press.

Durham, W. H. 1991. *Coevolution: Genes, Culture and Human Diversity.* Stanford, Calif.: Stanford University Press.

Gaulin, S. J. C., and Boster, J. Dowry as female competition. *American Anthropologist,* 92 (1990): 994–1005.

Gowaty, P. A. Evolutionary biology and feminism. *Human Nature,* 3 (1992): 217–249.

Gowaty, P. A., ed. 1997. *Feminism and Evolutionary Biology.* New York: Chapman and Hall.

Hartung, J. Polygyny and the inheritance of wealth. *Current Anthropology,* 23 (1982): 1–12.

Hartung, J. 1988. Deceiving down: Conjectures on the management of subordinate status. In Lockhard, J., and Paulhus, P. (eds.), *Self-Deception: An Adaptive Mechanism.* Englewood Cliffs, N.J.: Prentice Hall.

Hartung, J. Love thy neighbor: The evolution of in-group morality. *Skeptic,* 3, no. 4 (1995): 86–99.

Hausfater, G., and Hrdy, S. B., eds. 1984. *Infanticide: Comparative and Evolutionary Aspects.* Hawthorne N.Y.: Aldine de Gruyter.

Hrdy, S. B. 1981. *The Woman That Never Evolved.* Cambridge, Mass: Harvard University Press.

Hrdy, S. B. Raising Darwin's consciousness: Female sexuality and the prehominid origins of patriarchy. *Human Nature,* 8 (1997): 1–49.

Jones, O.D. Evolutionary Analysis in Law: An Introduction and Application to Child Abuse. *North Carolina Law Review*, 75 (1997): 1117–1242.

Konner, M. 1982. *The Tangled Wing: Biological Constraints on the Human Spirit*. New York: Holt, Rinehart and Winston.

Murphy, R. F. 1989. *Cultural and Social Anthropology: An Overture*. Englewood Cliffs, N.J.: Prentice Hall.

Packer, C., and Pusey, A. E. Divided we fall: Cooperation among lions. *Scientific American*, 276 (May 1997): 52–59.

Pinker, S. Why they kill their newborns. *New York Times Magazine*, (2 Nov. 1997): 53–54. New York: New York Times.

Richards, R. J. 1987. *Darwin and the Emergence of Evolutionary Theories of Mind and Behavior*. Chicago, Ill.: University of Chicago Press.

Ridley, M. 1997. *The Origins of Virtue: Human Instincts and the Evolution of Cooperation*. New York: Penguin Books.

Singh, D. Body shape and women's attractiveness: The critical role of waste-to-hip ratio. *Human Nature*, 4 (1993): 297–321.

Small, M. The evolution of female sexuality and mate selection in humans. *Human Nature*, 3(1991): 133–156.

Smuts, B. Male aggression against women: An evolutionary perspective. *Human Nature*, 3 (1992): 1–44.

Smuts, B. The evolutionary origins of patriarchy. *Human Nature*, 6 (1994): 1–32.

Trivers, R. L. 1981. Sociobiology and politics. In White, E. (ed.), *Sociobiology and Human Politics*, 1–44. Lexington, Mass: D.C. Heath & Co.

Voland, E. 1988. Differential infant and child mortality in evolutionary perspective: Data from late 17th to 19th century Ostfriesland (Germany). In Betzig, L., Borgerhoff Mulder, M., and Turke, P. (eds.), *Human Reproductive Behavior: A Darwinian Perspective*, 253–262. Cambridge, U.K.: Cambridge University Press.

Westermarck, E. 1894. *The History of Human Marriage*. London, U.K.: Macmillan.

Wetsman, A., and Marlow, F. How universal are preferences for female waist-to-hip ratios? Evidence from the Hadza of Tanzania. *Evolution and Human Behavior*, 20 (1999): 219–228.

Wilson, E. O. 1978. *On Human Nature*. Cambridge, Mass: Harvard University Press.

Wright, R. 1994. *The Moral Animal*. Toronto, Canada: Random House.

Credits

PART ONE
Opener: Courtesy NASA.

Chapter 1
Opener: M-SAT Ltd./Science Photo Library/Photo Researchers.
Figure 1.1: Andrew J. Martinez/Photo Researchers.
Figure 1.8A: Gary Retherford/Photo Researchers.
Figure 1.8B: Jim Steinberg/Photo Researchers.

Chapter 2
Opener: Frans Lanting/Photo Researchers.
Figure 2.1: © ARCHIV/Photo Researchers.
Figure 2.2: Modified from Moorehead, 1969.
Figure 2.3: From *Narrative of the Surveying Voyages of HMS Adventure and Beagle between the Years 1826 and 1836* by Robert FitzRoy (1839). H. Colburn, London.
Figure 2.4: From *The Life of Vertebrates* by J. Z. Young (1950). Reprinted by permission of Oxford University Press.
Figure 2.5: Courtesy New York Public Library.
Figures 2.6, 2.21: Modified from Brusca, R. C. and Brusca, G. J. (1990) *Invertebrates*, Sinauer Associates, Inc., Sunderland, MA.
Figure 2.7: Jeff Hunter/The Image Bank.
Figure 2.8: From Darwin's *The Structure and Distribution of Coral Reefs*, Part I of *The Geology of the Voyage of the 'Beagle.'* London, 1842.
Figure 2.10: M. Lustbader/Photo Researchers.
Figures 2.12A-C, E, F, 2.13A, B, E, 2.15, 2.17C, and 2.19A: Modified from Eaton, T. H., Jr. (1970) *Evolution*, W. W. Norton & Co., NY.
Figure 2.12D: After Gregory, W. K. (1951) in *Evolution Emerging*, Macmillan, with permission of the American Museum of Natural History.
Figures 2.13C, D, 2.16B, 2.17A, 2.18, and 2.27: After Colbert, E. H. and Morales, M. (1991) *Evolution of the Vertebrates*. John Wiley & Sons, NY.

Abridged citations can be found in complete form in the General Bibliography or at the ends of chapters.

Figure 2.14: After Orlov, J. (1961) *Traité de Paléontologie*, Masson et Cie.
Figures 2.16A, 2.20: After *Science and Creationism: A View from the National Academy of Sciences* (1999) National Academy Press, Washington, DC, with permission.
Figure 2.17B: From Ostrom, J. (1976) *Archaeopteryx* and the origin of birds. *Biological Journal of the Linnean Society 8*, 91–182.
Figure 2.19B: After Dunbar, C. O. (1949) *Historical Geology* John Wiley & Sons, NY.
Figure 2.19C: After Delevoryas, T. (1963) *Morphology and Evolution of Fossil Plants*. Holt, Rinehart and Winston, 1963.
Figure 2.22: After Haeckel, E. (1874) *Anthropogenie, order Entwickelungsgeschichte des Menschen*. Engelmann, Leipzig.
Figure 2.23: From *Embryos and Ancestors* by G. de Beer (1940). Reprinted by permission of Oxford University Press.
Figure 2.24: Birds from van Tyne, J. and Berger, A. J. (1965) *Fundamentals of Ornithology*. John Wiley and Sons, NY.
Figure 2.26 (top): Phil A. Dotson/Photo Researchers.
Figure 2.26 (bottom): John Mitchell/Photo Researchers.
Figure 2.26: Map after Conant, R. (1958) *A Field Guide to Reptiles and Amphibians*. Houghton Mifflin, Boston, MA.
Figure 2.28B: Map modified from Mayr, E. (1963). *Animal Species and Evolution*, Harvard University Press, Cambridge, MA.

Chapter 3
Opener: Dr. Gopal Murti/Science Photo Library/Photo Researchers.
Figures 3.2, 3.7, 3.9–3.10, 3.20–3.22: Modified from Alberts et al., (1994), *Molecular Biology of the Cell*, 3rd ed., Garland Publishing, New York, NY.
Figures 3.8, 3.11–3.13, 3.15–3.18: After Drlica, K. (1992) *Understanding DNA and Gene Cloning: A Guide for the Curious*, 2nd ed. New York: John Wiley & Sons.
Figure 3.19A: Stan Flegler/Visuals Unlimited.
Figure 3.28A: Andrew Syred/Science Photo Library/Photo Researchers.
Figure 3.28B: John Reader/Science Photo Library/Photo Researchers.
Figure 3.28: Modified from Sogin, M. L. and Silberman, J. D. 1998. Evolution of the protists and protistan parasites from the perspective of molecular systematics. *International Journal for Parasitology*, *28*, 11–20.

Chapter 4

Opener (left): Corbis-Bettmann.

Opener genealogy: From Mange, E. J. and Mange, A. (1999), *Basic Human Genetics*. Sinauer Associates, Sunderland, MA.

Figure 4.1: Modified from Simpson, G. G. and Beck, W. S. (1967) *Life: An Introduction to Biology* 2nd ed. Harcourt, Brace and World, New York, NY.

Figure 4.6: Modified from East, E. M. (1916) Studies on size inheritance in *Nicotiana. Genetics 1*, 164–176.

Figure 4.7: Modified from Lerner and Libby, 1976.

Figure 4.9: After Cavalli-Sforza and Bodmer, 1971.

Figure 4.10: Data from Boag and Grant, 1981.

Chapter 5

Opener (left): Art Wolfe/Photo Researchers.

Opener (right): Photo by T. H. Goldsmith.

Figure 5.1: Based on a gel by R. K. Koehn in Futuyma, D. (1986) *Evolutionary Biology*, 2nd ed. Sinauer Associates, Inc. Sunderland, MA.

Figure 5.2 (top): Bob & Clara Calhoun/Bruce Coleman, Inc.

Figure 5.2 (bottom): Bob Schorre/Bruce Coleman, Inc.

Figure 5.3: Adapted from Dickerson, R. E. (1971) *Journal of Molecular Evolution 1*, 26–45.

Figure 5.4: Adapted from Dayhoff, M. O. (ed.) (1972) *Atlas of Protein Sequence and Structure*. National Biomedical Research Foundation, Washington, D.C.

Figure 5.5: E. Hanumantha Rao/Photo Researchers.

Figure 5.6A: Vanessa Vick/Photo Researchers.

Figure 5.6B: Luis Castaneda/The Image Bank.

Figure 5.7A: Fletcher & Baylis/Photo Researchers.

Figure 5.7B: Joseph Van Os/The Image Bank.

Figure 5.8: Modified from Romer, A. S. (1949) *The Vertebrate Body*. W. B. Saunders, Philadelphia, PA.

Figure 5.9A, B: After Transeau, E. N., Sampson, A. C. and Tiffany, L. H. (1953) *Textbook of Botany*, Harper & Brothers, NY.

Figure 5.10: Photos by Erick Greene.

Figure 5.11: Field Museum/Photo Researchers.

Chapter 6

Opener (left): Pat & Tom Leeson/Photo Researchers.

Opener (right): Tom McHugh/Photo Researchers.

Figure 6.1: Modified from Wilson, 1975.

Figure 6.3: Photos by T. H. Goldsmith.

Figures 6.4, 6.5, 6.10: Drawings by Sarah Landry reprinted by permission of the publishers from *Sociobiology: The New Synthesis* by E. O. Wilson, Cambridge, Mass.: The Belknap Press of Harvard University Press, Copyright 1975, 2000 by the President and Fellows of Harvard College.

Figure 6.6: From *The Life of Vertebrates* by J. Z. Young (1950). reprinted by permission of Oxford University Press.

Figure 6.8: After Gilbert, S. F. (1991) *Developmental Biology*, 3rd ed., Sinaur Associates, Sunderland, MA.

Figure 6.9: Phil Farnes/Photo Researchers.

Figure 6.11: Stephen J. Krasemann/Photo Researchers.

Figure 6.12 (top): Gary Meszarus/MESZA/Bruce Coleman, Inc.

Figure 6.12 (bottom): David J. Boyle/Animals Animals.

Figure 6.13: Stephen Dalton/Animals Animals.

Figure 6.14: G. C. Kelley/Photo Researchers.

Figure 6.15: B. & C. Alexander/Photo Researchers.

PART TWO

Photo courtesy of Napoleon Chagnon.

Chapter 7

Opener (left): Gary Retherford/Photo Researchers.

Opener (right): J. P. Carin/Jacana/Photo Researchers.

Figure 7.1: After Wheeler, W. M. (1910) *Ants: Their Structure, Development and Behavior*. Columbia University Press, New York, NY.

Figures 7.2, 7.4, 7.5A: Reprinted by permission of the publishers from *The Insect Societies* by E. O. Wilson, Cambridge, Mass.: The Belknap Press of Harvard University Press, Copyright 1971 by the President and Fellows of Harvard College.

Figure 7.5B: Photo by T. H. Goldsmith.

Figure 7.6: Raymond A. Mendez/Animals Animals.

Figure 7.8: Modified from Lorenz, K. (1953) Die Entwicklung der vergleichenden Verhaltesforschung in den letzten 12 Jahren. *Zool. Anz. Suppl 16*, 36–58.

Figure 7.9: Based on von Frisch, 1967.

Chapter 8

Opener: Quido Noel/Liaison.

Figure 8.3: After Ewald, 1994.

Figure 8.8: Modified from Hahn et al., 2000.

PART THREE

Photo by Natalie Demong.

Chapter 9

Opener: fMRI Courtesy of John Gore.

Figure 9.1: Modified from Ramón y Cajal, S. (1933) Histology, 10th ed., Wood, Baltimore: MD.

Figure 9.15: After Penfield, W. and Rasmussen, T. (1950) *The Cerebral Cortex of Man: A Clinical Study of Localization of Function*. Macmillan, New York and Kandel et al., 1991.

Figures 9.16, 9.18, 9.20–9.22: Modified from Hubel, 1988.

Figure 9.26: Modified from Springer and Deutch, 1997.

Figure 9.27–9.29: After Kandel et al., 1991.

Chapter 10

Opener: Video microscopy courtesy of Paul Forscher.

Figures 10.2–10.4, 10.6: From Gilbert, S.F. (1991) *Developmental Biology*, 3rd ed., with permission of the publisher, Sinauer Associates, Sunderland, MA.

Figure 10.5: From Needham, J. (1936) *Order and Life*, Terry Lectures, Yale University Press, New Haven, CT.

Figure 10.7: Based on *From Egg to Adult*, 3rd in a series of reports from the Howard Hughes Medical Institute, 1992.

Figure 10.9: After Kirschfeld, K., 1969. Optics of the compound eye. In Reichardt, W. (ed) *Rendiconti S.I.F., XLIII*. Academic Press, NY.

Figure 10.10: From Halder et al., 1995 with permission of the publishers of *Science*.

Figure 10.13: Video microscopy courtesy of Paul Forscher.

Figure 10.14: After Hubel, 1988.

Figure 10.15: Biophoto Associates/Science Photo Library/Photo Researchers.

Figure 10.15A: Modified from drawings by Beth Johnston and Robert Warner.

Figure 10.16B: J. H. Carmichael, Jr./Bruce Coleman, Inc.

Chapter 11

Opener: George Holton/Photo Researchers.

Figure 11.1: Alan G. Nelson/Animals Animals.

Figures 11.2, 11.11: Modified from Tinbergen, N. (1965) *Animal Behavior* Life Nature Library, Time Inc., New York.

Figures 11.3, 11.5, 11.7, 11.8: After Gould and Marler, 1987.

Figures 11.4, 11.10: Nina Leen/LIFE Magazine © Time Inc.

Figures 11.9: Photos: French woman by Hans Hass; all others by I. Eibl-Eibesfeldt, from Eibl-Eibesfeldt, 1989.

Figure 11.12: Photo by Craig Stanford.

Figure 11.13: After Wrangham et al., 1994.

Figure 11.15: After Dunbar, 1992.

Figure 11.16: Photo by T. H. Goldsmith.

Figure 11.17: Adapted from Cummings, 1997.

PART FOUR
Opener: Photo by T. H. Goldsmith.

Chapter 12
Opener: John Reader/Science Photo Library/Photo Researchers.
Figure 12.1: Based on Sibley and Ahlquist (1987), Sibley et al. (1990), and Caccone and Powell (1989).
Figure 12.3: Based on Johanson and Edey, 1981.
Figure 12.4: After Jones et al., 1992 and Johanson and Edey, 1981.
Figure 12.5: After Brown, 1995.
Figure 12.6C: Modified from Wilson and Cann, 1992.
Figure 12.7: After Stoneking et al., 1986.
Figure 12.8: After Cavalli-Sforza et al., 1994.
Figure 12.9: After Diamond, 1992 and Cavalli-Sforza and Cavalli-Sforza, 1995.
Figures 12.10, 12.11: After Cavalli-Sforza and Cavalli-Sforza, 1995.

Chapter 13
Heading: Photo courtesy of Michael Huffman.
Figure 13.1A: Photo by T. H. Goldsmith.
Figure 13.1B: Jerry Ferrara/Photo Researchers.
Figure 13.2: Tim Davis/Photo Researchers.

Figure 13.3: © ASAP/Photo Researchers.
Figure 13.4 A,B: Photos courtesy of Tetsuro Matsuzawa.
Figure 13.4C: K & K Ammann/Bruce Coleman, Inc.
Figures 13.5–13.11: Photos by Frans deWaal.
Figure 13.12: Modified from Wrangham and Peterson, 1996.
Figures 13.13, 13.14: Modified from Rodseth et al., 1991.
Figure 13.15: Modified from Diamond, 1992.

Chapter 14
Heading: Photo courtesy of Stephen LeBlanc, from *Prehsitoric Warfare in the American Southwest*, 1999. University of Utah Press, Salt Lake City, UT.
Figures 14.2, 14.6: After Daly and Wilson, 1983.
Figure 14.3: Photo by Napoleon Chagnon.
Figure 14.4: Photo courtesy of Gülru Necipoglu, from *Architecture, Ceremonial, and Power: The Topkapi Palace in the Fifteenth and Sixteenth Centuries.* (1991) Architectural History Foundation, New York, NY and MIT Press, Cambridge, MA.
Figures 14.7–14.10: Data from Buss, 1999.
Figures 14.11–14.14: After Daly and Wilson, 1988.
Figure 14.15: Modified from Keeley, 1996.

Index